Smart Innovation, Systems and Technologies

Volume 326

The Smart Innovation, Systems and Technologies book series encompasses the topics of knowledge, intelligence, innovation and sustainability. The aim of the series is to make available a platform for the publication of books on all aspects of single and multi-disciplinary research on these themes in order to make the latest results available in a readily-accessible form. Volumes on interdisciplinary research combining two or more of these areas is particularly sought.

The series covers systems and paradigms that employ knowledge and intelligence in a broad sense. Its scope is systems having embedded knowledge and intelligence, which may be applied to the solution of world problems in industry, the environment and the community. It also focusses on the knowledge-transfer methodologies and innovation strategies employed to make this happen effectively. The combination of intelligent systems tools and a broad range of applications introduces a need for a synergy of disciplines from science, technology, business and the humanities. The series will include conference proceedings, edited collections, monographs, handbooks, reference books, and other relevant types of book in areas of science and technology where smart systems and technologies can offer innovative solutions.

High quality content is an essential feature for all book proposals accepted for the series. It is expected that editors of all accepted volumes will ensure that contributions are subjected to an appropriate level of reviewing process and adhere to KES quality principles.

Indexed by SCOPUS, EI Compendex, INSPEC, WTI Frankfurt eG, zbMATH, Japanese Science and Technology Agency (JST), SCImago, DBLP.

All books published in the series are submitted for consideration in Web of Science.

Vikrant Bhateja · Xin-She Yang ·
Jerry Chun-Wei Lin · Ranjita Das
Editors

Evolution in Computational Intelligence

Proceedings of the 10th International
Conference on Frontiers in Intelligent
Computing: Theory and Applications (FICTA
2022)

 Springer

Editors
Vikrant Bhateja
Department of Electronics Engineering
Faculty of Engineering and Technology
Veer Bahadur Singh Purvanchal University
Jaunpur, Uttar Pradesh, India

Jerry Chun-Wei Lin
Western Norway University of Applied
Sciences
Bergen, Norway

Xin-She Yang
School of Science and Technology
Middlesex University London
London, UK

Ranjita Das
Department of Computer Science
and Engineering
National Institute of Technology Agartala
Agartala, West Tripura, India

ISSN 2190-3018 ISSN 2190-3026 (electronic)
Smart Innovation, Systems and Technologies
ISBN 978-981-19-7515-8 ISBN 978-981-19-7513-4 (eBook)
https://doi.org/10.1007/978-981-19-7513-4

This Springer imprint is published by the registered company Springer Nature Singapore Pte Ltd.
The registered company address is: 152 Beach Road, #21-01/04 Gateway East, Singapore 189721,
Singapore

Organization

Chief Patron

Director, NIT Mizoram

Patrons

Prof. Saibal Chatterjee, Dean (Academics), NIT Mizoram
Dr. Alok Shukla, Dean (Dean RC), NIT Mizoram
Dr. P. Ajmal Koya, Dean (Faculty Welfare), NIT Mizoram
Dr. K. Gyanendra Singh, Dean (Students' Welfare), NIT Mizoram

General Chair

Dr. Jinshan Tang, College of Computing, Michigan Technological University, Michigan, US
Dr. Xin-She Yang, Middlesex University London, UK

Publication Chairs

Dr. Jerry Chun-Wei Lin, Western Norway University of Applied Sciences, Bergen, Norway
Dr. Peter Peer, Faculty of Computer & Information Science, University of Ljubljana, Slovenia

Prof. Vikrant Bhateja, Veer Bahadur Singh Purvanchal University, Jaunpur, Uttar Pradesh, India

Organising Chairs

Dr. Ranjita Das, Asst. Professor, Dept. of CSE, NIT Agartala
Dr. Sandeep Kumar Dash, Dept. of CSE, NIT Mizoram

Publicity Chairs

Dr. Sandeep Kumar Dash, Dept. of CSE, NIT Mizoram
Mr. Lenin Laitonjam, Dept. of CSE, NIT Mizoram
Ms. B. Sneha Reddy, Dept. of CSE, NIT Mizoram
Dr. Amit Kumar Roy, Dept. of CSE, NIT Mizoram
Dr. Ranjita Das, Asst. Professor, Dept. of CSE, NIT Mizoram

Advisory Committee

Aime' Lay-Ekuakille, University of Salento, Lecce, Italy
Annappa Basava, Department of CSE, NIT Karnataka
Amira Ashour, Tanta University, Egypt
Aynur Unal, Standford University, USA
Bansidhar Majhi, IIIT Kancheepuram, Tamil Nadu, India
Dariusz Jacek Jakobczak, Koszalin University of Technology, Koszalin, Poland
Dilip Kumar Sharma, IEEE U.P. Section
Ganpati Panda, IIT Bhubaneswar, Odisha, India
Jagdish Chand Bansal, South Asian University, New Delhi, India
João Manuel R. S. Tavares, Universidade do Porto (FEUP), Porto, Portugal
Jyotsana Kumar Mandal, University of Kalyani, West Bengal, India
K. C. Santosh, University of South Dakota, USA
Le Hoang Son, Vietnam National University, Hanoi, Vietnam
Milan Tuba, Singidunum University, Belgrade, Serbia
Naeem Hanoon, Multimedia University, Cyberjaya, Malaysia
Nilanjan Dey, TIET, Kolkata, India
Noor Zaman, Universiti Tecknologi, PETRONAS, Malaysia
Pradip Kumar Das, Professor, Department of CSE, IIT Guwahati
Roman Senkerik, Tomas Bata University in Zlin, Czech Republic
Sandeep Singh Sengar, Cardiff Metropolitan University, UK
Sriparna Saha, Associate Professor, Department of CSE, IIT Patna

Sukumar Nandi, Department of CSE, IIT Guwahati
Swagatam Das, Indian Statistical Institute, Kolkata, India
Siba K. Udgata, University of Hyderabad, Telangana, India
Tai Kang, Nanyang Technological University, Singapore
Ujjawl Maulic, Department of CSE, Jadavpur University
Valentina Balas, Aurel Vlaicu University of Arad, Romania
Yu-Dong Zhang, University of Leicester, UK

Technical Program Committee Chairs

Dr. Steven L. Fernandes, Creighton University, USA
Dr. Mufti Mahmud, Nottingham Trent University, Nottingham, UK

Technical Program Committee

A. K. Chaturvedi, Department of Electrical Engineering, IIT Kanpur, India
Abdul Rajak A R, Department of Electronics and Communication Engineering, Birla
Institute of Dr. Nitika Vats Doohan, Indore, India
Ahmad Al-Khasawneh, The Hashemite University, Jordan
Alexander Christea, University of Warwick, London, UK
Amioy Kumar, Biometrics Research Lab, Department of Electrical Engineering, IIT
Delhi, India
Anand Paul, The School of Computer Science and Engineering, South Korea
Anish Saha, NIT Silchar
Apurva A. Desai, Veer Narmad South Gujarat University, Surat, India
Avdesh Sharma, Jodhpur, India
Bharat Singh Deora, JRNRV University, India
Bhavesh Joshi, Advent College, Udaipur, India
Brent Waters, University of Texas, Austin, Texas, United States
Chhaya Dalela, Associate Professor, JSSATE, Noida, Uttar Pradesh, India
Dan Boneh, Computer Science Dept, Stanford University, California, USA
Dipankar Das, Jadavpur University
Feng Jiang, Harbin Institute of Technology, China
Gengshen Zhong, Jinan, Shandong, China
Harshal Arolkar, Immd. Past Chairman, CSI Ahmedabad Chapter, India
H R Vishwakarma, Professor, VIT, Vellore, India
Jayanti Dansana, KIIT University, Bhubaneswar, Odisha, India
Jean Michel Bruel, Departement Informatique IUT de Blagnac, Blagnac, France
Jeril Kuriakose, Manipal University, Jaipur, India
Jitender Kumar Chhabra, NIT, Kurukshetra, Haryana, India
Junali Jasmine Jena, KIIT DU, Bhubaneswar, India

Jyoti Prakash Singh, NIT Patna
K. C. Roy, Principal, Kautaliya, Jaipur, India
Kalpana Jain, CTAE, Udaipur, India
Komal Bhatia, YMCA University, Faridabad, Haryana, India
Krishnamachar Prasad, Department of Electrical and Electronic Engineering, Auckland, New Zealand
Lipika Mohanty, KIIT DU, Bhubaneswar, India
Lorne Olfman, Claremont, California, USA
Martin Everett, University of Manchester, England
Meenakhi Rout, KIIT DU, Bhubaneswar, India
Meenakshi Tripathi, MNIT, Jaipur, India
Mrinal Kanti Debbarma, NIT Agartala
M Ramakrishna, ANITS, Vizag, India
Mukesh Shrimali, Pacific University, Udaipur, India
Murali Bhaskaran, Dhirajlal Gandhi College of Technology, Salem, Tamil Nadu, India
Ngai-Man Cheung, Assistant Professor, University of Technology and Design, Singapore
Neelamadhav Padhi, GIET University, Odisha, India
Nilay Mathur, Director, NIIT Udaipur, India
Philip Yang, Price water house Coopers, Beijing, China
Pradeep Chouksey, Principal, TIT College, Bhopal, MP, India
Prasun Sinha, Ohio State University Columbus, Columbus, OH, United States
R K Bayal, Rajasthan Technical University, Kota, Rajasthan, India
Rajendra Kumar Bharti, Assistant Prof Kumaon Engg College, Dwarahat, Uttarakhand, India
S. R. Biradar, Department of Information Science and Engineering, SDM College of Engineering & Technology, Dharwad, Karnataka, India
Sami Mnasri, IRIT Laboratory Toulouse, France
Savita Gandhi, Professor, Gujarat University, Ahmedabad, India
Soura Dasgupta, Department of TCE, SRM University, Chennai, India
Sushil Kumar, School of Computer & Systems Sciences, Jawaharlal Nehru University, New Delhi, India
Ting-Peng Liang, National Chengchi University Taipei, Taiwan
V. Rajnikanth, EIE Deptt., St. Joseph's College of Engg., Chennai, India
Veena Anand, NIT Raipur
Xiaoyi Yu, National Laboratory of Pattern Recognition, Institute of Automation, Chinese Academy of Sciences, Beijing, China
Yun-Bae Kim, SungKyunKwan University, South Korea

Preface

This book is a collection of high-quality peer-reviewed research papers presented at the 10th International Conference on Frontiers in Intelligent Computing: Theory and Applications (FICTA-2022) held at National Institute of Technology, Mizoram, Aizawl, India during 18–19 June 2022, (Decennial Edition of FICTA Conference).

The idea of this conference series was conceived by few eminent professors and researchers from premier institutions of India. The first three editions of this conference: FICTA-2012, 2013 & 2014 were organized by Bhubaneswar Engineering College (BEC), Bhubaneswar, Odisha, India. The fourth edition FICTA-2015 was held at NIT, Durgapur, W.B., India. The fifth and sixth editions FICTA-2016 and FICTA-2017 were consecutively organized by KIIT University, Bhubaneswar, Odisha, India. FICTA-2018 was hosted by Duy Tan University, Da Nang City, Viet Nam. The eighth edition FICTA 2020 was held at NIT, Karnataka, Surathkal, India. The Ninth edition FICTA-2021 was held at NIT, Mizoram, Aizawal, India. All past editions of the FICTA conference proceedings are published by Springer. Presently, FICTA-2022 is the tenth edition of this conference series which aims to bring together researchers, scientists, engineers, and practitioners to exchange and share their theories, methodologies, new ideas, experiences, applications in all areas of intelligent computing theories and applications to various engineering disciplines like Computer Science, Electronics, Electrical, Mechanical, Bio-Medical Engineering etc.

FICTA-2022 had received a good number of submissions from the different areas relating to computational intelligence, intelligent data engineering, data analytics, decision sciences and associated applications in the arena of intelligent computing. These papers have undergone a rigorous peer-review process with the help of our technical program committee members (from the country as well as abroad). The review process has been very crucial with minimum 02 reviews each; and in many cases 3–5 reviews along with due checks on similarity and content overlap as well. This conference witnessed more than 500 submissions including the main track as well as special sessions. The conference featured many special sessions in various cutting-edge technologies of specialized focus which were organized and chaired by eminent professors. The total toll of papers included submissions received cross country along with many overseas countries. Out of this pool only 110 papers were given acceptance and segregated as two different volumes for publication under the proceedings. This volume consists of 55 papers from diverse areas of Evolution in Computational Intelligence.

The conference featured many distinguished keynote addresses in different spheres of intelligent computing by eminent speakers like: Dr. Xin-She Yang (Reader at Middlesex University London, UK) and Dr. Sumit K. Jha (Professor of Computer Science at the University of Texas-San Antonio (UTSA) US.) Dr. Xin-She Yang keynote lecture on "Nature-Inspired Algorithms: Insights and Open Problems" give an idea on Nature-inspired algorithms such as the particle swarm optimization, bat algorithm and firefly algorithm have been widely used to solve problems in optimization, data mining and computational intelligence. Also, Dr. Sumit K. Jha talk on the Trust in Artificial Intelligence received ample applause from the vast audience of delegates, budding researchers, faculty and students.

We thank the advisory chairs and steering committees for rendering mentor support to the conference. An extreme note of gratitude to Dr. Sandeep Kumar Dash (Head, Dept. of CSE, NIT Mizoram, Aizawl, India) and Dr. Ranjita Das (Dept. of CSE, NIT Mizoram, Aizawl, India) for providing valuable guidelines and being an inspiration in the entire process of organizing this conference. We would also like to thank Department of Computer Science and Engineering, NIT Mizoram, Aizawl, India who came forward and provided their support to organize the tenth edition of this conference series.

We take this opportunity to thank authors of all submitted papers for their hard work, adherence to the deadlines and patience with the review process. The quality of a refereed volume depends mainly on the expertise and dedication of the reviewers. We are indebted to the technical program committee members who not only produced excellent reviews but also did these in short time frames. We would also like to thank the participants of this conference, who have participated the conference above all hardships.

Jaunpur, Uttar Pradesh, India   Dr. Vikrant Bhateja
London, UK   Dr. Xin-She Yang
Bergen, Norway   Dr. Jerry Chun-Wei Lin
Agartala, India   Dr. Ranjita Das

Contents

About the Editors

Dr. Vikrant Bhateja is associate professor in Department of Electronics Engineering Faculty of Engineering and Technology, Veer Bahadur Singh Purvanchal University, Jaunpur, Uttar Pradesh, India. He holds a doctorate in ECE (Bio-Medical Imaging) with a total academic teaching experience of 19+ years with around 190 publications in reputed international conferences, journals and online book chapter contributions; out of which 35 papers are published in SCIE indexed high impact factored journals. Among the international conference publications, four papers have received "Best Paper Award". Among the SCIE publications, one paper published in *Review of Scientific Instruments (RSI) Journal* (under American International Publishers) has been selected as "Editor Choice Paper of the Issue" in 2016. He has been instrumental in chairing/co-chairing around 30 international conferences in India and abroad as Publication/TPC chair and edited 50 book volumes from Springer-Nature as a corresponding/co-editor/author on date. He has delivered nearly 20 keynotes, invited talks in international conferences, ATAL, TEQIP and other AICTE sponsored FDPs and STTPs. He has been Editor-in-Chief of IGI Global—*International Journal of Natural Computing and Research (IJNCR)* an ACM & DBLP indexed journal from 2017–2022. He has guest edited Special Issues in reputed SCIE indexed journals under Springer-Nature and Elsevier. He is Senior Member of IEEE and Life Member of CSI.

Xin-She Yang obtained his D.Phil. in Applied Mathematics from the University of Oxford. He then worked at Cambridge University and National Physical Laboratory (UK) as a Senior Research Scientist. Now he is Reader at Middlesex University London, Fellow of the Institute of Mathematics and its Application (IMA), and a Book Series co-Editor of the *Springer Tracts in Nature-Inspired Computing*. He was also the IEEE Computational Intelligence Society task force chair for Business Intelligence and Knowledge Management (2015 to 2020). He has published more than 25 books and more than 400 peer-reviewed research publications with over 73,000 citations, and he has been on the prestigious list of highly-cited researchers (Web of Sciences) for seven consecutive years (2016–2022).

Jerry Chun-Wei Lin received his Ph.D. from the Department of Computer Science and Information Engineering, National Cheng Kung University, Tainan, Taiwan in 2010. He is currently a full Professor with the Department of Computer Science, Electrical Engineering and Mathematical Sciences, Western Norway University of Applied Sciences, Bergen, Norway. He has published more than 400 research articles in top-tier journals and conferences, 12 edited books, as well as 33 patents (held and filed, three US patents). His research interests include data analytics, soft computing, artificial intelligence/machine learning, optimization, IoT, and privacy preserving and security technologies. He is the Editor-in-Chief of the *International Journal of Data Science* and *Pattern Recognition*. He has recognized as the most cited Chinese Researcher respectively in 2018, 2019, and 2020 by Scopus/Elsevier. He is the Fellow of IET (FIET), senior member for both IEEE and ACM.

Dr. Ranjita Das is currently working as an Assistant professor in the Department of Computer Science and Engineering, National Institute of Technology Agartala. She was previously working in NIT Mizoram as an Assistant Professor. She has total 11 years of Teaching experience. She did her Ph.D. from NIT Mizoram, M.Tech. from Tezpur University, and B.Tech. from NIT Agartala. Dr. Das's research covers wide areas related to Pattern recognition, Computational Biology, Information retrieval and image processing. She has published several research papers in reputed journals and conferences. She has been program chair and organizing chair of many international conferences and international workshops. She also serves as Reviewer, and Technical Program Committee Member for a number of IEEE and Springer journals and conferences. She won the best paper awards in FICTA-2021, IC4E2020, ICACCP-2019 & IEEE-INDICON 2015. She has been involved in sponsored research projects in the broad areas of Computational Biology and Machine Learning technologies, funded by the DBT and DST-SERB etc. She was the Board of Governor member of NIT Mizoram.

Chapter 1
SIFT Application Separates Motion Characteristics and Identifies Symbols on Tires

Nguyen Ha Huy Cuong, Doan Van Thang, Nguyen Trong Tung, Mai Nhat Tan, and Nguyen Thi Thuy Dien

Abstract Deep learning techniques have aided in the transfer of content-based image retrieval issues from manually constructed local features like scale-invariant feature transformations (SIFT) to features obtained from convolutional neural networks (CNNs). Existing image-based CNN features, which are taken directly from the entire picture, are not adequate for identifying small areas of overlap. These will have an impact on the performance of partial duplicate picture detection. In this work, we systematized the SIFT characteristics, compared them to other algorithms, and then employed it to separate the motion characteristics and recognize the characters on the tire tread. The experimental findings suggest that after enhancing the input data, the training process of digital characters may be identified with high accuracy.

1.1 Introduction

Identification of tire symbols and parameters is extremely important. Symbols and tire parameters can provide information about the tire type, production unit, wear level, maximum load, maximum pressure, heat resistance, speed limiter, etc. Each of these parameters affects a vehicle's performance and safety. Based on the extracted

N. H. H. Cuong
The University of Danang, Da Nang, Vietnam
e-mail: nhhcuong@sdc.udn.vn

D. Van Thang
Industrial University of Ho Chi Minh City, Ho Chi Minh, Vietnam
e-mail: doanvanthang@iuh.edu.vn

N. T. Tung (✉)
Dong A University, Da Nang, Vietnam
e-mail: tungqn@donga.edu.vn

M. N. Tan
Binh Duong Driving Test and Training Center, Binh Duong, Vietnam

N. T. T. Dien
The Quang Nam College, Quang Nam, Vietnam

features, we identify the tire symbols. Features must be extracted before content-based images can be analyzed. Matching natural photos is difficult due to factors such as lighting, scale, overlapping, viewing angle, and so on, which have a significant impact on accuracy. Some of the above challenges can be solved by David Lowe's image matching approach using the SIFT invariant feature. Face recognition has been made possible by convolutional neural networks (CNNs). Results show that CNN with dense SIFT performs better than conventional CNN and CNN with SIFT. We presented a novel method for improving SIFT matching accuracy, a crucial component in retrieval. To reduce noise interference, object export extracts compact CNN representations from several common feature regions. After that, a query-adapting method is proposed for selecting the appropriate CNN evidence to verify that each SIFT pair is matched. In this study [1], two different visual match verification functions are presented and evaluated; the author's optimized SIFT matches visible and short-wave infrared images.

1.2 Related Work

In [2], Wu and Zhou proposed a new end-to-end, coarse-to-fine deep synthesis model to reconstruct latent images from SIFT features. The designed deep phylogenetic model consists of two networks, the first of which attempts to learn the structural information of the latent image by converting SIFT features to local binary sample features (LBP), and the second of which attempts to reconstruct pixel values guided by the learned LBP; Zhao et al. [3] proposes a fast mosaic method based on scale-invariant property transformation (SIFT) to mosaic UAV images for monitoring plant growth. The process of dynamically determining the appropriate contrast threshold in the difference of space-to-Gaussian scale (DOG) based on the contrast character-istics of UAV images used to monitor plant growth; Dellinger et al. [4] developed SAR-SIFT, a special SIFT-like algorithm for SAR images. The algorithm includes key point detection as well as the computation of local descriptors. For the first time, a new gradient definition was introduced, which provides orientation and magni-tude that are consistent with noise; Chhabra et al. [5] proposes a content-based image retrieval (CBIR) system for extracting a feature vector from an image and retrieving content-based images one by one effectively. ORB detectors employ fast key-points, while descriptors employ descriptors. SIFT is a technique for analyzing images based on their orientations and scales; The improved algorithms PCA-SIFT, GSIFT, CSIFT, SURF, and ASIFT were compared by Wu et al. [6]. The results of the experiments show that each type has its own advantages. The best scaling and rota-tion algorithms are SIFT and CSIFT. CSIFT improves SIFT in terms of opacity and alignment change, but not in terms of brightness change. GSIFT performs best when dimming and changing brightness levels. ASIFT performs best when there is a link change. In various situations, PCA-SIFT is always second. SURF performs poorly in various situations but runs the fastest; In [7–9], authors Abu Doush and AL-Btoush

compared two approaches to the locally invariant feature descriptor to study the identification of banknotes intended to be classified as currency. Color and gray image
locally invariant feature descriptors (color SIFT approach) (gray SIFT approach). In
terms of processing time and accuracy, the evaluation results show that the color SIFT
method outperforms the gray SIFT method; In [10, 11], the author and collaborators
proposed using infrared rays to recognize human faces and provided a diagram of
the star window filter-SIFT (SWF-SIFT) to improve face recognition performance.
People in infrared can be identified by filtering out incorrect matches. Through tests
using a typical infrared human face database, the performance comparison between
SIFT and SWF-SIFT algorithm demonstrates the advantages of SWF-SIFT algorithm. To optimize the SIFT parameters to fit the recorded multi-wavelength image
sets, Author Zhu et al. [12] proposed the BP-SIFT image registration algorithm,
which treats crucial point matching of SIFT descriptors as a global optimization
problem and offers an optimal solution based on trust transfer (BP). With reasonable computational complexity, the test results show a significant improvement over
conventional SIFT-based matching. Although the SIFT descriptor has been shown
to outperform other existing local descriptors in image matching, it is not sufficiently discriminatory and powerful, particularly in the case of transforms. Multiple
mismatches can occur in affine and mirror. As a result, the authors of the study [13]
presented an enhancement to the SIFT descriptor for image matching and retrieval.
The steps in the proposed descriptor framework are as follows: Normalization of
the ellipse neighborhood, conversion to affine scale space A comparison of different
descriptors performed reveals that the current method outperforms existing methods;
Dou et al. [14] proposed a powerful image matching algorithm based on a combination of the wavelet transform and the scale-invariant feature transform (SIFT). First,
the corresponding discrete Wavelet transform is applied to the reference and sample
images to extract their low-frequency parts, and then Harris angle detection is used to
detect and match the interesting points in the samples, their low-frequency component
to determine the appropriate candidate region of the sample image in the reference
image using a *k-d* tree and a two-way matching strategy, extract SIFT features from
matching candidate regions and sample images. Finally, use the SIFT information to
generate match constraints and use them to find more precise matches. The experimental results show that the algorithm can improve matching accuracy. Meanwhile,
Rublee et al. [15] proposed ORB, a rotation invariant and noise resistant binary
descriptor based on BRIEF. ORB outperforms SIFT by two orders of magnitude in
a variety of real-world applications, including smartphone patch tracking and object
detection. High noise and local distortion in multimodal images reduce the accuracy
of scale-invariant feature-transform (SIFT) image matching. To address this issue,
the authors of the study [16] propose a new method based on the SIFT framework
that combines the phase consistency and gradient direction optimization strategy of
principal component analysis (PCA) with 8 decreasing latitude directions, which is
recommended; Authors Mistry and Banerjee in the study [17] where the comparison
of SIFT and SURF approaches is discussed. In terms of invariant rotation, blur, and

vertical transform, SURF outperforms SIFT. At different scale images, SIFT outperforms SURF. Because of the built-in image and box filters, SURF is three times faster than SIFT. SIFT and SURF are excellent at altering the lighting for images.

1.3   SIFT Characteristic Selection Model

Scale-invariant Feature Transform (SIFT), a common and very efficient technique based on invariant local features in the picture, was presented by David Lowe in 2004 and has since been modified. SIFT extracts key points, which are accompanied by descriptions and a vector with the key point as the origin.

The technique for obtaining invariant characteristics SIFT is approached via the pyramid filtering method, which involves the following stages:

Scale-space Extrema Detection: The derivative of the Gaussian function (DoG) will be used in this initial stage to discover probable feature sites. Typically, they are points with very little (invariant) reliance on magnification and rotation.

Key-point Localization: It will filter and extract the best collection of key points from the probable points listed above.

Orientation Assignment: Based on the gradient direction of the picture, each feature point will be allocated to one or more directions. Any subsequent processing procedures will be carried out on picture data that has been modified in relation to the given orientation, size, and position of each feature point. As a result, these operations will be consistent.

Key-point Descriptor: The local gradient directions are measured in a picture of a specific size near each feature point. They are then portrayed in a way that permits crucial phases of local shape distortion and brightness change to be depicted.

Identify Key Points

The Gaussian function, according to Koenderink (1984) and Lindeberg (1994), is the best function for representing the spatial measure of 2D images. As a result, an image's spatial measure will be defined as a do $L(x, y, \sigma)$ generated by convolution-alizing the original image $I(x, y)$ by a Gaussian function. The measured parameter of $G(x, y, \sigma)$ is variable.

$$L(x, y, \sigma) = G(x, y, \sigma) * I(x, y) \tag{1.1}$$

$L(x, y, \sigma)$: Scale-space function of the image I.
$G(x, y, \sigma)$: Gaussian scale variable (variable scale Gaussian).
$I(x, y)$: Input image.
* Is the convolution between x and y
and,

$$G(x, y, \sigma) = \frac{1}{2\pi\sigma^2} e^{-(x^2+y^2)/2\sigma^2} \tag{1.2}$$

The algorithm used to find highly invariant features is to find the local extrema of the derivative of the Gaussian function, abbreviated as Difference-of-Gaussian (DoG), denoted as $D(x, y)$. The difference between two adjacent spatial measures of an image is used to calculate this function, with the parameter difference being a constant k.

$$D(x, y, \sigma) = L(x, y, k\sigma) - L(x, y, \sigma) = (G(x, y, k\sigma) - G(x, y, \sigma)) * I(x, y) \tag{1.3}$$

The Gaussian function was chosen because it is a very efficient technique for computing L (as well as increasing image smoothness), and L always has to be computed a lot to describe the feature in the measurement space, and then D will be calculated simply with pixel matrix subtraction with a low implementation cost.

Furthermore, the derivative of the Gaussian function (DoG) can be used to produce a close approximation to Lindeberg's 1994 proposed standard-sized Laplace second derivative of the Gaussian function ($\sigma^2\nabla^2 G$). σ^2 demonstrates that in order for the measure invariant to be true, the second derivative must be normalized with the coefficient. He specifically stated that the maximum and minimum values of ($\sigma^2\nabla^2 G$) are the most stable (highly invariant) values when compared to other evaluation functions such as Gradient, Hessian, or Harris.

The relationship between D and ($\sigma^2\nabla^2 G$) is represented as follows:

$$\frac{\partial G}{\partial \sigma} = \sigma\nabla^2 G \tag{1.4}$$

So, can be calculated through partial approximation $\partial G / \partial \sigma$ at the closely related measurement parameters $k\sigma$ and σ:

$$\sigma\nabla^2 G = \frac{\partial G}{\partial \sigma} \approx \frac{G(x, y, k\sigma) - G(x, y, \sigma)}{k\sigma - \sigma} \tag{1.5}$$

Therefore: $G(x, y, k\sigma) - G(x, y, \sigma) \approx (k - 1)\sigma^2\nabla^2 G$.

Based on this formula, we can use the derivative of Gaussian (DoG) to approximate the Laplace second derivative of Gaussian when the measurement parameters deviate by a constant k. Because the coefficient $(k - 1)$ in the preceding equation is constant in all measurement spaces, it has no effect on the determination of extreme locations. When k is close to one, the error in approximating the second derivative approaches zero. The author's experimental results, on the other hand, show that the derivative approximation process has no effect on the detection of extreme positions. Even if k is chosen arbitrarily, for example, $k = 2$.

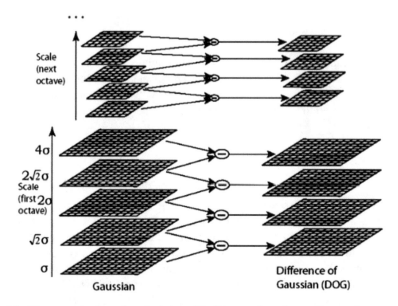

Fig. 1.1 Histogram simulates the calculation of DoG images from fuzzy adjacency images

As a result, the first step of the SIFT algorithm is to detect attractive points using a Gaussian filter at various scales and DoG images from various fuzzy adjacent images (Fig. 1.1).

Locating Attractions

Following step 1, a large number of potential points can be obtained as special points; however, some of them are unnecessary. The following step will remove points that have low contrast (are sensitive to noise), have fewer local features than others, or tend to be feature boundaries. This step is made up of three steps:

Neighborhood Interpolation Gives the Correct Location of the Potential Point

Neighborhood interpolation for the Difference of Gaussian function $D(x, y, \sigma)$ using the Taylor expansion:

$$D(X) = D + \frac{\partial D^T}{\partial X} X - \frac{1}{2} x^T \frac{\partial^2 D}{\partial^2 X} \tag{1.6}$$

where: D and its derivative are calculated at a potential point and $X = (x, y, \sigma)$ represents the distance from that point. Taking the derivative of the above function with the argument X and working toward 0 determines the position of the extreme point (Fig. 1.2):

$X = (x, y, \sigma)^T$: is the shift relative to the neighboring points of the sampling point.

The region containing the gravitational point is determined through $\hat{X}$.

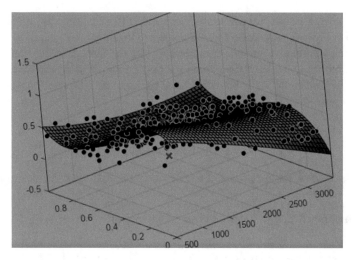

Fig. 1.2 Simulation using Taylor's extended formula for the DoG function

$$\hat{X} = -\frac{\partial^2 D^{-1}}{\partial X^2} \frac{\partial D}{\partial X} \tag{1.7}$$

If a dimension $\hat{X}$ has a value greater than 0.5, it has an extreme index that is not close to other potential points; it will be changed, and interpolation will replace its role with another point near it. Continue with the other sampling points.

Eliminate Points with Poor Contrast

The points that satisfy (0.5) are added to the best sample set, and the analysis is continued.

Used $D\left(\hat{X}\right)$ to get rid of unstable extremes (low contrast).

Substituting X into $D\left(\hat{X}\right)$ we get:

$$D\left(\hat{X}\right) = D + \frac{1}{2}\frac{\partial D^T}{\partial X}\hat{X} \tag{1.8}$$

If $D\left(\hat{X}\right) < 0.03$ then that sampling point will be rejected.

Remove the Redundancies Along the Boundary

When the position of the boundary is difficult to determine, using the DoG function has a strong impact on the boundary, and thus the potential points on the boundary are not invariant and noisy. In order to increase the stability of the points that will be chosen as special points, we will exclude potential difficult-to-locate points (i.e., the position is easy to change when there is noise due to being located at the boundary).

The image border will be unclear after applying the DoG function, and the main curvature will have a much larger value than the curvature along the border, so the special points along the same boundary must be removed. The solution is to use the value of a second-order Hessian matrix.

$$H = \begin{bmatrix} D_{xx} & D_{xy} \\ D_{xy} & D_{yy} \end{bmatrix} \tag{1.9}$$

H's eigenvalues are proportional to D's curvature, and its elements are D_{xx} and D_{yy}.

Determine the Direction for the Points of Attraction

After experimenting with various approaches to determining direction, the following method was discovered to produce the most consistent results. At each attractive point, the gradient direction graph in the vicinity of the gravitational point is computed. The formula determines the magnitude and direction of the gravitational points.

$$m(x, y) = \sqrt{(L(x + 1, y) - L(x - 1, y))^2 + (L(x, y + 1) - (L(x, y - 1))^2}$$
$$\theta(x, y) = \tan^- ((L(x, y + 1) - L(x, y - 1))/(L(x + 1, y) - (L(x - 1, y))) \tag{1.10}$$

In there:

$m(x, y)$: The magnitude of the orientation vector.

$\theta(x, y)$: The direction of the direction vector (expressed through the angle θ).

Immutable Local Feature Matching

To begin, in order to match the images, the corresponding set of keypoints from each image must be extracted using the steps outlined above. The matching will then be done on these key-point sets. The main step in the matching technique is to find a subset of keypoints in two images that match, and to do so, to find pairs of matching keypoints in the two images. The similar image region is the subset of **matching keypoints.**

1.4 The Experimental Results

SIFT Feature Extraction

The SIFT feature extraction process is performed according to the following steps:

Step 1: Read Images

The read (const string & filename, int flags) image read function reads an image from a specified file and returns a matrix of gray level values. The function returns an empty matrix if the image is not read. The following parameters are used in the function:

- Filename: The image to be loaded's name.
- Flags: Flags specify the color pattern of the loaded image; otherwise, it will read the image in its correct format.

  – CV_LOAD_IMAGE_ANYDEPTH: Returns a 16 or 32-bit image when the input image has the respective depth. Otherwise, the image is converted to 8 bits.
  – CV_LOAD_IMAGE_COLOR: Convert image to color.
  – CV_LOAD_IMAGE_GRAYSCALE: Convert to a grayscale image.

The show (const string& winname, InputArray mat) image display function displays an image in the specified window. The image will be displayed at its original size if the window is created with a flag, but it is limited by the screen resolution. Otherwise, the image will be reduced in size to fit the window. The following are the meanings of the function parameters:

- Winname: Name of the window.
- mat: Image matrix is displayed.

Using image processing transformations to recognize and match test images with train images (Fig. 1.3).

```
rotation_matrix = cv2.getRotationMatrix2D((num_cols/2, num_rows/2), 90, 1)
```

Step 2: Detect and Compute Feature Descriptions

In OpenCV, the following method allows to detect features:

Fig. 1.3 Read the preparation image for training and testing

```
cv2.xfeatures2d.SIFT_create()
```

```
train_keypoints, train_descriptor = sift.detectAndCompute(training_gray, None)
test_keypoints, test_descriptor = sift.detectAndCompute(test_gray, None)
```

In there:

- Training_gray, test_gray: The pixel matrix of the train image and the corresponding test image.
- Test_keypoint, train_keypoints: is a vector that stores the set of feature points for the training_gray and test_gray image matrices respectively.
- Test_descriptor, train_descriptor: describe the feature vectors for the training_gray and test_gray image matrices respectively.

Calculation results are displayed as shown in Fig. 1.4.

And on the train and test images, we find the corresponding number of keypoints as follows:

- The number of keypoints detected on train_image is 302.
- The number of keypoints detected on test_image is 91.

Step 3: Connect the Corresponding Feature Keypoints

- Create a Brute Force Matcher object for descriptor matching.

```
bf = cv2.BFMatcher(cv2.NORM_L1, crossCheck = False)
```

- Do a match between the SIFT descriptors of train_image and test_image.

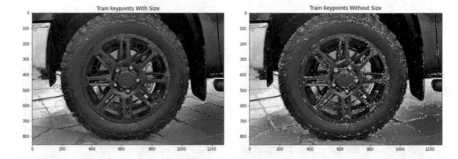

Fig. 1.4 Result using SIFT to find dimensioned and dimensionless features

```
matches = bf.match(train_descriptor, test_descriptor)
```

- Connect the distances between points with similar features.

```
matches = sorted(matches, key = lambda x : x.distance)
result = cv2.drawMatches(training_image, train_keypoints, test_gray, test_keypoints, m
atches, test_gray, flags = 2)
```

The number of matching keypoints of the two images train and test is 2568 points (Fig. 1.5).

The result of the feature detection process:

Character Recognition on Tires

Building a Character Image Dataset

Prepare a dataset consisting of original alphanumeric characters and changes in illumination, viewing angle, scaling, partial masking, etc. Images with sizes from 100 × 100px to 300 × 500px (Fig. 1.6).

The above dataset is divided into two datasets.

- Train set: Contains corresponding characters and numbers.
- Test set: Contains a few characters to check for accuracy.

After the data preparation is complete, each image is converted to black and white, each pixel has a value in the range [0…255], 0 for white, and 255 for black, respectively.

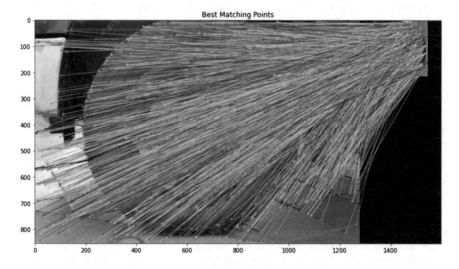

Fig. 1.5 Match the found features of two training (right) and test (left) images

Fig. 1.6 Alphanumeric and character data sets

Training and Testing Using SVM

- Feature scaling

Before putting data into training, it is necessary to perform data normalization. Data can come from many sources, with units and components having large variances. To calculate, we need to bring all the data to a common standard, the components have values in the same range like [0, 1] or [−1, 1].

To bring the value between [0, 1] [0, 1], we use the rescaling method with the following formula:

$$x' = \frac{x - \min(x)}{\max(x) - \min(x)}$$

Applied to our data, each pixel has a value in the range 0 … 255, so we have $\min(x) = 0$, $\max(x) = 255$, from which we there is a formula to normalize the data (Fig. 1.7).

1.5 Conclusion

The tire identification gave 2568 feature points for the matching results of the two images based on the experimental results. The process of identifying numbers on tires has also been done using two datasets, the Train set, and Test set, and yields accurate results. However, due to the machine's limited configuration to perform the simulation, the experiment on the input data set is not diverse, which affects the algorithm's efficiency when executing. The next step in the research will be to experiment on a large number of input datasets and improve the data classification in the algorithm before putting it through training.

In the future, we continue to study deep learning models for data classification, reinforcement learning methods, controlled learning, …. research to improve the

(a)   (b)

Fig. 1.7 **a** Evaluating the training model with the evaluators, **b** character recognition on tires

quality of classification models, we will focus on improving the speed of model training, experimenting on many large data sets and parallelizing the convolutional neural network model. In addition, we research and apply deep learning to the problem of resources allocation in cloud computing.

References

1. Sima, A.A., Buckley, S.J.: Optimizing SIFT for matching of short wave infrared and visible wavelength images. Remote Sens. **5**(5) (2013). https://doi.org/10.3390/rs5052037
2. Wu, H., Zhou, J.: Privacy Leakage of SIFT features via deep generative model based image reconstruction. IEEE Trans. Inf. Forensics Secur. **16** (2021). https://doi.org/10.1109/TIFS.2021.3070427
3. Zhao, J. et al.: Rapid mosaicking of unmanned aerial vehicle (UAV) images for crop growth monitoring using the SIFT algorithm. Remote Sens. **11**(10) (2019). https://doi.org/10.3390/rs11101226
4. Dellinger, F., Delon, J., Gousseau, Y., Michel, J., Tupin, F.: SAR-SIFT: a SIFT-like algorithm for SAR images. IEEE Trans. Geosci. Remote Sens. **53**(1) (2015). https://doi.org/10.1109/TGRS.2014.2323552
5. Chhabra, P., Garg, N.K., Kumar, M.: Content-based image retrieval system using ORB and SIFT features. Neural Comput. Appl. **32**(7) (2020). https://doi.org/10.1007/s00521-018-3677-9
6. Wu, J., Cui, Z., Sheng, V.S., Zhao, P., Su, D., Gong, S.: A comparative study of SIFT and its variants. Meas. Sci. Rev. **13**(3) (2013). https://doi.org/10.2478/msr-2013-0021
7. Doush, I.A., Btoush, S.A.: Currency recognition using a smartphone: comparison between color SIFT and gray scale SIFT algorithms. J. King Saud Univ. Comput. Inf. Sci. **29**(4) (2017). https://doi.org/10.1016/j.jksuci.2016.06.003
8. Yang, J. et al.: SIFT-aided path-independent digital image correlation accelerated by parallel computing. Opt. Lasers Eng. **127** (2020). https://doi.org/10.1016/j.optlaseng.2019.105964

9. Connie, T., Al-Shabi, M., Cheah, W.P., Goh, M. : Facial expression recognition using a hybrid CNN-SIFT aggregator. In: Lecture Notes in Computer Science (including subseries Lecture Notes in Artificial Intelligence and Lecture Notes in Bioinformatics), vol. 10607 (2017). https://doi.org/10.1007/978-3-319-69456-6_12

10. Tan, C., Wang, H., Pei, D.: SWF-SIFT approach for infrared face recognition. Tsinghua Sci. Technol. **15**(3) (2010). https://doi.org/10.1016/S1007-0214(10)70074-2

11. Zhang, G., Zeng, Z., Zhang, S., Zhang, Y., Wu, W.: SIFT Matching with CNN Evidences for Particular Object Retrieval. Neurocomputing **238** (2017). https://doi.org/10.1016/j.neucom.2017.01.081

12. Zhu, Y., Cheng, S., Stanković, V., Stanković, L.: Image registration using BP-SIFT. J. Vis. Commun. Image Represent. **24**(4) (2013). https://doi.org/10.1016/j.jvcir.2013.02.005

13. Liao, K., Liu, G., Hui, Y.: An improvement to the SIFT descriptor for image representation and matching. Pattern Recogn. Lett. **34**(11) (2013). https://doi.org/10.1016/j.patrec.2013.03.021

14. Dou, J., Qin, Q., Tu, Z.: Robust image matching based on the information of SIFT. Optik **171** (2018) https://doi.org/10.1016/j.ijleo.2018.06.094

15. Rublee, E., Rabaud, V., Konolige, K., Bradski, G.: ORB: an efficient alternative to SIFT or SURF (2011). https://doi.org/10.1109/ICCV.2011.6126544

16. Liu, H., Luo, S., Lu, J., Dong, J.: Method for fused phase and PCA direction based on a SIFT framework for multi-modal image matching. IEEE Access **7** (2019). https://doi.org/10.1109/ACCESS.2019.2953539

17. Mistry, D., Banerjee, A.: Comparison of feature detection and matching approaches: SIFT and SURF. GRD J. Glob. Res. Dev. J. Eng. **2** (2017)

Chapter 2
Comparative Study on Different Approaches for Understanding the Privacy Policies

Souvik Maitra and Dwijen Rudrapal

Abstract Privacy policies are the policies designed by businesses to interact with the user's personal information. In this age of digital world, every website is dealing with user's sensitive information which is mandatory in nature for using the content of the websites. But the fact is that most of the users do not read privacy policies due to the poor readability and time consuming task. This sharing of information is a potential threat to the user considering the information may reveal to any unwanted environment. Prior research proposed various approaches to summarize or extract key contents from the privacy policies to get the gist with least time. In this work, we report a detail review of prior researches on understanding of privacy policies. We also review the evaluation procedure of proposed approaches by surveying recent evaluations. At the end, we highlighted the current research challenges in this domain.

2.1 Introduction

Privacy policy is a document [1, 12] published in digital or physical form by an organization that represents which personal information of the user is collected, managed, utilized shared and protected by the organization. These are the legal documents which follows legal clauses while handling the user's information. Organizations present various terms and conditions in the privacy policies for utilization of their products and services. Hypothetically, the terms and conditions are to be read by the users and agreed if they are acceptable to the users before using the products or services. But in reality, the privacy policies include a large amount of text and legal clauses [9] that they are not easy to understand. Often, the texts in the policies are presented in complex manner with small fonts or complex sentences that users lost interest to these documents. Thus, most of the times, users share personal information to the organizations to access a product or service knowingly or unknowingly. This sharing of information poses a threat towards the user's privacy. In the last decade, many promising research works have been conducted to benefit user under-

S. Maitra (✉) · D. Rudrapal
National Institute of Technology Agartala, Agartala, Tripura, India
e-mail: nascar.20296@gmail.com

© The Author(s), under exclusive license to Springer Nature Singapore Pte Ltd. 2023
V. Bhateja et al. (eds.), *Evolution in Computational Intelligence*, Smart Innovation,
Systems and Technologies 326, https://doi.org/10.1007/978-981-19-7513-4_2

standability by making privacy policy documents short and specific. The direction of those works intends to extract the important and relevant sentences about the terms and conditions and presents to the users as per their requirements. The outcomes of previous researches make the user well aware before sharing of personal information that which crucial information is going to be shared and the associated threats. All the previous research works mainly focus on two approaches. One, summarization-based approaches to extract the required points by processing the whole document. Two, question-answer-based approach to present a specific answer in the Boolean form (Yes/No) or statement in reply of the query or question placed by user.

In this survey paper, we reviewed the previous promising research works in this domain by comparing the techniques and evaluation methodologies followed. Primarily, we have investigated the corpora and techniques used in various proposed approaches for understanding the privacy policies at minimum effort. Then, different evaluation procedures are discussed to establish the quality of the content for understanding the policies. We have highlighted limitations of various proposed techniques and noted research outcomes as well as challenges in this domain. The paper is organized as follows: Sect. 2.2 describes various corpora used and experimented by promising research works. Section 2.3 describes different state-of-the-art approaches followed by review of evaluation methodologies in Sect. 2.4. In Sect. 2.5 reports various research challenges followed by the conclusion of survey work in Sect. 2.6.

2.2   Data set

A lion share of the previous research work on privacy policies mainly targeted English corpora of variable number of policies ranging from the lowest in [12] with 35 privacy policies to a million number of policies used in the work by [2]. The nature of the policies is also domain independent covering the area of mobile applications, e-commerce, insurance, private companies and many more. The data set experimented in [14] is having 76 numbers of policies. The policies are collected by downloading from the web which belong to major related to a wide range of business areas. The data set OPP-115 used in [1, 6, 13] is an annotated data set containing 115 privacy policies. In addition, [13] utilizes 35,000 more privacy policies along with another data set containing privacy-related questions, both of the data sets being inherited from the PriBot project [6]. For [6], 130,326 policies of android applications were used in the process taken from the PlayDrone project [15]. The work by [17] uses a corpus of 400 privacy policies of different companies chosen randomly from a list by New York Stock Exchange, Nasdaq and AMEX. The data set used in work by [1] is a collection of 35 policies of different applications from Google play store. And the PolicyQA being the main focus of the work contains 25,017 comprehension style examples extracted from OPP-115.

The researchers have also built an interface [2] to track changes in the evolution of privacy policies over time.

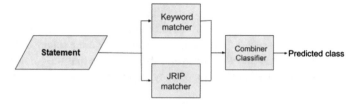

Fig. 2.1 Schematic components of the two-level system

2.3 Various Approaches for Understanding the Privacy Policies

In the last decade, researchers have proposed various approaches to understand the privacy policies in the simplest way considering the volume of text, complexity and time. The proposed approaches can be categorized in two classes. One, based on the summarization of the content and other, based on the question-answer approach. Summarization approaches are based on conventional summarization techniques formerly applied on traditional text to extract the gist of the policy information. Question-answer-based approaches focuses on the specific queries of the users and retrieval of the most suitable relevant information from the policy information. The following subsections describe various techniques under each approach.

2.3.1 Summarization-Based Approaches

The main goal of the summarization-based approaches is to provide the user with a short version of the typical lengthy policy. The system generates a summary of a policy rather than the full text. The works in this domain primarily use machine learning (ML) or deep learning (DL) techniques. The work in [17] summarizes privacy policy content to inform users about how their data are being accessed. The approach is applicable on any online privacy policy. It provides a graphical at-a-glance representation of the privacy policies. The work proposed in [14] summarizes a policy by extracting the sentences that address the key points. The approach uses two-level hierarchical structures as depicted in Fig. 2.1 to extract sentences. In the first level, keywords are extracted based on manually drafted patterns provided by human domain experts as well as based on the set of rules derived from the text patterns. In the second level, a machine learning-based classifier is applied to categorize relevant and irrelevant sentences to generate summary.

The system proposed in [17] is a semi-automatic approach focused on privacy factors like email, third party sharing, etc. in policies. Keywords selected through manual process and then text snippets are generated for each privacy factor using classifications models.

The proposed approach in [2] concludes that many privacy policies may not be disclosing all present third parties and an alternate explanation has been assumed as specific technologies may not be named because they are explained in the third parties instead of the first party's. The work also suggests that the implementation of the "notice and choice" model for privacy policy regulation is not maintained properly.

The research work by [7] focused on riskiest sentences in each privacy policy and generates summary by minimizing information redundancy. The work incorporates domain information predicted by the classifier and compares risk and coverage-focused content selection mechanism to form a summary. The generated summary consists of sentences that acquire the highest risk score.

2.3.2   Question-Answer-Based Approaches

The other way of providing a solution to understanding these lengthy privacy policies is a question-answering (QA) approach. Closed QA systems [4, 8] to open QA systems [4, 18] along with privacy assistants capable of engaging in dialogues with users have been developed to help the users understand about the privacy of their data and information. The closed QA system refers to providing answers for a pre-defined set of questions that users generally have in mind while reading a privacy policy. Open QA system refers to formulating answers from large-scale unstructured documents for questions.

2.3.2.1   Set of Questionnaire

In-depth analysis of the approaches based on question-answer to understand the privacy policies reveals that the researchers considered different sets of questions/points in their work. These questions can be broadly categorized into nine categories [12]. The categories are:

1. **First Party Collection/Use**: This category mainly addressed what information is being collected, why the information is being collected and how the information is being collected and used.
2. **Third Party Sharing/Collection**: Which user data is being shared, for what purpose it is being shared and how it is shared with the third parties are the queries addressed in this category.
3. **Data Security**: This category primarily focus on the measures those are being taken to keep the user's data safe. So, how secure the information with the organization is the question included in this category.
4. **Data Retention**: This category enquires about the time duration for which user's data will be retained with the organization.

5. **User Choice/Control**: What control options do the users have over their data can be found in this category.
6. **User Access, Edit and Deletion**: Can a user access his/her data, if so, then how it can be accessed and other facilities like modifying the user data or even deleting the information are the lookouts of this category.
7. **Policy Change**: This category contains the information whether the policy contains any context about informing the user if changes are made in the current privacy policy.
8. **International and Specific Audiences**: Any applicable clauses present in the policy for specific users are addressed in this category.
9. **Other**: All other general topics that are not covered by any of the previous categories are addressed by this category.

The questions elicitation were either done through professional crowd workers or skilled personnel. To avoid inadvertent biases, the contents of the privacy policies are kept hidden from the crowd workers or the skilled personnel. The crowd workers cite questions that come to their mind after looking at the website or the applications. And in contrast, experts are recruited to identify answers to the cited questions. In many cases, the experts provided answers along with evidence from the policies supporting them.

2.3.2.2   Proposed Approaches

The work proposed in [12] is built on closed QA approach where long text spans are extracted from the policies to answer questions. The approach is experimented on the OPP-115 corpus [16]. The set of questions prepared by crowdsourcing. To avoid inadvertent biases while crowdsourcing questions, the content of the privacy policies was kept hidden from the crowdsourcing task. Legal experts were hired to provide answers for the elicited questions. Each question was characterized using 3 sets of features based on SVM along with Bag-of-words (SVM-BOW), length of question in words (SVM-BOW+LEN) and parts-of-speech (SVM-BOW+LEN+POS). The experimental setup tried to evaluate the model's ability to provide answers to an isolated question. The proposed approach also used GloVe [10] and BERT [5] for encoding and fine tuning the system performance.

Recognizing the limitations of a closed QA system, the proposed approach [13] has developed an open QA system, where the user can ask any question for which the system tries to return a best matching answer from the privacy policy document. The open QA system has an upper hand over the closed QA system as users may be concerned about privacy issues that are not regularly discussed. The closed domain system addresses the concern and provides a solution for the same. It divides the policy documents into segments and then the segments are ranked based on BM25 scores of the segments with the user's query. The segment with the highest rank is considered as the answer. A length reduction strategy based on BM25 was used to further shorten the replies by the assistant.

The work proposed in [6, 13] is based on both open QA system and closed QA system to present answers of the questions from privacy policies. The machine learning-based classifiers developed for closed QA system works on both supervised and unsupervised stages. Domain-specific word vectors are built in the supervised stage and the neural networks having an upper hand over traditional techniques are utilized for the unsupervised stage. The proposed approaches show that the use of generic word vectors significantly improves the classification results. Generated machine learning-based models predict the answer for a specific question or query from the input policy content and further classify based on each privacy attribute. Embedding of the words also has a significant contribution in the open QA field where it addresses free-form questions and answers from the policies.

The approach proposed in [1] extracts shorter text span from the privacy policies to present the answers of the questions quickly. The proposed model trained on word embedding of policy content by using fastText [3] approach. The performance of the model is experimented by pre-training the model on SQuAD [11]. Performance shows that the proposed model outperforms over the experimented models. The experiments also indicate that training BERT on a larger corpus produces an increase in performance of the system. Table 2.1 presents a brief summary of reviewed research works with key aspects.

Table 2.1 Summary of the prior research works on different aspects

Work	Approach	Data set	Techniques	Key features	Evaluation metrics
[14]	Summarization	Company policies	Two-level ensembly hierarch	Keyword matching	Precision, recall and F measure
[13]	QA based	Company policies	Deep neural model	Bag-of-words	Precision and MPP
[17]	Summarization	Company policies	Machine learning	Bag-of-words	Precision
[12]	QA based	Mobile apps	Neural network	Bag-of-words, Part-of-speech	Precision
[1]	QA based	OPP-115	Neural network	BiDAF, BERT	$F1$ measure
[7]	Summarization	Company policies	Neural network	Bag-of-words	Precision, recall and F measure
[6]	Summarization	OPP-115	Neural network	Bag-of-words	Precision, recall and F measure

2.4 Evaluation Methodologies

Evaluation of summary or answers of questions is a complex task. Different persons may have different understanding, context, views and directions regarding a topic. So, it is very common that summary of a topic or answer of a question may vary person to person. However, most of the content will be similar in different human generated summaries or answers for a specific question. Comparison and evaluation of summaries or answers generated by different approaches are a challenging task due to the absence of standard annotated data set. Prior research works have followed various evaluation metrics and justified their proposed approach performance.

Evaluation of the system using the $F1$ scores is a popular technique. Yet in the early works, it was not much in practice. A tenfold cross-validation was used to evaluate the performance of the system proposed in [14] that identified sentences belonging to the stated 5 risk categories. The precision scores were comparatively higher in comparison to recall (which exhibits lesser accuracy for the system). Abstractive models being difficult to evaluate, mostly have to rely on manual analysis. A set of questions prepared manually was used to evaluate the system built using BM25 as described in the work by [13]. Evaluation of the deep neural model was done based on question-answer pairs generated. Mean Reciprocal Rank (MRR) was used to state the accuracy of both the approaches along with top-1 precision only for the deep neural model.

Evaluation method in [17] employed a three step process. Firstly, it was manually checked for 50 privacy policies followed by comparison with some alternative tools (P3P, ToS;DR, Privee). Lastly, feedback from the users using the PrivacyCheck browser extension was also considered for the evaluation process. $F1$ score being the popular evaluation metric finds its place assessing several models. A sentence-level $F1$ is calculated and the average of maximum $F1$ from each $n-1$ subset is considered for evaluating the proposed model in the work by [12]. For a better understanding of the system's performance recall and precision was also computed. Similarly, $F1$ was accompanied with exact match of the tokens to evaluate the accuracy of the model in [1]. The work by [7] used micro-$F1$ and macro-$F1$ to denote the ability of the system to identify the risk sentences in the policies. The approach evaluated the content of the generated summaries using ROUGE metrics. The work proposed in [6] evaluated by using a chart containing precision, recall, $F1$ and Top-1 precision.

2.5 Research Challenges

In-depth study of the previous promising research works reveals that there are many aspects yet unsolved which pose challenges to develop a robust system for quick understanding of privacy policies. First of all, the length of the privacy policy documents has increased over time. On top of that, more than 20Secondly, newer privacy factors have come up which need to be incorporated into the models for better performance. Thirdly, the lack of a gold standard data set remains to be a big limitation in this research domain. Most of the models developed till now experimented on OPP-115 annotation scheme which needs to be upgraded to keep up with the changing trends in creating privacy policy legislation. Lastly, the most important setback till now is that while constructing the summary which points need to be included and which do not. This is because a certain point may be important to one person but it may not have much importance to another. This implies that we need a person specific summary approach in future to overcome this limitation.

2.6 Conclusion

Studying the works can always bring refinement in future approaches to come. Evidences acquired from the previous promising researches state that the users mostly search for specific categories of risks in the privacy policies. This review work made a comprehensive study of the recent proposed research for understanding the privacy policies quickly and easily. The approaches are mainly summarization and question-answer based. The current work also reviewed the evaluation methodologies and matrices employed to asses the performance of proposed systems. As the contributions in this field have increased, it has become evident that better word representation can have a greater impact on the performance of the system. In case of supervised summarization methods, a risk-focused and coverage-focused content selection mechanism performs better. At the of the detail review of recent research works, this paper highlighted the research challenges in this domain.

Scopes for betterment always persist as for summarization-based methods the classifier design can be improved to reduce redundancy and aim for a nearly perfect system. By enhancing the productivity of the annotators to identify the risk-focused areas can remarkably help in annotating larger data sets for training abstractive models. QA-based systems developed so far have not yet been robust. With an average precision average, the closed QA systems can only answer pre-defined questions. As the open QA approaches can definitely address more user queries than the closed QA approaches, the scope of works on open QA systems is still remaining unexplored and can significantly contribute to the improvement of the system.

References

1. Ahmad, W.U., Chi, J., Tian, Y., Chang, K.W.: Policyqa: a reading comprehension dataset for privacy policies. arXiv Preprint (2020). arXiv:2010.02557
2. Amos, R., Acar, G., Lucherini, E., Kshirsagar, M., Narayanan, A., Mayer, J.: Privacy policies over time: curation and analysis of a million-document dataset. In: Proceedings of the Web Conference 2021, pp. 2165–2176 (2021)
3. Bojanowski, P., Grave, E., Joulin, A., Mikolov, T.: Enriching word vectors with subword information. Trans. Assoc. Comput. Linguist. 5, 135–146 (2017)

4. Caballero, M.: A brief survey of question answering systems. Int. J. Artif. Intell. Appl. **12**, 01–07 (2021)
5. Devlin, J., Chang, M.W., Lee, K., Toutanova, K.: Bert: pre-training of deep bidirectional transformers for language understanding. In: Burstein, J., Doran, C., Solorio, T. (eds.) NAACL-HLT (1), pp. 4171–4186. Association for Computational Linguistics (2019). https://dblp.uni-trier.de/db/conf/naacl/naacl2019-1.html
6. Harkous, H., Fawaz, K., Lebret, R., Schaub, F., Shin, K.G., Aberer, K.: Polisis: automated analysis and presentation of privacy policies using deep learning. In: 27th {USENIX} Security Symposium ({USENIX} Security 18), pp. 531–548 (2018)
7. Keymanesh, M., Elsner, M., Sarthasarathy, S.: Toward domain-guided controllable summarization of privacy policies. In: NLLP@ KDD, pp. 18–24 (2020)
8. Lende, S., Raghuwanshi, M.: Closed domain question answering system using NLP techniques (2016)
9. McDonald, A.M., Cranor, L.F.: The cost of reading privacy policies. Isjlp **4**, 543 (2008)
10. Pennington, J., Socher, R., Manning, C.D.: Glove: global vectors for word representation. In: EMNLP, vol. 14, pp. 1532–1543 (2014)
11. Rajpurkar, P., Zhang, J., Lopyrev, K., Liang, P.: Squad: 100,000+ questions for machine comprehension of text. arXiv Preprint (2016). arXiv:1606.05250
12. Ravichander, A., Black, A.W., Wilson, S., Norton, T., Sadeh, N.: Question answering for privacy policies: combining computational and legal perspectives. arXiv Preprint (2019). arXiv:1911.00841
13. Sathyendra, K.M., Ravichander, A., Story, P.G., Black, A.W., Sadeh, N.: Helping users understand privacy notices with automated query answering functionality: an exploratory study. Technical report, Carnegie Mellon University (2017)
14. Tomuro, N., Lytinen, S., Hornsburg, K.: Automatic summarization of privacy policies using ensemble learning. In: Proceedings of the Sixth ACM Conference on Data and Application Security and Privacy, pp. 133–135 (2016)
15. Viennot, N., Garcia, E., Nieh, J.: A measurement study of google play. In: The 2014 ACM International Conference on Measurement and Modeling of Computer Systems, pp. 221–233 (2014)
16. Wilson, S., Schaub, F., Dara, A.A., Liu, F., Cherivirala, S., Leon, P.G., Andersen, M.S., Zimmeck, S., Sathyendra, K.M., Russell, N.C., et al.: The creation and analysis of a website privacy policy corpus. In: Proceedings of the 54th Annual Meeting of the Association for Computational Linguistics (vol 1: Long Papers), pp. 1330–1340 (2016)
17. Zaeem, R.N., German, R.L., Barber, K.S.: Privacycheck: automatic summarization of privacy policies using data mining. ACM Trans. Internet Technol. (TOIT) **18**(4), 1–18 (2018)
18. Zhu, F., Lei, W., Wang, C., Zheng, J., Poria, S., Chua, T.: Retrieving and reading: a comprehensive survey on open-domain question answering. CoRR (2021). https://arxiv.org/abs/2101.00774

Chapter 3
A Framework for the Development of Intrusion Detection System Using Deep Learning

Madhab Paul Choudhury and J. Paul Choudhury

Abstract An intrusion detection system (IDS) is a system that monitors network traffic for suspicious activity and issues alerts when such activity has been discovered. Any malicious venture or violation is normally reported either to an administrator or collected centrally using a security information and event management (SIEM) system. In this paper, an effort has been made to apply multivariate statistical tools on the available data items for eliminating redundant items and to form a cumulative data item. Under multivariate statistical tool, factor analysis and principal component analysis have been used. The particular multivariate statistical tool has been chosen based on the minimum value of value of standard deviation and coefficient of variation using the estimated data based on factor analysis and on principal component analysis. Thereafter, deep learning-based neural network has to be applied on the cumulative data item as formed using the selected multivariate statistical tool. *K*-means clustering algorithm has to be applied on the estimated data based on deep learning based neural network to get optimum number of clusters. Based on the clusters as formed and available network type data, confidence matrix has to be formed which can indicate the accuracy of the proposed system. Finally certain test data (containing network parameters) have been taken and the proposed model has to be applied with the test data indicating whether the prediction made by the proposed model is correct or not. KDD cup 99 data set have been taken from UCI machine repository data have been used here.

M. P. Choudhury
NIT Jamshedpur, Jamshedpur, Jharkhand, India

J. Paul Choudhury (✉)
Kalyani Government Engineering College, Kalyani, West Bengal, India
e-mail: jnpckgec@gmail.com

Narula Institute of Technology, Kolkata, India

© The Author(s), under exclusive license to Springer Nature Singapore Pte Ltd. 2023   25
V. Bhateja et al. (eds.), *Evolution in Computational Intelligence*, Smart Innovation,
Systems and Technologies 326, https://doi.org/10.1007/978-981-19-7513-4_3

3.1   Introduction

Insertion of malicious activity in a network is called threat. It is necessary to analyze the security ecosystem to identify any malicious activity that can hamper the activities of the network. If a threat is detected, then mitigation efforts must be enacted to properly neutralize the threat before it can exploit any present vulnerability.

Vinayakumar et al. [1] have proposed the holistic approach to obtain a real-time intrusion detection system using the deep learning technique. Here the information regarding the current network traffic attacks have been furnished. The authors have made an effort to scale the performance of a Network anomaly detection between KDD 99 and varied KDD cup 99 data sets.

Moradi and Zulkernine [2] have proposed a neural network based system for Intrusion Detection and Classification of attacks to solve a multi class problem. Here a three class has been described and that can be extended to cases with more attack types. For that reason a neural network with three layers has been used. Under Matlab, neural network toolbox has been used for MLP networks.

In identifying important features for intrusion detection using support vector machines and neural networks. Andrew Sung and Mukkamala [3] have proposed an intrusion detection model used by DARPA. It has identified important features of intrusion detection system which has produced more appropriate result.

Hidalgo-Espinoza et al. [4] have applied artificial neural networks with deep learning approaches using Intrusion Detection Evaluation Dataset (CICIDS2017) in the Canadian Institute for Cybersecurity (CIC). The objective of building the dataset is to perform a massive amount of different controlled hacking attacks from different parts of the world to the dedicated servers. Here in the deep learning model, two different ARN architectures have been used as a shallow network for the usage of a back propagation algorithm (BPA) and a deep network formed by artificial neural network (ANN) architectures as a two-layered auto encoder and the back propagation algorithm (BPA).

A multi-dimensional approach toward intrusion detection system have been suggested by Manoj Ramesh Chandra Thakur and Sanyal [5]. Here, a multi-dimensional approach toward intrusion detection has been proposed. Network and system usage parameters like source and destination IP addresses, source and destination ports, incoming and outgoing network traffic data rate and number of CPU cycles per request have been divided into multiple dimensions. Hongyu Liu and Lang [6] have performed a survey comprising of machine learning and deep learning based models in the field of intrusion detection.

From the literature review, it has been found that neural network, support vector has been used in intrusion detection system [2, 4, 5]. Deep neural network has been used in [4]. But, no work has been done to find out the type of malicious attack in the network based on the available parameters of the network. The performance of the proposed model has not been examined. That is the reason for doing this type of work. The research work in the field of soft computing has been done in [7–9].

Here, in this paper an effort has been made to apply multivariate statistical tools on the available data items for eliminating redundant items and to form a cumulative data item. Under multivariate statistical tool, factor analysis and principal component analysis have been used. The particular multivariate statistical tool has been chosen based on the minimum value of value of standard deviation and coefficient of variation using the estimated data based on factor analysis and principal component analysis. Thereafter, deep learning based neural network has been used on the cumulative data based on selected multivariate statistical tool to get the estimated data. K-means clustering algorithm has to be applied on the estimated data based on deep learning based neural network to get optimum number of clusters. After that, based on clusters as formed and available network type data, confidence matrix has to be formed. The performance of confidence matrix has to be compared with the confidence matrix as formed without the application deep learning based neural network to verify the superiority of the deep learning based neural network model as compared to without the application deep learning based neural network model. Finally, the proposed model has to tested with certain test data indicating the prediction is correct or not.

This paper has been organized as follows. Necessity of network intrusion detection system and objective of the paper is furnished in Sect. 3.1. Theory of proposed model has been described in Sect. 3.2. The contribution of multivariate statistical tool (FA and PCA) has been discussed in Sects. 3.3.1 and 3.3..2. The evaluation of the performance of multivariate statistical tool (FA and PCA) has been furnished in Sect. 3.3.3. The formation of a number of clusters has been discussed in Sect. 3.3.4. The number of clusters indicate the number of the network attack types which have been occurred already. To improve the performance of the total proposed system deep learning based neural network model has been used as discussed in Sect. 3.3.5. The formation of a number of clusters has been discussed using estimated data based on deep learning based neural network model in this section. Testing of the proposed model using new network data has been performed in Sect. 3.3.6. The evaluation of the performance of the proposed model has been discussed in Sect. 3.4. Finally the views of the proposed model has been furnished in Sect. 3.5. The flow diagram of the proposed contribution has been furnished in Fig. 3.1 which has been placed at the end of paper.

3.2   Multivariate Statistical Tool

Multivariate statistical analysis [10] is considered a useful tool for evaluating the significance of anomalies in relation to both any individual variable and the mutual influence of variables on each other [10]. In basic terms, multivariate analysis aims to identify spatial correlations between groups of elements in a complex system and reduce a multidimensional data set to more basic components. Principal component analysis (PCA), factor analysis (FA), and cluster analysis (CA) are some of the most widely used multivariate analysis techniques. The result of a multivariate

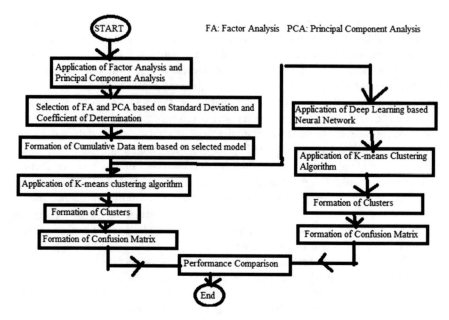

Fig. 3.1 Flow diagram of proposed contribution (snapshot)

analysis is an array of data in which elements are grouped as associations by means of their correlation coefficients or other measures of association.

3.2.1 Factor Analysis

Step 1:   Standardize the dataset.
Step 2:   Calculate the correlation matrix for the features in the dataset.
Step 3:   Calculate the eigenvalues and eigenvectors for the covariance matrix.
Step 4:   Sort eigenvalues and their corresponding eigenvectors.
Step 5:   Pick k eigenvalues and form a matrix of eigenvectors.
Step 6:   Transform the original matrix.

3.2.2 Principal Component Analysis (PCA)

Principal component analysis (PCA) is a statistical procedure that is used to reduce the dimensionality. It uses an orthogonal transformation to convert a set of observations of possibly correlated variables into a set of values of linearly uncorrelated variables called principal components. It is often used as a dimensionality reduction technique.

Steps Involved in the PCA

Same as factor analysis in step 2. Calculate the covariance matrix for the features in the dataset.

3.3 Implementation

3.3.1 Factor Analysis

Step 1: The KDD Cup 99 dataset have been taken from UCI machine repository [11] which have been used as input data set.

Step 2: Data have been standardized by deleting redundant rows and unnecessary columns to get revised dataset (matrix).

Step 3: The correlation coefficient among various network parameters (columns) of the dataset.

Step 4: Now the Eigen value and Eigen vector have been computed based on correlation coefficient matrix.

Step 5: Now it is necessary to calculate the percentage contribution of Eigen value among all Eigen values. It is known that if the Eigen value percentage value is less than 2.5 then those corresponding datasets will be omitted.

Step 6: The major factors have been formed as the product of the square root of the Eigen value with the square of Eigen vector.

Step 7: The sum of each major factors will be computed as row wise to get the cumulative factor value.

Step 8: Now based on the cumulative factor value and column item, a relation has been formed to produce the actual data source value. Using the actual values and the relation as formed, the estimated data value based on factor analysis has been produced.

3.3.2 Principal Component Analysis

Step 1: The covariance matrix has been computed using various network characters (columns) of the dataset as produced in step-2 (Sect. 3.3.1) of factor analysis.

Step 2: Now the Eigen values and Eigen vectors have been computed based on covariance matrix.

Step 3: Steps as narrated in step-5 to step-8 of factor analysis (Sect. 3.3.1) have been repeated to compute the estimated data value based on Principal component analysis.

Table 3.1 Comparative study of factor analysis and principal component analysis

Model	Factor analysis	Principal component analysis
Mean	4538	7841.8
Standard deviation	8937.2	11,999
Coefficient of variation	0.5078	0.6536

3.3.3 Comparison on the Performance of Factor Analysis and Principal Component Analysis

A comparative study has been made between estimated data based on factor analysis and on principal component analysis using the value of standard deviation and coefficient of determination. Lesser value of standard deviation and that of coefficient of determination indicates the superiority of the model. From Table 3.1, it has been observed that factor analysis is preferable as compared to principal component analysis. Therefore factor analysis has been chosen as selected multivariate statistical tool. The estimated data based on factor analysis has to be used in the next step.

3.3.4 Contribution Using Clustering Algorithm

Step 1: K-means algorithm has been used on the estimated data value based on factor analysis as narrated in step-3 to step-8 of Sect. 3.3.1.

Step 2: It has been observed that the number of clusters as 6 gives minimum intra cluster distance. Here, it is to note that euclidean distance has been used as intra cluster distance. The said number of clusters as 6 has been used for further processing. It can be mentioned that the cluster center value defines the network label which specifies the type of network attack which has been occurred in the network.

Step 3: The cluster centers of 6 clusters are as follows:-
Cluster 1 as 12.5, cluster 2 as 3741.3, cluster 3 as 68.6, cluster 4 as 41.9, cluster 5 as 121.25, cluster 6 as 7.5.

Step 4: The estimated data value based on factor analysis has been placed in respective cluster as formed in step-3 (Sect. 3.3.4) based on minimum distance with respective cluster center.

Step 5: From actual data it has been observed that out of 25 normal network level, 12 numbers belong to cluster 4, 5 numbers belong to cluster 6, 1 belong to cluster 5, 3 numbers belong to cluster 3 4 numbers belong to cluster 1.
Out of 25 back levels, 25 belongs to cluster 2.
Out of 51 ipsweep levels, 22 belong to cluster 6, 11 numbers belong to 3, 18 numbers belong to cluster 1.

Out of 51 nmap levels, 28 numbers belong to cluster 1, 13 numbers belong to cluster 3, 10 numbers belong to cluster 6.

Out of 51 smurf levels, 45 numbers belong to cluster 5 and 6 numbers belong to cluster 1.

Out of 46 neptune levels, 31 numbers belong to cluster 4 and 15 numbers belong to cluster 1, respectively.

Step 6:   Based on the maximum number of cluster allocation, the selected cluster number has been mapped with the network level data types. Therefore it can be mentioned that cluster 4 is meant for 'neptune' type network, cluster 2 for 'back', cluster 3 for 'normal' network type, cluster 6 for 'ipsweep' type, cluster 1 for 'nmap' type, cluster 5 for 'smurf' type and cluster 4 for 'neptune' type.

Step 7:   Based on mapping of data items with network type along cluster number, confusion matrix has been formed based on certain parameters. The parameters used are as follows:-

Accuracy = (TP + TN)/(TP + FP + TN + FN), Recall = TP/(TP + FN), Precision = TP/(TP + FP), Specificity = TN/(FP + TN), F-Measure = (2 * precision * recall)/(precision + recall).

Where TP stands for True Positive, TN stands for True Negative, FP stands for False Positive, and FN stands for False Negative, respectively.

3.3.5   Contribution of Deep Learning Based Neural Network

Step 1:   In order to improve the performance of the system, deep learning based neural network has been employed on the estimated data based on factor analysis as computed in step to get the estimated data value as computed in in step-3 to step-8 of Sect. 3.3.1.

Step 2:   K-means algorithm has been used on the estimated data value based on deep learning based neural network model as computed in previous step (step-1 of Sect. 3.3.5).

Step 3:   It has been observed that the number of clusters as 6 gives minimum intra cluster distance. The said number of clusters as 6 has been used for further processing.

Step 4:   The cluster centers of 6 clusters are as follows:
Cluster 1 as 16.2, cluster 2 as 3745.7, cluster 3 as 70.4, cluster 4 as 42.45, cluster 5 as 123.25, cluster 6 as 8.2.

Step 5:   The estimated data value based on deep learning based neural network has been placed in respective cluster as formed in step-4(Sect. 3.3.5) based on minimum distance with respect to cluster center.

Step 6:   From actual data it has been observed that out of 25 normal network level 12 numbers belong to cluster 6, 11 numbers belong to cluster 4, 1 belong to cluster 3, 1 numbers belong to cluster 1.

Out of 25 back levels 23 belongs to cluster 2, 1 number belong to cluster 4 and 1 belong to cluster 5.

Out of 51 ipsweep levels, 51 numbers belong to cluster 4.

Out of 51 nmap levels, 26 numbers belong to cluster 3, 25 numbers belong to cluster 4.

Out of 51 smurf levels, 26 numbers belong to cluster 3 and 25 numbers belong to cluster 3.

Out of 19 neptune levels, 16 numbers belong to cluster 1, 3 numbers belong to cluster 6, respectively.

Step 7: Based on the maximum cluster allocation the selected cluster number has been mapped with the network level data types. Therefore it can be mentioned that cluster 6 is meant for 'normal' type network, cluster 2 for 'back' network type, cluster 4 for 'ipsweep' type, cluster 3 for 'nmap' type, cluster 1 for 'smurf' type and cluster 1 for 'neptune' type.

Step 8: Based on mapping of data items with network type along cluster number, confusion matrix has been formed based on certain parameters. The confusion matrix is furnished in Table 3.2. The parameters used in confusion matrix are as follows:

Accuracy $= (TP + TN)/(TP + FP + TN + FN)$.

Recall $= TP/(TP + FN)$.

Precision $= TP/(TP + FP)$.

Specificity $= TN/(FP + TN)$.

F-Measure $= (2 * \text{precision} * \text{recall})/(\text{precision} + \text{recall})$.

where TP stands for True Positive, TN stands for True Negative, FP stands for False Positive, and FN stands for False Negative respectively.

Step 9: A comparison is made based on the confusion matrix as formed using estimated data based on deep learning based neural network and that of original data (without using deep learning based neural network). The performance of confusion matrix based on two methods has been furnished in Table 3.2.

Step 10: From Table 3.2, it has been observed that based on five parameters (accuracy, recall, precision, specificity and F-measure) out of 30 cases, deep learning technique is preferred by 18 cases and the model without learning is preferred by 11 cases and in one case both the models give same performance. Therefore, it can be concluded that the performance of the model has been improved through the application of deep learning based neural network model.

Table 3.2 Comparative statistics of confusion matrix

Network type	Accuracy with deep learning	Accuracy without learning	Recall with deep learning	Recall without learning	Precision with deep learning	Precision without learning	Specificity with deep learning	Specificity without learning	F-measure with deep learn	F-measure without learning
Normal	0.94	0.7799	0.75	0.444	0.667	0.279	0.961	0.8296	0.706	0.3427
Back	0.987	1	1	1	0.92	0.48	0.985	1	0.958	0.6486
ipsweep	0.811	0.8069	0.579	0.431	1	0.6875	1	0.9338	0.733	0.529
nmap	0.779	0.7118	0.565	0.549	0.509	0.394	0.842	0.758	0.5355	0.4569
smurf	0.8112	0.9588	0.646	0.882	0.608	0.978	0.459	0.992	0.6264	0.9275
neptune	0.8196	0.858	0.333	0.6739	0.842	0.7209	0.9143	0.9167	0.4773	0.6966

3.3.6 Testing of Network Attack Using New Network Parameter Values

Step 1:   A set of new network parameter values have been taken. The formula for calculating estimated data based on factor analysis (as computed in step-3 to step-8 of Sect. 3.3.1) has been applied on the network parameter values to get the estimated data based on factor analysis.

Step 2:   Now, the deep learning based neural network model has been applied on the estimated data based on factor analysis as computed in previous step. The estimated data based on deep learning based neural network has to be compared with the cluster center as formed in step 4 of Sect. 3.3.5. Cluster center value with respect to cluster data type has been furnished in Table 3.4.

Step 3:   The distance vector value has been computed between the deep learning based neural network model and six cluster centers as formed in step 3 of Sect. 3.3.5. The minimum distance indicates the type of deep learning based neural network data value is similar to the characteristic of the selected cluster. The selection of cluster center has been furnished in Table 3.3. In Table 3.3, tested data has been furnished in column 2, type of tested data in column 3. Column 4 of Table 3.3 represents distance of tested data as furnished in column 2 from cluster center1. The distances of tested data as furnished in column 2 of Table 3.3 from cluster 2, cluster 3, cluster 4, cluster 5, and cluster 6 have been furnished in column 5, column 6, column 7, column 8, column 9, respectively. Based on the distance values as furnished in column 4, column 5, column 6, column 7, column 8, and column 9, minimum distance value has to be selected. From the minimum distance value, the cluster number has to be selected based on selected cluster heading as furnished in row 1 of Table 3.3. The selected cluster number has been furnished in column 10 of Table 3.3. The type of selected cluster as decided in step 7 of Sect. 3.3.5, has been furnished in column 11 of Table 3.3. In this way, Table 3.3 has been filled up.

Step 4:   Step-4.From Table 3.3, it is evident that data no 1 with data set value as 39,333 (as furnished in row 2), belongs to back network category, selected cluster is cluster 2 which belongs to back category. Similarly data no 4 with data value 69.2 belongs to normal category has been selected as cluster no 3 which belongs to normal category. Further for data no. 5 with data value as 201.6 belongs to normal category has been selected as cluster no. 5 which belongs to smurf category. Accordingly for all data value, predicted cluster no with predicted cluster data type has been found out.

Table 3.3 Data set versus selected cluster no

No.	Data set	Data type original	Distance cluster1	Distance cluster2	Distance cluster3	Distance cluster4	Distance cluster5	Distance cluster6	Selected cluster No.	Type predicted
1	39,333	back	39,316	35,587.3	39,262.6	39,290.55	39,209.75	39,324.8	2	back
2	39,348	back	39,331.8	35,602.3	39,262.6	39,277.6	39,224.75	39,398.8	2	back
3	21	neptune	4.8	3,724.7	49.4	21.45	102.25	12.8	1	nmap
4	69.2	normal	53	3,676.5	1.2	26.75	54.05	61	3	normal
5	201.6	normal	185.4	3,544.1	131.2	159.14	78.35	193.4	5	smurf
6	65.2	normal	49	3,680.5	5.2	22.75	57.75	57	3	normal
7	4.2	ipsweep	12	3,741.5	66.2	38.25	119.05	4	6	ipsweep
8	10.4	ipsweep	5.79	3,735.3	60	32.05	112.85	2.2	6	ipsweep
9	12.1	nmap	4.1	3,733.6	58.3	30.35	111.15	3.9	6	ipsweep
10	133.6	smurf	117.4	3,612.1	63.19	91.15	10.35	125.4	5	smurf

Table 3.4 Initial data type, predicated data type and evaluation of prediction (correct or not)

Data No.	Initial data type	Predicted data type based on cluster allocation	Evaluation (correct predication yes/no)
1	back	back	yes
2	back	back	yes
3	neptune	nmap	no
4	normal	normal	yes
5	normal	smurf	no
6	normal	normal	yes
7	ipsweep	ipsweep	yes
8	ipsweep	ipsweep	yes
9	nmap	ipsweep	no
10	smurf	smurf	yes

3.4 Result

From Table 3.3, original data type and predicted data type has been ascertained. Table 3.4 has been created based on the original data type and predicted data type which has indicated whether correct prediction has been made or not.

From Table 3.4, it is evident that out of 10 data items, seven cases have been predicted correctly whereas three cases could not predict correctly. Accordingly based on any data set, type of network attack can be ascertained in advance. In that case any preventive measure can be taken by the network administrator so that type of network attack can be avoided.

3.5 Conclusion

KDD cup 99 dataset have been taken from UCI machine repository data [11]. Here in this data set, type of network attack with respect to various network parameters have been furnished. It is to note that if there is no network attack, type of network is assigned as normal. This means that when there is no network attack, normal network transmission occurs. If any network attack occurs, type of network has been defined as that network type (back, Neptune, smurf, ipsweep, etc.).

Based on the network parameter values, type of network attack can be ascertained using this procedure. It is expected that similar type of network attack can be ascertained with other similar network data sets using this procedure.

Deep learning based neural network model plays an important role in the research field. Ordinary neural network which normally consists of an input layer, an output layer and a number of hidden layers work much better in terms of performance as compared to other unsupervised models. Whereas deep learning based neural network consists of quite huge number of hidden layers and as a result the performance of it is much better as compared to ordinary neural network. Multiple deep networks have been used to improve the performance of intrusion detection. The performance of deep learning models is much better as compared to other machine learning models. The performance of deep learning neural network has been shown better as compared to without the application of deep learning based neural network in this proposed model.

References

1. Vinayakumar, R., Alazab, M., Somani, K.P., Poornachandran, P., Al-Nemrat, A., Venkatraman, S.: Deep learning approach for intelligent intrusion detection system. IEEE Access (2019)
2. Moradi, M., Zulkernine, M.: A neural network based system for intrusion detection and classification of attacks, pp. 148–04. https://citeseerx.ist.psu.edu/viewdoc/download?doi=10.1.1. 457.455&rep=rep1&type=pdf
3. Sung, A.H., Mukkamala, S.: Identifying important features for intrusion detection using support vector machines and neural networks. In: Proceedings of the 2003 Symposium on Applications and the Internet (SAINT'03). IEEE Computer Society 2003, New Mexico Institute of Mining and Technology (2003)
4. Hidalgo-Espinoza, S., Chamorro-Cupuerán, K., Chang-Tortoler, O.: Intrusion detection in computer systems by using artificial neural networks with deep learning approaches. AIRCC Publishing Corporation (2020). https://arxiv.org/ftp/arxiv/papers/2012/2012.08559.pdf
5. Thakur, M.R., Sanyal, S.: A multi-dimensional approach towards intrusion detection system. https://arxiv.org/ftp/arxiv/papers/1205/1205.2340.pdf
6. Liu, H., Lang, B.: Machine learning and deep learning methods for intrusion detection systems: a survey. Appl. Sci. MDPI (2019). file:///C:/Users/Admin/Downloads/applsci-09-04396%20(1).pdf
7. Singh, D., Paul Choudhury, J., De, M.: A comparative study of meta heuristic model to assess the type of breast cancer disease. IETE J. Res. (2020). https://doi.org/10.1080/03772063-2020. 1775139. Taylor & Francis Online, https://www.tandfonline.com/doi/abs/10.1080/03772063. 2020.1775139?journalCode=tijr20#.X7qJOASky1M.gmail
8. Burman, M., Paul Choudhury, J., Biswas, S.: Automated skin disease detection using multiclass PNN. Int. J. Innov. Eng. Technol. 14(4), 19–24 (2019). ISSN: 2319-1058. https://doi.org/10. 21172/ijet.144.03
9. Quraishi, M.I., Hasnat, A., Paul Choudhury, J.: Selection of optimal pixel resolution for landslide susceptibility analysis within Bukit Antarabangsa, Kuala Lampur, by using image processing and multivariate statistical tools. EURASIP J. Image Video Process. 2017, 1–12 (2017). Springer Open, Thompson Reuter. https://doi.org/10.1186/s13640-017-0169-2
10. Multivariate statistical analysis, https://www.sciencedirect.com/topics/earth-and-planetary-sciences/multivariate-statistical-analysis#:~:text=Multivariate%20statistical%20analysis%20is%20considered,of%20variables%20on%20each%20other
11. KUDD CUP 99 data set, http://kdd.ics.uci.edu/databases/kddcup99/kddcup99.html

Chapter 4
Localization and Classification of Thoracic Abnormalities from Chest Radiographs Using Deep Ensemble Model

Satya Vandana Nallam, Neha Raj, Madhuri Velpula, and Srilatha Chebrolu

Abstract In computer vision applications, object detection is an important task. Detection helps in identifying the objects in an image and localizing them. Detection has many applications including human recognition, vehicle parking, self-driving cars, and X-ray analysis. X-rays are helpful in diagnosis and treatment. Target recognition from images helps the radiologists to find insightful information. Initial identification of an abnormality can help in proper treatment. In this work, the objective is to find the abnormalities of chest radiographs from the VinDr-CXR data set. Considering different thoracic abnormalities, this work identifies various abnormalities in the X-ray. The proposed model needs to identify critical findings and localize the abnormalities by their locations. Combining deep learning models helps in reducing misdiagnosis. The architectural design of the proposed model includes two pipelines and an ensemble method. In the first pipeline, YOLOv5 and EfficientNet are used. In second pipeline, the Faster R-CNN model is used. Through the Weighted box fusion method, the fused predictions are created from pipeline results. The final detection results illustrate confidence scores of abnormalities found in particular locations. Abnormalities with little probabilities also help in early prediction of abnormalities.

4.1 Introduction

A chest X-ray (CXR) examination uses a small dose of ionizing radiation to produce radiographs. A chest radiograph is mostly preferred in the medical assessment of patients. Localization, image-level prediction, segmentation, and classification tasks are performed on CXRs [1]. CXR also helps in identifying the position of peripherally inserted central catheter [2] and detection of lung boundary [3]. Several thoracic diseases are examined by CXR scans [4]. Evaluation of the chest wall helps to diag-

S. V. Nallam (✉) · N. Raj · M. Velpula · S. Chebrolu
Department of Computer Science and Engineering, National Institute of Technology,
Tadepalligudem, Andhra Pradesh, India
e-mail: satyavandananallam@gmail.com

S. Chebrolu
e-mail: srilatha.chebrolu@nitandhra.ac.in

nose medical conditions. Several thoracic diseases are examined by CXR scans [4]. Evaluation of the chest wall helps to diagnose medical conditions. A chest radiograph is a projection of the chest used to diagnose health conditions. Abnormalities in chest X-ray signs different diseases. Manual examination by radiologists is challenging to find all abnormalities in detail and it takes more time. Deep learning models can locate critical findings in chest X-rays. Deep convolutional neural models help in recognizing abnormal chest radiographs [5]. Object detection locates the affected area. Computer-aided diagnosis is used to screen chest radiographs and reduces workload [6]. It helps in reducing false positives through convolutional networks. Computer-aided solutions localize the findings contains a portion of an image that needs to filter. Supervised learning models are used which are trained on predefined weights and classify thoracic diseases.

Radiologists check the lungs, chest wall, pleura, and mediastinum from each X-ray image [7]. Abnormality is found in any of the regions. CXR is used for the detection of many abnormalities. Very complicated conditions of diseases will be known using X-ray [8]. Misdiagnosis interpretation of X-ray is the main challenge. In this paper, YOLOv5, EfficientNet, and Faster R-CNN are used to detect abnormalities. And Weighted box fusion method combined the results to get final detection results. Rest of the paper is organized as follows. Section 4.2 discusses related work, Sect. 4.3 discusses proposed approach, Sect. 4.4 experimental work, and Sect. 4.5 concludes the paper.

4.2 Related Work

Computer-aided solutions such as deep learning models like DenseNet [9, 10] will improve the quality of diagnosis. DenseNet architecture contains five dense blocks with convolutional layers. These layers are subsequently connected and have batch normalization, pooling which helps in the reduction of dimensions. This neural network takes resized pooling layer based upon the input size and gives an output of all the abnormality classes. DenseNet helps in multi-scale feature extraction [11]. The memory cost taken by the feature maps is reduced by using shared memory allocations. In the classification process, it addresses the multi-label problem by taking loss functions and also additionally uses weights to the function of cross-entropy. Class categorization can be avoided by computing the label gradients of the data set from the current image. It gives the details of all images to each network and uses normalization to solve many variations of images by ensuring proper alignment of images and adjusting the brightness, intensities of images, or any other contrast factors.

EfficientNet [12, 13] is an efficient network used to scale up the model performance. In the task of image classification, EfficientNet gives good accuracy and reduces false positives by balancing network resolution, depth, and width. It has a compound scaling method that takes a fixed coefficient and scales the resolution, depth, and width uniformly with the coefficient. EfficientNet learns from complex features to capture patterns. It determines predictions and gives final results.

Abnormality detection includes both detection and classification. You Only Look Once (YOLO) [14] detects the abnormalities and is preferred for real-time object detection because of its accuracy and speed. It consists of backbone, neck, and head parts. YOLOv5 uses Cross-Stage Partial Networks (CSPDarknet) [15] for extraction of features as a backbone network. CSPDarknet uses a cross-stage hierarchy and has low computations.YOLOv5 uses Path Aggregation network (PANet) [16] to create feature pyramids [17]. PANet used bottom-up path augmentation to propagate information and feature pooling for prediction. YOLO layer is used as a head model to construct class probabilities. Optimization is done using sigmoid and ReLU activation functions.

Combination of YOLOv5 and Residual Network (ResNet) [18] used in critical findings of abnormalities. YOLO is a one-stage detector that improves the speed of detection and detects multi-targets. It uses a multi-scale prediction method and detects the abnormal region. YOLOv5 predicts bounding boxes which are weighted with probability expected through one forward propagation. ResNet [19] is used for target classification. They used ResNet50 for classification. ResNet50 model was initialized on ImageNet data [20]. It has two dense layers and used ReLU as activation.

Faster R-CNN [21] is a unified convolutional network for object detection. In Faster R-CNN, high-quality region proposals are generated by training with the region proposal network. It outputs bounding box offsets and corresponding class labels. Mask R-CNN [22] is for dense object detection. Mask R-CNN is an extended form of Faster R-CNN with a segmentation mask which is the annotation of an object detected. It results in bounding box object detection. Detectron originates from Mask R-CNN. Detectron is a fast detection system having models for instance segmentation. The instance segmentation task detects and localizes founded objects. Detectron2 [23] is a library of Facebook AI [24] which is an improved version of the detectron with more flexibility.

Data augmentation is a technique for the diversity of data by adding modified data which is the main part of training. Augmentation includes HorizontalFlip, ShiftScaleRotate, and random brightness contrast [25]. These are transformations of the image. In Faster R-CNN, to get feature extraction in a convolutional way, feature pyramid network (FPN) is used [17]. Detectron2 has different models including Faster-RCNN-R50-FPN. It provides many simple models for pre-processing of images and loading of models such as Default Predictor [26] which is an end-to-end predictor for an individual image. It customizes training behaviour and supports evaluation through the DatasetEvaluator interface [26].

Weighted Boxes Fusion (WBF) method [27] ensembles different predictions from different models to generalize better. Each prediction of object detection contains bounding boxes for the objects detected by the model. Each bounding box is predicted with some confidence score. WBF method takes the predictions in the descending order of their confidence scores. For each prediction taken, it looks for overlapping objects detected by considering the Intersection over Union (IoU) metric with a certain threshold. If any detected objects are found at the same location, now boxes are combined to form a new bounding box. The confidence score is calculated by

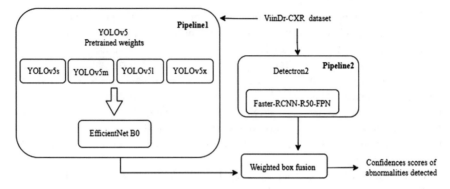

Fig. 4.1 Block diagram of ensembled model using YOLOv5, EfficientNetB0, Faster R-CNN

taking an average of confidence scores of predicted boxes in that cluster. And the coordinates of the fused bounding box is calculated from the weighted summation of coordinates of overlapping bounding boxes. Varying the weights of models taken will give better results. WBF method creates fused bounding boxes returning the locations of objects.

4.3 Ensembled Model Using YOLOv5, EfficientNetB0, Faster R-CNN

The design of the proposed model is shown in Fig. 4.1. Two pipelines are used in proposed model. YOLOv5 and EfficientNet are used in pieline1. A Faster R-CNN model from the detectron2 library is used in pipeline2. Each pipeline runs on the test data set to give predictions. These predictions are ensembled through the Weighted box fusion method. Details of pipelines are as follows.

In pipeline 1, two neural network models are used. YOLOv5 is used for object detection and EfficientNet is used for the classification task. YOLOv5s, YOLOv5m, YOLOv5l, and YOLOv5x are backbones used for training. YOLOv5 is also trained with the backbone ResNet101. ResNet101 is a feature extraction network that replaces typical bilinear interpolation upsampling with deconvolution to extract deeper features. Images are trained on dimensions 512 * 512 * 3. EfficientNet is the second neural network that was deployed. EfficientNet models outperform existing CNNs in terms of accuracy and efficiency. And for classification, EfficientNet B0 is applied. B0 is a mobile-sized architecture with a million parameters that can be trained. In this, the Non-Maximum Suppression (NMS) [28] method is used. NMS works by iteratively merging highly overlapping detections. One bounding box with a high IoU score is selected from overlapping detected entities.

Detectron2 library is used as pipeline 2. Detectron is a fast detection system having models for instance segmentation. Images are trained on dimensions 256 *

256 * 3. The Detectron2 APIs come in handy for inferring with an existing model and training a built-in model on a custom data set. It has a model-zoo API which consists of a collection of functions to create a model. Convolutional neural networks have a good performance in image analysis [29]. Faster-RCNN-50-FPN-3x model from the library is used. Feature Pyramid Network (FPN) employs a backbone that includes standard convolutional and box prediction techniques. It accepts a single-scale image of any size as input and outputs correspondingly sized feature maps at several levels. An instance of configuration is created with get-cfg configuration. Detectron2 engine has Defaultpredictor which is an end-to-end predictor that runs on a given configuration for an image.

WBF ensembles the predictions and constructs bounding boxes from different models. WBF method fuses the abnormality bounding boxes at overlapping positions. This approach ranks the predictions by their confidence scores in descending order. Overlapping bounding boxes are found by using the IoU metric. If more than the one identified object is located at the same place, the boxes are joined to generate a new bounding box. It creates the abnormality bounding box with minimal coordinates by a weighted sum of overlapping detected boxes from predictions.

4.4   Experimental Work

Each abnormality class is corresponding with a confidence score of being abnormal and coordinates of bounding boxes. Each critical prediction includes class id, confidence score, coordinates of bounding boxes, i.e. xmin, ymin, xmax, and ymax. The objective is to find all the different possible abnormalities for each chest radiograph. For images with no findings, the prediction will be class 14 of confidence score 1 with a bounding box of 1-pixel size.

Two pipelines are used for predictions with GPU runtime. In pipeline 1, a combination of YOLOv5 and EfficientNet is used. Images are trained on dimensions 512 * 512 * 3. A computer vision tool Albumentations is used to perform augmentations of data. Data is augmented by rotating, vertical, and horizontal transformations. Training of YOLO is based on pre-trained weights. Hyperparameters used for YOLOv5 are object confidence threshold = 0.005, IoU threshold for NMS = 0.45. EfficientNet-B0 model is used. Hyperparameters used for EfficientNet-B0 include batch size = 16, classification threshold = 0.003751 and different IoU thresholds used are 0.4 and 0.0001.

In pipeine2, the detectron2 library is used. Images are trained on dimensions 256 * 256 * 3. Object detection model Faster-RCNN-R50-FPN-3X is used. Hyperparameters used are base learning rate = 0.00025, batch size = 4 and max no. of iterations = 10,000. Predictor runs with model Faster R-CNN and hyperparameters to get detection bounding boxes.

Two pipelines give detection abnormalities with locations. WBF method is used with the results of two pipelines. It uses bounding boxes coordinates of predictions to get fused bounding boxes. Hyperparameters used for WBF are IoU threshold for finding overlapping boxes = 0.6, skip box threshold = 0.01, weights taken for YOLOv5-EfficientNet = 2, Faster R-CNN = 1. The confidence score is calculated as an average confidence score of overlapping bounding boxes of predictions.

4.4.1 Data Set

In this work, the VinDr-CXR data set [30] is analysed for classifying and localizing abnormalities. It contains chest X-ray images. Chest radiographs are provided by Hospital 108 and the Hanoi Medical University Hospital in Vietnam. Data set consists of 18,000 chest X-ray images. Experienced radiologists annotated 14 types of abnormalities in chest radiographs. 14 radiographic findings in (class id, abnormality name, detail) format are as follows. (0, Aortic enlargement, enlargement of largest artery aorta), (1, Atelectasis, the partial collapse of an area of lung), (2, Calcification, irregular thickened nodular edges), (3, Cardiomegaly, enlargement of cardiac diameter), (4, Consolidation, compressed lung tissue), (5, ILD, space around air sacs of lungs), (6, Infiltration, white spots in lungs), (7, Lung Opacity, grey areas in lungs), (8, Nodule, mass with small density), (9, Other lesions, oval growth appear), (10, Pleural effusion, excessive fluid surround lungs), (11, Pleural thickening, bulkiness in top of lungs), (12, Pneumothorax, the gap between lung and chest wall), (13, Pulmonary fibrosis, damage of lung tissue), (14, No Finding, without having any abnormalities).

4.4.2 Results and Analysis

Table 4.1 gives the results for ten experiments conducted using the proposed model on VinDr-CXR data set. These experiments are conducted by varying the hyperparameters as follows. Hyperparameters include optimizer, batch size and weight in pipeline 1, and learning rate, batch size and weight in pipeline 2. The result was obtained after the ensemble of these predictions from 2 pipelines with the WBF method. Among all the experiments conducted, Experiment 7 has achieved the highest mAP score of 0.249.

Each chest radiograph can have more than one abnormality. Figure 4.2 shows abnormality results of different chest X-ray images with their detection boxes. Each detection box in the result refers to different abnormalities found. Each bounding box specifies one particular abnormality detected. Detection has a class id and probability for abnormality found. Each colour specifies one particular abnormality detected among the 14 abnormalities. Table 4.2 describes the abnormalities of Fig. 4.2. It has the details of the number of abnormalities and number of areas of abnormalities in each chest X-ray. And the probabilities of each detected abnormality are mentioned in column 4 of Table 4.2. For Fig. 4.2a, Abnormality ID: % of abnormality value is mentioned as 11:6%, 11:2%, 0:26%, 3:52%, 13:2%, 10:2% in Table 4.2.

Table 4.1 mAP detection scores obtained by conducting experiments using the proposed model with various values set for hyperparameters

Experiment No.	Pipeline 1			Pipeline 2			Score
	Optimizer	Batch size	Weight	Learning rate	Batch size	Weight	
1	AdamW	8	1	0.002	4	2	0.233
2	SGD	16	3	0.003	32	2	0.246
3	Adam	4	2	0.00025	16	1	0.244
4	AdamW	24	1	0.00035	8	2	0.230
5	Adam	32	2	0.0004	24	3	0.210
6	SGD	64	1	0.0003	8	1	0.233
7	AdamW	16	2	0.0025	4	1	0.249
8	SGD	16	2	0.0037	4	3	0.230
9	AdamW	24	3	0.0004	16	2	0.244
10	Adam	32	3	0.002	32	1	0.203

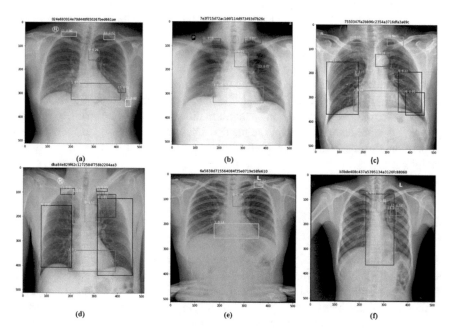

Fig. 4.2 Detected bounding boxes of abnormalities and their corresponding confidence scores

This value represents Pleural thickening with class id 11 is detected with 6% and 2% probabilities in different areas, Aortic enlargement with class id 0 is detected with 26% probability, Cardiomegaly with class id 3 is detected with 52% probability, Pulmonary fibrosis with class id 13 is detected with 2% probability and

Table 4.2 Description of obtained abnormalities of Fig. 4.2

Image label	No. of abnormalities	No. of areas	Abnormality ID: % of abnormality
Figure 4.2a	5	6	11:6%, 11:2%, 0:26%, 3:52%, 13:2%, 10:2%
Figure 4.2b	4	5	11:16%, 11:7%, 0:31%, 13:2%, 3:6%
Figure 4.2c	5	9	11:1%, 11:4%, 11:5%, 0:28%, 5:13%, 5:22%, 5:1%, 2:1%, 3:6%
Figure 4.2d	6	8	11:1%, 11:3%, 9:1%, 0:52%, 5:1%, 5:3%, 13:1%, 3:1%
Figure 4.2e	3	3	11:2%, 0:19%, 3:14%
Figure 4.2f	3	4	9:1%, 0:1%, 13:1%, 13:2%

Pleural effusion with class id 10 is detected with 2% probability. Similarly, for Fig. 4.2b–f the Abnormality ID: % of abnormality values are as mentioned in Table 4.2.

The mean average precision score (mAP) compares the ground-truth bounding box with the detected bounding box and gives the accuracy. Precision measures the percentage of our predictions which are correct. Precision is the ratio of true positives to the total positive results. Recall measures how well we find all the actual detections. A recall is the ratio of true positives to the total positive detections. AP is finding the area under the curve of precision-recall. The average of the APs for all classes is the mAP. With the 40% IoU threshold, the mAP score obtained is 0.249. Detected bounding boxes show the region of abnormalities found. Abnormalities of little probabilities are also detected in the results. These probabilities also show a great impact on treatment. Based on probabilities, further action will be taken by the radiologists.

4.5   Conclusion

A chest radiograph is preferred to examine health conditions. During the manual screening, invisible abnormalities in chest X-rays may lead to severe effects. Manual diagnosis is critical to finding the abnormalities and takes more time. Computer-based technology helps radiologists in identifying critical areas. Target detection helps in finding abnormalities. Deep learning models help to find the interpretation of chest X-rays and localize the abnormality area. VinDr-CXR considers 14 different types of thoracic abnormalities to be found from chest radiographs. The proposed model detects abnormalities using the deep learning technique.

An ensembled model is proposed for the localization of thoracic abnormalities from chest radiographs. Exploring multiple models gives an effective improvement in detection tasks. The model is based on 2 pipelines. One is a combination of object detection algorithms YOLOv5 and EfficientNet. The second pipeline is based on the

Detectron2 library with the Faster-RCNN-50-FPN-3x model. These pipelines are ensembled through the WBF method. By combining results from different predictions, optimized bounding boxes are formed. These bounding boxes signs abnormalities detected at different areas showing corresponding confidence scores. Hence, early prevention can be taken based on confidence scores.

References

1. Çalli, E., Sogancioglu, E., van Ginneken, B., van Leeuwe, K.G., Murphy K.: Deep learning for chest X-ray analysis: a survey. Med. Image Anal. (2021)
2. Yu, D., Zhang, K., Huang, L., Zhao, B., Zhang, X., Guo, X., Li, M., Gu, Z., Fu, G., Hu, M., Ping, Y., Sheng, Y., Liu, Z., Hu, X., Zhao, R.: Detection of peripherally inserted central catheter (PICC) in chest X-ray images: a multi-task deep learning model. Comput. Methods Programs Biomed. **197** (2020)
3. Candemir, S., Antani, S.: A review on lung boundary detection in chest X-rays. Comput. Assist. Radiol. Surg. (2019)
4. Nguyen, N.H., Pham, H.H., Tran, T.T., Nguyen, T.N., Nguyen, H.Q.: VinDr-PCXR: An open, large-scale chest radiograph dataset for interpretation of common thoracic diseases in children. medRxiv https://arxiv.org/abs/2203.10612 (2022)
5. Tang, Y.X., Tang, Y.B., Peng, Y., Yan, K., Bagheri, M., Redd, B.A., Brandon, C.J., Lu, Z., Han, M., Xiao, J., et al.: Automated abnormality classification of chest radiographs using deep convolutional neural networks. NPJ Digit. Med. **3**(1), 1–8 (2020)
6. Lin, C., Zheng, Y., Xiao, X., Lin, J.: CXR-RefineDet: single-shot refinement neural network for chest X-Ray radiograph based on multiple lesions detection. J. Healthc. Eng. (2022)
7. Nguyen, N.H., Nguyen, H.Q., Nguyen, N.T., Nguyen, T.V., Pham, H.H., Nguyen, T.N.M.: A clinical validation of VinDr-CXR, an AI system for detecting abnormal chest radiographs. arXiv preprint arXiv:2104.02256 (2021)
8. Guendel, S., Ghesu, F.C., Grbic, S., Gibson, E., Georgescu, B., Maier, A., Comaniciu, D.: Multi-task learning for chest X-ray abnormality classification on noisy labels. arXiv preprint arXiv:1905.06362 (2019)
9. Huang, G., Liu, Z., Van Der Maaten, L., Weinberger, K.Q.: Densely connected convolutional networks. In: Proceedings of the IEEE Conference on Computer Vision and Pattern Recognition, pp. 4700–4708 (2017)
10. Pleiss, G., Chen, D., Huang, G., Li, T., van der Maaten, L., Weinberger, K.Q.: Memory-efficient implementation of DenseNets. CoRR abs/1707.06990, 1707.06990 (2017)
11. Hwang, J.J., Liu, T.L.: Pixel-wise deep learning for contour detection. arXiv preprint arXiv:1504.01989 (2015)
12. Tan, M., Le, Q.: EfficientNet: rethinking model scaling for convolutional neural networks. In: International Conference on Machine Learning, PMLR, pp. 6105–6114 (2019)
13. Xu, R., Lin, H., Lu, K., Cao, L., Liu, Y.: A forest fire detection system based on ensemble learning. Forests **12**(2), 217 (2021)
14. Thuan, D.: Evolution of YOLO algorithm and YOLOv5: the state-of-the-art object detection algorithm (2021)
15. Wang, C.Y., Liao, H.Y.M., Wu, Y.H., Chen, P.Y., Hsieh, J.W., Yeh, I.H.: CSPNet: a new backbone that can enhance learning capability of CNN. In: Proceedings of the IEEE/CVF Conference on Computer Vision and Pattern Recognition Workshops, pp. 390–391 (2020)
16. Liu, S., Qi, L., Qin, H., Shi, J., Jia, J.: Path aggregation network for instance segmentation. In: Proceedings of the IEEE Conference on Computer Vision and Pattern Recognition, pp. 8759–8768 (2018)

17. Lin, T.Y., Dollár, P., Girshick, R., He, K., Hariharan, B., Belongie, S.: Feature pyramid networks for object detection. In: Proceedings of the IEEE Conference on Computer Vision and Pattern Recognition, pp. 2117–2125
18. Luo, Y., Zhang, Y., Sun, X., Dai, H., Chen, X.: Intelligent solutions in chest abnormality detection based on YOLOv5 and ResNet50. J. Healthc. Eng. (2021)
19. He, K., Zhang, X., Ren, S., Sun, J.: Deep residual learning for image recognition. In: Proceedings of the IEEE Conference on Computer Vision and Pattern Recognition, pp. 770–778 (2016)
20. Elsayed, S., Brinkmeyer, L., Schmidt-Thieme, L.: End-to-end image-based fashion recommendation. arXiv preprint arXiv:2205.02923 (2022)
21. Ren, S., He, K., Girshick, R., Sun, J.: Faster R-CNN: towards real-time object detection with region proposal networks. Adv. Neural Inf. Process. Syst. **28** (2015)
22. He, K., Gkioxari, G., Dollár, P., Girshick, R.: Mask R CNN. In: Proceedings of the IEEE International Conference on Computer Vision, pp. 2961–2969 (2017)
23. Wu, Y., Kirillov, A., Massa, F., Lo, W.Y., Girshick, R.: Detectron2. https://github.com/facebookresearch/detectron2 (2019)
24. Geramifard, A.: Conversational AI Efforts within Facebook AI applied research. In: Proceedings of the 2nd ACM Multimedia Workshop on Multimodal Conversational AI (2021)
25. Shorten, C., Khoshgoftaar, T.M.: A survey on image data augmentation for deep learning. J. Big Data **6**(1), 1–48 (2019)
26. Yagüe, F.J., Diez-Pastor, J.F., Latorre-Carmona, P., Osorio, C.I.G.: Defect detection and segmentation in X-ray images of magnesium alloy castings using the detectron2 framework. arXiv preprint arXiv:2202.13945 (2022)
27. Solovyev, R., Wang, W., Gabruseva, T.: Weighted boxes fusion: ensembling boxes from different object detection models. Image Vis. Comput. **107**, 104–117 (2021)
28. Qiu, S., Wen, G., Deng, Z., Liu, J., Fan, Y.: Accurate non-maximum suppression for object detection in high-resolution remote sensing images. Remote Sens. Lett. **9**(3), 237–246 (2018)
29. Pasa, F., Golkov, V., Pfeiffer, F., Cremers, D., Pfeiffer, D.: Efficient deep network architectures for fast chest X-ray tuberculosis screening and visualization. Sci. Rep. **9**(1), 1–9 (2019)
30. Nguyen, H.Q., Lam, K., Le, L.T., Pham, H.H., Tran, D.Q., Nguyen, D.B., Le, D.D., Pham, C.M., Tong, H.T., Dinh, D.H., et al.: VinDr-CXR: an open dataset of chest X-rays with radiologist's annotations. arXiv preprint arXiv:2012.15029 (2020)

Chapter 5
Particulate Matter Concentration Estimation from Images Based on Convolutional Neural Network

Anju S. Mohan and Lizy Abraham

Abstract Air pollution is a critical environmental issue that causes severe health risks. Accurate estimation of air pollutants can facilitate air pollution control and alert the public. $PM_{2.5}$ (particulate matter with diameters less than 2.5 μm) is hazardous, and its estimation requires high-cost sensors. This paper proposes an image-based $PM_{2.5}$ estimation model, a low-cost alternative. The proposed method is a convolutional neural network (CNN) architecture for estimating $PM_{2.5}$ concentration from images. The experiments conducted on the Beijing dataset show promising results, and therefore our proposed image-based CNN model can be used to estimate $PM_{2.5}$ concentrations.

5.1 Introduction

Air pollution is a significant threat to humanity. World Health Organization (WHO) statistics [1] show that 9 out of 10 people breathe air that contains high levels of pollutants, and 90% of people live in areas where air pollution exceeds WHO standards. Air pollution can affect human health and cause adverse environmental effects leading to global climate change, acid rain, and ozone depletion. Airborne particulate matter (PM) with less than 2.5 μm aerobic diameter ($PM_{2.5}$) is the most dangerous among the various air pollutants. $PM_{2.5}$ can even penetrate human lungs and blood vessels, causing severe health problems such as heart attacks, stroke, lung cancer, and other pulmonary diseases. Due to rapid industrialization, $PM_{2.5}$ concentration is increasing worldwide at an alarming rate, especially in developing countries, which affirms the need for accurate monitoring. However, monitoring stations are very few and sparsely distributed. The sophisticated $PM_{2.5}$ sensors used in monitoring stations

A. S. Mohan (✉) · L. Abraham
Department of Electronics and Communication Engineering, LBS Institute of Technology for Women, APJ Abdul Kalam Technological University, Thiruvananthapuram, Kerala, India
e-mail: anjusmohan@lbsitw.ac.in

L. Abraham
e-mail: lizyabraham@lbsitw.ac.in

© The Author(s), under exclusive license to Springer Nature Singapore Pte Ltd. 2023
V. Bhateja et al. (eds.), *Evolution in Computational Intelligence*, Smart Innovation, Systems and Technologies 326, https://doi.org/10.1007/978-981-19-7513-4_5

are based on gravimetric and other optical methods. Due to high setup and maintenance costs, it is impossible to set up monitoring stations almost everywhere. To monitor $PM_{2.5}$ pollution, we need alternate low-cost methods.

In recent years, advancements in artificial intelligence and computer vision have enabled automatic feature extraction from images and solved many problems in multidisciplinary domains. The presence of $PM_{2.5}$ in the air can reduce visibility [2], which can be easily captured in an image and correlate to particulate matter concentration. With low-cost cameras and affordable smartphones available, the acquisition of outdoor images is effortless nowadays. So, image-based methods have been explored over the last few years as an alternate method for estimating particulate matter concentration.

5.2 Literature Survey

The existing approaches for $PM_{2.5}$ concentration estimation using images generally fall under two categories: feature extraction from image-based and deep learning-based. Image features affected by PM pollution are extracted from images in image feature-based techniques, and some algorithm is used to get PM concentration from these features.

Wang et al. [3] studied the amount of degradation in the images and analyzed $PM_{2.5}$ concentration from measured light extinction. Six image features such as transmission, sky color and smoothness, whole and local image contrast, and image entropy are extracted by Liu et al. [4] and integrated with the sun's position and other weather conditions. The extracted features are then fed to a Support Vector Regression (SVR) model for $PM_{2.5}$ index estimation. In the method proposed by Li et al. [5], the depth map and transmission matrix are computed parallelly using deep convolutional neural fields (DCNF) and dark channel prior (DCP), respectively, and then combined to yield the $PM_{2.5}$ estimate. $PM_{2.5}$ monitoring from images taken by smartphone cameras is proposed by Liu et al. [6]. A learning-based model extracts features, and a haze model estimates $PM_{2.5}$ concentration. Liaw et al. [7] selected ROI from images automatically using proper image processing methods; then, an average pixel value is extracted from this ROI and fed to a simple linear regression model to get the final $PM_{2.5}$ value.

In the last few years, deep learning-based methods have been explored, and researchers have attempted convolutional neural networks (CNNs) to quantify $PM_{2.5}$ from pictures. Zhang et al. [8] built a CNN model to categorize images based on their $PM_{2.5}$ indices. A negative log-log ordinal classifier is utilized instead of softmax, and a revised rectified linear as the activation function. For image-based $PM_{2.5}$ level analysis, Chakma et al. [2] use transfer learning with the VGG-16 model. Two primary transfer learning algorithms, CNN finetuning, and random forest with CNN features are employed to classify the photos into three classes based on their $PM_{2.5}$ levels. To extend this study, a regression model with image features obtained by CNN and two weather parameters, wind speed and humidity, merged using an SVR, is proposed by Bo et al. [9] to estimate $PM_{2.5}$ concentration.

The work put forward by Ma et al. [10] uses a deep hybrid CNN model to classify smartphone photos based on pollution levels into three classes: Good, Moderate, and Severe. The model consists of two-channel deep CNN with skip connections. The image is fed to one network, while the dark channel map is the second channel's input, and at the end, the feature maps are combined. Rijal et al. [11] extracted $PM_{2.5}$ concentrations using an ensemble of pre-trained CNN models from outdoor images. The preliminary $PM_{2.5}$ concentration values obtained from ResNet50, VGG-16, and Inception-v3 are merged utilizing a feed-forward neural network to get the final $PM_{2.5}$ concentration.

Inspired by the spectacular success of CNNs in the computer vision field, we propose an end-to-end CNN model for estimating $PM_{2.5}$ from images in this paper. The paper is organized as follows: Firstly, the CNN model is introduced briefly, followed by the proposed model architecture. Then model development, training, and testing are elaborated. Finally, the experiment results are discussed with performance metrics and plots.

5.3 Methodology

An explicit end-to-end CNN architecture capable of extracting both low-level and high-level features is used in this study.

CNN is a multi-layered neural network inspired by the human brain and visual system. CNN is commonly used for classification problems [12], and it consists of two stages: the feature extraction stage and the classification stage. Features relevant to the task are extracted from the images through convolutional and pooling layers. The kernels convolve with the image regions and extract the features. The kernel weights are shared to extract low-level features from the image. Activation functions are applied after the convolution operation to accelerate the convergence of CNN. Commonly used activation functions are 'ReLU', 'Tanh', and 'Softmax'. The network's spatial feature and parameter size are reduced by applying a pooling layer, which does the downsampling operation. The second stage is the classification stage, with fully-connected dense layers. For regression applications, a continuous real value must be obtained from the images. Therefore some modifications are made to the classical CNN to fit the regression problem. The methodology flow diagram of the proposed study is depicted in Fig. 5.1.

The CNN model used in this study is shown in Fig. 5.2. It is a 5-layer architecture with two convolutional layers and three fully-connected dense layers (FC5, FC6, and FC7). Extraction of $PM_{2.5}$ concentration values from RGB images is a regression problem; hence, the last fully-connected layer (FC7) applies linear activation with a single neuron. The input to the model is outdoor images of dimensions $150 \times 150 \times 3$. The structure of the proposed CNN network is summarized in Table 5.1. By feeding images directly to the CNN model, a lot of time and effort can be saved by

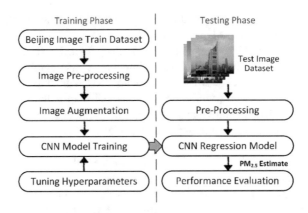

Fig. 5.1 Methodology flow diagram

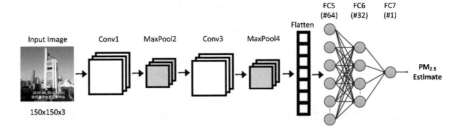

Fig. 5.2 Proposed CNN model for the estimation of PM$_{2.5}$ values from images

eliminating the tedious task of extracting features like dark channel maps, entropy, etc., from the image. The presence of PM$_{2.5}$ in the atmosphere can cause reduced visibility due to particle scattering of light rays. This visual information can be extracted using a CNN model, and then these extracted features can be mapped to PM$_{2.5}$ concentration values. The CNN model can automatically extract the relevant visual features from the image, improving accuracy. Therefore the CNN model can perform better than traditional machine learning models, where the features have to be extracted manually and fed to the model.

5.4　Experiments and Results

This section introduces the image dataset used in this study, the training and testing process, and the experimental results and discussion.

Table 5.1 Network structure of the proposed CNN model

Layer	Input size	Proposed model
Conv1	$150 \times 150 \times 3$	#kernels = 64, kernel size: (3×3), 'valid', ReLU, stride =1
MaxPool2	$148 \times 148 \times 64$	kernel size: (2×2), stride = 2, 'valid'
Conv3	$74 \times 74 \times 64$	#kernels = 32, kernel size: (3×3), 'valid', ReLU, stride =1
MaxPool4	$72 \times 72 \times 32$	kernel size: (2×2), stride = 2, 'valid'
FC5	$36 \times 36 \times 32$	#neurons = 64, ReLU
FC6	2,654,272	#neurons = 32, ReLU
FC7	33	#neuron = 1, Linear

5.4.1 Dataset

A Beijing-based image dataset [4] collected by Yi Zou is used to train and test the proposed model. The dataset consists of 327 single-scene images taken at the Beijing Television Tower in Beijing, China, at almost 7 am daily from January to December 2014. The dataset was downloaded from the figshare repository.[1] The ground truth $PM_{2.5}$ values corresponding to each image are obtained from the U.S. consulate's website[2] in Beijing city. The hourly $PM_{2.5}$ data for 2014 is downloaded and manually mapped to each image corresponding to the date and time of capture. Figure 5.3 illustrates the histogram plot, and Fig. 5.4 shows the kernel density estimation (KDE) curve of the $PM_{2.5}$ concentrations in the Beijing dataset. The PM values range from 3 to 477 $\mu g/m^3$.

5.4.2 Image Pre-processing

The image dataset consists of 327 images with varying resolutions, so we first resize the images to $256 \times 256 \times 3$ RGB images. Then, the images are normalized, making the pixel values between 0 and 1. Since the number of images is significantly less, image augmentation is done to improve model performance. Commonly used augmentation methods like zooming, rotation, shearing, and brightness adjustment are insignificant for regression tasks since they change pixel values. In our work, we have used only two augmentation methods: horizontal flip and random cropping. From the 256×256 dimension image, 150×150 image patches are cropped out, and horizontal flipping is also applied. Sample images obtained after applying augmentation to a single image are shown in Fig. 5.5.

---

[1] https://figshare.com/articles/figure/Particle_pollution_estimation_based_on_image_analysis/160 3556/2.

[2] http://www.stateair.net/web/historical/1/1.html.

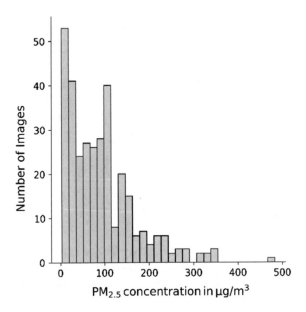

Fig. 5.3 Histogram of PM$_{2.5}$ concentrations in Beijing dataset

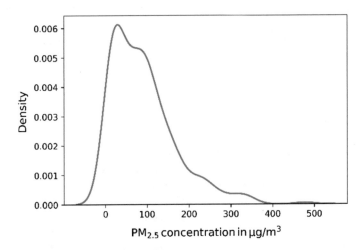

Fig. 5.4 KDE curve of PM$_{2.5}$ concentrations in Beijing dataset

5.4.3 Evaluation Metrics

The performance of the proposed CNN model is evaluated using two commonly used metrics for regression: Root Mean Squared Error (RMSE) and R-squared (R^2) value. The mathematical expressions of RMSE and R^2 are given in Eqs. (5.1) and (5.2), respectively,

Fig. 5.5 Image augmentation: **a** original image, **b** augmented images

$$\text{RMSE} = \sqrt{\frac{\sum_{i=1}^{n}(y_{\text{pi}} - y_{\text{ai}})^2}{n}} \tag{5.1}$$

$$R^2 = 1 - \frac{\sum_{i=1}^{n}(y_{\text{pi}} - y_{\text{ai}})^2}{\sum_{i=1}^{n}(y_{\text{ai}} - \overline{y_{\text{a}}})^2} \tag{5.2}$$

where y_{ai} is the actual ground truth $PM_{2.5}$ concentration value of the ith image, y_{pi} is the estimated $PM_{2.5}$ concentration value of the ith image, and $\overline{y_{\text{a}}}$ is the average of actual $PM_{2.5}$ values.

5.4.4 Experimental Results and Discussion

The experiments were conducted using an Intel Core i7-9750H CPU and an NVIDIA GeForce GTX 1650 graphics processing unit (GPU) with 16GB of RAM. The proposed model was implemented through Keras with TensorFlow backend in python language. The model is trained on 80% of images selected randomly, while 20% are used for testing. The batch size is 5 images, and adam is used as the optimizer with a learning rate of 0.00001. The number of epochs is 50, but early stopping is provided to avoid overfitting. The loss function used is the mean squared error.

The proposed CNN model is evaluated using the Beijing dataset and computed the RMSE and R^2 scores. Table 5.2 summarizes the results. Liu et al. [4] used the same Beijing dataset and achieved an RMSE = 38.28 and $R^2 = 0.7$. Our model's performance shows an improvement of almost 8 μg/m^3 in RMSE and a slight improvement

Table 5.2 Performance evaluation of the proposed CNN model using the Beijing dataset

Model	R^2 score	RMSE
Liu et al. [4]	0.70	38.28
VGG16Net	0.59	39.53
XceptionNet	0.59	39.49
Proposed CNN model	**0.76**	**30.17**

The best results are highlighted in bold

in R^2 value. Rijal et al. [11] have used a different dataset from Beijing with 1460 images and used an ensemble of pre-trained CNN models to attain an RMSE of 49.37 and R^2 of 0.684. The other methods in the literature use different datasets, so we cannot use them for comparison directly. Instead, we have used two transfer learning approaches for performance comparison, VGG16Net [13] and XceptionNet [14]. These models are pre-trained on the ImageNet dataset and built for classification problems with 1000 classes. For suiting $PM_{2.5}$ estimation, the last FC layer with 1000 neurons is replaced with one neuron, and linear activation is applied. Since these pre-trained models are trained for classification purposes and the model's weights are also learned for classification tasks, the experimental results reveal that these models are not suitable for regression tasks. Our proposed CNN model outperforms all three comparison models.

The regression results obtained for the Beijing dataset are shown in Fig. 5.6 with the line of best fit. For lower values of $PM_{2.5}$, the actual and estimated concentration values are correlated, and the points are almost distributed around the best-fit line. But with higher concentration values, the points drift farther from the best-fit line. This may be because the Beijing dataset under study is not uniformly distributed. The number of images with high values of $PM_{2.5}$ is limited. Only 8 out of 327 images have $PM_{2.5}$ values above 300, while 201 images have concentrations below 100.

Figure 5.7 shows some sample images with their actual $PM_{2.5}$ values and the estimated values using the CNN model. The first figure, Fig. 5.7a, shows three examples that the model correctly predicts. The model fails in the $PM_{2.5}$ estimation of some images, as evident from Fig. 5.7b. These errors are mainly due to the change in weather conditions. Pollution estimation from images is very different from classification and recognition applications, making it a challenging and emerging research field.

5.5 Conclusion and Future Scope

This paper proposes a CNN-based approach to estimate the particular matter concentration using RGB images without expensive sensors. The CNN model analyses the visibility changes in the image and calculates the $PM_{2.5}$ concentrations. The architecture contains two convolutional layers and max-pooling layers for feature extraction.

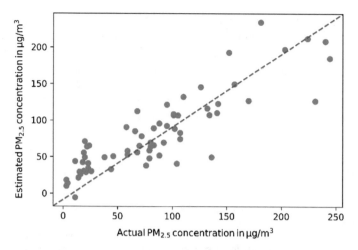

Fig. 5.6 Actual versus estimated PM$_{2.5}$ concentration scatter plot

Fig. 5.7 Sample images **a** correctly predicted, **b** failed predictions

Then a regression model with three fully-connected layers extracts $PM_{2.5}$ values. The proposed model is evaluated on the Beijing dataset with 327 single scene images. Experimental results show that the proposed CNN model is valid as a low-cost $PM_{2.5}$ pollution monitoring method.

In the future, we'll explore other CNN models and hybrid methods. To minimize estimation error, meteorological parameters can be included. In addition, more images with diverse scenarios and uniform PM distribution will be collected.

References

1. WHO: 9 out of 10 people worldwide breathe polluted air, but more countries are taking action. Saudi Med. J. **39**, 641–643 (2018)
2. Chakma, A., Vizena, B., Cao, T., Lin, J., Zhang, J.: Image-based air quality analysis using deep convolutional neural network. In: Proceedings-International Conference on Image Processing (ICIP), pp. 3949–3952. IEEE Computer Society (2018). https://doi.org/10.1109/ICIP.2017.829 7023
3. Wang, H., Yuan, X., Wang, X., Zhang, Y., Dai, Q.: Real-time air quality estimation based on color image processing. In: 2014 IEEE Visual Communications and Image Processing Conference (VCIP), pp. 326–329 (2015). https://doi.org/10.1109/VCIP.2014.7051572
4. Liu, C., Tsow, F., Zou, Y., Tao, N.: Particle pollution estimation based on image analysis. PLoS One **11**, e0145955 (2016). https://doi.org/10.1371/journal.pone.0145955
5. Li, Y., Huang, J., Luo, J.: Using user generated online photos to estimate and monitor air pollution in major cities. In: ACM International Conference Proceeding Series, pp. 11–15. Association for Computing Machinery (2015). https://doi.org/10.1145/2808492.2808564
6. Liu, X., Song, Z., Ngai, E., Ma, J., Wang, W.: $PM_{2.5}$ monitoring using images from smartphones in participatory sensing, 630–635 (2015)
7. Liaw, J.J., Huang, Y.F., Hsieh, C.H., Lin, D.C., Luo, C.H.: $PM_{2.5}$ concentration estimation based on image processing schemes and simple linear regression. Sensors **20** (2020). https://doi.org/10.3390/s20082423
8. Zhang, C., Yan, J., Li, C., Rui, X., Liu, L., Bie, R.: On estimating air pollution from photos using convolutional neural network. In: MM 2016—Proceedings of the 2016 ACM Multimedia Conference. pp. 297–301. Association for Computing Machinery, Inc, New York (2016) https://doi.org/10.1145/2964284.2967230
9. Bo, Q., Yang, W., Rijal, N., Xie, Y., Feng, J., Zhang, J.: Particle pollution estimation from images using convolutional neural network and weather features. In: Proceedings—International Conference on Image Processing (ICIP), pp. 3433–3437. IEEE Computer Society (2018). https://doi.org/10.1109/ICIP.2018.8451306
10. Ma, J., Li, K., Han, Y., Yang, J.: Image-based air pollution estimation using hybrid convolutional neural network. In: Proceedings—International Conference on Pattern Recognition, pp. 471–476. Institute of Electrical and Electronics Engineers Inc. (2018). https://doi.org/10.1109/ICPR.2018.8546004
11. Rijal, N., Gutta, R.T., Cao, T., Lin, J., Bo, Q., Zhang, J.: Ensemble of deep neural networks for estimating particulate matter from images. In: 2018 3rd IEEE International Conference on Image, Vision and Computing (ICIVC), pp. 733–738. Institute of Electrical and Electronics Engineers Inc. (2018). https://doi.org/10.1109/ICIVC.2018.8492790
12. Krizhevsky, A., Sutskever, I., Hinton, G.E.: ImageNet classification with deep convolutional neural networks. Commun. ACM **60**, 84–90 (2017). https://doi.org/10.1145/3065386

13. Simonyan, K., Zisserman, A.: Very deep convolutional networks for large-scale image recognition (2015)
14. Chollet, F.: Xception: deep learning with depthwise separable convolutions. In: Proceedings of the IEEE Conference on Computer Vision and Pattern Recognition, pp. 1800–1807 (2017). https://doi.org/10.1109/CVPR.2017.195

Chapter 6
Brain Tumor Prediction from MRI Images Using an Ensemble Model Based on EfficientNet-B2, B4, and ResNet34 Architectures

Jatin Singh, Govind Prasad Lakhotia, Aerva Shiva, and Srilatha Chebrolu

Abstract Brain is an exceptionally complex and crucial organ of the human body which has a life-threatening disease associated with it, i.e., brain tumor. It is one among the world's top causes of death. Because of the complicated structure of the brain and the comparable nature of many types of brain tumor disease, manual detection of brain tumors is a difficult task. The classification of medical images using Deep Learning techniques has benefited much attention in past years. It has shown optimistic results toward improving accuracy in classification of brain tumors based on Magnetic Resonance Imaging (MRI). The goal of this research is to develop an integrated model which explores a number of Convolutional Neural Networks (CNN), namely, EfficientNet-B2, EfficientNet-B4, and ResNet34 for distinguishing brain tumor MRI image sets. Tests are performed on brain tumor dataset BraTS 2021 containing four types of MRI images, i.e., (i) Inversion Recovery with Fluid Attenuation, (ii) T1-weighted pre-contrast, (iii) T1-weighted post-contrast, and (iv) T2-weighted. Proposed deep ensembled network has demonstrated satisfactory performance and efficiency on the dataset.

6.1 Introduction

The growth of atypical cells in the brain, which can adversely disrupt the central nervous system, is described as a brain tumor. This growing mass of cells can interrupt the brain's smooth working. Many types of tumors make the brain tissue enlarge over time, which leads to damage of brain cells. However, if brain tumor detection is made as early as possible, it will significantly increase the chances of treatment and survivability of the patients. Magnetic Resonance Imaging (MRI) is a widely utilized method in medicine as it produces high-quality images. These images prove

J. Singh (✉) · G. P. Lakhotia · A. Shiva · S. Chebrolu
Department of Computer Science and Engineering, National Institute of Technology, Tadepalligudem, Andhra Pradesh, India
e-mail: jatinhmu@gmail.com

S. Chebrolu
e-mail: srilatha.chebrolu@nitandhra.ac.in

© The Author(s), under exclusive license to Springer Nature Singapore Pte Ltd. 2023
V. Bhateja et al. (eds.), *Evolution in Computational Intelligence*, Smart Innovation, Systems and Technologies 326, https://doi.org/10.1007/978-981-19-7513-4_6

61

beneficial for the automatic disease analysis field by facilitating the visualization of complex brain structure and hence provide comprehensive information about it. Detecting the tumor manually using a significant number of MRI images is a time taking, work intensive, and prone to error task.

Deep learning architectures are proven to be more efficient and accurate in prediction of brain tumor. The following are a few examples of deep learning-based brain tumor classification and segmentation work. In [13], Min Jiang et al., have introduced DDU-net for segmentation of brain tumor using MRI images. DDU-net is a combination of dual-stream decoding CNN (DDNet) and U-net [19]. The focus point of DDU-net is edge feature learning. DDU-net has been found to achieve better prediction w.r.t Dice coefficient [20] on BraTS 2017 [12] and BraTS 2018 [18]. Emre Dandil et al. [5] have introduced a model for identification of brain tumors from pseudo brain tumors. For this purpose, Magnetic Resonance Spectroscopy (MRS) data has been analyzed. The proposed model is based on Long Short-Term Memory (LSTM) [21] and Bidirectional LSTM (Bi-LSTM) [9] networks. For training and testing, MRS signals from brain tissue in the INTERPRET database [25] are used. The proposed model differentiate pseudo brain tumor from the following: i) glioblastoma brain tumor, ii) diffuse astrocytoma brain tumor and iii) metastatic brain tissue. This model classifies pseudo brain tumor from normal brain tissue as well. This method gives better results compared to 1D-Convolutional Neural Network (CNN) [1]. Jindong Sun et al. [23] have introduced a multi-pathway architecture for brain tumor detection. A 3D fully connected CNN (FCN) is used in this architecture. The authors have performed MRI image resizing or normalization and then used 3D dilated convolution for feature extraction. The dataset used for evaluation of this model is BraTS 2019 [26] and is found to achieve a better segmentation score. Karayegen et al. [15] have proposed a semantic segmentation method using CNN. It is performed on a 3D Brain Tumor Segmentation image dataset. The dataset contains four different types of images i.e. (i) T1, (ii) T1C, (iii) T2, and (iv) Flair. This model works on 3D images of the brain where it contrasts real and predicted labels.

In this paper, an ensembled model based on Residual Network (ResNet) [10], EfficientNet [24], LSTM deep learning models is proposed. The proposed model predicts the presence of Brain Tumor for unknown cases in BraTS 2021 [2]. The remaining paper is laid out as follows: (i) Sect. 6.2 contains the related work, (ii) Sect. 6.3 contains the proposed approach, (iii) Sect. 6.4 contains the experimental results, and analysis and (iv) Sect. 6.5 contains the conclusion.

6.2   Related Work

This section provides a quick overview of each of the architectures used in this work. The architecture of CNN is motivated by the neuron connectivity pattern in the Brain. It's a simple deep learning technique that takes an input image. It allocates importance by learning parameters related to numerous image aspects, and differentiates them. Filters are used to pull out features from the images and form a feature map. Pooling operation is performed on the feature map which reduces its dimension and hence reduces the number of parameters to learn. There are some optional improvement techniques used by CNNs for better performance. First is a dropout [17, 22], this technique works by randomly selecting the neurons and ignoring them at each update of the training phase hence reducing the computation cost and preventing the problem of overfitting. The second is batch normalization [11], this technique works by introducing extra layers in the CNN which performs the standardizing and normalizing of input coming from the previous layer thus making it faster, stable, and overcoming the situation of vanishing gradient descent which is a situation when gradients of the loss function approach zero thus makes parameter learning difficult. Another way of overcoming the vanishing gradient problem is to use simpler activation functions [7] like Rectified Linear Unit (ReLU) which doesn't cause small derivatives.

The EfficientNet architecture is one of the powerful CNN architectures. It is a scaling method which uniformly scales (i) width, (ii) depth, and (iii) resolution using compound coefficients instead of randomly scaling up these factors. The intuition behind the compound scaling method is that if the input image is large, then the network requires more number of layers and channels. It increases the receptive field and captures more fine-grained patterns. There are seven EfficientNet architectures i.e. (i) B1, (ii) B2, (iii) B3, (iv) B4, (v) B5, (vi) B6, and (vii) B7 in the family scaled using baseline EfficientNet-B0 using different compound coefficients. EfficientNet-B2 contains 9.2 million parameters and 1 billion flops. It gives top-1 accuracy of 80.1% and top-5 accuracy of 94.9% for ImageNet [6] dataset. EfficientNet-B4 contains 19 million parameters and 4.2 billion flops. It gives top-1 accuracy of 82.9% and top-5 accuracy of 96.4% for ImageNet.

ResNet is another famous CNN architecture. Its main idea is based on identity shortcut / shortcut path / skip connection. To avoid the vanishing gradient descent problem, it skips a few layers and hence makes the construction of a network with up to thousands of convolutional layers possible. ResNet is classified based on the number of layers present, for example ResNet34 [8]. It is an architecture that is 34 layers deep and contains 34 billion flops having an top-1 error rate of 25% and the top-5 error rate of 7.7% for ImageNet.

Recurrent Neural Network (RNN) [21] is another frequently used neural network when the output is dependent on multiple previous states. It is a class of ANN. In this, the output of the previous phase is supplied into the following step as input. LSTM [27] is the advancement of RNN which has three gates, i.e., input, output and forget gate. Long-term dependencies can be learned via LSTM. It is able to overcome the vanishing gradient problem as well by freeing up those memory locations that are not needed for the prediction of the final classification labels.

The architecture presented in this paper is based on three pipelines using ResNet, EfficientNet, and LSTM models. The proposed approach is applied on BraTS 2021. For each case, a set of MRI images are given and these MRI images fall under one of the four categories, i.e., (i) FLAIR, (ii) T1w, (iii) T1wCE, and (iv) T2w. For each case, in the training data, the target value is specified as the presence or absence of the O[6]-methylguanine-DNA methyltransferase (MGMT) promoter methylation. The objective of the proposed architecture is to predict the existence or non-existence of MGMT for an unknown case. The size of the training dataset used in model training plays a crucial role in determining its performance. The number of MRI images available for training the model can be increased by performing augmentation of the original dataset. It includes flip, rotation, height, width, brightness change, etc. in images. Albumentations [4] is a computer vision tool that is used to perform the augmentation of images. This process boosts the performance of the model by providing more relevant data required for better training. Images are needed to be resized and normalized as per the pipeline's requirement. Normalization of images makes the pixel values between 0 and 1. To do resizing, OpenCV [3] is used which is a great library for image processing and computer vision related work.

6.3 Ensembled Approach: EfficientNet-B2, EfficientNet-B4, ResNet34, and LSTM Models

The proposed approach is a deep learning ensembled model consisting of three pipelines.The proposed architecture is depicted as shown in Fig. 6.1. The proposed model is trained on BraTS 2021. The weighted average of output from the three pipelines is considered to be the predicted MGMT value. In the following subsections, each pipeline is described in detail.

6.3.1 Pipeline 1

This pipeline classifies each case into one of the two classes, either presence or absence of MGMT promoter methylation. This pipeline consists of a 2D convolutional (Conv2d) layer that has four input channels (IC) and three output channels (OC) which take the input of size 4 * 512 * 512, extracts features, and forms a feature map using three kernels. Hence an output of size 3 * 256 * 256 is generated. It uses EfficientNet-B4 which has three IC and two OC. The components of EfficientNet-B4 architecture are as shown in Fig. 6.1. It consists of (i) Conv2d layer that extracts features from the given input, (ii) 2D batch normalization (BatchNorm2d) layer which reduces Internal Covariate Shift (ICS) to enhance the learning rate (LR) of the network, (iii) Sigmoid-Weighted Linear Unit (SiLU) activation function which is defined as $x\sigma(x)$ where x is input and $\sigma(x)$ is sigmoid activation function, (iv) 2D adaptive average pooling (SelectAdaptivePool2d) which selects stride and kernel size automatically based on the output size mentioned and the input size received, (v) linear layer which is a single layer feed-forward network, and (vi) softmax activation which gives the probability distribution for presence and absence of MGMT promoter methylation.

6.3.2   Pipeline 2

This pipeline uses MRI images of only one type, i.e., T1wCE. It randomly chooses 64 MRI images of type T1wCE and forms a resultant input of size 256 * 256 * 64. This pipeline is making use of ResNet34 which has one IC that's why the input images are reshaped to a size of 1 * 256 * 256 * 64. It uses ResNet34 which has one IC and one OC. It takes an input of size 1 * 256 * 256 * 64 and gives a single value as output. Figure 6.1 shows the components of the ResNet34 architecture. It contains (i) a 3D convolutional (Conv3d) layer that applies a 3D filter to the input data and outputs a 3D feature map, (ii) 3D batch normalization (BatchNorm3d) layer which minimizes ICS, (iii) 3D max pooling (MaxPool3d) layer which downsampled the feature map by calculating the maximum value for patches of the feature map, (iv) ReLU activation function, (v) 3D adaptive average pooling (AdaptiveAvgPool3d) layer that selects stride and kernel size automatically based on the output size mentioned and the input size received, and (vi) a linear layer. The sigmoid activation is applied on the output of ResNet34 and the probability of the presence of MGMT is determined.

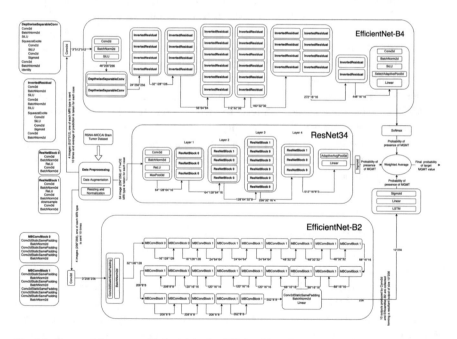

Fig. 6.1 Proposed deep ensembled model using EfficientNet-B2, B4, and ResNet34 for brain tumor prediction

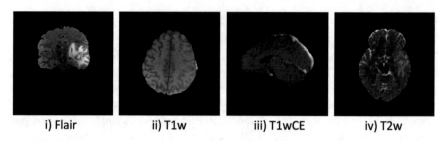

i) Flair   ii) T1w   iii) T1wCE   iv) T2w

Fig. 6.2 Predicted MGMT value for selected case: 0.596

6.3.3 Pipeline 3

In this pipeline, ten images are selected from each MRI type. The pipeline is fed an input of size 4 * 256 * 256 ten times. This pipeline consists of Conv2d which takes an input of size 4 * 256 * 256 and gives an output of size 3 * 256 * 256. It uses EfficientNet-B2 which has three IC and 256 OC hence it takes input of size 3 * 256 * 256 and gives output of size 256. The components of EfficientNet-B2 architecture are as shown in Fig. 6.1. It consists of (i) Conv2d with static same padding (Conv2dStaticSamePadding) which applies padding of zeros at right and bottom of the input received, and performs convolution operation on it, (ii) BatchNorm2d layer which minimizes ICS, and (iii) linear layer. Ten outputs of size 256 produced by EfficientNet-B2 are fed as input to the LSTM model with 2 hidden layers. It has 256 IC and 32 OC and hidden state features, producing an output of size 32. At last, there is a linear layer which gives the probability of MGMT presence.

The three pipelines are used to make individual predictions and finally, the weighted average is taken to give a final probability for the presence of brain tumor, i.e., MGMT promoter methylation.

6.4 Experimental Results and Analysis

The Google Colab environment is used for establishing the complete experimental setup. Graphics Processing Unit (GPU) is employed for training pipelines on BraTS 2021. In pipeline 1, the model used is EfficientNet-B4. The images are preprocessed to 512 * 512 dimensions and cases are split into five folds using the stratified k-fold technique. The hyper parameters are experimented with values as follows: (i) number of epochs = 10, (ii) weight decay = 10^{-6}, (iii) LR = 0.0001, (iv) batch size = 4, and (v) number of workers = 4. In pipeline 2, the model used is ResNet34. The images are preprocessed to 256 * 256 dimensions and cases are split into five folds using the stratified k-fold technique. The hyper parameters are experimented with values as follows: (i) number of epochs = 15, (ii) LR is 0.0001 for first 10 epochs and 0.00005 for last 5 epochs, and (iii) batch size = 8.

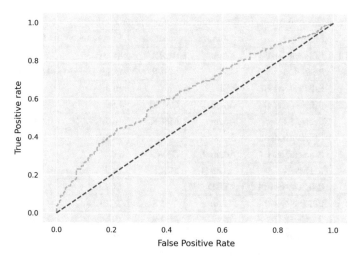

Fig. 6.3 ROC curve for the training data

In pipeline 3, the model used is EfficientNet-B2. The images are preprocessed to 256 * 256 dimensions and cases are split into five folds using the stratified k-fold technique. Different augmentation techniques used are flip, shift, scale, rotate. The hyper parameters are experimented with values as follows: (i) number of epochs = 15, (ii) LR = 0.0004, (iii) batch size = 8, and (iv) number of workers = 8.

The Area Under the Receiver Operating Characteristic Curve (ROC-AUC) is the evaluation metric employed. The ROC-AUC shown in Fig. 6.3 gives the ROC-AUC score of 0.636 for training data. Table 6.1 shows the experimental results obtained by fine-tuning the hyper parameters: (i) Optimizer, (ii) Scheduler, and (iii) Loss function. For optimizer, Adaptive moment estimation (Adam) [16], Stochastic Gradient Descent (SGD), Asynchronous Stochastic Gradient Descent (ASGD), and Resilient Backpropagation (Rprop) are chosen. For scheduler, ReduceLROnPlateau (RLRP), MultiStepLR (MSLR), CosineAnnealingWarmRestarts (CAWR), and CosineAn-nealingLR (CALR) are chosen. For loss function, TaylorCrossEntropyLoss (TCEL), BinaryCrossEntropyWithLogits (BCEL), LabelSmoothing (LS), and FocalLoss (FL) are chosen for conducting experiments. Among all the experiments conducted, Experiment 10 has obtained the highest ROC-AUC score. The best result obtained from Kaggle submissions is as follows : (i) Public score = 0.634 and (ii) Private score = 0.609 [14]. Figure 6.2 shows the predicted MGMT value as 0.596 for a selected case.

The performance of the proposed architecture is compared with ResNet34, EfficientNet-B2, and EfficientNet-B4 on the BraTS 2021 dataset. Table 6.2 shows the comparison results. The performance of individual models is low in comparison with the proposed ensemble architecture of ResNet34, EfficientNet-B2, and EfficientNet-B4 models.

Table 6.1 Experimental results by fine-tuning the hyper parameters

	Model	Optimizer	Scheduler	Loss function	ROC-AUC Score
Experiment 1	EfficientNet-B4	Adam	RLRP	TCEL	
	ResNet34	Adam	MSLR	BCEL	0.59676
	EfficientNet-B2	SGD		BCEL	
Experiment 2	EfficientNet-B4	Adam	CAWR	LS	
	ResNet34	Adam	MSLR	BCEL	0.60853
	EfficientNet-B2	SGD		BCEL	
Experiment 3	EfficientNet-B4	Adam	CAWR	LS	
	ResNet34	Adam	MSLR	BCEL	0.60849
	EfficientNet-B2	ASGD		BCEL	
Experiment 4	EfficientNet-B4	Adam	CALR	TCEL	
	ResNet34	Adam	MSLR	BCEL	0.59425
	EfficientNet-B2	ASGD		BCEL	
Experiment 5	EfficientNet-B4	Adam	CALR	TCEL	
	ResNet34	Adam	MSLR	BCEL	0.59387
	EfficientNet-B2	Rprop		BCEL	
Experiment 6	EfficientNet-B4	Adam	CAWR	FL	
	ResNet34	Adam	MSLR	BCEL	0.58643
	EfficientNet-B2	Rprop		BCEL	
Experiment 7	EfficientNet-B4	Adam	CAWR	FL	
	ResNet34	Adam	MSLR	BCEL	0.58647
	EfficientNet-B2	SGD		BCEL	
Experiment 8	EfficientNet-B4	Adam	CAWR	FL	
	ResNet34	SGD	MSLR	BCEL	0.56837
	EfficientNet-B2	SGD		BCEL	
Experiment 9	EfficientNet-B4	Adam	CAWR	LS	
	ResNet34	SGD	MSLR	BCEL	0.60866
	EfficientNet-B2	SGD		BCEL	
Experiment 10	EfficientNet-B4	Adam	CAWR	LS	
	ResNet34	SGD	MSLR	BCEL	0.60883
	EfficientNet-B2	ASGD		BCEL	

Table 6.2 Comparison of proposed architecture with the State-of-the-Art models

Architecture	ROC-AUC score
ResNet34	0.565
EfficientNet-B4	0.587
EfficientNet-B2	0.589
Ensembled architecture	0.609

6.5 Conclusion

In this paper, an ensemble model is introduced which automatically predicts brain tumor from MRI images. The model is based on three pipelines which are making use of some well known architectures like EfficientNet, ResNet. The experiment results obtained shows that these pipelines can be used to build a classifier which can efficiently predict the brain tumor presence. The specific type of brain tumor associated with the dataset used is Glioblastoma. Glioblastoma is a most common brain tumor type found in adults and with the most serious form, the average survival duration is less than a year. So to predict the presence of such deadly disease, this model can be proved helpful as it eliminates the need for human assistance and related complications. The ROC-AUC score obtained by the proposed model is 0.597 on the BraTS 2021 dataset. The architecture of this model is so simple that it can be used in real-time situations. Similarly, this model can detect other diseases from MRI images if supplementary training is provided.

This model can therefore help patients with brain tumors to get appropriate diagnoses and treatments.These pre-surgery strategies have the ability to improve the management, survival, and prospects of patients with brain tumors.

References

1. Albawi, S., Mohammed, T.A., Al-Zawi, S.: Understanding of a convolutional neural network. International Conference on Engineering and Technology (ICET) pp. 1–6 (2017)
2. Baid, U., Ghodasara, S., Mohan, S., et al.: The RSNA-ASNR-MICCAI BraTS 2021 benchmark on brain tumor segmentation and radiogenomic classification. arXiv:2107.02314 (2021)
3. Bradski, G.: The OpenCV Library. Dr. Dobb's J. Softw. Tools (2000)
4. Buslaev, A., Iglovikov, V.I., Khvedchenya, E., Parinov, A., Druzhinin, M., Kalinin, A.A.: Albumentations: fast and flexible image augmentations. Information 125 (2020)
5. Dandil, E., Karaca, S.: Detection of pseudo brain tumors via stacked LSTM neural networks using MR spectroscopy signals. Biocybernetics Biomed. Eng. **41**(1), 173–195 (2021)
6. Deng, J., Dong, W., Socher, R., Li, L.J., Li, K., Fei-Fei, L.: ImageNet: a large-scale hierarchical image database. In: 2009 IEEE Conference on Computer Vision and Pattern Recognition, pp. 248–255 (2009)
7. Ding, B., Qian, H., Zhou, J.: Activation functions and their characteristics in deep neural networks. In: Chinese Control And Decision Conference (CCDC), pp. 1836–1841 (2018)
8. Gao, M., Qi, D., Mu, H., Chen, J.: A transfer residual neural network based on ResNet-34 for detection of wood knot defects. Forests **12**(2), 212 (2021)
9. Graves, A., Fernández, S., Schmidhuber, J.: Bidirectional lstm networks for improved phoneme classification and recognition. In: Duch, W., Kacprzyk, J., Oja, E., Zadrożny, S. (eds.) Artificial Neural Networks: Formal Models and Their Applications—ICANN 2005, pp. 799–804. Springer, Berlin Heidelberg, Berlin, Heidelberg (2005)
10. He, K., Zhang, X., Ren, S., Sun, J.: Deep residual learning for image recognition. arXiv:1512.03385 (2015)
11. Ioffe, S., Szegedy, C.: Batch normalization: Accelerating deep network training by reducing internal covariate shift. arXiv:1502.03167v3 (2015)
12. Isensee, F., Kickingereder, P., Wick, W., Bendszus, M., Maier-Hein, K.H.: Brain tumor segmentation and radiomics survival prediction: contribution to the BRATS 2017 challenge. CoRR abs/1802.10508 (2018)

13. Jiang, M., Zhai, F., Kong, J.: A novel deep learning model DDU-net using edge features to enhance brain tumor segmentation on MR images. Artif. Intell. Med. **121**, 102180 (2021)
14. Kaggle: Submission Notebook for Brain tumor prediction on BraTS 2021 dataset https://kaggle.com/code/jatinhmu/experimentbraintumor?scriptVersionId=94039956
15. Karayegen, G., Aksahin, M.F.: Brain tumor prediction on MR images with semantic segmentation by using deep learning network and 3D imaging of tumor region. Biomed. Signal Process. Control **66**, 102458 (2021)
16. Kingma, D.P., Ba, J.: Adam: a method for stochastic optimization. arXiv:1412.6980v9 (2014)
17. Labach, A., Salehinejad, H., Valaee, S.: Survey of dropout methods for deep networks. arXiv:1904.13310 (2019)
18. Myronenko, A.: 3D MRI brain tumor segmentation using autoencoder regularization. arXiv:1810.11654v3 (2018)
19. Ronneberger, O., Fischer, P., Brox, T.: U-net: Convolutional networks for biomedical image segmentation. arXiv:1505.04597v1 (2015)
20. Shamir, R.R., Duchin, Y., Kim, J., Sapiro, G., Harel, N.: Continuous dice coefficient: a method for evaluating probabilistic segmentations. arXiv:1906.11031v1 (2019)
21. Sherstinsky, A.: Fundamentals of recurrent neural network (RNN) and long short-term memory (LSTM) network. Physica D: Nonlinear Phenom. **404**, 132306 (2020)
22. Srivastava, N., Hinton, G., Krizhevsky, A., Sutskever, I., Salakhutdinov, R.: Dropout: a simple way to prevent neural networks from overfitting. J. Mach. Learn. Res. **15**(56), 1929–1958 (2014)
23. Sun, J., Peng, Y., Guo, Y., Li, D.: Segmentation of the multimodal brain tumor image used the multi-pathway architecture method based on 3D FCN. Neurocomputing **423**, 34–45 (2021)
24. Tan, M., Le, Q.V.: Efficientnet: rethinking model scaling for convolutional neural networks. arXiv:1905.11946 (2020)
25. Tate, A.R., Underwood, J., Acosta, D.M., Julià-Sapé, M., Majós, C., Moreno-Torres, A., Howe, F.A., van der Graaf, M., Lefournier, V., Murphy, M.M., Loosemore, A., Ladroue, C., Wesseling, P., Luc Bosson, J., Cabañas, M.E., Simonetti, A.W., Gajewicz, W., Calvar, J., Capdevila, A., Wilkins, P.R., Bell, B.A., Rémy, C., Heerschap, A., Watson, D., Griffiths, J.R., Arús, C.: Development of a decision support system for diagnosis and grading of brain tumours using in vivo magnetic resonance single voxel spectra. NMR Biomed. **19**(4), 411–434 (2006)
26. Wang, F., Jiang, R., Zheng, L., Meng, C., Biswal, B.: 3D U-Net based brain tumor segmentation and survival days prediction. arXiv:1909.12901v2 pp. 131–141
27. Yu, Y., Si, X., Hu, C., Zhang, J.: A review of recurrent neural networks: LSTM cells and network architectures. Neural computation **31**(7), 1235–1270 (2019)

Chapter 7
Automatic Abnormality Detection in Musculoskeletal Radiographs Using Ensemble of Pre-trained Networks

Raksha Verma, Sweta Jain, S. K. Saritha, and Shubham Dodia

Abstract Musculoskeletal disability (MSDs) defined as the injuries that affect the movement or musculoskeletal system of the human body. Over the worldwide, it is the second most cause of physical disability. Musculoskeletal disability worsens over time and can result in long-term discomfort and severe disability. As a result, early detection and diagnosis of these anomalies is essential. But the diagnosis process is very time consuming, error prone and required diagnostic professional. Deep learning algorithms have recently been applied in medical imaging that provides a robust platform with very reliable outcomes. The development of Computer Aided Detection (CAD) system extensively speed up the diagnosis process. In this paper, a weighted ensemble model has been proposed, which is the combination of three pre-trained models (DenseNet169, MobileNet, and XceptionNet). The weighted ensemble model is tested on MURA dataset, a large public dataset provided by Stanford ML Group. Our model achieved a cohen's kappa score 0.739 with precision of 0.885 and recall of 0.854, which is higher than many existing approaches such as densenet169 and ensemble200 model.

7.1 Introduction

Musculoskeletal disorders are often characterized by pain and limitations in mobility, dexterity, and general level of functioning, reducing people's ability to work. More than 1.7 billion people suffer from these disorders [1]. Around 150 conditions are categorized as musculoskeletal conditions because they have an impact on an individual's locomotor system [2]. TThey range from those that happen abruptly and are transient, such as fractures, sprains, and strains, to those that last a lifetime and are associated with permanent functional limitations and impairment. Conditions affecting the musculoskeletal system includes: ankylosing spondylitis, osteoporosis,

R. Verma · S. Jain · S. K. Saritha
Maulana Azad National Institute of Technology, Bhopal, Madhya Pradesh, India

S. Dodia (✉)
National Institute of Technology Karnataka, Surathkal, Karnataka, India
e-mail: shubham.dodia8@gmail.com

osteopenia, and related fragility fractures, traumatic fractures, sarcopenia, muscles, back and neck pain.

The reliability of diagnosis of bone abnormalities is depending on both the expertise of the specialist and the imaging quality. Due to the low quality and high noise levels of medical images, computer-aided detection (CADs) can help us to overcome these shortcomings of anomalies. There is a lot of scope of work in this research area [3]. A specific difficulty for medical image analysis is that disability commonly arise at small location of regions within a high-resolution image that may lead to a problem of missing or misdiagnosing radiological examination. These errors have major consequences and result in delayed treatment, a poorer long-term prognosis, an increase in treatment costs and time, and can lead to patient becoming handicapped [4]. To enhance the treatment method, a CAD system is required that could detect the abnormality or fracture early.

In recent years, Artificial Intelligence strategies have found their application in medical imaging, with promising results. The importance of the most recent deep learning approaches has grown in a variety of fields [5]. Deep learning approaches, particularly convolutional neural networks (CNN), have attracted the interest of researchers for medical image classification and segmentation applications [6]. For the effective use of the deep learning solution huge volumes of labeled training data are required. To detect the musculoskeletal abnormalities in X-ray images, Stanford University provides the MURA dataset of radiographs, which contains seven different category of upper extremity.

In this paper, The MURA abnormality detection approach is developed using three prominent pre-trained deep learning models (DenseNet169, MobileNet, and XceptionNet) followed by ConvNet and DenseNet layers. X-ray images are collected from a variety of sources. The raw input images typically contain poor X-ray image quality, and varying image sizes [7]. The images must be preprocessed before they can be fed to the model. The contributions of this work are as follows:

- Data preprocessing for the images was performed for better clarity.
- The second step includes developing the proposed ensemble architecture that uses the preprocessed data for training and validation of the model.
- In the next step, the illustration of the models performance on the test data set is shown.
- The last step includes the comparison between our model with that of the existing approaches such as densenet169 model [8] and ensemble200 model [9].

The rest of the work is organized as follows: Sect. 7.2 covers research literature in computer-aided identification of anomalies. Section 7.3 illustrates the proposed methodologies followed by experimental findings in Sects. 7.4 and 7.5 have a conclusion.

7.2 Related Works

This section will give a brief description of the studies that happened previously in the field of bone abnormality detection and classification. Pranav et al. [8] designed a 169-layer DenseNet architecture to detect and localize abnormalities in the MURA dataset. The model obtained an AUROC of 0.929, with cohen's kappa 0.70, 0.815 sensitivity, and 0.887 specificity. Among all seven MURA categories, the model only give comparable results on finger and wrist with that radiologist based on cohen's kappa score. Govind et al. [10] used pre-trained networks DenseNet-169, DenseNet-201, and InceptionResNetV2 to evaluated the performance on the humerus and finger category among all categories, contain MURA dataset. The DenseNet-201 and InceptionResNetV2 models detected abnormalities in the humerus radiographs with 95% CI. The DenseNet-201 model gave an accuracy of 88.19% on humerus and 76.57% on finger. Whereas, the InceptionResNetV2 model gave an accuracy of 86.46% on humerus and 77.66% on finger. Anna et al. [4] developed a 169-layer convolutional neural network to calculate the percentage of abnormality for each category of the MURA dataset. Model shows best results on wrist (0.942), hand (0.862) and shoulder (0.735) in term of kappa score. Dennis and Peter [9] developed the ensemble200 model which scored overall 0.66 Cohen Kappa score. The best cohen kappa score obtained on wrist is 0.7408 and the lowest is obtained on hand is 0.5844. A. F. M. SAIF et al. [11] proposed the capsule network architecture for detecting the abnormality in musculoskeletal radiographs that trained only on 50% of whole MURA dataset. The model scored 10% higher kappa score of 0.705 as compared to the 169-layer densenet model. Minliang He et al. [1] introduced a new calibrated ensemble of three baseline networks (ConvNet, ResNet, and DenseNet). The model gives an overall performance of Accuracy of 0.87, AUC of 0.93, and Cohen's kappa of 0.74. on the publicly available MURA dataset. Stanford ML Group provide only training and validation dataset. The exact amount of test dataset is not given in any of the research paper.

7.3 Proposed Work

Musculoskeletal Disability (MSDs) defined as the injuries that affect the movement or musculoskeletal system of the human body (i.e., discs, muscles, blood vessels,tendons, ligaments, nerves, etc.). Various researchers worked on MURA abnormality classification. But, each research shows some limitations like kappa score is low and variability of kappa score between different categories is high. Therefore, in this research the work is focused to overcome these challenges, in which first we evaluate the performance of some pre-trained models. Then design the weighted ensemble model to improve the results of pre-trained model. The ensemble model improves the kappa score of the model as well as kappa score variability in between the different anatomy of the upper extremities. It is given that the performance of the best radiologist lies between 0.73 and 0.78 in Cohen's kappa metrics [1]. Basically, the raw image data is not understood by the model if fed directly to the system. To make the image system understandable, some preprocessing methods are performed. To construct the weighted ensemble model, we employed three pre-trained models named MobileNet, XceptionNet, and DenseNet169, all of which were trained on ImageNet, the world's largest dataset [12]. These models, as well as the final ensembled model, were developed, trained, and evaluated performance in term of the metrics (Accuracy, AUROC, Cohen's kappa, precision and recall) [9]. The ensemble model results are then compared to the existing models. The model's training was fine-tuned by freezing parts of the models layers.

The details of the proposed ensemble model shown in Fig. 7.1 can be described as follows: Input X-ray images are given for preprocessing. Then these preprocessed images are fed to classification model. The output layer of each pre-trained model is stacked in such a way that, for each image three values will be generated. Now we apply the mean function on these values resulting in the mean of these three values in the range between 0 and 1. Finally, if the mean value is greater than 0.5 then it belongs to abnormal class and if the mean value is less than 0.5 then it belongs to normal class.

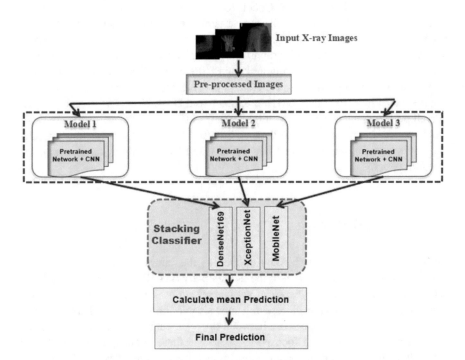

Fig. 7.1 The workflow of proposed Weighted Ensembled model

7.3.1 Dataset Description

Researchers develop different Deep Learning approaches based on public and private datasets. There are several public datasets of joint abnormalities like musculoskeletal (Hand, knee, left hand), chest, etc. computer science, medicine, and radiology department of Stanford university compiled seven upper extremities (Hand, Elbow, Finger, Forearm, Humerus, Shoulder, and Wrist) datasets of a total of 40,561 musculoskeletal radiographs images collected from the Picture Archive and Communication System (PACS) of Stanford Hospital in 2018 [8].

Figure 7.2 contains sample abnormal and normal images from the original MURA dataset. Table 7.1 is the detailed explanation of the dataset that was used for our research [13]. The test dataset is not publicly available, so to evaluate the models performance the train data is split into 80:20 ratio manually.

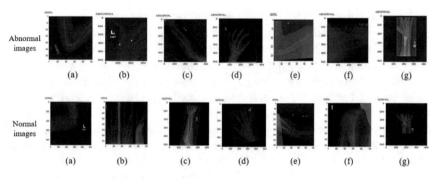

Fig. 7.2 Sample radiographic images for abnormal and normal from the MURA dataset. **a** Elbow, **b** Finger, **c** Forearm, **d** Hand, **e** Humerus, **f** Shoulder, **g** Wrist

Table 7.1 Number of normal and abnormal images of upper extremity

Extremity name	Training images		Validation images		Total images
	Abnormal	Normal	Abnormal	Normal	
Elbow	2006	2925	230	235	5396
Finger	1968	3138	247	214	5567
Forearm	661	1164	151	150	2126
Hand	1484	4059	189	271	6003
Humerus	599	673	140	148	1560
Shoulder	4168	4211	278	285	8942
Wrist	3987	5765	295	364	10411
Total No. of images	14873	21935	1530	1667	40005

7.3.2 Data Augmentation and Preprocessing

The MURA datasets [13] of bone abnormality are found in a wide range of diverse images in multiple formats and sizes. For better classification, the dataset is preferred to be in the same format and resolution. So, the use of image preprocessing techniques becomes important to improve the image quality. Also, data augmentation like the rotation of image up to $10°$, width shift, height shift, horizontal flip to enlarge the dataset was peformed. After the data augmentation all the images were normalized in the same range of 0–1. Preprocessing techniques are applied to prevent our model from overfitting.

7.3.3 Experimental Setup

One important stage here is to choose between various Deep Learning model architectures in order to discover the most suited one. Deep Neural Networks (DNN) and Convolutional Neural Networks (CNN) are most used recent architectures. Therefore, a combination of CNN and DNN architecture was chosen as it gives fast training ability and eliminates the propagation of errors [4]. To implement the ensemble model, selection of three pre-trained models named DenseNet169, XceptionNet, and MobileNet was done. The reason behind the selection of these models is the size of the model and the number of parameters used [14]. During the model development phase, the very first layer is the input layer that contains the shape parameter value equal to the input image size of 224 * 224 with 3 channels. Instead of using all the layers, fine-tuning of some of the layers to increase the model performance was done (freezing top 350 layers out of 595 layers in DenseNet169, 116 out of 132 in XceptionNet, and 70 out of 96 in MobileNet). Then each model is connected with 2 convolution layers followed by the global average pooling layer that takes 4D input and gives 2D output. This 2D output is now given as input for the following dense layers. All Dense layers except the output layer have ReLU as an activation function. The output layer of all models predicts some result that would be stored in the stack. The new dataset that is created by stack is test image data-size x number of models, (i.e., number of test image x 3). Now, we use the reduce mean function that will give a single output across each image that is shown in Fig. 7.1. The last layer contains only 1 neuron and sigmoid activation function as our problem is binary classification. So the output layer gives output in the range between 0 or 1.

Other training parameters include binary cross-entropy as the loss function and the stochastic gradient descent (SGD) optimizer with a learning rate of 0.01. Batch size of 32 was used for training the model. Evaluation of the model is done on the test dataset and it gives satisfying results. Figure 7.3 shows the variation of training and validation accuracy of proposed model during training.

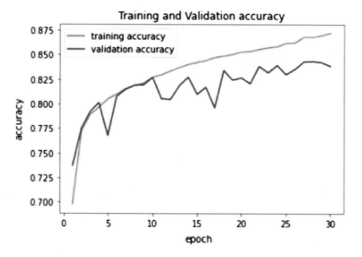

Fig. 7.3 Variation of validation accuracy of proposed ensemble model with epochs

7.3.4   Evaluation Metrics

We choose the five basic criteria to assess the performance of our architecture, namely Kappa, AUC ROC, accuracy, precision, and recall. The mathematical paradigms for the performance indicators are as follows (in Eqs. (7.1), (7.2), (7.3) and (7.4)):

$$\text{Kappa Score}(k) = (P_o - P_e)/(1 - P_e) \tag{7.1}$$

$$\text{Accuracy} = (TP + TN)/(TP + TN + FP + FN) \tag{7.2}$$

$$\text{Recall} = TP/(TP + FN) \tag{7.3}$$

$$\text{Precision} = TP/(TP + FP) \tag{7.4}$$

where P_o is Observational Agreement and P_e is Expected Agreement, TP, TN, FP, and FN represents True Positive, True Negative, False Positive, and False Negative are the parameter of the confusion matrix.

7.4 Result Analysis

7.4.1 Our Model Performance

On the basis of performance metrics we evaluate the performance of each pre-trained network and ensemble network. The evaluation is done for each category as well as overall model performance. Tables 7.2, 7.3, 7.4 and 7.5 show the results of each network.

Comparison of our proposed model was done with the existing state-of-the-art approaches like Raj et al. [8] results and Dennis and Peter [9] results, to determine the level of success of the proposed ensembled architecture. After performing comparison as shown in Table 7.6, it is found that overall our proposed model gives comparative better performance. Our model overcomes the challenges found till now. On Elbow study our model give 0.718 cohen's kappa score, which is higher than compared paper. Similarly on Finger, Forearm and Humerus, our model gives 0.675, 0.771, and 0.748 kappa score with precision is 0.849, 0.883 and 0.866 and recall

Table 7.2 Evaluation of DenseNet169 model on each category

Study part	Accuracy	Cohen Kappa Score	Precision	Recall	AUC value
Elbow	0.815	0.630	0.863	0.865	0.884
Finger	0.772	0.545	0.842	0.814	0.835
Forearm	0.774	0.549	0.895	0.742	0.871
Hand	0.780	0.536	0.810	0.730	0.819
Humerus	0.837	0.673	0.837	0.893	0.886
Shoulder	0.723	0.447	0.708	0.899	0.789
Wrist	0.854	0.702	0.925	0.847	0.915
Overall	0.801	0.600	0.837	0.830	0.860

Table 7.3 Evaluation of XceptionNet model on each category

Study part	Accuracy	Cohen Kappa Score	Precision	Recall	AUC Value
Elbow	0.847	0.694	0.892	0.830	0.908
Finger	0.714	0.431	0.797	0.749	0.805
Forearm	0.801	0.602	0.870	0.735	0.870
Hand	0.761	0.482	0.840	0.640	0.792
Humerus	0.833	0.667	0.827	0.879	0.889
Shoulder	0.705	0.412	0.696	0.838	0.804
Wrist	0.842	0.675	0.912	0.783	0.891
Overall	0.784	0.565	0.821	0.778	0.850

Table 7.4 Evaluation of MobileNet model on each category

Study part	Accuracy	Cohen Kappa Score	Precision	Recall	AUC value
Elbow	0.841	0.681	0.911	0.830	0.896
Finger	0.755	0.516	0.861	0.749	0.837
Forearm	0.817	0.635	0.943	0.755	0.901
Hand	0.772	0.503	0.872	0.656	0.829
Humerus	0.830	0.659	0.858	0.857	0.895
Shoulder	0.739	0.479	0.740	0.888	0.815
Wrist	0.847	0.685	0.900	0.797	0.895
Overall	0.800	0.597	0.857	0.788	0.862

Table 7.5 Evaluation of ensemble network on each category

Study part	Accuracy	Cohen Kappa Score	Precision	Recall	AUC value
Elbow	0.859	0.718	0.854	0.860	0.861
Finger	0.839	0.675	0.849	0.798	0.840
Forearm	0.888	0.771	0.883	0.853	0.887
Hand	0.850	0.678	0.927	0.690	0.871
Humerus	0.873	0.748	0.866	0.879	0.873
Shoulder	0.890	0.717	0.921	0.927	0.860
Wrist	0.885	0.765	0.866	0.877	0.883
Overall	0.869	0.739	0.885	0.854	0.870

Table 7.6 Compare our results to that of paper [8] results and paper [9] results in terms of cohen's kappa (precision, recall)

Study part	Raj et al. [8]	Dennis and Peter [9]	Proposed model results
Elbow	0.710 (0.674,0.745)	0.617 (0.8182, 0.8804)	0.718 (0.854, 0.860)
Finger	0.389 (0.332,0.446)	0.653 (0.7870, 0.9239)	0.675 (0.849, 0.798)
Forearm	0.737 (0.707, 0.766)	0.6954 (0.7753, 1.000)	0.771 (0.883, 0.853)
Hand	0.851 (0.830, 0.871)	0.5844 (0.7778, 0.9703)	0.678 (0.927, 0.690)
Humerus	0.600 (0.558, 0.642)	0.5995 (0.7595, 0.8824)	0.748 (0.866, 0.879)
Shoulder	0.729 (0.697, 0.760)	0.6597 (0.8367, 0.8283)	0.717 (0.921, 0.927)
Wrist	0.931 (0.922, 0.940)	0.7408 (0.8581, 0.9500)	0.765 (0.866, 0.877)
Overall	0.705 (0.700, 0.710)	0.66 (0.81,0.92)	0.739 (0.885, 0.854)

is 0.798, 0.853, and 0.879 greater than the compared results. Whereas, on Hand, Shoulder and Wrist our models performance is lower than Raj et al. [8] results, but, more than Dennis and Peter [9] results. The overall performance of our model is higher than both the model.

7.5   Conclusion

This paper presents the utilization of weighted ensemble model in improving the overall performance of deep learning models in anticipating anomaly or normality in Musculoskeletal radiographs. The model processes a radiographic image and detects whether the image is normal of abnormal. We trained our models on the MURA dataset that consists of radiographic studies of the upper extremity including the shoulder, humerus, elbow, forearm, wrist, hand and finger. The evaluated results show that the weighted ensemble model outperformed than pre-trained models. The model predicts the label of the radiographic : 0 for normal and 1 for abnormal, along with the probability of prediction of abnormality. The best models were selected for forming the weighted ensemble model that consisted of DenseNet169, XceptionNet, and MobileNet models. The weighted ensemble model achieved the highest Cohen kappa score of 0.739 with precision of 0.885 and recall of 0.854. Future works will be focused on improving the current results, and a promising application will be to extend the idea of abnormality detection for public use.

References

1. He, M., Wang, X., Zhao, Y.: A calibrated deep learning ensemble for abnormality detection in musculoskeletal radiographs. Sci. Rep. **11**, 1–11 (2021)
2. Library, W., Services, H.: Styles for bibliographic citations : guidelines for WHO-produced bibliographies. World Health Organization (1988)
3. Xi, P., Guan, H., Shu, C., Borgeat, L., Goubran, R.: An integrated approach for medical abnormality detection using deep patch convolutional neural networks. Vis. Comput. **36**, 1869–1882 (2020)
4. Solovyova, A., Solovyov, I.: X-Ray bone abnormalities detection using MURA dataset. *ArXiv Preprint* ArXiv:2008.03356 (2020)
5. Kong, J., Xu, Y. Yu, H.: Deep transfer learning for abnormality detection. In: Proceedings Of The 4th International Conference On Crowd Science And Engineering, pp. 233–237 (2019)
6. Sage, A., Badura, P.: Intracranial hemorrhage detection in head CT using double-branch convolutional neural network, support vector machine, and random forest. Appl. Sci. **10**, 7577 (2020)
7. Nigam, B., Nigam, A., Jain, R., Dodia, S., Arora, N., Annappa, B.: COVID-19: automatic detection from X-ray images by utilizing deep learning methods. Expert Syst. Appl. **176**, 114883 (2021)
8. Rajpurkar, P., Irvin, J., Bagul, A., Others Mura: Large dataset for abnormality detection in musculoskeletal radiographs. ArXiv Preprint ArXiv:1712.06957. (2017)

9. Banga, D., Waiganjo, P.: Abnormality detection in musculoskeletal radiographs with convolutional neural networks (ensembles) and performance optimization. ArXiv Preprint ArXiv:1908.02170 (2019)
10. Chada, G.: Machine learning models for abnormality detection in musculoskeletal radiographs. Reports **2**, 26 (2019)
11. Saif, A., Shahnaz, C., Zhu, W., Ahmad, M.: Abnormality detection in musculoskeletal radiographs using capsule network. IEEE Access **7**, 81494–81503 (2019)
12. Deng, J., Dong, W., Socher, R., Li, L., Li, K., Fei-Fei, L.: Imagenet: a large-scale hierarchical image database. In: 2009 IEEE Conference On Computer Vision And Pattern Recognition, pp. 248–255 (2009)
13. https://stanfordmlgroup.github.io/competitions/mura/
14. Chollet, F., et al.: https://github.com/fchollet/keras (GitHub, 2015)

Chapter 8
A Naïve Approach for Mutation Detection Using Color Encodings and String Matching Techniques

Masabattula Teja Nikhil, Kunisetty Jaswanth, Mutyala Sai Sri Siddhartha, D. Radha, and Amrita Thakur

Abstract Genome sequencing is one of the key areas of research that helps analyze the genome and contributes to the diagnosis of various diseases like cancer, Ebola virus, covid, etc. There are various genome sequencing techniques available for detecting the mutations and for sequencing a gene including whole-genome sequencing methods. There are also various deep learning techniques like mutations identification and disease classification. Mutation detection in the genome can also be achieved with sequence alignment techniques such as the Needleman Wunsch algorithm and smith-waterman algorithm. In this paper, a new approach of mutation identification technique has been proposed using color encoding, hamming distance, and Levenshtein distance in identifying the cancerous mutation in the gene. Research says that most of the mutations (>50%) that lead to cancer are due to the mutations in the TP53 gene which is present in chromosome 17. So, the TP53, a coding gene that is responsible for the production of P53 protein has been taken as a reference gene for the identification of cancer. A visual mutation identification technique using color encodings is also added which makes the user easily identify the mutation visually. The implementation of the sequencing techniques has been done using python programming.

8.1 Introduction

Genome sequencing is basically a powerful technique that can solve various mysteries about the genome. There are also deep learning techniques proposed like [1] to identify the mutation classification of diseases based on the mutations. In

M. T. Nikhil · K. Jaswanth · M. S. S. Siddhartha · D. Radha (✉)
Department of Computer Science and Engineering, Amrita School of Engineering, Amrita Vishwa Vidyapeetham, Bengaluru, India
e-mail: d_radha@blr.amrita.edu

A. Thakur
Department of Chemistry, Amrita School of Engineering, Amrita Vishwa Vidyapeetham, Bengaluru, India
e-mail: t_amrita@blr.amrita.edu

V. Bhateja et al. (eds.), *Evolution in Computational Intelligence*, Smart Innovation, Systems and Technologies 326, https://doi.org/10.1007/978-981-19-7513-4_8

general, genome sequencing techniques include alignment algorithms as well as string matching algorithms. The reasons for getting cancer can be due to changes in the structure of the protein [2] or may be due to mutations in the cancer-related gene. Research has also been done to identify the rare genetic mutations by analyzing the genotype and phenotype of the disease causing gene [3]. This paper concentrates on String matching algorithms. String matching techniques can be of two types exact and approximate string matching, Exact string matching [4] techniques will not allow any mistakes in string and return true only if the string is matched exactly (100% matching), Approximate string matching techniques can allow a certain number of mistakes and returns true if the strings are matched such that the dissimilarities between the strings are less. The allowable number of mistakes which means that the strings may need not be matched exactly (100%). Coming to genome sequencing which works with gene data, approximate string matching techniques are good to go for, as the gene will never be consistent and will not be matched with the reference gene exactly all the time. Taking the advantage of approximate string matching techniques the number of mutations can be identified and followed by the type of mutation (Insertion/ Deletion/ Substitution) and further according to the results obtained, one can also identify the type of substitution mutation (Missense/ Nonsense/ Silence).

It is already known that a genome is made up of a sequence of nucleotides (*A*, *T*, *C*, *G*) and a genome consists of two types of regions, a coding region which is responsible for the production of proteins, a non-coding region that regulates the production of proteins. A sequence of three nucleotides that together form a unit of genetic code in a DNA or RNA molecule is called codons. Any coding gene starts with a start codon mostly ATG and ends with a stop codon (TGA, TAA, TAG). Using the start and stop codons and length of the reference gene one can identify the coding gene from a particular chromosome. In this paper, a coding gene called the tumor suppressor gene is taken as a reference gene. A tumor suppressor gene directs the production of a protein that is a part of the system that regulates cell division. The tumor suppressor protein plays a key role in keeping cell division in check. When mutated, a tumor suppressor gene is unable to do its job and uncontrolled cell growth may occur as a result. There are various tumor suppressor genes in the human genome which include the TP53 gene, BRCA1, and BRCA2. This paper has taken the TP53 gene as a reference coding gene which is present on the short arm of chromosome 17 and contributes 2.5–3% of the whole DNA.

8.2 Related Works

Paper [1] has implemented and analyzed the involvement of various cancer-associated genes like BARD1, CHEK2, ATM, BRCA2, BRCA1, and RAD51 using a deep learning model called Siamese neural network. The work presented in paper [1] helped in making valid predictions on TP53 mutation. In this study, Hamming distance is used to extract the mutated TP53 gene from chromosome 17 using a reference gene. In Paper [5], a novel system has been proposed for the detection of mutations in the TP53 gene which is a tumor suppressor gene. The proposed system utilizes Content addressable memory (CAM) along with hamming distance for mutation detection and specifying the location of the mutation.

It is not like always there exists a reference gene, sometimes one must generate the coding gene using their properties like length, protein encodings, and start and stop codons. Paper [6] has proposed a new technique of identifying the protein-coding genes in genomic sequences using a set of encoded proteins and reviewed the computational pipelines that are used to generate the reference protein-coding gene sets. Paper [7] worked on pattern matching with compressed genomic data which helps in decreasing computation cost in matching the strings (genes). The compression of gene data means converting the DNA data into ordinary characters and this research attempts to identify the known patterns in the compressed gene without decompressing it. Pattern matching algorithms are generic algorithms that can be used to find the mutations in any type of genes. Alignment algorithms unlike pattern matching can also be used for finding the mutations in the gene as presented in paper [8] which works with Leukemia using a dynamic short distance pattern matching algorithm.

8.3   System Model

Figure 8.1 depicts the block diagram of the model in which the input to the model is the two DNA sequences reference and mutated TP53 genes which is then converted into color encodings further the observed mutations can be classified as substitution or frameshift mutations. The two algorithms that are used to analyze the mutations are hamming distance and Levenshtein distance.

8.3.1   Hamming Distance

Hamming distance is one of the approximate string-matching techniques, generally used to find the number of mismatches between the strings. In this paper, Hamming distance is used to calculate the number of substitution mutations in the gene. Hamming distance can be used for extracting the mutated gene from the whole genome assuming that the mutations in the genome are substitutions.

8.3.2   Levenshtein Distance

Levenshtein distance also known as edit distance is an approximate string-matching technique used for calculating the number of mutations by using a dynamic program-ming approach [10]. In this paper, edit distance, which is stored in the form of a

Fig. 8.1 Block diagram of the proposed model

matrix, is used to find the total number of frameshift and substitution mutations, unlike hamming distance which calculates only the substitution mutations. The type of mutation can be identified by backtracking the obtained distance matrix.

8.3.3 Color Encodings

Before analyzing the mutated gene with the approximate string-matching techniques, it is required to find whether the mutation in the gene is a substitution or a frameshift and whether the mutations are wide apart or occurred consecutively. Color encoding helps in completing this task by assigning different colors to each nucleotide and converting the nucleotide sequence into an image. In this method each nucleotide Adenine (A), Cytosine (C), Thymine (T), Guanine (G) is assigned with different colors like Adenine as blue, cytosine as red, Guanine as green, and Thymine as yellow as shown in Fig. 8.2 and then the nucleotide sequence is reconstructed such that it forms a perfect square matrix and followed by converting into an image with each pixel representing each nucleotide with their corresponding colors. The dimension of the image is dependent on the length of the coding gene as shown in Eq. (8.1).

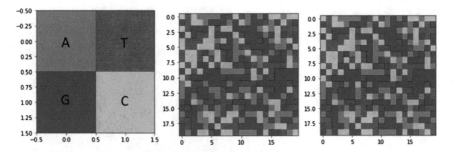

Fig. 8.2 Color encodings of reference and mutated genes

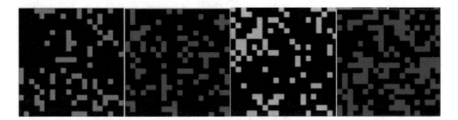

Fig. 8.3 Combined color encodings of individual nucleotides

$$\text{Let} m = \sqrt{\text{gene length}}$$

$$\text{dimensions of image} = m * m \qquad (8.1)$$

The pixelated image of the nucleotide sequence is shown in Fig. 8.2. The left figure indicates the color codes of four nucleotides A, T, G, and C, the middle and the right one represents the color encodings for the reference and mutated genes respectively.

Once the pixelated images are generated, the next step is the identification of mutations. This can be done using the bitwise "and" operation between the color-encoded matrices containing only one nucleotide which is shown in Fig. 8.3.

Once the color encodings of individual nucleotides are generated then the final matrix can be obtained by adding all the color encodings in Fig. 8.4. The black dots in Fig. 8.4 represents that the nucleotide has been mutated.

8.4 Implementation

Implementation has been done using the python programming language. The flow of the algorithm is shown in Fig. 8.5.

Fig. 8.4 Final combined
color encoded matrix

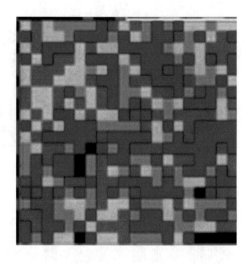

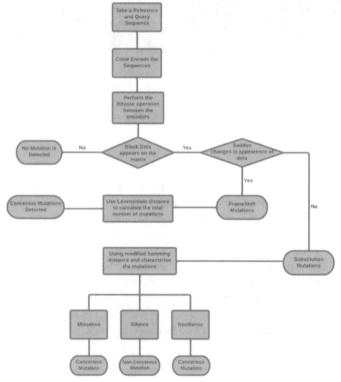

Fig. 8.5 Flowchart of the proposed algorithm

Sequencing Algorithm: Sequence traversal algorithm

```
1: Input: Reference TP53 Gene, Mutated TP53 gene
2: Output:
3. Reference_gene - - - > Color coded reference gene
4. Mutated_gene - - - > Color coded mutated gene
5. imageshow(bitwise_And(Color coded reference gene, Color coded mutated gene))
6. If((isSubstitutions()==true) & (isframeShift()==false)) :
7.    (num_mutations,indexposition) = Hamming_Distance(Reference gene, mutated gene)
8.    for i in indexposition:
9:        if (protein_encoding(codon_pair(reference_gene[i])) == protein_encoding(codon_pair(mutated_gene[i]))):
10:            mutation.append("silence mutation")
11:        elif (protein_encoding(codon_pair(reference_gene[i])) != protein_encoding(codon_pair(mutated_gene[i]))):
12:            mutations.append("missense/nonsense mutation(depends on changes takes place due to protein codings)")
13: else:
14:    num_mutations = Leveinshtein_Distance(Reference gene, mutated gene)
15:    selection = select_index(top substitutions)
16:    for i in selection:
17:        if (protein_encoding(codon_pair(reference_gene[i])) == protein_encoding(codon_pair(mutated_gene[i]))):
18:            mutation.append("silence mutation")
19:        elif (protein_encoding(codon_pair(reference_gene[i])) != protein_encoding(codon_pair(mutated_gene[i]))):
20:            mutations.append("missense/nonsense mutation(depends on changes takes place due to protein codings)")
```

Fig. 8.6 A blueprint of sequence algorithm

The model takes the reference TP53 [14] which is taken from the sequence bank (Gene ID: 7157) and the query gene as the input and then converts them into a color encoding matrix. The obtained color encodings of the genes are then combined using the bitwise operator which results in a new color matrix with mutations identified as black dots. If the black dots occur at different points separated by a distance, then the mutation is concluded as substitution mutation and if there is a sudden increase in the black dots sequentially then the mutations are concluded as frameshift mutations. If the mutations in the gene are substitutions, then hamming distance is used to calculate the number of mismatches and their position in the gene. In case of frameshift mutation, Levenshtein distance is used to calculate the number of mutations. Once the mutations are observed they will be encoded into proteins and followed by classifying the mutation into missense, silence, or nonsense mutation. The logic of the sequencing algorithm is as mentioned in Fig. 8.6.

8.4.1   Tumor Suppressor Gene

TP53 is a coding gene that codes for a tumor protein called P53 [9], also known as the guardian of the genome. TP53 is a tumor suppressor gene, and the name indicates that it helps in regulating the uncontrollable growth of the cells which could lead to the formation of a tumor. There are many stages of cancer, but any type of cancer starts with the formation of a tumor which is due to the piling of dead cells. The TP53 is a regulatory gene that subjects to the mutations in most cancerous diseases. A mutation in TP53 may lead to various diseases such as breast cancer, bone cancer, leukemia, and soft tissue cancers also called sarcomas. A deep understanding of the mutations and functions of TP53 is described in [11–13]. There are many string-matching techniques that can be applied to various genes to find the mutations in the gene. So, in this paper TP53 is taken as a reference gene on which pattern matching is applied. Generally, the TP53 gene is located in the short arm of chromosome 17 which contributes only about 2.5–3% of whole human DNA.

8.5 Results

The reference taken in the proposed work is the TP53 gene and during the implementation, a small segment of the TP53 gene has been taken as input shown in Fig. 8.7. Using this reference gene, a mutated gene has been extracted from a genome with a larger length than the reference gene which is shown in Fig. 8.8 and then using the color encodings, the corresponding algorithm has been selected which is the hamming distance in this case, and the results are as shown in Fig. 8.9.

Figure 8.9 mentioned below indicates that the mutation in the gene is only a substitution. As the mutations in the gene are only substitutions, hamming distance is the best searching algorithm to identify the mutation, and the results obtained are shown in Fig. 8.10.

With the same reference gene, if the mutated gene is changed such that it also consists of substitutions along with insertions and deletions, then the mutated gene is given as input to the model. The results obtained representing the mutations in the gene for the corresponding gene are as shown in Figs. 8.11 and 8.12.

The sudden increase in the sequential changes in the gene indicates that the mutation in the gene is not only the substitution but frameshift as well. As the mutations

```
CTCAAAAGTCTAGAGCCACCGTCCAGGGAGCAGGTAGCTGCTGGGCTCCGGGGACACT
TTGCGTTCGGGCTGGGAGCGTGCTTTCCACGACGGTGACACGCTTCCCTGGATTGGGTA
AGCTCCTGACTGAACTTGATGAGTCCTCTCTGAGTCACGGGCTCTCGGCTCCGTGTATTT
TCAGCTCGGGAAAATCGCTGGGGCTGGGGGTGGGGCAGTGGGGACTTAGCGAGTTTGG
GGGTGAGTGGGATGGAAGCTTGGCTAGAGGGATCATCATAGGAGTTGCATTGTTGGGA
GACCTGGGTGTAGATGATGGGGATGTTAGGACCATCCGAACTCAAAGTTGAACGCCTA
GGCAGAGGAGTGGAGCTTTGGGGAACCTTGAGCCGGCCTAAAGCGTACTTCTTTGCACA
TCCACCCGGT
```

Fig. 8.7 Reference gene taken from sequence bank

```
CTCAAAAGTCTAGAGCCACCGTCCAGGGAGCAGGTAGCTGCTGGGCTCCGGGGACACT
TTGCGTTCGGGCTGGGAGCGTGCTTTCCACGACGGTGACACGCTTCCCTGGATTGGGTA
AGCTCCTGACTGAACTTGATGAGTCCTCTCTGAGTCACGGGCTCTCGGCTCCGTGTATTT
TCAGCTCGGGAAAATCGCTGGGGCTGGGGGTGGGGCAGTGGGGACTTAGGGAGTTTGG
GGGTGAGTGGAATGGAAGCTTGGCTAGAGGGATCATCATAGGAGTTGCATTGTTGGGA
GACCTGGGTGTAGATGATGGGGACGTTAGGACCATCCGAACTCAAAGTTGAACGCCTA
GGCAGAGGAGTGGAGCTTTGGGGAACCTTGAGCCTGCCTAAAGCGTACTTCTTTGCACA
TCCACCCGGT
```

Fig. 8.8 Mutated reference gene with only substitutions

Fig. 8.9 Substitution
mutations identified in the
gene

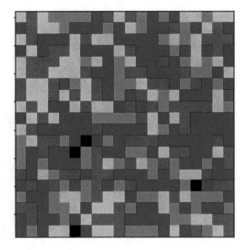

Substitution mutation detected at index position 226
Substitution mutation detected at index position 245
Substitution mutation detected at index position 316
Substitution mutation detected at index position 385
Mutation at 226 can cause cancer as it is a missense mutation.(not much sever)
Mutation at 245 cannot be responsible for cancer as it is a silence mutation
Mutation at 316 can cause cancer as it is a missense mutation.(not much sever)
Mutation at 385 can cause cancer as it is a missense mutation.(not much sever)

Fig. 8.10 Output when the mutations are only substitution

CTCAAAAGTCTAGCGCCACCGTCCAGGGAGCAGGTAGCTGCTGGGCTCCGAGGACACTT
TGCGTTCGGGCTGGGAGCGTGCTTCCACGACGGTGACACGCTTCCCTGGATTGGGTAAG
CTCCTGACTCGAACTTGATGAGTCCTCTCTGAGTCACGGGCTCTCGGCTCCGTGTATTTT
CAGCTCGGGAAAATCGCTGGGGCTGGGGGTGGGGCAGTGGGGACTTAGCGAGTTTGGG
GGTGAGTGGGATGGAAGCTTGGCTAGAGGGATCATCATAGGAGTTGCATTGTTGGGAG
ACCTGGGTGTAGATGATGGGGATGTTAGGACCATCCGAACTCAAAGTTGAACGCCTAG
GCAGAGGAGTGGAGCTTTGGGGAACCTTGAGCCGGCCTAAAGCGTACTTCTTTGCACAT
CCACCCGGT

Fig. 8.11 Mutated reference gene with substitutions along with insertions and deletions

in the gene are not an only substitution, Levenshtein distance is the best searching algorithm to identify the mutation, and the results obtained are shown in Fig. 8.13.

The results computed from the proposed algorithm are represented in the Table 8.1. There are three total outcomes for each mutation type, from the table it is evident that if the mutation type is substitution, then missense and non-sense mutations can create a severity of cancer while silence mutations doesn't. In case of Frame Shift mutations, irrespective of occurrence of Substitution mutations the severity of cancer

Fig. 8.12 Substitutions and frameshift mutations identified

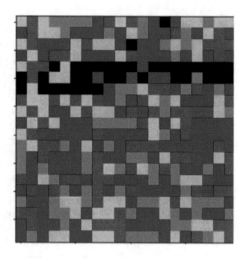

Substitution mutation detected at index position 13
Substitution mutation detected at index position 50
Deletion mutation detected at index position 83
Insertion mutation detected at index position 128
Mutation at 4000 can cause cancer as it is a missense mutation.(High severity)
Mutation at 5000 cannot be responsible for cancer as it is a silence mutation

Fig. 8.13 Output when the mutations are substitutions and frameshift

Table 8.1 Result analysis

Mutations		Outcome possibilities				
Substitution	Missense	Y/N	N	Y	N	Y
	Non-sense	Y/N	N	N	Y	Y
	Silence	Y/N	Y/N	Y/N	Y/N	Y/N
Frame shift mutations		Y	N	N	N	N
Severity		Highly severe	No cancer	Severe	Severe	Severe

is very high. Color encodings in our model helps in drawing the above conclusions so easily because of visualization as mutations are shown in black pixels.

8.6 Conclusion

Pattern matching techniques, especially the approximate pattern matching techniques have a great significance in identifying the mutations. Color encoding in genome sequencing is very useful for the visualization of mutations and diagnostics of cancer or any genetic disease that occurs frequently. Hamming distance is useful only when all the gene mutations are substitutions and fail to detect the frameshift mutations. Levenshtein distance is used to find all types of mutations in the gene either substitution or frameshift. It depends on the color encodings whether to select the hamming distance or edit distance algorithm. The analysis says that whenever the mutation is a frameshift mutation then it is sure that the severity of the cancer is high (highly prone to cancer) and when the mutation is a substitution then the severity depends on the type of substitution. If the substitution is a silence mutation, then there is no chance of getting cancer (not prone to cancer) and whenever the mutation is a nonsense or missense mutation then it is said to be severe (prone to cancer). Color encodings helps in quick identification of occurrence of cancer and its severity. Various genome sequencing techniques are available for detecting the mutations [15] in the human genome and there are various approaches to analyze the human genome by using graphs with the size of the node representing the number of genes that are associated [16] with the disease.

References

1. Lakshmi Anand, C., Krishnan, N.P.K.: Design and development of a diagnostic system for early prediction of P53 mutation causing cancer from microscopic biopsy images. In: 2020 Fourth International Conference on I-SMAC (IoT in Social, Mobile, Analytics and Cloud) (I-SMAC), pp. 435–440 (2020)
2. Bipin Nair, B.J., Anju, K.J., Jeevakumar, A.: Tobacco smoking induced lung cancer prediction by Lc-Micrornas secondary structure prediction and target comparison, In: 2017 2nd International Conference for Convergence in Technology (I2CT), pp. 854–857 (2017)
3. Bipin, B.J.: Identifying genetic mutation rare genetic disorder by analyzing characteristics of genotype-phenotype by implementing Apriori algorithm, **11**, 262–265 (2020)
4. Joshi, O.S., Upadhvay. B.R., Supriya, M.: Parallelized advanced Rabin-Karp algorithm for string matching. In: 2017 International Conference on Computing, Communication, Control and Automation (ICCUBEA), pp. 1–5 (2017)
5. Ibrahim, M., Mujahid, O., Rehman, N.U., Qazi, A., Ullah, Z., Fouzder, T.: An FPGA-based accelerated mutation detection system for the tumor suppressor gene. IEEE Access **9**, 164542–164550 (2021)
6. Harrow, J., Nagy, A., Reymond, A., et al.: Identifying protein-coding genes in genomic sequences. Genome Biol. **10**, 201 (2009)
7. Keerthy, A.S., Manju Priya, S.: Pattern Matching in compressed genomic data. In: 2nd International Conference on Communication and Electronic Systems (ICCES) (2017)
8. Ananya, B., Prabisha, A., Kanjana, V.: Novel approach to find the various stages of chronic myeloid leukemia using dynamic short distance pattern matching algorithm. In: 2018 3rd International Conference for Convergence in Technology (I2CT), pp. 1–5 (2018)
9. Kumar Das, J., Ghosh, S., Kumar Rout, R., Pal Choudhury, P.: A study of P53 gene and its regulatory genes network. In: 2018 8th International Conference on Cloud Computing, Data Science and Engineering (Confluence), pp. 14–19 (2018)
10. Clarke, W.A., Ferreira, H.C.: Leveinshtein distance-based coding for synchronous, fixed length decoding windows in the presence of insertions/deletions. SAIEE Afr. Res. J. **97**(2), 146–150 (2016)

11. Ryan, K.M., Phillips, A.C., Vousden, K.H.: Regulation and function of the p53 tumor suppressor protein. Curr. Opin. Cell Biol. **13**(3), 332–337 (2001)
12. Liu, J., Zhang, C., Hu, W., Feng, Z.: Tumor suppressor p53 and its mutants in cancer metabolism. Cancer Lett. **356**(2), 197–203 (2015)
13. Olivier, M., Hollstein. M., Hainaut, P.: TP53 mutations in human cancers: origins, consequences, and clinical use. Cold Spring Harb. Perspect. Biol. (2010)
14. TP53 tumor protein p53 [Homo sapiens (human)], protein-coding gene, Gene ID: 7157
15. Martins, P., Abbasi, M.: End to end distance measurement algorithms in biology sequences. In: 2020 15th Iberian Conference on Information Systems and Technologies (CISTI), pp. 1–6 (2020)
16. Doke, A.R., Garla, N., Radha, D.: Analysis of human gene-disease association as a social network. Int. J. Adv. Trends Comput. Sci. Eng. **8**(4) (2019)

Chapter 9
Ablation Study of Indian Automatic Vehicle Number-Plate Recognition Model Trained Over Synthetic Dataset

Pranav Gaur, Arpon Basu, Pritam P. Shete, Abhilash Bhardwaj, and Dinesh M. Sarode

Abstract Modern automatic vehicle number-plate recognition models require a large corpora of *labelled* examples as training dataset. Recent advances in data-generation methods have addressed this challenge by leveraging auto-encoders (Yang et al. in 2015 8th International Symposium on Computational Intelligence and Design (ISCID). pp. 465–469, 2015 [1]) and adversarial networks (Kukreja et al. in 2020 4th International Conference on Electronics, Communication and Aerospace Technology (ICECA). pp. 1190–1195, 2020 [2]); (Han et al. in Appl. Sci. 10:2780, 2020 [3]); Liu et al. in Proceedings of the 25th ACM International Conference on Multimedia (MM '17). Association for Computing Machinery, New York, NY, USA, pp. 1618–1626, 2017 [4]) to model the latent space of vehicle number-plate images based on unlabeled real images. As per our knowledge, however, there is no systematic study assessing efficacy of such methods for Indian vehicle number plates. While the usage of deep learning models for generation of training data has been reported with success, it further limits the interpretability of the overall solution by adding another deep neural network in the ANPR system pipeline (Linardatos et al. in Entropy 23:18, 2021[5]). The objective of this work, therefore, is to address these challenges by reporting promising evidences from an ablation study focused on understanding the impact of a simple image-generation control parameter like text font over the performance of convolutional recurrent neural network-based (CRNN) Indian vehicle number-plate recognition model. The design of our study is guided by initial experiments that suggested that merely exposing CRNN to variations in vehicle number text font alone in the synthetic training data cannot provide

P. Gaur (✉) · P. P. Shete · A. Bhardwaj · D. M. Sarode
Computer Division, Bhabha Atomic Research Centre, Bombay, Maharashtra, India
e-mail: pranav@barc.gov.in

P. P. Shete
e-mail: ppshete@barc.gov.in

A. Bhardwaj
e-mail: abhilashb@barc.gov.in

D. M. Sarode
e-mail: dinesh@barc.gov.in

A. Basu
Indian Institute of Technology, Bombay, Maharashtra, India

© The Author(s), under exclusive license to Springer Nature Singapore Pte Ltd. 2023
V. Bhateja et al. (eds.), *Evolution in Computational Intelligence*, Smart Innovation, Systems and Technologies 326, https://doi.org/10.1007/978-981-19-7513-4_9

generalization over real images as we observed 100% mean word-error rates (WERs) and mean character-error rates (CER) regardless of the size of training data. We have, therefore, maintained a uniform range of rotation, translation, and Gaussian blurring across all of our reported experiments in the ablation study. We have created a test dataset of 100 real images drawn from an ANPR deployment site camera and 350 images under unconstrained environment setting available from the Kaggle platform. We observed that the lack of text font variety severely affected the performance of CRNN, rendering a mean WER and CER of 100% on test data, whereas increasing the number of fonts to even 10 classes enhanced the performance to upto 10% mean WER and 1.62% mean CER on constrained dataset and around 50% mean WER and 15.5% mean CER on unconstrained dataset. Our results show that it is possible to achieve a substantial performance headstart using conventional synthetic image-generation methods alone.

9.1 Introduction

Vehicle identification plays a key role in root-cause analysis in event of road accidents, crime forensics, law enforcement [6]. Manual logging of vehicle details is not feasible in most cases because of the scale and nature of traffic as well as the economics of labor. ANPR systems address this by automating this process by leveraging advances in computer vision techniques. Given the variety of vehicles, image backgrounds, dynamics of vehicle movement, conventional computer vision algorithms fail to generalize beyond rigid test environment [7]. Computer vision technologies based on deep learning have proved to be promising for ANPR applications [8, 9]; however, it requires large amounts of labeled training dataset for delivering a decent performance [10].

In order to alleviate scarcity of *labelled* training dataset, unsupervised learning approaches based on auto-encoders and generative adversarial networks have been reported with promising results [1–4]. Modern ANPR systems typically employ two (and sometimes three) cascaded neural network models for vehicle number-plate localization followed by text prediction [11]. Adding neural network-based unsupervised methods for dataset generation, while delivering performance, hinders explainability of the overall ANPR solution [5]. Conventional synthetic dataset generation methods, on the other hand, with very few control parameters like text rotation, translation, image noise, and text font, have proved to be effective for ANPR [12] and even for the more general problem of optical character recognition (OCR) [13], while at the same time facilitating interpretability of the effect of data-generation strategy on model performance. While there has been a considerable work in direction of ANPR system and algorithm development for Indian ANPR, a systematic study of effect of such training dataset development strategies on ANPR model performance is still missing. Owing to the wild variation in number-plate shape, size, text font, image background, image noise and blur as shown in Fig. 9.1, the effectiveness of artificial data-generation methods in adapting to real-image domain is still not established.

Fig. 9.1 Indian vehicle number plates

The *contribution of our work* is that we have demonstrated that even for challenging image domains like Indian vehicle number plates, simple synthetic data-generation methods can facilitate a performant baseline number-plate recognition model which can then be fine-tuned over real images or images generated with deep neural networks like GANs and auto-encoders. The design of our experiments is inspired by [10]; however, we have kept this study limited to programmatically generated synthetic datasets. We observed that our results were in agreement with those reported by [13] in terms of the effectiveness of synthetic number-plate data over real images. The paper is organized as follows: we report our methodology for defining the problem based on real-image test data, evaluation metric, and selection of hyper-parameters for training CRNN across these ablation experiments. We then report our the experimental results in detail. We further highlight the observed issues in the discussion section. The conclusion section summarizes the insights we gained from this activity and highlights the challenges we look forward to address in future.

9.2   Methodology

We have adopted a combination of *constrained* and *unconstrained* test datasets for assessing the performance of the developed models during the ablation study. Constrained dataset is collected under controlled lighting, number-plate style, and vehicle movement conditions, whereas the unconstrained dataset has no such assumptions. We have used standard OCR evaluation metric for evaluation of the ANPR model. The initial experiments focused purely on frontal images as shown in Fig. 9.2, exposing CRNN to variations in the text fonts alone. These initial experiments helped us identify the importance of text rotation, text translation, and Gaussian noise as

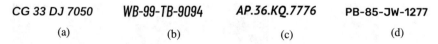

CG 33 DJ 7050   **WB-99-TB-9094**   **AP.36.KQ.7776**   **PB-85-JW-1277**

(a)   (b)   (c)   (d)

Fig. 9.2 Some samples from the synthetic frontal image dataset

(a)   (b)

(c)   (d)

Fig. 9.3 Some test image samples: **a**, **b** are from constrained dataset and **c**, **d** from the unconstrained dataset

prerequisites for CRNN before an ablation study involving text font can provide meaningful insights. Following sections describe various aspects of our methodology in more detail.

9.2.1   Test Dataset

We have acquired dataset from IP cameras installed at the entrance of our ANPR deployment site. The images are captured when a vehicle arrives in a region-of-interest (ROI) at the deployment site. The ROI is decided based on operational constraints imposed by the local authority utilizing the ANPR solution for law enforcement. We have manually cropped full-size images to contain only the vehicle number plates. In order to maximize the diversity in text in the test dataset, we have ensured that each test image contains a unique vehicle number. As we focused on number-plate text only, we have avoided experimenting with number-plate background and number-plate text color combinations, keeping number-plate background as white and number-plate text as black. In order to assess generalizabilty of the best-performing model against extreme text rotations, perspective transformation, text background, we identified and annotated a publicly available unlabeled dataset at Kaggle. Some representative images from constrained dataset and unconstrained dataset are shown in Fig. 9.3.

9.2.2   Synthetic Data Generation

Our initial experiments did not include text rotation, text translation, and Gaussian noise, experimenting only with text font drawn from Google Fonts Library. We generated random vehicle numbers and utilized text-to-image utilities to generate corresponding number-plate images with various text fonts. In subsequent experiments, we augmented data generation with uniform variations in rotation, translation, and noise

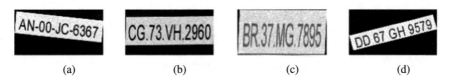

(a)   (b)   (c)   (d)

Fig. 9.4 Some samples from the synthetic image dataset

parameters utilizing OpenCV [14]. The text fonts were selected randomly to prevent aiding CRNN with those matching the fonts in test images. The size of training data as well as the validation was increased progressively as we increased the number of text fonts so as to maintain a uniform ratio of number of images per font across all experiments. The target scenario for ANPR deployment required addressing both 1 and 2-line vehicle number plates; therefore, we maintained an equal proportion of 1 and 2-line number-plate samples across our training and validation dataset. Some representative images are shown in the Fig. 9.4.

9.2.3 Evaluation Metric and Model training

Character and word-error rate are standard metric for evaluation of optical character recognition systems [15]; as ANPR is a special case of OCR, we have used the same evaluation metric in this work. For model development, we leveraged our earlier work on CRNN [16], which has worked well over publicly available number-plate recognition datasets. The model control parameters like batch size, LSTM step size, feature-extractor architecture, weight decay, and learning rate were kept fixed across all experiments.

9.3 Experiments

We have generated synthetic data in a progressive and controlled manner and analyzed its effect on the performance of CRNN model over real-image test dataset. We initially performed experiments to assess if variation in text font alone is sufficient to develop a performing CRNN model by feeding a large sample size of 1.28M samples with 10 fonts per text. As shown in the Table 9.1, however, regardless of the number of fonts used, CRNN failed to generalize beyond synthetic dataset achieving close to 99% validation accuracy on synthetic data but 100% WER on real images. These observations motivated us to experiment with other control parameters like text rotation, translation, and image noise to arrive at a reasonable starting point where we can keep the other parameters fixed and vary text font alone. As can be observed in experiment-3, adding rotation, translation, and noise resulted in improved CER.

Based on above experiments, we designed our ablation study, by uniformly randomizing the text rotation, text translation, and image-noise parameters across all subsequent experiments while experimenting with the number of text fonts exposed

Table 9.1 Initial experiments to assess significance of text font

Index	Training data	Validation data	CER and WER
1	128K single-line frontal images with 1 font class	12.8K single-line frontal images	100, 100
2	1.28M single-line frontal images with 1 image per font per text	128K single-line frontal images with 1 image per font per text	100, 100
3	100K single-line images with 1 font class, random text rotation, translation, and Gaussian noise	20K single-line images with 1 font class and random text rotation, translation, and Gaussian noise	66.29, 100

Table 9.2 Ablation study of CRNN performance with respect to number of text fonts

Index	Training data	Validation data	CER and WER
4	50K, 50K single-line and two-line images with 5 font classes	10K, 10K single-line and two-line images with 5 font classes	18.38, 98.57
5	50K, 50K single-line and two-line images with with test-image background over 5 fonts	10K, 10K single-line and two-line images with test-image background over 5 fonts	10.05, 71.43
6	100K, 100K single-line and two-line images with 5 font classes	20K, 20K single-line and two-line images with 5 font classes	5.22, 41.43
7	100K, 100K single-line and two-line images with 10 font classes	20K, 20K single-line and two-line images with 10 font classes	3.80, 28.57
8	200K, 200K single-line and two-line images with 10 font classes	40K, 40K single-line and two-line images with 10 font classes	1.62, 10.0

to CRNN during training as shown in Table 9.2. As an evidence of how matching data distribution can significantly improve network performance, we also performed experiment-5 which is unrelated to the current study but sets the motivation for generative neural networks which attempt to generate *realistic* images. Specifically, in experiment-5, for training dataset of identical size as experiment-4, we achieved a better performance because synthetic number plates were embedded in a background sample taken from test images. We further evaluated the generalization performance of model developed in experiment-8 over our unconstrained dataset, over which, we achieved 50% mean WER and 15.5% mean CER.

9.4   Discussion

The experiments demonstrated a strong correlation between text-font variations and CRNN performance given the model is also exposed to random variations in text rotation, translation, and image noise. We further analyzed the quality of predictions and identified multiple issues, as discussed in this section.

Poor performance on 2-line number-plate images: We observed that while CRNN could generalize over single-line number-plate test images, it did not show such progression over 2-line number-plate images. We suspect that the data distribution between the synthetic 2-line training images is considerably different from that of the 2-line test images. A practical work around to address multiline text could be to train CRNN over OCR datasets [13] instead and inferring over the resulting OCR model 1-text line at a time and combining the result [17].

Validation error is an unreliable proxy for test error: During the course of model training runs over multiple epochs, we observed that the best model with respect to validation dataset may not be the best model on test dataset. For instance, in experiment-7, the validation accuracy of the best-performing model (on test data) is 0.01% lower than the model with 40% lower test accuracy. In order to avoid such pitfalls, we had to evaluate all models checkpoints over the test data and select the one with least WER.

9.5   Conclusion

In this work, we have attempted to provide evidences for the proposition that using synthetic dataset based on the conventional image-generation methods alone, a significant performance head start over the number-plate recognition task can be achieved. We have described our process of development of a baseline model for Indian number-plate recognition which achieves 10% mean WER on constrained test dataset and 50% mean WER on an unconstrained dataset. For the unconstrained dataset, it achieved a 15.5% mean CER implying that only 1 in every 6 number-plate letters were predicted wrongly, which in case of Indian vehicle numbers imply a maximum of 2 letters per vehicle number plate. Given that the model has not been exposed to any real image, the desired performance can be very easily achieved with fine-tuning the model on few real images. We also observed that the model could predict well on synthetic 2-line number-plate images but failed to generalize over real 2-line number-plate images, a challenge we wish to investigate in future.

Acknowledgements Authors would like to thank colleagues from high-performance computing section, Computer Division, BARC for providing model training infrastructure and security section, BARC, for providing access to on-field cameras for test-image collection.

References

1. Yang, R., Yin, H., Chen, X.: License plate detection based on sparse auto-encoder. In: 2015 8th International Symposium on Computational Intelligence and Design (ISCID), vol. 2, pp. 465–469 (2015)
2. Kukreja, V., Kumar, D., Kaur, A., Geetanjali, Sakshi: GAN-based synthetic data augmentation for increased CNN performance in vehicle number plate recognition. In: 2020 4th International Conference on Electronics, Communication and Aerospace Technology (ICECA), pp. 1190–1195 (2020). https://doi.org/10.1109/ICECA49313.2020.9297625
3. Han, B.-G., Lee, J.T., Lim, K.-T., Choi, D.-H.: License plate image generation using generative adversarial networks for end-to-end license plate character recognition from a small set of real images. Appl. Sci. **10**, 2780 (2020). https://doi.org/10.3390/app10082780
4. Liu, W., Liu, X., Ma, H., Cheng, P.: Beyond human-level license plate super-resolution with progressive vehicle search and domain priori GAN. In: Proceedings of the 25th ACM International Conference on Multimedia (MM '17), pp. 1618–1626, Association for Computing Machinery, New York, NY, USA (2017). https://doi.org/10.1145/3123266.3123422
5. Linardatos, P., Papastefanopoulos, V., Kotsiantis, S.: Explainable AI: a review of machine learning interpretability methods. Entropy **23**, 18 (2021). https://doi.org/10.3390/e23010018
6. Gurney, R., Rhead, M., Lyons, V., Ramalingam, S.: The effect of ANPR camera settings on system performance. In: 5th International Conference on Imaging for Crime Detection and Prevention (ICDP 2013), pp. 1–6 (2013). https://doi.org/10.1049/ic.2013.0276
7. Lubna, Mufti, N., Shah, S.A.A.: Automatic number plate recognition: a detailed survey of relevant algorithms. Sensors **21**, 3028 (2021). https://doi.org/10.3390/s21093028
8. Naren Babu, R., Sowmya, V., Soman, K.P.: Indian car number plate recognition using deep learning. In: 2019 2nd International Conference on Intelligent Computing, Instrumentation and Control Technologies (ICICICT), pp. 1269–1272 (2019). https://doi.org/10.1109/ICICICT46008.2019.8993238
9. Zherzdev, S., Gruzdev, A.: LPRNet: License plate recognition via deep neural networks. arXiv. cs.CV1806.10447 (2018)
10. Wu, C., Xu, S., Song, G., Zhang, S.: How many labeled license plates are needed? arXiv. cs.CV1808.08410 (2018)
11. Guo, J., Liu, Y.: License plate localization and character segmentation with feedback self-learning and hybrid binarization techniques. IEEE Trans. Veh. Technol. **57**(3), 1417–1424 (2008). https://doi.org/10.1109/TVT.2007.909284
12. Wang, H., Li, Y., Dang, L.-M., Moon, H.: Robust Korean license plate recognition based on deep neural networks. Sensors **21**, 4140 (2021). https://doi.org/10.3390/s21124140
13. Jaderberg, M., Simonyan, K., Vedaldi, A., Zisserman, A.: Synthetic data and artificial neural networks for natural scene text recognition. Workshop on Deep Learning, NIPS (2014)
14. Grady, B.: In *The OpenCV Library*, Dr.Dobbs Journal., https://www.drdobbs.com/open-source/the-opencv-library/184404319. Cited 31 Mar 2022
15. Neudecker, C., Baierer, K., Gerber, M., Clausner, C., Antonacopoulos A., Pletschacher, S.: A survey of OCR evaluation tools and metrics. In: The 6th International Workshop on Historical Document Imaging and Processing, pp. 13–18. Association for Computing Machinery, New York, NY, USA, (2021). https://doi.org/10.1145/3476887.3476888
16. Pal, S., Shete, P.P.: Recognizing low quality vehicle license plates using image based sequence recognition. In: 2019 10th International Conference on Computing, Communication and Networking Technologies (ICCCNT), pp. 1–5 (2019). https://doi.org/10.1109/ICCCNT45670.2019.8944845
17. Du, Y., et al.: PP-OCRv2: Bag of tricks for ultra lightweight OCR system. arXiv preprint arXiv:2109.03144 (2021)

Chapter 10
A Self-explainable Face Anti-spoofing Solution Based on Depth Estimation

Shyam Sunder Prasad, Naval Kishore Mehta, Ankit Shukla, Pranav Mahajan, Arshdeep Singh, Sumeet Saurav, and Sanjay Singh

Abstract The human face is one of the most widely available biometric methods of identification and verification. In the age of Industry 4.0, one can find digital cameras everywhere, making a face recognition-based digital identity system much more viable. The face is vulnerable to spoofing attacks because it is the most accessible and commonly used biometric information among all biometric modalities. Most deep learning-based face anti-spoofing solutions have poor generality, and none of them provides a rationale for their results. We propose an explainable AI model that not only classifies spoof and live faces but also explains why they are different using mid-level features. Our proposed end-to-end face anti-spoofing network extracts the depth map and classifies the face input.

10.1 Introduction

The face recognition market is expected to reach 8.5 billion US dollars between 2021 and 2026, rising at a compound annual growth rate of 21.71% [1]. Face recognition solutions based on AI platforms are becoming increasingly popular among companies, hospitals, and schools these days. The face's uniqueness is one of the primary reasons it is used in applications such as access control, facial recognition, and person identification, and the majority of systems rely on AI technology and machine learning. Face recognition systems have made significant advances in past years, but they are still vulnerable to spoofing. Spoofing is the process of gaining

S. S. Prasad (✉) · N. K. Mehta · S. Saurav · S. Singh
Academy of Scientific and Innovative Research (AcSIR), Pilani, Rajasthan, India
e-mail: shyam.ece56@gmail.com

S. Singh
e-mail: sanjay@ceeri.res.in

S. S. Prasad · N. K. Mehta · A. Shukla · S. Saurav · S. Singh
CSIR-Central Electronics Engineering Research Institute (CSIR-CEERI), Pilani, Rajasthan, India

P. Mahajan · A. Singh
BITS Pilani, Goa Campus, Goa, India

© The Author(s), under exclusive license to Springer Nature Singapore Pte Ltd. 2023   103
V. Bhateja et al. (eds.), *Evolution in Computational Intelligence*, Smart Innovation,
Systems and Technologies 326, https://doi.org/10.1007/978-981-19-7513-4_10

unauthorized access to a system by impersonating the real source of authentication. Face detection in devices can be easily spoofed using presentation attacks, so a real-time face anti-spoofing system is essential in addition to face detection. The majority of past anti-spoofing research has been approached as a simple end-to-end binary classification problem. There are two major flaws in deep learning-based face anti-spoofing models with binary supervision. Firstly, they are end-to-end types; for example, CNN with softmax loss may discover arbitrary cues (such as screen bezels) rather than more reliable cues such as depth estimation, skin detail loss, and colour distortion. When these cues disappear during testing in real-world scenarios, models tend to fail when distinguishing between spoof and live faces, causing them to fail to generalize appropriately, and the resulting discrepancy is the model's interpret-ability or explainability. Models trained with binary supervision will produce a binary decision that humans may or may not be able to interpret. We present a self-explainable face anti-spoofing method in this research by predicting the depth map of a given input face. Our method provides the reason for its decision using mid-level features depth map. We use auxiliary supervision to nudge the models to learn explainable spatial and temporal information in the data, and we use our domain knowledge of the important distinctions between real and fake images to provide the auxiliary information. From a structural standpoint, we know that live faces have depth because they are three-dimensional, but faces in print or replay attacks are flat. In the temporal domain, we should use videos or a stack of contiguous frames as input, which will include chromatic and depth related information. As a result, we propose a 3D CNN-based architecture with skip connections for estimating the depth map of input video and classifying it as live or spoof. Camera quality is a critical factor in determining quality of spoof faces and can directly affect the performance of the model; therefore, we trained and tested our model on two widely accepted databases, Replay Attack [2] and SiW [3]. It is clear that with advancements in camera technology and spoofing systems, we have also achieved good results on the SiW [3] dataset, which was recently collected and has many PIE on depth-to-camera variations.

The rest of this work is structured as follows. The next section highlights the prior works. Section 10.3 goes over database specifics, while Sect. 10.4 goes over pre-processing details. In Sect. 10.5, we explain the architecture, and in Sect. 10.6, we report on the experimentation and results. Section 10.7 contains the conclusions and future directions.

10.2 Prior Work

Presentation attacks (PA) are divided into 13 categories [4] , including print attack—showing the victim's picture , replay attack—playing a face video on the device, and mask attack—wearing a printed mask 2D/3D, among others. Face anti-spoofing methods [5, 6] have been developed to identify PA before a face image is identified in order to prevent fake facial verification. Face anti-spoofing algorithms have recently gained popularity, and they have improved their performance by identifying presentation attacks [3, 7, 8].

Very early solutions for face anti-spoofing based on just temporal cues include work by Pan et al. [9] and Patel et al. [10] which take into account eye blinking as well as work by Kollreider et al. [11] and Shao et al. [12] which track the motion of mouth and lip movement to detect spoof vs live faces. Such methods are vulnerable to replay attack. Few works prior to deep learning methods propose utilizing Haralick features [13], motion mag [14], and optical flow [15].

Face anti-spoofing systems, which are similar to face verification systems, use RGB image and video as its standard input. Previously, handcrafted features were used in texture-based anti-spoofing techniques, such as LBP [2], HoG [16] and SURF [17] were used to find the differences between live and spoofed faces. Traditional classifiers like SVM and LDA come in play for such systems . These methods are vulnerable to illumination variations, and thus, changing the input domain to HSV or YCbCr colour space [17, 18] or Fourier spectrum [19] improved the performance of these systems. In [20], for the first time, CNN was employed for face anti-spoofing, and the results in intra-database testing were impressive. As a result of this work, many CNN-based solutions for addressing face anti-spoofing as a binary classification task have been developed.

Recently introduced GFA-CNN [8] tries to make the existing CNN approaches more generalizable by using total pairwise confusion (TPC) loss and fast domain adaptation (FDA) for pre-processing to mitigate negative effects due to domain changes. Deep learning methods which treat face anti-spoofing as a binary classification tend to overfit, thus [21] propose to improve the performance by synthesizing 3D virtual samples. Gan et al. [22] present a 3D CNN to discriminate between real and fake faces, but they approach the problem as a binary classification task. Xu [23] introduces a network that combines LSTM units and CNN.

With binary supervision, there are two major concerns with deep learning-based face anti-spoofing models. Firstly, they are end-to-end in nature, such as CNN with softmax loss. They might discover arbitrary cues (like screen bezel) instead of more faithful cues like depth estimation, skin detail loss, colour distortion. When those cues vanish during testing in real-world circumstances, models will be unable to discriminate between spoof and live faces, resulting in poor generalization. Secondly, the issue is interpret-ability or explainability of the model. Models trained with binary supervision will output a binary decision which may not be necessarily interpret-able by humans. Thus by using auxiliary supervision, we can nudge the models to learn explainable spatial and temporal information in the data.

Wang et al. [24] propose a meta-learning-based adaptor learning algorithm for learning a better adaptor with unsupervised adaptor loss by using data from multiple source domains during the training phase. Li et al. [25] proposed a nine-layer CNN architecture that utilizes end-to-end learning and several color space models to deliver complimentary features that directly assess the raw input face image's corresponding output class. Supervised CNN architectures have been proposed by Atoum et al. [26] and Liu et al. [3] which discover static spoofing cues under pixel-wise supervision.

To distinguish between spoof and live faces, Atoum et al. used a depth map of the face as an auxiliary input and create a representation by combining depth and patch from a single frame. Liu et al. used rPPG signals as extra supervisory cues for utilizing temporal information. They did not use any depth information in the temporal domain and instead learned the appropriate rPPG signals using a basic RNN. In our work, we use a 3D CNN autoencoder architecture which also takes into account the depth information in temporal domain for auxiliary supervision.

3D Dense Face Alignment (3DDFA) [27] methods are used to create pseudo ground truth depth maps which are to be used for auxiliary supervision. Liu et al. [3] used Pose-Invariant DeFA methods [28] for generating depth maps for supervision. Newer methods such as 3DDFA , SuperFAN [29], and PRNet [30] give better depth maps from RGB face images.

10.3 Database Details

SiW (Spoof in wild) [3] database is the most recent database for anti-spoofing that is publicly available. There are 4478 live, print, and display attack videos of 165 subjects in this database. This collection contains 8 live and 20 spoof videos recorded in four sessions for each subject. Subjects in various sessions display a variety of expressions and poses while standing at various distances from the camera.

The **Replay Attack** [2] is one of the most widely used anti-spoofing dataset. It includes 1300 video clips of 50 different subjects in various lighting conditions. All of the videos were created by presenting a photo or video clip of the same person for at least 9 s, or by having a real person attempt to access a laptop via the built-in webcam.

10.3.1 3D Dense Face Alignment

We use 3D DFA method [27] for generating pseudo ground truth depth maps. 3DDFA uses a combination of Cascade regression and CNNs where CNN acts as a regressor. Cascade regression can be formally written as

$$\mathbf{f}^{n+1} = \mathbf{f}^n + R^n(F(\mathbf{I}, \mathbf{f}^k)) \qquad (10.1)$$

At the nth iteration, the shape parameter $\mathbf{f}^n$ is upgraded by the regression function $\mathbf{R}$ on the shape indexed feature $\mathbf{F}$, which is reliant on both the image $\mathbf{I}$ and the current parameter $\mathbf{f}^n$. The regression Reg has an essential "feedback" property in that its input feature $F(\mathbf{I}, \mathbf{f}^n)$ by its output $\mathbf{f}$ that also update after each iteration. This property allows an array of weak regression to be cascaded to gradually minimize the alignment error. The regressor R is replaced with a CNN in 3D DFA.

To accomplish pose-free face alignment, 3D DFA uses a dense 3D Morphable Model instead of the conventional landmark detection framework. The 3D Morphable Model (3DMM) is a mathematical model that explains the 3D face space with PCA is described as follow,

$$S = \overline{S} + A_{id}\alpha_{id} + A_{exp}\alpha_{exp} \tag{10.2}$$

where S is 3D face, $\overline{S}$ is the mean shape, A_{id} is the main (principal) axis trained for 3D face with neutral facial expression and α_{id} is a shape parameter, A_{exp} is the main (principal) axis that are taught on the differences between expression and neutral scans and α_{exp} is the expression parameter. After the 3D face is built, scale orthographic projection may be used to project it onto the image plane.

10.4   Preprocessing

For depth estimation network, we first generate the pseudodepth map M Fig. 10.1a of live RGB videos I (stacked 5 frames) which acts as the ground truth using 3DDFA [27] method. For spoof videos (e.g., print attack , display attack), we are assuming depth map is consisting of all RGB value 0,0,0 i.e., a black image Fig. 10.1b. This idea comes from the assumption that both types of videos whether print attack or replay attack come from plain surface where the depth is zero. Since our focus is to estimate depth only for live faces, that is why the background in case of live videos has been assumed as black or zero. We used Dlib [?] to detect faces form input video frame.

Classification network consists of a flatten layer(output of depth estimation) at the start and series of five dense layers followed by ReLU activation layer except for the last dense layer where softmax activation function is used. Classification network input is the output of depth estimation network. Our model is combination of two network one is 3DCNN based, and other one is 3DCNN autoencoder (depth estimation followed by classifier network) based. 3DCNN network takes n (number of frames, i.e., depth) continuous frames from the input video. Number of frames (depth) can be fixed with different depths, and in our case, we fixed number of frames 5. In prior work, most of them covered only spatial domain information; some of them are 2DCNN-based network; from single frame, it is very difficult to predict whether it is Live/Spoof.

10.5   Explanation of Architecture

Our proposed 3D CNN network for depth estimation and classification, as shown in Figure 10.2, is divided into two blocks: the first estimates the depth map of a given input image sequence, and the second classifies the estimated depth image into the live and spoof categories.

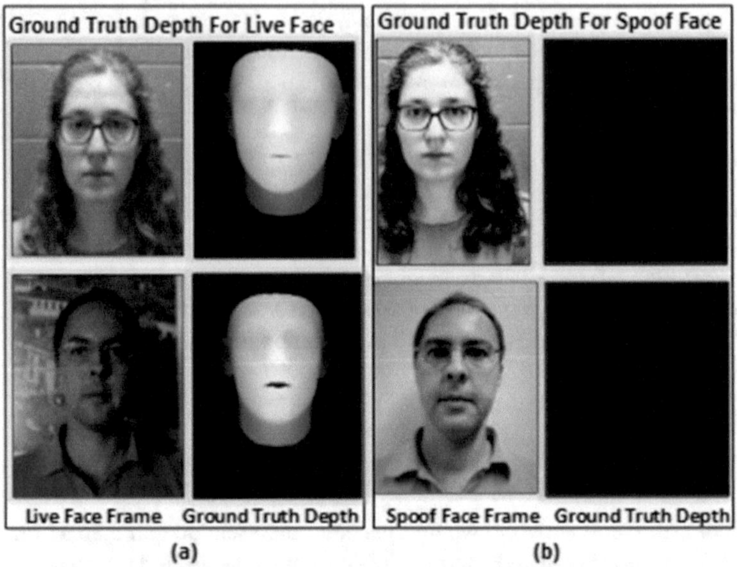

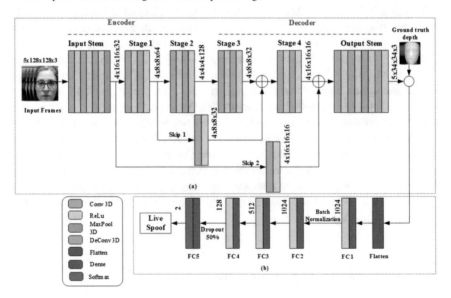

Fig. 10.1 Examples of the depth estimation of SiW and Replay Attack datasets obtained by 3DDFA method. In the image **a** live face frame. In sub-figure **b** spoof face frame. All the depth masks of the live and spoof faces shown in ground truth depth sub-figures

Fig. 10.2 Proposed end-to-end architecture for live and spoof classification using depth estimation approach

Depth estimation network consists of an encoder block followed by a decoder block with skip connections. There are three stages to the encoder block: input stem, stage 1, and stage 2. The decoder block has three stages as well: stage 3, stage 4, and the output stem. To ensure feature reuse from encoder to decoder layers, the output of the input stem is concatenated with the output of stage 4 via a skip connection (skip 1), and similarly, the output of stage 1 is concatenated with the output of stage 3 via a skip connection (skip 2). The rest of the interconnections between the stages are depicted in Figure 10.2a.

As illustrated in Fig. 10.2b, the classification network input is flatten output of the depth estimation layer, and the network consists of 5 dense layers(FC1-FC5) followed by a ReLU activation layer, with the exception of the last dense layer, which uses the softmax activation function to detect the live and spoof attack from the input frames.

10.6   Experimentation and Results

Our training strategy is split into two steps. The first step is to train the depth estimation block, which converts the input RGB video **I** (stacked 5 frames) to the corresponding depth map **M**. The second step is to train the classifier to differentiate between live and spoof video.

10.6.1   Depth Estimation Model Training

After the preparation of the training data, we train the 3D CNN-based depth estimator network. The model is trained for 1000 epochs with a batch size of 64, and the learning rate is set to 0.0001. We employ the mean squared error (MSE) loss and the Adam optimizer. We have trained and tested the models with SiW and Replay Attack datasets. The SiW dataset includes 90 training subjects and 75 testing subjects. We split the SiW training dataset into 70:30 ratio, with 70% of the data used to train the depth estimation network and 30% used to train the classification network. The Replay Attack has distinct training, validation, and test sets. The train set is used to train the depth estimation network, while the validation set is used to train the classification network.

10.6.2   Classifier Model Training

After our depth estimation network has been successfully trained, we combine the classification and depth estimation blocks to create an end-to-end depth estimation and classification system. While training the classification block, the depth estimator

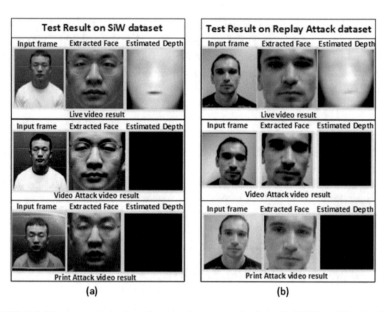

Fig. 10.3 Model-generated depth estimation feature results from the SiW and Replay Attack datasets on live video, video attack, and print attack

network's parameters were set to non-trainable. The Adam optimizer and cross-entropy loss were utilized, with a batch size of 64 and a learning rate of 0.0001. We train the network for 50 epochs using both datasets individually.

10.6.3 Results

We tested our model on the test sets provided by SiW and Replay Attack datasets, and we evaluated the network's performance by three parameters: False Acceptance Rate (FAR), False Rejection Rate (FRR), and Half Total Error Rate (HTER). FAR means the number of samples accepted as true that were false in nature, whereas FRR indicates the number of samples rejected by the model or classified as false that were true in nature. In other words, FRR is the rate at which a system fails to recognize legitimate people, whereas FAR is the rate at which a system accepts illegitimate people. HTER is the arithmetic mean of FRR and FAR.

Results of our network on SiW and Replay Attack are presented in Tables 10.1 and 10.2 respectively. The accuracy of each type of attack, as well as the FAR and FRR, is assessed. For real faces, FRR is calculated, while for spoof faces, FAR is calculated. As demonstrated by the results, we achieved good accuracy for both datasets, particularly in the case of video attack, which is 99%. Also the mid-level

Table 10.1 Test results on SiW database

Database	Accuracy (%)	FAR/FRR
SiW-Live	97.35	0.02719
SiW-PA	97.86	0.02143
SiW-VA	99.02	0.00979
HTER		0.01874

Table 10.2 Test results on Replay Attack database

Database	Accuracy (%)	FAR/FRR
RA-Live	96.28	0.03718
RA-PA	99.12	0.00883
RA-VA	99.01	0.00992
HTER		0.02038

features means the estimated depth map for a given input face is shown in Fig. 10.3. It is evident that our network estimates a fair quality depth map for live attacks while generating a blank image for both types of attacks.

10.7 Conclusion

In this paper, we proposed an end-to-end face anti-spoofing model that can detect both print and display attacks simultaneously. The main point of this paper is that it explains the output using mid-level features and auxiliary information, which makes it explainable. A depth map has been used as supplemental data. The work presented here is limited to detecting print and display attacks; however, other PAs such as 3D masks, makeup, and partial attacks can be addressed in future. The model's accuracy and generalization can also be improved when more auxiliary information is used, such as rPPG signals.

References

1. Song, Z., Nguyen, K., Nguyen, T., Cho, C., Gao, J.: Spartan face mask detection and facial recognition system. In: Healthcare, Multidisciplinary Digital Publishing Institute, vol. 10, p. 87 (2022)
2. Chingovska, I., Anjos, A., Marcel, S.: On the effectiveness of local binary patterns in face anti-spoofing. In: 2012 BIOSIG-Proceedings of the International Conference of Biometrics Special Interest Group (BIOSIG), pp. 1–7. IEEE (2012)

3. Liu, Y., Jourabloo, A., Liu, X.: Learning deep models for face anti-spoofing: Binary or auxiliary supervision. In: Proceedings of the IEEE Conference on Computer Vision and Pattern
   Recognition, pp. 389–398 (2018)
4. Liu, Y., Stehouwer, J., Jourabloo, A., Liu, X.: Deep tree learning for zero-shot face anti-
   spoofing. In: Proceedings of the IEEE/CVF Conference on Computer Vision and Pattern
   Recognition, pp. 4680–4689 (2019)
5. Frischholz, R.W., Dieckmann, U.: Biold: a multimodal biometric identification system. Computer 33(2), 64–68 (2000)
6. Frischholz, R.W., Werner, A.: Avoiding replay-attacks in a face recognition system using
   head-pose estimation. In: 2003 IEEE International SOI Conference. Proceedings (Cat. No.
   03CH37443), pp. 234–235. IEEE (2003)
7. Jourabloo, A., Liu, Y., Liu, X.: Face de-spoofing: Anti-spoofing via noise modeling. In: Proceedings of the European Conference on Computer Vision (ECCV), pp. 290–306 (2018)
8. Tu, X., Ma, Z., Zhao, J., Du, G., Xie, M., Feng, J.: Learning generalizable and identity-
   discriminative representations for face anti-spoofing. ACM Trans. Intell. Syst. Technol. (TIST)
   11(5), 1–19 (2020)
9. Pan, G., Sun, L., Wu, Z., Lao, S.: Eyeblink-based anti-spoofing in face recognition from a
   generic webcamera. In: 2007 IEEE 11th International Conference on Computer Vision, pp
   1–8. IEEE (2007)
10. Patel, K., Han, H., Jain, A.K.: Cross-database face antispoofing with robust feature represen-
    tation. In: Chinese Conference on Biometric Recognition, pp 611–619. Springer (2016)
11. Kollreider, K., Fronthaler, H., Faraj, M.I., Bigun, J.: Real-time face detection and motion
    analysis with application in "liveness" assessment. IEEE Transactions on Information Forensics
    and Security 2(3), 548–558 (2007)
12. Shao, R., Lan, X., Yuen, P.C.: Deep convolutional dynamic texture learning with adaptive
    channel-discriminability for 3d mask face anti-spoofing. In: 2017 IEEE International Joint
    Conference on Biometrics (IJCB), pp. 748–755. IEEE (2017)
13. Agarwal, A., Singh, R., Vatsa, M.: Face anti-spoofing using haralick features. In: 2016 IEEE 8th
    International Conference on Biometrics Theory, pp. 1–6. Applications and Systems (BTAS),
    IEEE (2016)
14. Bharadwaj, S., Dhamecha, T.I., Vatsa, M., Singh, R.: Face anti-spoofing via motion magnifi-
    cation and multifeature videolet aggregation. Tech. Rep. (2014)
15. Bao, W., Li, H., Li, N., Jiang, W.: A liveness detection method for face recognition based on
    optical flow field. In: 2009 International Conference on Image Analysis and Signal Processing,
    pp. 233–236. IEEE (2009)
16. Komulainen, J., Hadid, A., Pietikäinen, M.: Context based face anti-spoofing. In: 2013 IEEE
    Sixth International Conference on Biometrics: Theory, pp. 1–8. Applications and Systems
    (BTAS), IEEE (2013)
17. Boulkenafet, Z., Komulainen, J., Hadid, A.: Face antispoofing using speeded-up robust features
    and fisher vector encoding. IEEE Sign. Process. Lett. 24(2), 141–145 (2016)
18. Boulkenafet, Z., Komulainen, J., Hadid, A.: Face anti-spoofing based on color texture analysis.
    In: 2015 IEEE International Conference on Image Processing (ICIP), pp 2636–2640. IEEE
    (2015)
19. Li, J., Wang, Y., Tan, T., Jain, A.K.: Live face detection based on the analysis of fourier spectra.
    In: Biometric Technology for Human Identification, SPIE vol. 5404, pp. 296–303 (2004)
20. Yang, J., Lei, Z., Li, S.Z.: Learn convolutional neural network for face anti-spoofing. arXiv
    preprint arXiv:1408.5601 (2014)
21. Guo, J., Zhu, X., Xiao, J., Lei, Z., Wan, G., Li, S.Z.: Improving face anti-spoofing by 3d virtual
    synthesis. In: 2019 International Conference on Biometrics (ICB), pp. 1–8. IEEE (2019)
22. Gan, J., Li, S., Zhai, Y., Liu, C.: 3d convolutional neural network based on face anti-spoofing.
    In: 2017 2nd International Conference on Multimedia and Image Processing (ICMIP), pp. 1–5.
    IEEE (2017)
23. Xu, Z., Li, S., Deng, W.: Learning temporal features using lstm-cnn architecture for face anti-
    spoofing. In: 2015 3rd IAPR Asian Conference on Pattern Recognition (ACPR), pp. 141–145.
    IEEE (2015)

24. Wang, J., Zhang, J., Bian, Y., Cai, Y., Wang, C., Pu, S.: Self-domain adaptation for face anti-spoofing. arXiv preprint arXiv:2102.12129 (2021)
25. Li, T., Lian, Z.: A novel face anti-spoofing method using multiple color space models. In: Twelfth International Conference on Graphics and Image Processing (ICGIP 2020), International Society for Optics and Photonics, vol. 11720, p. 1172021 (2021)
26. Atoum, Y., Liu, Y., Jourabloo, A., Liu, X.: Face anti-spoofing using patch and depth-based CNNs. In: 2017 IEEE International Joint Conference on Biometrics (IJCB), pp. 319–328. IEEE (2017)
27. Zhu, X., Liu, X., Lei, Z., Li, S.Z.: Face alignment in full pose range: a 3d total solution. IEEE Trans Pattern Anal Mach Intell **41**(1), 78–92 (2017)
28. Jourabloo, A., Liu, X.: Pose-invariant face alignment via CNN-based dense 3D model fitting. Int. J. Comput. Vis. **124**(2), 187–203 (2017)
29. Bulat, A., Tzimiropoulos, G.: Super-fan: Integrated facial landmark localization and super-resolution of real-world low resolution faces in arbitrary poses with gans. In: Proceedings of the IEEE Conference on Computer Vision and Pattern Recognition, pp. 109–117 (2018)
30. Feng, Y., Wu, F., Shao, X., Wang, Y., Zhou, X.: Joint 3d face reconstruction and dense alignment with position map regression network. In: Proceedings of the European Conference on Computer Vision (ECCV), pp. 534–551 (2018)

Chapter 11
An Approach of Hate Speech Identification on Twitter Corpus

Kavita Kumari⊙ and Anupam Jamatia⊙

Abstract In recent times, the Internet and social media are very well-known and popular among people. Usage of social media is increased exponentially during the last few years globally, and it allows people to engage with one another and share ideas, thoughts, opinions, etc. Every day, massive amounts of data are disseminated at breakneck speed via social media platforms, reaching a massive audience. Furthermore, the ability to write anonymous posts and comments makes expressing and spreading hate speech even easier. To improve the users' experience, social media sites are attempting to remove hateful remarks. In this research work, the main focus is on developing automated hate speech and offensive language detection models. Started with traditional hate speech and offensive language identification approaches, than reaches to advanced hate speech recognition methods for social media. This brings a need for integrated datasets and the hate speech prediction method. This paper describes the study on hate speech and offensive content identification in English language by using the various approaches based on machine learning algorithms (Support vector machine, decision tree, and so on) and NLP, along with the features used for the classification problem. The models were tested on the HASOC (2021) datasets and concluded that the ensemble model perform better than other algorithms with the test dataset for different tasks. Results and analysis part of this paper offers researchers a comprehensive picture of approaches.

K. Kumari (✉) · A. Jamatia
Department of Computer Science and Engineering, National Institute of Technology, Agartala, India
e-mail: jmdkavita87@gmail.com

© The Author(s), under exclusive license to Springer Nature Singapore Pte Ltd. 2023 115
V. Bhateja et al. (eds.), *Evolution in Computational Intelligence*, Smart Innovation, Systems and Technologies 326, https://doi.org/10.1007/978-981-19-7513-4_11

11.1 Introduction

Social networking Web sites like Twitter and Facebook became a major medium for individuals and especially youths for social interaction and news consumption. Currently, 24.45 million are Twitter users in India.[1] Twitter is a medium to know about current news and what is going on in the country. But, at the same time, it gives a platform to debate with the people which results in hate speech and mental stress. The first challenge in controlling hate speech is to define its parameters. People use social media to share their thoughts and experiences, and issues like nationality, gender, race, castes, and other personal qualities are frequently discussed. The public may differ about the mortality of particular religious doctrines and the wisdom of a country's foreign policy, and netizens want them to be allowed to discuss those topics on social media like Twitter and Facebook. Natural language processing plays a vital role in analyzing and processing hate speech of social media. Before, much research has been done to predict hate speech on Twitter. As pointed out in a recent survey [6], various terms have been used in the literature to describe phenomena with overlapping characteristics such as toxicity, hate speech, and cyberbullying. But, it is a big problem to tackle abusive language on social media nowadays. Various machine learning-based methods like supervised learning, semi-supervised, and deep learning are available to predict toxic language. There has been notable research into the exploration of offensive language in different languages, but most of the research is carried out in English. The primary goal of this research work is to investigate the role of social media platforms in promoting intolerance and test the limits of AI systems that can detect hate speech, to investigate the application of these automated AI systems that can greatly aid in controlling hate speech on social media platforms, and to ensure that these systems are fair to all ethnic groups, ages, and genders.

The rest of this paper is arranged as follows. In Sect. 11.2, sum up various previous research problems and shared tasks composed under the scope of hate speech and offensive language detection. In Sect. 11.3, corpus collection is described with the problem statement of this paper. The ideas behind data preprocessing, feature engineering, and the training of the model are explained in Sect. 11.4. The analysis with its results of solving tasks of machine learning classification models is described in Sect. 11.5. Section 11.6, explaining about the error of the models which it misclassified. At last, Sect. 11.7 is the conclusions of this paper along with the challenges faced during the experiments of HASOC corpus and shared task.

11.2 Related Work

In the previous years, there has been published research on hate and offensive content identification as the term abusive and offensive language detection appears as

[1] https://www.statista.com/statistics/242606/number-of-active-Twitter-users-in-selected-countries/.

a multiclass classification problem [3], and research of this has typically focused on different types of finding class like hate, profane, bullying [17], and offensive tweets prediction problems. In the past, most research areas are focused on task: cyberbullying [2], aggression and trolling [11], sexism [24], hate speech [7], and etc. Corpus for these tasks has been collected from different types of social media platforms like Instagram [28], Twitter [4], Facebook [11], and so on. Schmidt and Wiegand [21] and Fortuna and Nunes [6] conducted surveys in which they extracted features from the text such as sentiment and linguistic features, and used several lexical resources to label hate tweet. According to Malmasi and Zampieri [12], the SVM model predicts the posts with three types of extracted features for these tests that are word skip-grams, surface n-gram, and brown cluster. They gave accuracy scores and decide the lexical axis for the difference between profane and hate speech with the maximum number of corpus. Malmasi and Zampieri [12] and Zampieri et al. [26] said that the rise of abusive language in online chats and user reviews has come out as an important problem in the environment as well as society. Ramakrishnan et al. [19] gave an accuracy score of 83.14% for predicting the hate tweets on the real test for the labels of offensive and non-offensive with additional training data. The model's concept has also been questioned. Models trained on one dataset tend to perform well only when tested on the same dataset, as Karan and Snajder [10] have demonstrated. After experimenting with cross-domain training and testing, Karan and Snajder [10]decided to use the LSVM model with minimal features and to preprocess for answerability. User-specific metadata was also used as a feature by Waseem and Hovy [24] claim that knowing your gender improves your performance, whereas Unsvag and Gamback [23] claim that knowing your network is more important. The dataset was published by Davidson et al. [3] in a row text format and claim to have gathered this information from Twitter using a HateBase33[2] dictionary of hate words and phrases. They annotated the tweets into the three classifications using a crowdsourcing technology (Figure-Eight44,[3] formerly CrowdFlower) and 24,802 tweets in English in a corpus (5.77% labeled as hate speech, 77.43% as offensive, and 16.80% as neither).

11.3 Corpus

In this research study, only, English tweets were included in the dataset used for the experiment. The datasets have been collected from the HASOC,[4] 2021 shared task which is in CSV (Comma-separated values) format, contains 3000 posts for the training data and 844 tweets were used as the test dataset. Noticeably, from Table 11.1 that datasets are not equally divided in all labels, the percentage and number of tweets are shown in Table 11.1, and the examples related to various classes are also

---

[2] https://hatebase.org/.

[3] https://appen.com/.

[4] https://hasocfire.github.io/hasoc/2021/organizers.html.

Table 11.1 Corpus distribution

Classes	No. of posts	Percentage
NOT	1046	35.19
HOF	1954	64.81
NONE	1046	35.19
PRFN	944	30.74
HATE	525	17.73
OFFN	485	16.33

Table 11.2 Tweet example of all classes

Classes	Example
HOF	@Sweeeet_FA Wee twat. How can she even joke about that?
NOT	@Warix_gay Can't expect God to do all the work
PRFN	@milkydonah fuck off this is you
OFFN	@dxnprivv Stop being a twat then
HATE	@bananapixelsuk That's why the whole thing is a load of crap. Corporate bollocks
NONE	JEALOUSY A BITCH IT LOOK LIKE

shown in Table 11.2. Another thing to consider is that the dataset was sampled at a time when India was dealing with the second wave of covid-19. As a result, major posts and tweets on social media are related to covid-19. Moreover, posts were also related to the post-election violence in West Bengal in India.

In HASOC shared task, there are two subtasks given to predict the offensive language. Subtask 1A is a binary classification of predicting tweets into two classes, namely HOF (Hate and Offensive) and NOT (Non-Hate and Offensive). HOF is made up of offensive, hateful, and profane tweets whereas NOT represents non-biased posts. Subtask 1B is a fine-grained classification of predicting tweets with 4 different classes that are hate speech (HATE), offensive (OFFN), profane (PRFN), and neutral (NONE).

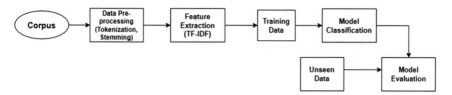

Fig. 11.1 Proposed framework

11.4   Experiments

11.4.1   Data Preprocessing

In this method, the first datasets are cleaned using a tweet preprocessing library.[5] Twitter uses different types of symbols like URLs, retweets, hashtags (#), emojis, smileys, mentions, and reserved words which have been removed with the help of the preprocessor library. This preprocessed data also exclude the stopwords with the use of the NLTK6 library in English while tokenizing the texts for the feature extraction. Clean data increase the efficiency and accuracy of the machine learning model. Both subtasks and unseen data use same preprocessing. The tokenization [25] helps in clarifying the meaning of the text by scanning the sequence of the words. Wordnet Lemmatization [16] is used to build structured semantic relationships within words. To estimate any models, corpus collection is the primary step after this, data pre-processing technique is used for cleaning the datasets along with the features of converting raw data into numerical characteristics. And, train the various types of machine learning models on the training dataset to do verification of various models from the test datasets, to confirm that models were trained effectively as shown in the proposed framework in Fig. 11.1.

11.4.2   Feature Extraction

Feature Engineering is the key step for text classification. Depending on the model, TF-IDF is used for the conversion of text data into numerical form. TF-IDF [20] is a technique for determining the meaning of words in a phrase, and it overcomes the limitations of the bag of words technique, which is useful for text classification and assisting machines in reading numbers.

$$\text{TF}(t, d) = \text{Total of } t \text{ in } d \text{ / Total number of words in } d \qquad (11.1)$$

$$\text{IDF}(t) = \log(N/(\text{df} + 1)) \qquad (11.2)$$

---

[5] https://pypi.org/project/tweet-preprocessor/.

$$TF - IDF = TF * IDF = TF(t, d) * \log(N/(df + 1)) \qquad (11.3)$$

where, d = document (group of words), t = term (word), corpus = the whole set of documents, df = occurrence of t in N documents, N = corpus count.

11.4.3 Machine Learning Algorithms

The machine learning, deep learning, and ensemble methodology-based (Max voting) classifiers for the English corpus for Subtask 1A and Subtask 1B have been developed, namely—SVM [22], RF [13], DT [18], KNN [27], NB [1], ensemble [5], and LSTM [8] + GLOVE [15]. The scikit-learn [14] library has been used for the implementation. Global vectors for word representation (Glove) are an unsupervised learning system that aims to construct word embeddings by aggregating global word co-occurrence matrices from the corpus. DT is an algorithm that uses the CART algorithm (Classification and regression tree algorithm) to construct a tree. SVM's main aim is to find the best line or decision boundary that can divide n-dimensional space into classes so that fresh data points can be readily placed in the correct category in the future. NB is based on the Bayes theorem and predicts the classification problem based on the probability of the object. RF is an algorithm that is built from subsets of data, and the final output is based on average or majority rating, overfitting problem is avoided. KNN is a non-parametric algorithm, which means it makes no assumptions about the data it's working with. Ensemble learning is a combination of different ML models to make an optical predictive model. LSTM enhances performance by memorizing and finding patterns in pertinent information. The followings are the parameters that are used in the algorithms:

1. For TF-IDF vectorizer, max_features = 2000 and min_df = 5 used.
2. In KNN, max_samples = 0.5 and max_features = 0.5.
3. In RepeatedKFold, n_splits = 10 and n_repeats = 10 have been used.
4. For LSTM, batch size of 64 and 1 epoch for training data.

11.5 Results and Analysis

After performing the experiments of the classification model, NB (Naive Bayes), DT (Decision tree), SVM (Support vector machine), KNN (K-nearest neighbors), and RF (Random forest) techniques, the outputs of these methods are shown in Table 11.3, and the performance of models are evaluated in terms of macro F_1 score and accuracy [9]. The chosen models were then optimized on the training set, before being tested on the test dataset. F_1 score is the harmonic mean of P and R, and accuracy is defined as the total number of correct predictions divided by the total number of original instances.

Table 11.3 Experiments results on train and test set for two tasks

Models	Train data				Test data			
	Task 1A		Task 1B		Task 1A		Task 1B	
	F1	Acc	F1	Acc	F1	Acc	F1	Acc
DT	96.36	96.70	93.28	93.73	69.11	72.16	45.57	51.30
SVM	91.56	92.43	88.03	89.27	71.59	74.41	**53.05**	60.78
RF	64.17	74.26	94.49	94.83	52.68	67.42	50.48	61.97
NB	77.54	80.57	68.02	70.47	67.40	71.80	49.82	57.11
KNN	96.60	96.93	36.68	54.43	67.81	74.41	31.64	48.82
Ensemble	95.63	96.72	92.56	95.21	**71.68**	**76.78**	52.09	**63.51**
LSTM	81.93	85.11	79.56	75.67	69.12	76.66	55.63	58.25

$$F_1 = 2 * R * P / (R + P) \tag{11.4}$$

$$Accuracy = (TP + TN)/(TP + FP + TN + FN) \tag{11.5}$$

TP, FN, TN, FP, P, and R are true positives, false negatives, true negatives, false positives, precision, and recall, respectively.

11.5.1   Train Data

The dataset was made available as a pre-tagged training dataset (containing 3000 posts) and an un-tagged test dataset (containing 844 samples). After performing the data preprocessing methods, each model in the ensemble was run through a repeated 10-Fold cross-validation process, with the metrics from each fold aggregated to get a median for each model's performance on the dataset. Firstly, the model is trained by using the trained datasets through which gets outputs in terms of accuracy and F_1 score as shown in Table 11.3, and for two different tasks, the TF-IDF feature has been applied to the datasets. It was also observed that models of task 1A with the highest F_1 score of 96.60% and accuracy of 96.93%, KNN is the best model, and for task 1B with F_1 score of 94.49% and accuracy of 94.83%, RF is good on the train datasets in comparison to any other single classifier model.

11.5.2   Test Data

After the models had been trained, they were tested on 844 unseen tweets, and the macro F_1 score is an evaluation metric of all highly imbalanced corpus.

11.5.2.1 Task1A

The best single ensemble model of task 1A is linear support vector of 71.59% of F_1 score and accuracy of 74.41% which is slightly different from the ensemble model. Small decimal difference of values in ensemble method which makes the best model for task 1A with F_1 score of 71.88% and accuracy of 76.88% which is bold in Table 11.3. The ensemble model has a recall of 50.70% on offensive tweets and 90.14% on non-offensive tweets of task 1A on the test data. LSTM trained model with epochs = 1 was used to predict the offensive posts on the unseen data for task 1A. The dense layer uses the 'softmax' activation function to allow for speedy convergence. The F_1 score and accuracy of train data for task1A are 81.93% and 85.11%, respectively. The F_1 score and accuracy of test data for task 1A are 69.12% and 76.66%, respectively.

11.5.2.2 Task1B

The corpus of task 1B is highly imbalanced as shown in Table 11.1. In this task, different particular classifiers showed a significant discrepancy in two classes; thus, ensemble model performed better among all individual ensemble classifiers with the highest accuracy of 63.51% and F_1 score of 52.09%. The precision and recall of the ensemble model of various tweet labels such as HATE (54.26%, 33.55%), OFFN (50%, 12.12%), and PRFN (72.59%, 87.96%), respectively, which shows the high percentage difference of prediction of hate words.

11.6 Error Detection

This section provides a brief qualitative examination of the misclassifications in each task and speculates on possible causes. Some models were unable to successfully categorize posts into various classes due to the under-representation of specified labels, and this is because of the appearance of highly marginalized hate and offensive content (HATE label), profane (PRFN label) where very few posts could be correctly classified. Previously, Davidson et al., Ramakrishnan et al., and so on [4, 19] talk about the errors of the model that is, the classifiers find difficulties to identify offensive and hate posts that lack obscenity for example "@KnabeWolf DAMN let me get what you got lmao," "@RobsBlob @EvokeDistrict @evokeartistry np pimp" as shown in Table 11.4. Due to an extremely skewed data set, some labels misclassify most that as PRFN, OFFN, and HATE such as "@BBCNews Let the dog deal with the wanker once he's unarmed ...," "@KanganaTeam This is such a vile, xenophobic and uneducated comment... struggling to believe someone thinks like this, let alone posted this?! Daylight #Islamophobia and it should be stopped. #Elections2021". The classifier also had trouble identifying political dialogue, since

Table 11.4 Model prediction of actual versus predicted class

Tweets	Actual class	Predicted class
@Healey17 Waistcoat wanker at it again, no lingard either	PRFN	HATE
@BBCNews Let the dog deal with the wanker once he's unarmed	HOF	NOT
@RobsBlob @EvokeDistrict @evokeartistry np pimp	OFFN	NONE

it may have picked up on trends of words like 'ResignModi', 'ModiTerrorist', 'Liberals', 'ElectionResult', 'BengalBurning', 'BengalViolence', and 'bjp4india' which comes under the offensive label.

11.7   Conclusion

In this project, machine learning-based models were preferred to predict hate and vilify language. In the Internet community, the occurrence of hate speech and derogatory language that has gained attention in the previous years is considered a social challenge. Offensive language like hate, hurtful, derogatory, and toxic content is directed from one user to others. As a consequence, many social media platforms monitor users' posts, and this brings a method to verify or identify suspicious posts automatically. In this paper, the HASOC dataset has been used to detect hate speech and offensive content in English. Different types of data preprocessing and word embedding method like TF-IDF have been used on the corpus. Effective prediction systems are used to reduce abusive and offensive content, and classification models were experimented to predict hate speech as well as offensive tweets to verify the datasets like NB, DT, LSVM, and so on in this research work. Later, the best results were obtained by the ensemble model with F_1 score of 71.68% for task 1A. Outputs of these experiments show that dealing with offensive and hate content on social media platforms is a major challenge. The datasets of this research work give better results when it is trained with advanced machine learning and deep learning techniques and also lead a life in a safe online environment.

References

1. Abbas, M., Memon, K.A., Jamali, A.A., Memon, S., Ahmed, A.: Multinomial naive Bayes classification model for sentiment analysis. IJCSNS Int. J. Comput. Sci. Netw. Secur. **19**(3), 62 (2019)
2. Dadvar, M., Trieschnigg, D., Ordelman, R., Jong, F.D.: Improving cyberbullying detection with user context. In: European Conference on Information Retrieval, pp. 693–696. Springer (2013)
3. Davidson, T., Warmsley, D., Macy, M., Weber, I.: Automated hate speech detection and the problem of offensive language. In: Proceedings of the International AAAI Conference on Web and Social Media, vol. 11 (2017)

4. Davidson, T., Warmsley, D., Macy, M., Weber, I.: Automated hate speech detection and the problem of offensive language. In: Proceedings of the International AAAI Conference on Web and Social Media, vol. 11, pp. 512–515 (2017)
5. Dietterich, T.G., et al.: Ensemble learning. Handb. Brain Theory Neural Netw. **2**(1), 110–125 (2002)
6. Fortuna, P., Nunes, S.: A survey on automatic detection of hate speech in text. ACM Comput. Surveys (CSUR) **51**(4), 1–30 (2018)
7. Founta, A.M., Djouvas, C., Chatzakou, D., Leontiadis, I., Blackburn, J., Stringhini, G., Vakali, A., Sirivianos, M. and Kourtellis, N.: Large scale crowdsourcing and characterization of twitter abusive behavior. In: Twelfth International AAAI Conference on Web and Social Media (2018)
8. Hochreiter, S., Schmidhuber, J.: Long short-term memory. Neural Comput. **9**(8), 1735–1780 (1997)
9. Hossin, M., Sulaiman, M.N.: A review on evaluation metrics for data classification evaluations. Int. J. Data Min. Knowl. Manag. Process **5**(2), 1 (2015)
10. Karan, M., Šnajder, J.: Cross-domain detection of abusive language online. In: Proceedings of the 2nd Workshop on Abusive Language Online (ALW2), pp. 132–137 (2018)
11. Kumar, R., Ojha, A.K., Zampieri, M., Malmasi, S.: Proceedings of the first workshop on trolling, aggression and cyberbullying (trac-2018). In: Proceedings of the First Workshop on Trolling, Aggression and Cyberbullying (TRAC-2018) (2018)
12. Malmasi, S., Zampieri, M.: Detecting hate speech in social media. arXiv preprint arXiv:1712.06427 (2017)
13. Nugroho, K., Noersasongko, E., Fanani, A.Z., Basuki, R.S., et al.: Improving random forest method to detect hatespeech and offensive word. In: 2019 International Conference on Information and Communications Technology (ICOIACT), pp. 514–518. IEEE (2019)
14. Pedregosa, F., Varoquaux, G., Gramfort, A., Michel, V., Thirion, B., Grisel, O., Blondel, M., Prettenhofer, P., Weiss, R., Dubourg, V., et al.: Scikit-learn: machine learning in python. J. Mach. Learn. Res. **12**, 2825–2830 (2011)
15. Pennington, J., Socher, R., Manning, C.D.: Glove: global vectors for word representation. In: Proceedings of the 2014 Conference on Empirical Methods in Natural Language Processing (EMNLP), pp. 1532–1543 (2014)
16. Plisson, J., Lavrac, N., Mladenic, D., et al.: A rule based approach to word lemmatization. In: Proceedings of IS, vol. 3, pp. 83–86 (2004)
17. Pohjonen, M., Udupa, S.: Extreme speech online: an anthropological critique of hate speech debates. Int. J. Commun. **11**, 19 (2017)
18. Priyam, A., Abhijeeta, G.R., Rathee, A., Srivastava, S.: Comparative analysis of decision tree classification algorithms. Int. J. Curr. Eng. Technol. **3**(2), 334–337 (2013)
19. Ramakrishnan, M., Zadrozny, W., Tabari, N.: UVA wahoos at SemEval-2019 task 6: hate speech identification using ensemble machine learning. In: Proceedings of the 13th International Workshop on Semantic Evaluation, pp. 806–811 (2019)
20. Ramos, J., et al.: Using tf-idf to determine word relevance in document queries. In: Proceedings of the First Instructional Conference on Machine Learning, vol. 242, pp. 29–48. Citeseer (2003)
21. Schmidt, A., Wiegand, M.: A survey on hate speech detection using natural language processing. In: Proceedings of the Fifth International Workshop on Natural Language Processing for Social Media, 3 April 2017, Valencia, Spain, pp. 1–10. Association for Computational Linguistics (2019)
22. Ukil, A.: Support vector machine. In: Intelligent Systems and Signal Processing in Power Engineering, pp. 161–226. Springer (2007)
23. Unsvåg, E.F., Gambäck, B.: The effects of user features on twitter hate speech detection. In: Proceedings of the 2nd Workshop on Abusive Language Online (ALW2), pp. 75–85 (2018)
24. Waseem, Z., Hovy, D.: Hateful symbols or hateful people? Predictive features for hate speech detection on Twitter. In: Proceedings of the NAACL Student Research Workshop, pp. 88–93 (2016)
25. Webster, J.J., Kit, C.: Tokenization as the initial phase in NLP. In: COLING 1992 Volume 4: The 14th International Conference on Computational Linguistics (1992)

26. Zampieri, M., Malmasi, S., Nakov, P., Rosenthal, S., Farra, N., Kumar, R.: Semeval-2019 task 6: identifying and categorizing offensive language in social media (offenseval). arXiv preprint arXiv:1903.08983 (2019)
27. Zhang, S., Li, X., Zong, M., Zhu, X., Cheng, D.: Learning k for KNN classification. ACM Trans. Intell. Syst. Technol. (TIST) **8**(3), 1–19 (2017)
28. Zhong, H., Li, H., Squicciarini, A.C., Rajtmajer, S.M., Griffin, C., Miller, D.J., Caragea, C.: Content-driven detection of cyberbullying on the Instagram social network. In: IJCAI, vol. 16, pp. 3952–3958 (2016)

Chapter 12
A Fuzzy C-Means Based Clustering Protocol to Improve QoS of Multi Sink WSNs

Shashank Singh and Veena Anand

Abstract WSNs contain many sensors which are interconnected by wireless links. The data is collected by sensors and sent to the sink through a single path or multiple paths. The placement of the sink affects the performance of the network. The sink deployment and sensor to sensor routing are crucial in wireless sensor networks. In the proposed work, the network is divided into clusters using Fuzzy-C means, and cluster centers are found by using improved Fuzzy-C means. Each node is assigned to a cluster by calculating the distance between every center and node. The cluster with the least center to node distance is assigned to the node. Cluster heads are elected according to metrics residual energy, the richness of connectivity which is used to select the node with high neighbor density, and also the average distance of that node to all other nodes in the cluster. In this proposed work, we deployed multiple sinks to collect the data from the cluster head. Each node is aware of its nearest base station. This protocol ensures network reliability and energy efficiency. We perform simulations on MATLAB and compare the performance of the proposed scheme with other existing state of art protocols like LEACH-FC (Lata et al. IEEE Access 8:66013–66024, 2020) and UCMRP (Adnan et al. IEEE Access 9:38531–38545, 2021).

12.1 Introduction

A wireless sensor network (WSN) is a network in which many geographically distributed wireless sensors are connected to sense the physical or environmental conditions. In wireless sensor networks, the nodes can communicate using wireless links. The data is transferred through one or more nodes and a gateway. The data can also be transferred between one or more wireless sensor networks [1]. There

S. Singh (✉) · V. Anand
Department of CSE, National Institute of Technology, Raipur, Chhattisgarh 492010, India
e-mail: ssingh.phd2018.cse@nitrr.ac.in

V. Anand
e-mail: vanand.cs@nitrr.ac.in

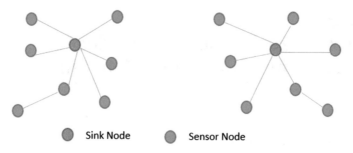

Sink Node   Sensor Node

Fig. 12.1 Multiple sinks

is limited energy in wireless sensor nodes. So, we should not allow the node to frequently transmit the data to prolong the lifetime of the network, and we should not allow the direct communication of nodes with base stations. An efficient method can be used which divides the network into groups called clustering. It involves the collection of sensor nodes into groups and electing cluster heads for all clusters. In this way, only some nodes are responsible for transmitting the data and those nodes are also changed periodically to save energy consumption. The cluster head is responsible for collecting the data and routing it to the base station [2].

A major challenge in this method is to decide the appropriate cluster head for the cluster. Election of cluster head is done by different parameters like residual energy which is the energy that is remaining with the sensor, richness of connectivity which is used to select the node with high neighbor density, and also the average distance of that node to all other nodes in the cluster, etc. [3]. Use of the second cluster head, that sends the data collected to the cluster head and the cluster head sends the data to the base station. The end goal is to expand the system's lifetime and efficiency. Another method is the deployment of multiple sinks shown in Fig. 12.1. In multiple sink networks, major factors are sink number and optimal placement of the sinks.

The placement of the sink is referred to as an optimization problem. Generally, only one sink is deployed and every node forwards the data in a single hop to that particular sink this increases the load on the sink and this creates an energy hole problem [4]. To solve this problem, we can place multiple sinks in the network which is discussed in Sect. 12.4.

A common factor for research on wireless sensor networks is energy efficiency. Wireless sensor nodes are powered using batteries. The node is dead if the energy is exhausted. Only in some cases batteries can be replaced or recharged. Even if it is possible, these operations are expensive, slow, and network performance is also decreased. Various protocols are proposed to slow down the exhaustion of battery energy, energy harvesting-based WSNs (EHWSNs) are one of the solutions to solve this problem by using the ability to take energy from the environment [5]. Energy harvesting can exploit different sources of energy, such as wind, magnetic fields, temperature variations, solar power, and mechanical vibrations. Continuously providing energy, and storing it for future use, energy harvesting subsystems enable WSN nodes to last potentially forever. To solve all the discussed problems, we have proposed an enhanced Fuzzy-C Means for clustering in WSNs having multiple sinks.

Table 12.1 Table representing the summary of the latest related works

References	Proposed methods	Gap identified
[6]	Uniform connectivity based clustering	Improvement can be done in the reliability of WSNs
[7]	Cluster-Tree based Energy-Efficient Data Gathering based on Fuzzy logic	Needs to focus on performing a real-time implementation
[8]	Fuzzy logic with competition radius	Needs to focus on some other performance parameters of WSN
[9]	Mobility based cluster head selection	Needs to focus on the reliability of WSNs
[10]	A probabilistic approach is also performed by LEACH and BCF	Improvement can be done in other QoS parameters
[11]	Fuzzy-C means (FCM) and moth-flame optimization method (MFO)	Needs to focus on some other application areas
[12]	A centralized approach based on LEACH-FUZZY	Improvement in routing is required for large-scale networks
[13]	Energy and traffic-aware sleep-awake	Improvement can be done in other QoS parameters
[14]	Fuzzy C-means	Improvement can be done in other QoS parameters

The remainder of the paper is organized as Sect. 12.2 conducts a brief survey of some of the clustering algorithms with their pros and cons. In Sect. 12.3, the proposed methodology. Section 12.4 provides the details of the system model. Then, Sect. 12.5 presents the simulation results, and Sect. 12.6 concludes the paper.

12.2   Related Work

Significant research has been conducted in the past to increase efficiency and QoS measures. Several methods have been proposed. The following (Table 12.1) are some of the most important and current works that address network longevity and energy consumption challenges.

12.3   Methodology

12.3.1   Optimal Number of Clusters

To divide the network into clusters, we need to know the optimal number of clusters. The efficiency of the protocol decreases if we do not divide the network into optimal parameters that decide the optimal number of clusters. The optimal number of clusters is given by Eq. (12.1).

$$K_{\text{opt}} = \sqrt{\frac{2N E_{\text{fs}} A}{\pi E_{\text{mp}} N - 4E_{\text{elec}}}} \tag{12.1}$$

where K_{opt} is the optimal number of clusters, N is the number of nodes, E_{fs} is energy for free space routing, A is the area of the network, E_{mp} energy for multipath routing, and E_{elec} is energy for transmission and Receiving.

12.3.2 Initialization of the Clusters Centroid

Fuzzy-C means clustering is a type of unsupervised learning, which is used when you have data without defined groups or categories. In Fuzzy-C means, membership values are assigned to each data point and this membership value indicates to which cluster the data point belongs [12].

The Algorithm Used to Find Cluster Centers and Clustering

- Decide the number of clusters needed
- Assign random coefficients to the data points for their membership in the cluster
- Repeat until the stopping condition is achieved, the stopping condition is changed between the coefficients in two iterations is less than a threshold value:

  - Calculate the centroid of the cluster
  - Calculate the coefficients of data points for being in the cluster.

The mean of all points is the centroid is calculated by Eq. (12.2), it shows the intensity of belonging to the cluster.

$$C_k = \frac{\sum_x W_k(x)^m x}{\sum_x W_k(x)^m} \tag{12.2}$$

where m is the degree and w_k is the weight. The weight of the data point is given by Eq. (12.3).

$$W_{ij} = \frac{1}{\sum_{k=1}^{c} \left(\frac{\|x_i - c_j\|}{\|x_i - c_k\|} \right)^{\frac{2}{m-1}}} \tag{12.3}$$

The principal objective of the Fuzzy-C means is to optimize the objective. The function is given by Eq. (12.4).

$$\arg \min C \sum_{i=1}^{n} \sum_{j=0}^{c} W_{ij}^m \|x_i - c_j\|^2 \tag{12.4}$$

In MATLAB, we use Fuzzy-C means by Eq. (12.5):

$$[\text{centers}, \ U] \ = \ \text{FCM (data, } N_c) \tag{12.5}$$

Here, centers relate to the centers of the clusters, and U gives a matrix that gives the membership values of the data points to the clusters. Data is the set of data points, and N_c is the number of clusters.

We assign a node to a cluster by calculating the distance between the node and all the cluster centers. The cluster with minimum distance is assigned as the cluster to the node.

12.3.3   Election of Cluster Head

The formation of clusters is performed at the moment of network deployment. In the proposed work, we do not form predefined cluster formation before starting a wireless sensor network. The deployment is performed randomly and several nodes in each cluster are not pre-defined and cluster formation depends on the initial geographic positions. Each node is aware of its neighbor. Cluster head selection is based on the communication delay and energy consumption. To achieve this, a weight is assigned to each node based on remaining energy, the entity of neighboring nodes, and distance to the neighboring node. The weight of each sensor node is given based on Eq. (12.6).

$$W_i = (N_i \cdot c \cdot E_i) \big/ D_i \tag{12.6}$$

where c is the coefficient, N_i is the Number of neighbors of sensor node i, E_i is the Residual Energy of sensor node i, D_i is the number of neighbors, and W_i is the weight of sensor node i.

12.3.4   Energy Dissipation

The energy dissipation will be different for cluster heads and normal nodes by using Eq. (12.7). Many constants are used to determine the energy dissipated like $E_{TX}, E_{RX}, \varepsilon_{mp}, \varepsilon_{fs}$, and E_{DA}. E_{TX} is the energy constant for transmission. E_{RX} is the energy constant for receiving. ε_{mp} is the energy constant for multipath. ε_{fs} is the energy constant for free space routing. E_{DA} is an energy constant for data aggregation. We will take a heterogeneous network which means all nodes do not have the same energy and based on the factor of heterogeneity, several heterogeneous nodes are decided [15].

$$E_{TX}(L, d) = \begin{cases} LE_{elec} + L\varepsilon_{fs}d^2, \ d \leq d_0 \\ LE_{elec} + L\varepsilon_{mp}d^4, \ d \geq d_0 \end{cases} \tag{12.7}$$

$$E_{RX}(L) = LE_{elec}$$

12.4 System Model

12.4.1 Sink Placement Strategy

In our simulation scenario, we have 100 nodes and these nodes are deployed in a 100 × 100 m² area. To place the sinks in the network, we have to partition the network area. The main aim is to deploy the sink in the partition to increase the efficiency of the network and decrease energy consumption. If the sink is placed in a low-density area, then the nodes near the sinuses create more energy and result in a loss of network reliability. But, the network we generated is random so we cannot predict the place where there can be a high density of nodes [7].

So, we partition the network into grids for each grid (shown in Fig. 12.2) we can deploy a sink. The nodes that are nearer to a particular sink can be assigned to that sink.

We deployed five sinks in our network. The position of the sinks is shown in Table 12.2.

12.4.2 Network Assumptions

Simulation scenarios for the protocol are:

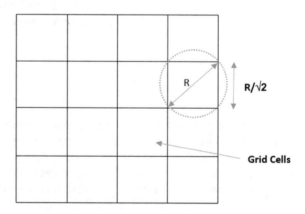

Fig. 12.2 Grid cell structure

Table 12.2 Sink placement

Sink name	Position
Sink1	(25,25)
Sink2	(25,75)
Sink3	(50,50)
Sink4	(25,75)
Sink5	(75,0.75)

Table 12.3 Simulation parameters

Parameter	Value
Surface area of interest	$100 \times 100 \text{ m}^2$
Number of sensor nodes	100
Sensor node deployment	Random
Packet size	4000 Bytes
The initial energy of the sensor nodes	0.5–1.5 J
E_{TX}	50 Nj/bit
E_{RX}	50 Nj/bit
E_{fs}	10 pj/bit/m^4
Coefficient c	0.4

- Randomly distributed sensor nodes are present in a square region.
- Sensor nodes are heterogeneous and have a unique identification number throughout the network, the energy of other nodes is limited. The location of the node does not change after the deployment.
- The base stations are deployed according to strategy (Position mentioned in Table 12.2) and their locations have not changed.
- Single-hop or multi-hop communication is done.

12.4.3   Simulation Parameters

See Table 12.3.

12.5   Performance Evaluation

Figure 12.3 shows the random distribution of nodes in a $100 \times 100 \text{ m}^2$ area. The blue color nodes are nodes with low energy and the green color nodes are nodes with high energy and X is the base station. Figure 12.4 shows the state of the network after 1200 rounds. The star nodes are the cluster heads of the cluster. Figure 12.5 shows the number of alive nodes after each round which describes the reliability of the protocol.

The proposed scheme was compared with some other schemes like LEACH-FC [12] and UCMRP [10]. It is found that the proposed scheme performs better in terms of network lifetime.

Figure 12.6 shows the variation of energy dissipated with the number of clusters. We can observe from the graph that the energy dissipated is being reduced and Fig. 12.7 shows the residual energy to the number of rounds. The graph is linear which shows that there is no sudden decrease in energy in any particular round.

Fig. 12.3 Random
distribution

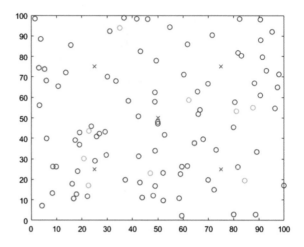

Fig. 12.4 After1200 rounds

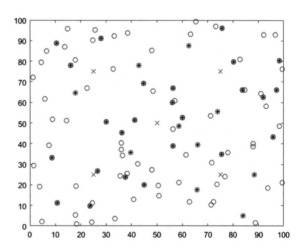

Figure 12.8 shows the comparison of residual energy of the proposed scheme and other state of art schemes like LEACH-FC [12] and UCMRP [10], and it is found that the performance of the proposed scheme is better than existing schemes in terms of energy efficiency.

Figures 12.9 and 12.10 show the FND and HND for the proposed scheme and compared with the other two existing schemes, it is observed that the proposed scheme has better stability than the other two.

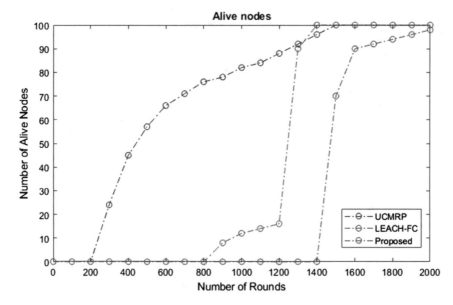

Fig. 12.5 Number of alive nodes per rounds

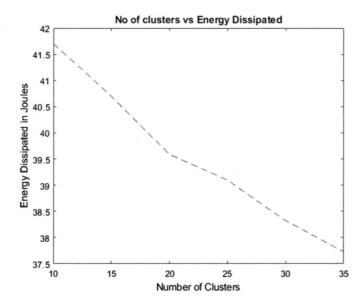

Fig. 12.6 Energy versus no. of clusters

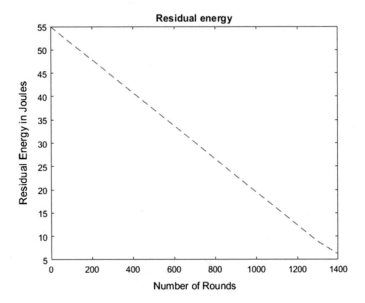

Fig. 12.7 Residual energy versus rounds

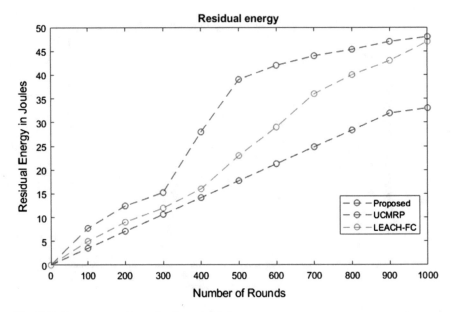

Fig. 12.8 Energy comparison of various protocols

Fig. 12.9 Comparison of
FND

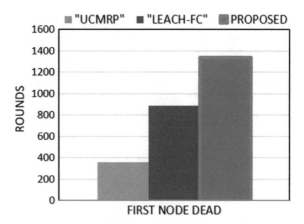

Fig. 12.10 Comparison of
HND

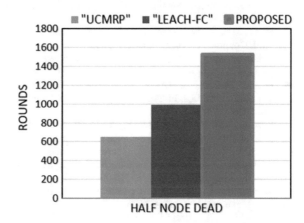

12.6 Conclusion

The proposed scheme addresses the issue of energy efficiency and lifetime in the
wireless sensor networks as the wireless sensors have limited energy.

   In our proposed scheme, the reliability of the network is increased as the first node
is dead after 1200 rounds. The energy consumption is also decreased because we have
taken the factors like neighbor distance, and the number of neighbors into consideration, with multiple sinks. Deployment of multiple sinks increases the efficiency
of the network and decreases the transmission delay. In events like disasters, recognition time is very important, so in networks related to these applications, multiple
sinks must be used. In this way, we have increased the reliability, and efficiency
and decreased the energy consumption of the protocol in comparison with other
mentioned existing state of art protocols.

In the future, we can also improve the sink placement strategy or harvest the energy from sources like solar energy, mechanical energy, etc., to make the network more efficient.

Acknowledgements We are thankful to the faculty members of the National Institute of Technology, Raipur (Chhattisgarh), and colleagues of SRMCEM (Lucknow), India for their motivation and continuous support. Our special vote of thanks to Prof. Ajay Kumar Srivastava for his valuable suggestions and contributions.

References

1. Kumar, S., Duttagupta, S., Rangan, V.P., et al.: Reliable network connectivity in wireless sensor networks for remote monitoring of landslides. Wireless Netw. **26**, 2137–2152 (2020). https://doi.org/10.1007/s11276-019-02059-7
2. Fu, C., Jiang, Z., Wei, W., Wei, A.: An energy balanced algorithm of LEACH protocol in WSN. Int. J. Comput. Sci. **10**(1), 354–359 (2013)
3. Oladimeji, M.O., Turkey, M., Dudley, S.: HACH: Heuristic algorithm for clustering hierarchy protocol in wireless sensor networks. Appl. Soft Comput. **55**, 452–461 (2017)
4. Das, D., Rehena, Z., Roy, S., Mukherjee, N.: Multiple-sink placement strategies in wireless sensor networks. In: 2013 5th International Conference on Communication Systems and Networks (COMSNETS 2013) (2013)
5. Bozorgi, S.M., Shokouhi Rostami, A., Hosseinabadi, A.A.R., Balas, V.E.: A new clustering protocol for energy harvesting-wireless sensor networks. Comput. Electr. Engi. (2017)
6. Sood, T., Sharma, K.: LUET: a novel lines-of uniformity based clustering protocol for heterogeneous-WSN for multiple-applications. J. King Saud Univ. Comput. Inf. Sci. (2020). https://doi.org/10.1016/j.jksuci.2020.09.016
7. Karunanithy, K., Velusamy, B.: Cluster-tree based energy efficient data gathering protocol for industrial automation using WSNs and IoT. J. Ind. Inf. Integr. 100156 (2020). https://doi.org/10.1016/j.jii.2020.100156
8. Singh, J., Yadav, S.S., Kanungo, V., Yogita, Pal, V.: A node overhaul scheme for energy efficient clustering in wireless sensor networks. IEEE Sens. Lett. **5**(4), 1–4, Art no. 7500604. https://doi.org/10.1109/LSENS.2021.3068184
9. Umbreen, S., Shehzad, D., Shafi, N., Khan, B., Habib, U.: An energy-efficient mobility-based cluster head selection for lifetime enhancement of wireless sensor networks. IEEE Access **8**, 207779–207793 (2020). https://doi.org/10.1109/ACCESS.2020.3038031
10. Adnan, M., Yang, L., Ahmad, T., Tao, Y.: An Unequally clustered multi-hop routing protocol based on fuzzy logic for wireless sensor networks. IEEE Access **9**, 38531–38545 (2021). https://doi.org/10.1109/ACCESS.2021.3063097
11. Fei, W., Hexiang, B., Deyu, L., Jianjun, W.: Energy-efficient clustering algorithm in underwater sensor networks based on Fuzzy C means and moth-flame optimization method. IEEE Access **8**, 97474–97484 (2020). https://doi.org/10.1109/ACCESS.2020.2997066
12. Lata, S., Mehfuz, S., Urooj, S., Alrowais, F.: Fuzzy Clustering algorithm for enhancing reliability and network lifetime of wireless sensor networks. IEEE Access **8**, 66013–66024 (2020). https://doi.org/10.1109/ACCESS.2020.2985495
13. Shagari, N.M., Idris, M.Y.I., Salleh, R.B., Ahmedy, I., Murtaza, G., Shehadeh, H.A.: heterogeneous energy and traffic aware sleep-awake cluster-based routing protocol for wireless sensor network. IEEE Access **8**, 12232–12252 (2020). https://doi.org/10.1109/ACCESS.2020.2965206

14. Hassan, A.A.-H., Shah, W.M., Habeb, A.-H.H., Othman, M.F.I., Al-Mhiqani, M.N.: An improved energy-efficient clustering protocol to prolong the lifetime of the WSN-based IoT. IEEE Access **8**, 200500–200517 (2020). https://doi.org/10.1109/ACCESS.2020.3035624
15. Bongale, A.M., Nirmala, C.R., Bongale, A.M.: Hybrid cluster head election for wsn based on firefly and harmony search algorithms. Wireless Pers. Commun. **106**, 275–306 (2019). https://doi.org/10.1007/s11277-018

Shashank Singh is a research scholar in the Department of Computer Science and Engineering at the National Institute of Technology, Raipur, India. His research interest focuses on Wireless Sensor Networks.

Dr. Veena Anand is Asst. Professor in the Department of Computer Science and Engineering at the National Institute of Technology, Raipur, India. Her research interest focuses On Wireless Sensor Networks and Mobile Adhoc Network.

Chapter 13
Streamflow Forecasting Using Novel ANFIS-GWO Approach

Gopal Krishna Sahoo, Niharika Patel, Debiprasad Panda, Shaswati Mishra, Sandeep Samantaray, and Deba Prakash Satapathy

Abstract One of the most crucial features in ensuring reliable and resilient water resource management and planning is accurate streamflow forecasting. It is critical for hydropower generation, flood management, and agricultural planning. Due to the sheer importance of streamflow forecasting, in this study the Grey Wolf optimization algorithm (GWO) is used to optimize the Adaptive Neuro-Fuzzy Inference System (ANFIS), and the model performance is compared to that of the standalone ANFIS model. Considering the importance of rainfall in streamflow forecasting, these parameters are used as input to the model for forecasting the streamflow in Mahanadi River, India. To assess the efficacy of our suggested model WI, NSE and MSE were used and it is found that ANFIS-GWO model outperforms than ANFIS model in streamflow forecasting.

G. K. Sahoo · D. Panda · S. Samantaray (✉) · D. P. Satapathy
Department of Civil Engineering, OUTR Bhubaneswar, Bhubaneswar, Odisha, India
e-mail: sandeep1139_rs@civil.nits.ac.in

D. P. Satapathy
e-mail: dpsatapathy@cet.edu.in

N. Patel
Department of Civil Engineering, GIET University, Gunupur, Odisha, India

S. Mishra
Department of Philosophy, Utkal University, Bhubaneswar, Odisha, India

© The Author(s), under exclusive license to Springer Nature Singapore Pte Ltd. 2023   141
V. Bhateja et al. (eds.), *Evolution in Computational Intelligence*, Smart Innovation, Systems and Technologies 326, https://doi.org/10.1007/978-981-19-7513-4_13

13.1 Introduction

The Mahanadi River is one of India's most significant rivers, flowing through the states of Chhattisgarh and Odisha. This basin's streamflow forecasting is critical for integrated water resource development as well as flood prevention and relief, exploitation, optimal scheduling, and scientific management [1–7]. As numerous variables influence the streamflow in Mahanadi River, several approaches have been utilized by the researchers to anticipate it's streamflow over years. Since machine learning models are capable of effectively modelling various nonlinear interdependencies between input and output values, these techniques are utilized for streamflow forecasting for this study.

Sharma et al. [8] for Chickasaw Creek watershed, USA, compared ANFIS with LSPC and concludes that when rain gauge stations are scarce, ANFIS may be a preferable option. Sanikhani and Kisi [9] inspected the combination of ANFIS with grid partition and ANFIS with sub-clustering for the monthly streamflow estimation for Firat-Dicle Basin, Turkey and suggested that the ANFIS-SC model marginally outperforms the ANFIS-GP model. Firat and Turan [10] checked the applicability of ANFIS for monthly river flow forecasting by considering time series of river flow data of Goksu River, Turkey and tested it against autoregressive methods and feed forward neural networks, and found that ANFIS outperforms other models. Pramanik and Panda [11] used upstream river flow data to estimate the downstream river flow in Mahanadi River, India by using ANN and ANFIS techniques and found that ANFIS outperformed the ANN model. Firat and Gungor [12] proved the superiority of the ANFIS model over ANN and multiple regression models by considering antecedent river flow in the Menderes River, Turkey. Adnan et al. [13] in the first part of his research found that ANFIS-SC outperformed ANFIS-GP, GRNN, RBNN, and FFNN and in the second part of the study they proved that periodic ANFIS-SC outperforms the periodic RBNN model for predicting monthly streamflow of the Gilgit river basin. Considering temperature and antecedent streamflow as input parameters. Jaafari et al. [14] estimated the landslide susceptibility using ANFIS optimized with biogeography-based optimization and grey wolf optimization for Tehri Garhwal district, Uttarakhand, India by using landslide locations and their causative factors as input variables.

Dehghani et al. [15] optimized ANFIS with GWO algorithm to forecast hydropower generation in Dez basin considering dam inflow, rainfall, and hydropower time-series data and found that GWO-ANFIS is superior to the ANFIS model. Dehghani et al. [16] demonstrated that accurate estimation of 5-min to 10-days ahead influent flow rate can be obtained from the developed ANFIS-GWO models compared to the ANFIS models by considering Influent time-series data from a wastewater treatment plant situated at Isfahan city, Iran. Maroufpoor et al. [17] validated the ANFIS-GWO model against ANFIS-FCM, ANFIS-SC, SVR, and ANN for soil moisture content estimation using clay content, organic matter, dielectric constant and soil bulk density as input parameters for Dehgolan plain, Iran, and results indicated that ANFIS-GWO was the finest of all the models. Madvar et al. [18]

forecasted river flow in three forecast horizons, i.e. long-term (Annual), mid-term (Monthly and Weekly), and short-term (Daily) for Karun III dam, Iran using daily inflow data. Best model for forecasting in long term is ANFIS-GWO, for mid-term is ANFIS-GWO and ANFIS-GA, and for short term is ANFIS-DE, ANFIS-FFA, ANFIS-GA, and ANFIS-GWO.

13.2 Study Area

The Mahanadi is the most significant river flowing through the state of Odisha and drains a total area of 1,41,589 Km^2 (covering Chattisgarh, Odisha, Jharkhand, and Maharastra) Before emptying into the Bay of Bengal. The central India hills, the Eastern Ghats and the Maikal range bounds the basin from northern, eastern and southern, and from the western direction, respectively. The basin is located between $80° 30'$ and $84° 50'$ E longitudes and $19° 20'$ and $23° 35'$ N latitudes. The altitude of the basin varies from 30 to 200 m above the mean sea level. The mean yearly precipitation across the basin varies from 1200 to 1400 mm, more than 90% of the precipitation occurs from June to October, i.e. during the monsoon season. In the Mahanadi catchment, the coldest and the hottest months are December and May, respectively. The basin experiences a minimum diurnal range during July and August and maximum during February and March. The basin can be classified into four physiographic regions: the Eastern Ghats, the northern plateau, the coastal plain, and the erosional lowlands of the central tableland. The first two are mountainous areas. The delta area, which is exceptionally fertile, is located on the coastal plain. The central tableland in the basin's core interior is covered by the river and its tributaries (Fig. 13.1).

13.3 Methodology

13.3.1 ANFIS

Jang [19] combined artificial neural network (ANN) and fuzzy logic to develop a correlation between input and output space by utilizing the learning capability of the ANN model and fuzzy reasoning. ANFIS assigns and adjusts membership functions using ANN's training capability. The primary distinction between the types of fuzzy inference Systems (FIS) is how the consequent (or "then") component of the ANFIS model is determined. The most commonly used FISs are Takagi–Sugeno, Mamdani, and Tsukamoto FIS [20, 21]. For determining the consequent part of the ANFIS model, we used the Takagi–Sugeno FIS. Two fuzzy if–then rules are used in the Takagi–Sugeno FIS:

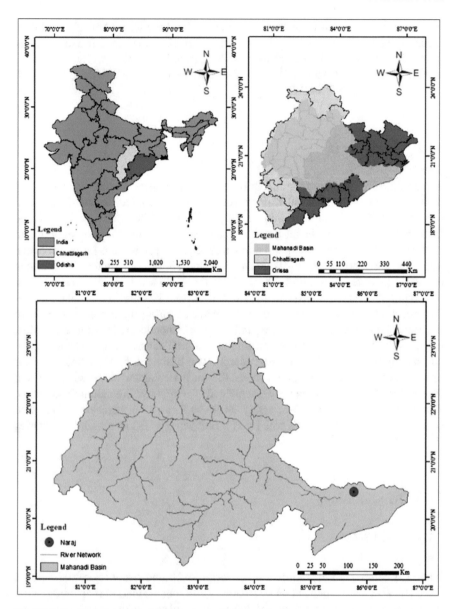

Fig. 13.1 Study area showing Naraj station in Mahanadi Basin

$$\text{Rule1 : IF } y \text{ is } U_1 \text{ and } z \text{ is } V_1 \text{ THEN } f_1 = m_1 * y + n_1 * z + o_1 \qquad (13.1)$$

$$\text{Rule2 : IF } y \text{ is } U_2 \text{ and } z \text{ is } V_2 \text{ THEN } f_2 = m_2 * y + n_2 * z + o_2 \qquad (13.2)$$

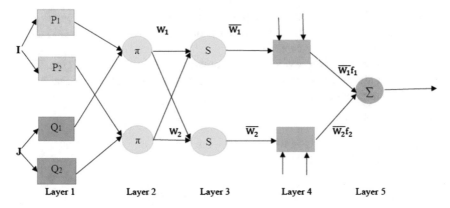

Fig. 13.2 Architecture of ANFIS

where the inputs are y and z and the membership functions of the inputs are U and V, the output is f, and m, n, and o are ANN-determined linear parameters in the model's consequent part. ANFIS model consists of five layers with output f and input being y and z. The first layer of the ANFIS structure fuzzifies the input variables (y and z), and every rule's firing strength is computed by the second layer. The firing strength normalization of each rule occurs in the third layer and the rule's output is specified in the fourth layer, by adding the fourth layer's outputs the fifth layer computes the fuzzy system's total output (Fig. 13.2).

13.3.2   Grey Wolf Optimization (GWO)

GWO algorithm developed by Mirjalili et al. [22] optimizes the model based on the social hierarchy of grey wolves and their hunting methods.

Social hierarchy

The optimum solution is alpha wolves (α), who are the pack leaders. The alpha is primarily in charge of making decisions about waking time, sleeping location, hunting, and so on. Although the pack's strongest wolf is not always the alpha, but he or she is the best at managing the pack.

Beta ranks second in the grey wolf hierarchy. The betas help the alpha with pack decisions and other responsibilities. He/she is most likely the greatest prospect of being the alpha wolf in the scenario when one of the pack's alpha dies or gets extremely old.

Delta wolves must submit to betas and alphas, but rule the omegas. This group includes elders, sentinels, scouts, hunters, and caretakers. Elders are experienced wolves who were once alpha or beta. Sentinels defend and ensure the pack's safety. Scouts are responsible for alerting the pack in case of any danger and patrolling the region. When hunting animals and supplying sustenance for the pack, hunters assist the alphas and betas. Finally, those in charge of caring for the pack's weak, ailing, and injured wolves are the caretakers.

The most inferior grey wolves are the omegas, it serves as a scapegoat. These wolves must always be subordinate to the other dominating wolves. Omegas are the last which are permitted to eat.

Encircling the prey

The prey is encircled by grey wolves before hunting.

For mathematical modeling:

$$\vec{Q} = \left| \vec{R}.\overrightarrow{A_P}(i) - \vec{Y}(i) \right|, \ \vec{A}(i+1) = \vec{A}(i) - \vec{C}.\vec{Q} \ \vec{C} = 2\vec{a}.\vec{r_1} - \vec{a}, \ \vec{R} = 2.\vec{r_2}$$
(13.3)

where coefficient vector is represented by C and R. A_P represents the location vector of the pray. i is the current iteration. Grey wolf's location vector is D, a is co-efficient vector that linearly decreases from 0 to 2, and r_1 and r_2 varying in the range [0, 1] are random vectors.

Hunting

The alpha, beta, and delta wolves are thought to have a greater awareness of the possible location of the prey and can thus better estimate the prey's location. As a result, the additional search agents use the following formulas to update the position of the prey based on the three primary wolves:

$$\overrightarrow{Q_\alpha} = \left| \vec{R_1}.\vec{A_\alpha} - \vec{A} \right|, \ \overrightarrow{Q_\beta} = \left| \vec{R_2}.\vec{A_\beta} - \vec{A} \right|, \ \overrightarrow{Q_\delta} = \left| \vec{R_3}.\vec{A_\delta} - \vec{A} \right|,$$
(13.4)

$$\vec{A_1} = \vec{A_\alpha} - \vec{C_1}.\vec{Q_\alpha}, \ \vec{A_2} = \vec{A_\beta} - \vec{C_2}.\vec{Q_\beta}, \ \vec{A_3} = \vec{A_\delta} - \vec{C_3}.\vec{Q_\delta}$$
(13.5)

$$\vec{A}(i+1) = \frac{\vec{A_1} + \vec{A_2} + \vec{A_3}}{3}$$
(13.6)

Attacking the prey (exploitation)

The grey wolves attack the prey when it stops moving. The value of $\vec{a}$ decreases (i.e. from 2 to 0) when the pray approaches, $\vec{C}$ is a random value that depends on 'a' and is in the interval $[-2a, 2a]$, the next position of the search agent is between its present position and the prey's position. When $|C| < 1$ the wolves attack the prey.

Search for the prey (exploration)

To find a fitter pray the grey wolves diverge if the random value of $\vec{C}$ exceeds the interval $[-1, 1]$ (i.e. $|C| > 1$). This allows the GWO algorithm to explore globally by emphasizing exploration. $\vec{R}$ favours exploration by providing random weights to the pray and contains random values in the range [0, 2]. Random weight is provided for the prey by this component to stochastically deemphasize ($R < 1$) or emphasize ($R > 1$) the prey's effect in determining the span. During optimization it enables GWO to exhibit more random behaviour, avoiding local optima and promoting exploration.

13.3.3 Performance Evaluation

For this research, 30 years (1992–2021) rainfall and streamflow data were collected from Central Water Commission, Bhubaneswar, from which the first 21 years of data (1992–2012) were considered for training and the last 9 years of data (2013–2021) were considered for testing purpose. For assessing the model performance mean squared error (MSE), willmott index (WI), and nash sutcliffe efficiency (NSE) were used [23–31].

$$\text{MSE} = \frac{1}{N} \sum_{i=1}^{N} (S_P - S_o)^2 \tag{13.7}$$

$$\text{WI} = 1 - \left[\frac{\sum_{i=1}^{N} (S_o - S_P)^2}{\sum_{i=1}^{N} \left(|S_P - \overline{S_o}| + |S_O - \overline{S_O}| \right)^2} \right] \tag{13.8}$$

$$\text{NSE} = 1 - \left[\frac{\sum_{k=1}^{N} (S_p - S_0)^2}{\sum_{i=1}^{N} (S_0 - \overline{S_o})^2} \right] \tag{13.9}$$

where, S_P = Predicted variables, S_O = Observed variables, $\overline{S_O}$ = Mean observed variables.

13.4 Results and Discussion

For each of the five input combinations, ANFIS and ANFIS-GWO models were developed. The models were then compared using MSE, WI, and NSE values. For both training and validation data, MSE, WI, and NSE values are calculated. For the validation data, MSE, WI, and NSE values are used to assess the actual performance. In terms of WI and NSE values, the ANFIS-GWO model has the greatest model efficiency for the training data, whereas ANFIS has the lowest efficiency. Among all the models, the fourth model (ANFIS/F-4 model) has the highest WI and NSE model efficiency, as well as the lowest MSE values. Those are 0.9502, 0.9477, and 7.3285, respectively. Similarly, the ANFIS-GWO model with input case four yields the best MSE, WI, and NSE values. It is observed that in case of ANFIS-GWO, WI values for the developmental stage of models F-I, F-II, F-III, and F-IV were 0.9936,

0.9942, 0.9959, and 0.9502, whereas during the testing period. It was 0.9707, 0.9729, 0.9741 and 0.9753. WI values close to one indicate that the model performed well (Fig. 13.3; Table 13.1).

The comparative plot of observed and estimated streamflow for Naraj gauge station is presented in Fig. 13.4. It can be observed from the Fig. 13.4 that the ANFIS-GWO model outperformed the conventional ANFIS model for estimation of streamflow. The ANFIS models estimate the maximum peak as 5064.856 m^3/s, instead of the observed 5423.919 m^3/s, whereas the ANFIS-GWO models provided a forecast equal to 5324.661 m^3/s.

Figure 13.5 presents the frequency distribution "histogram" plot of the predicted and actual stream flow. The presented graphical presentation can assist in a better

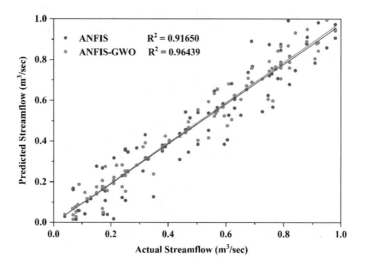

Fig. 13.3 Scatter plot between actual and predicted streamflow

Table 13.1 Performance of ANFIS and ANFIS-GWO model

Technique	Input combination	Model name	WI	MSE	NSE	WI	MSE	NSE
			Training			Testing		
ANFIS	P_1	F-I	0.9461	8.8864	0.9425	0.9237	11.5543	0.9187
	P_{t-1}	F-2	0.947	8.249	0.9449	0.9248	10.6094	0.9201
	P_{t-2}	F-3	0.9495	7.88	0.946	0.9287	10.116	0.9215
	P_{t-3}	F-4	0.9502	7.3285	0.9477	0.9296	9.2769	0.9223
ANFIS-GWO	P_1	F-I	0.9936	2.781	0.9903	0.9707	5.287	0.9664
	P_{t-1}	F-2	0.9942	2.3472	0.9921	0.9729	4.689	0.9678
	P_{t-2}	F-3	0.9959	1.93	0.9934	0.9741	4.0221	0.9692
	P_{t-3}	F-4	0.9964	1.006	0.9947	0.9753	3.27	0.9709

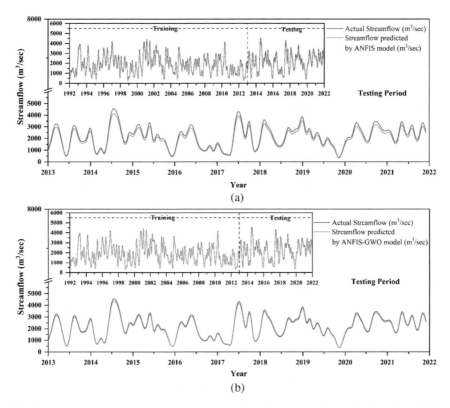

Fig. 13.4 Comparison between observed versus predicted Streamflow by **a** ANFIS and **b** ANFIS-GWO models

understanding of the prediction performance of the model. The figure clearly shows that the ANFIS-GWO model outperforms the standalone ANFIS model.

13.5 Conclusion

The current study evaluated the ANFIS-GWO and ANFIS models for streamflow forecasting in the Mahanadi River utilizing different lag tie of rainfall as input parameters and by using WI, MSE and WI performance matrices. The findings revealed that the ANFIS-GWO model outperformed than the ANFIS model, with WI = 0.9964, NSE = 0.9947 and MSE = 1.006, respectively. In this work, the significance of the ANFIS-GWO model in streamflow forecasting is proven. This research's present findings can be enhanced by integrating more hydrological and climatological data for the proposed model and testing with other input combinations for various time series data.

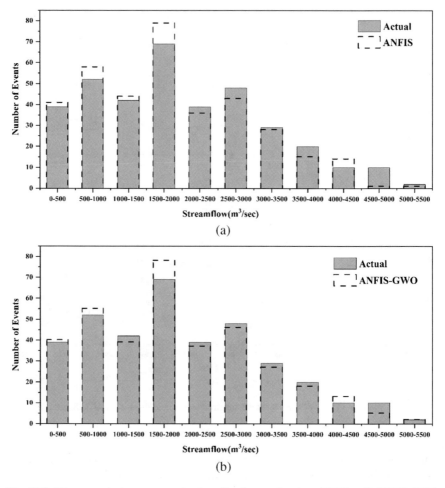

Fig. 13.5 Histogram plot between actual and predicted streamflow by **a** ANFIS and **b** ANFIS-GWO models

References

1. Sahoo, A., Samantaray, S., Ghose, D.K.: Prediction of flood in Barak River using hybrid machine learning approaches: a case study. J. Geol. Soc. India **97**(2) (2021a)
2. Sahoo, A., Samantaray, S., Bankuru, S., Ghose, D.K.: Prediction of flood using adaptive neuro-fuzzy inference systems: a case study. In: Smart Intelligent Computing and Applications, pp. 733–739. Springer, Singapore (2020)
3. Sahoo, A., Singh, U.K., Kumar, M.H., Samantaray, S.: Estimation of flood in a river basin through neural networks: a case study. In: Communication Software and Networks, pp. 755–763. Springer, Singapore (2021b)
4. Samantaray, S., Sahoo, A.: Estimation of flood frequency using statistical method: Mahanadi River basin, India. H$_2$Open J. **3**(1), 189–207 (2020)

5. Sahoo, A., Samantaray, S., Paul, S.: Efficacy of ANFIS-GOA technique in flood prediction: a case study of Mahanadi River basin in India. H_2Open J. **4**(1), 137–156 (2021c)
6. Samantaray, S., Das, S.S., Sahoo, A., Satapathy, D.P.: Evaluating the application of meta-heuristic approaches for flood simulation using GIS: a case study of Baitarani River Basin, India. J. Mater. Today: Proc. (2021a)
7. Samantaray, S., Sahoo, A., Agnihotri, A.: Assessment of flood frequency using statistical and hybrid neural network method: Mahanadi River Basin, India. J. Geol. Soc. India **97**(8), 867–880 (2021b)
8. Sharma, S., Srivastava, P., Fang, X., Kalin, L.: Performance comparison of adaptive neuro fuzzy inference system (ANFIS) with loading simulation program C++ (LSPC) model for streamflow simulation in El Niño Southern Oscillation (ENSO)-affected watershed. J. Expert Syst. Appl. **42**, 2213–2223 (2015)
9. Sanikhani, H., Kisi, O.: River flow estimation and forecasting by using two different adaptive neuro-fuzzy approaches. J. Water Resources Manage. **26**, 1715–1729 (2012)
10. Firat, M., Turan, M.E.: Monthly river flow forecasting by an adaptive neuro-fuzzy inference system. J. Water and Environ. J. **24**, 116–125 (2010)
11. Pramanik, N., Panda, R.K.: Application of neural network and adaptive neuro-fuzzy inference systems for river flow prediction. J. Hydrological Sci J. **54**(2), 247–260 (2009)
12. Firat, M., Gungor, M.: River flow estimation using adaptive neuro fuzzy inference system. J. Math. Comput. Simul. **75**, 87–96 (2007)
13. Adnan, R.M., Yuan, X., Kisi, O., Yuan, Y., Tayyab, M., Lei, X.: Application of soft computing models in streamflow forecasting. J. Water Manage. **172**(3), 123–134 (2019)
14. Jaafari, A., Panahi, M., Pham, B.T., Shahabi, H., Bui, D.T., Rezaie, F., Lee, S.: Meta optimization of an adaptive neuro-fuzzy inference system with grey wolf optimizer and biogeography-based optimization algorithms for spatial prediction of landslide susceptibility. J. Catena. **175**, 430–445 (2018)
15. Dehghani, M., Madvar, H.R., Hooshyaripor, F., Mosavi, A., Shamshirband, S., Zavadskas, E.K., Chau, K.W.: Prediction of hydropower generation using grey wolf optimization adaptive neuro-fuzzy inference system. J. Energies **12**, 289 (2019)
16. Dehghani, M., Seifi, A., Madvar, H.R.: Novel forecasting models for immediate-short-term to long-term influent flow prediction by combining ANFIS and grey wolf optimization. J. Hydrol. **576**, 698–725 (2019)
17. Maroufpoor, S., Maroufpoor, E., Haddad, O.B., Shiri, J., Yaseen, Z.M.: Soil moisture simulation using hybrid artificial intelligent model: hybridization of adaptive neuro fuzzy inference system with grey wolf optimizer algorithm. J. Hydrol. **575**, 544–556 (2019)
18. Madvar, H.R., Dehghani, M., Memarzadeh, R., Gharabaghi, B.: Short to long-term forecasting of river flows by heuristic optimization algorithms hybridized with ANFIS. J. Water Resour. Manage. **35**, 1149–1166 (2021)
19. Jang, J.S.R.: ANFIS: adaptive-network-based fuzzy inference system. J. IEEE Trans. Syst. Man Cybern. **23**, 3 (1993)
20. Samantaray, S., Sumaan, P., Surin, P., Mohanta, N.R., Sahoo, A.: Prophecy of groundwater level using hybrid ANFIS-BBO approach. In: Proceedings of International Conference on Data Science and Applications, pp. 273–283. Springer, Singapore (2022b)
21. Samanataray, S., Sahoo, A.: A comparative study on prediction of monthly Streamflow using hybrid ANFIS-PSO approaches. KSCE J. Civ. Eng. **25**, 4032–4043 (2021)
22. Mirjalili, S., Mirjalili, S.M., Lewis, A.: Grey wolf optimizer. J. Adv. Eng. Softw. **69**, 46–61 (2014)
23. Samantaray, S., Das, S.S., Sahoo, A., Satapathy, D.P.: Monthly runoff prediction at Baitarani river basin by support vector machine based on Salp swarm algorithm. J. Ain Shams Eng. J. **13**(5) (2022a)
24. Samantaray, S., Ghose, D.K.: Prediction of S12-MKII rainfall simulator experimental runoff data sets using hybrid PSR-SVM-FFA approaches. J. Water Climate Change **13**(2), 707–734 (2022)

25. Samantaray, S., Sahoo, A.: Prediction of suspended sediment concentration using hybrid SVM-WOA approaches. J. Geocarto Int. 1–27 (2021b)

26. Samantaray, S., Ghose, D.K.: Modelling runoff in an arid watershed through integrated support vector machine. J. H_2Open J. **3**(1), 256–275 (2020a)

27. Samantaray, S., Sahoo, A., Ghose, D.K.: Assessment of sediment load concentration using SVM, SVM-FFA and PSR-SVM-FFA in arid watershed, India: a case study. KSCE J. Civil Eng. **24**(6), 1944–1957 (2020)

28. Samantaray, S., Ghose, D.K.: Assessment of suspended sediment load with neural networks in arid watershed. J. Inst. Eng. (India): Series A **101**(2), 371–380 (2020b)

29. Samantaray, S., Sahoo, A.: Modelling response of infiltration loss toward water table depth using RBFN, RNN, ANFIS techniques. Int. J. Knowl.-Based Intell. Eng. Syst. **25**(2), 227–234 (2021)

30. Samantaray, S., Ghose, D.K.: Dynamic modelling of runoff in a watershed using artificial neural network. In: Smart Intelligent Computing and Applications, pp. 561–568. Springer, Singapore (2019)

31. Sarkar, B.N., Samantaray, S., Kumar, U., Ghose, D.K.: Runoff is a key constraint toward water table fluctuation using neural networks: a case study. In: Communication Software and Networks, pp. 737–745. Springer, Singapore (2021)

Chapter 14
Streamflow Forecasting Using Machine Learning Approach: A Case Study

Aiswarya Mishra, Narottam Nayak, Shaswati Mishra, Debiprasad Panda, Sandeep Samantaray, and Deba Prakash Satapathy

Abstract The ability to anticipate monthly streamflow accurately is critical for proper planning and management of water resources. The capability of three distinct data-driven approaches for forecasting the streamflow in Rushikulya River Basin of Odisha was investigated in this study and for this purpose artificial neural network (ANN), adaptive neuro fuzzy inference system (ANFIS), and support vector machine (SVM) models were used. The performance of these models is compared to observed data in training and validation sets. The model with the optimum fit for streamflow prediction has been chosen, which includes three antecedent flow values. In order to get a more precise assessment of the results of the three models, four standard statistical performance indicators such as mean squared error (MSE), Willmott index (WI) and Nash–Sutcliffe efficiency coefficient (NSE) were used. The results demonstrate that the performance of the machine learning models in terms of various evaluation criteria did not change considerably over the training and validation period; the performance of these soft computing methods in streamflow prediction was acceptable. In streamflow predictions, ANN and ANFIS models were outperformed by the SVM model, according to a comprehensive examination of overall performance.

A. Mishra · D. Panda · S. Samantaray (✉) · D. P. Satapathy
Department of Civil Engineering, OUTR Bhubaneswar, Bhubaneswar, Odisha, India
e-mail: sandeep1139_rs@civil.nits.ac.in

D. P. Satapathy
e-mail: dpsatapathy@cet.edu.in

N. Nayak
Land Development and Municipal Engineering, WSP Consultants India Private Limited, Bengaluru, India

S. Mishra
Department of Philosophy, Utkal University, Bhubaneswar, Odisha, India

© The Author(s), under exclusive license to Springer Nature Singapore Pte Ltd. 2023
V. Bhateja et al. (eds.), *Evolution in Computational Intelligence*, Smart Innovation, Systems and Technologies 326, https://doi.org/10.1007/978-981-19-7513-4_14

14.1 Introduction

Modelling of streamflow processes is critical for water resource planning, operation, and management. Accurate modelling is also critical for reducing the effect of droughts and floods. As a result, several hydrological models have been created to represent this intricate process. Despite the fact that technological advancements have substantially improved streamflow monitoring using low-cost sensors, isolated locations, and developing nations are still largely unserved by such low-cost equipment. Researchers have used several physical and empirical approaches to estimate streamflow at various temporal resolutions and geographical scales for decades. Nonetheless, the majority of these models are only useful at a basin, and they are depended on the geomorphological characteristics of the area in which they were used. That's why, the use of soft computing models has shown significant promise in a variety of hydrological and water resource subfields.

Besaw et al. [1] designed and evaluated two recurrent ANN models to anticipate streamflow in Northern Vermont's sub-basins. Climate data of a different basin was used to forecast the climate data of this basin. Because the model input consists of time lags of locally recorded meteorological data for recurrent flow prediction, it was discovered that time-lagged streamflow improves forecasts. The shift from one basin to another in terms of drainage is accounted for by a scaling ratio on the basis of a link between basin drainage area and bankfull discharge. For the smaller streams studied, hourly streamflow forecasts outperformed those based on daily data. Jeong and Kim [2] employed two types of ANN models: ENN and SNN to anticipate ensemble streamflow prediction (ESP) and enhance the modelling of rainfall-runoff process; the research was carried out South Korea's Daecheong multifunctional dam. Due to the current ESP system's strong probabilistic forecasting accuracy, assessed on the basis of hit rate, half-Brier score and average hit score. ENN proved to be the best model among these. Kagoda et al. [3] illustrated the use of RBF-ANN for forecasting 1-day streamflow in the Luvuvhu River in South Africa. This model has proven to be useful for predicting streamflow in regions where full climate data is unavailable such as developing countries. Makwana and Tiwari [4] compared performances of classical ANN model and wavelet ANN model for intermittent streamflow forecast in a semiarid watershed of Gujarat, India. In the first method testing data inside the training range was considered, while testing data outside the training range was evaluated in the second approach. The WANN model provided considerably greater outcomes than the ANN model. Even when testing data is beyond the training range, the WANN model was found to be adequate for streamflow prediction. Danandeh Mehr et al. [5] demonstrated that consecutive station prediction models may be used as an alternative to traditional streamflow forecast in irregular rain gauge catchments. Monthly streamflow records from two sites on Turkey's Oruh River were used in the study. FFBP was utilised to model prediction scenarios, and highest accuracy was obtained with 1 month lagged streamflow data. Then, RBF and GRNN models were used for 1 month ahead records and the results were compared with FFBP, which showed RBF as the best model in the study domain. Ali and Shahbaz [6] demonstrated daily river streamflow prediction using appropriate set of rainfall patterns and then developing an effective ANN(MLP) model for it. The research region was Pakistan's Upper Jhelum River Basin (UJRB). It was found that an ANN-based technique is an effective and helpful way to predict streamflow. The created MLP models may be utilised and adapted for various hydrology-related tasks. Wagena et al. [7] conducted real-time streamflow forecast by incorporating an ARMA model, a stochastic model (ANN), process-based model (SWAT-VSA), and a Bayesian ensemble model that utilises the ARMA, ANN, and SWAT-VSA results. Although, SWAT-VSA and ANN produced better streamflow prediction among the individual models; ANN had the highest predicting capability during the prediction phase. The ARMA model required less data but couldn't handle complex simulations. It was concluded that no model was found to be ideal for streamflow prediction in different scenarios.

Dalkiliç and Hashimi [8] investigated the capabilities of ANN, ANFIS, and wavelet NN models for forecasting of daily streamflow in the region of Büyük Menderes Basin, Greece. The WNN model had the greatest performance of all the models since it used the strategy of dividing the dataset into sub-series and removing their noises. Yaseen et al. [9] utilised the ANFIS-FFA model to estimate monthly streamflow in Malaysia's Pahang River Basin. When the hybrid model's findings were compared with the traditional ANFIS model. The ANN-FFA model excelled the standalone ANFIS model because FFA improved ANFIS' predicting accuracy. Dariane and Azimi [10] created a data-driven model for streamflow forecasting with binary GA as the input selection method. Fuzzy C-means (FCM)-ANFIS and Subtractive (Sub)-ANFIS were the two types of ANFIS models employed. Wavelet transformations were also used in conjunction with the models. The research region was the Iranian basins of Lighvan and Ajichai. When ANFIS was paired with GA and wavelet transform, significant improvements were obtained. Adnan et al. [11] assessed the monthly streamflow forecast accuracy of the M5 regression tree (M5RT), ANFIS-GA, and ANN-GA models in the Neelum and Kunhar Rivers area of Pakistan. The ANFIS-GA and ANN-GA models beat the M5RT model, and periodicity was shown to improve the model's prediction performance. Anusree and Varghese [12] for Karuvannur River Basin, Thrissur, India calculated daily streamflow using ANN, ANFIS, and MNLR by considering precipitation data from nine rain gauge stations. Simulation input vectors included various time delays of antecedent precipitation and flow combinations. Dariane and Azimi [13] used monthly hydrological data from the Ajichai Basin in Iran for evaluating the performance of ANN-BP and ELM training methods of RBF-ANN. Because ELM is ineffective with numerous input variables, an input selection approach such as wavelet transform, GA, or SSA was used. To produce the projected flow, a hybrid model is utilised, in which the ANN-BP and ANN-ELM model outputs were used as inputs to the ANFIS model. The findings demonstrate the relevance of input selection as well as the ascendancy of SSA and ELM over BP and wavelet transforms. Khadangi et al. [14] incorporated ANN-RBF and ANFIS to forecast daily river flow in Mahabad River at Mahabad Dam station, Iran. On the basis of the results obtained, it was observed that ANFIS model had higher accuracy and was more suitable. Hadi and Tombul [15] examined the potency of SVM, ANFIS, and ANN in daily streamflow prediction with an autoregressive model, for basins with varying physical features. A comparison of accuracy was made across three basins in Turkey's Seyhan River Basin. Among all of the models discussed, ANN proved to be the most superior. Khazaee Poul et al. [16] adopted MLR, ANN, ANFIS, and KNN for monthly flow estimation in St. Clair River between Canada and US. These models were evaluated individually and coupled with wavelet approach, respectively. Results indicated that ANFIS was the most accurate model. Zamani Sabzi et al. [17] proposed using ARIMA, ANN, ANN-ARIMA, and ANFIS for monthly and daily prediction of streamflow on Elephant Butte Reservoir, New Mexico. The most accurate model turned out to be ANFIS.

Meng et al. [18] used M-EMD SVM, EMD-SVM, WA-SVM, SVM, and ANN models to forecast monthly streamflow in Wei River Basin in China. The results of this research demonstrated that the suggested modified EMD-SVM model was more accurate than a single SVM model in forecasting strong non-stationary streamflow. Maity et al. [19] evaluated SVR and Box–Jenkins approach models for monthly streamflow forecasting in Mahanadi River, Orissa, India. For parameter calibration and model construction, LS-SVM was utilised. In the prediction of monthly streamflow, SVR beat the Box–Jenkins technique. Jajarmizadeh et al. (2014) compared the outputs of SVM with SWAT to estimate monthly streamflow, in the Roodan watershed in Iran. (Gamma test) GT-SVM approach was utilised to find the optimal input combinations, and regression (Reg-SVM) was employed to evaluate the gamma test's capabilities. The GT-SVM model's performance was then compared with LLR model's performance. In terms of streamflow prediction, SWAT and GT-SVM were shown to offer good results. Noori et al. (2011) investigated the influence of evaluation techniques such as forward selection (FS), gamma test (GT), and principle component analysis (PCA) approaches on the performance of SVM for monthly streamflow prediction. Following that the GT-SVM and PCA-SVM models were contrasted with PCA-ANN model. Result clearly showed that using evaluating techniques increased the accuracy of classical SVM model, and PCA-SVM was the best performing model. Rauf et al. (2018) used three ANNs and four SVR models to forecast streamflow in the Upper Indus River. The results of three distinct optimization strategies, including, and back propagation algorithms, conjugate gradient, and Broyden–Fletcher–Goldfarb–Shannon were compared. These ANNs were compared against four different types of SVR models, including sigmoid, radial basis function, polynomial, and linear kernels. The Broyden–Fletcher–Goldfarb–Shannon-ANN model outperformed all others, whereas the radial basis function kernel-based SVR model predicted stream flows with greater accuracy than the other kernels. The objective of the study is to predict stream flow prediction in Rushikulya River Basin, India.

14.2 Study Area

The Rushikulya River Basin, located in the Odisha districts of Ganjam and Kandhamal and stretching between 19.07 and 20.19 north latitude and 84.01–85.06 east longitude, is taken into consideration in this paper. It is one of the state's most important river basins, with an 8963-km catchment area. The Rushikulya River Basin is formed by the Rushikulya River which originates in the Daringbadi Hills of the Eastern Ghats range. Badanadi, Dhanei, Ghodahado, Padma, and Baghua are some of its important tributaries. In Ganjam district, it drains into the Bay of Bengal. The Rushikulya River has an average annual water discharge of 1800 million m^3, with discharge being extremely seasonal owing to monsoon conditions (Fig. 14.1).

14.3 Methodology

14.3.1 ANN

Artificial neural networks (ANNs) are the systems that process information, having the capacity to emulate the neural system of humans by modelling a structure that maps complicated nonlinear correlations processes that are intrinsic among numerous governing factors. Basically, it is a type of nonlinear regression model that includes a collection of weights to perform a mapping of the input–output. A feed-forward neural network comprises an input layer, one or more hidden layers, and an output layer. This technique has been proven to be quick and effective in modelling complicated relationships among variables, even in noisy contexts, and has been used to tackle a number of real-world issues. Because of its benefits, ANNs have been widely employed in a variety of real-world applications, such as time series forecasts. The output node of an ANN model is shown below:

$$\gamma_n = \sum_{j=1}^{J} \omega_{jk} \phi \left(\sum_{i=1}^{n} \omega_{ij} \lambda_i \right) \qquad (14.1)$$

where the weights of the connections whose values are tuned during training are represented by ω_{ij} and ω_{jk}, and γ represents sigmoidal function; The number of hidden and input layers are j and n, respectively, and model input variable is λ_i. . Readers interested in a more extensive discussion of the generic features of ANNs might see Bishop (1995) (Fig. 14.2).

14.3.2 ANFIS

Jang (1993) for measuring any real continuous function proposed ANFIS as an universal estimator. The FIS is classified on the basis of consequent (or "then") component into three categories, i.e. Mamdani, Tsukamoto, and Sugeno. Sugeno FIS

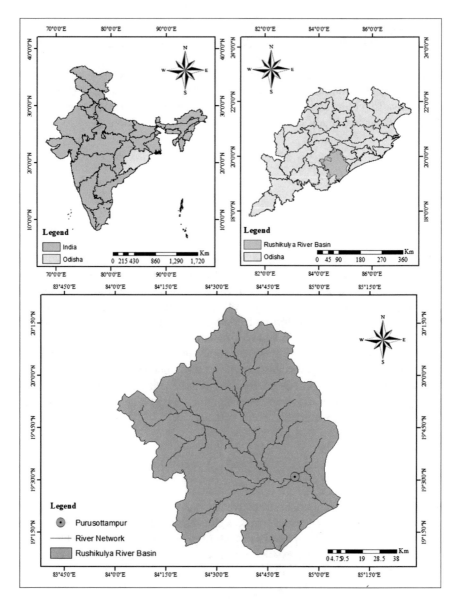

Fig. 14.1 Study area: Purusottampur station, Rushikulya River Basin

is used in this study to determine the consequent part. The hybrid learning algorithm of ANFIS consists of back propagation, gradient descent, and least-squares methods. In ANFIS system, the output is y and the inputs are y_1 and y_2.

For first order Sugeno FIS the if–then rules are represented by

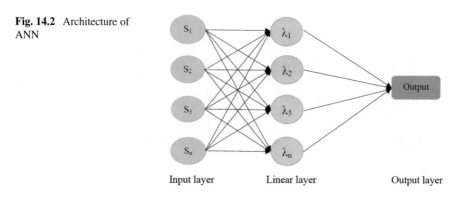

Fig. 14.2 Architecture of ANN

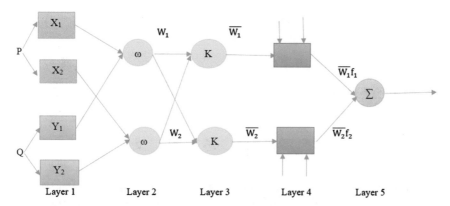

Fig. 14.3 Architecture of ANFIS

$$\text{Rule1 : If } y_1 \text{ is } P_1 \text{ and } y_2 \text{ is } Q_1, \text{ then } \gamma_1 = l_1 y_1 + m_1 y_2 + n_1, \quad (14.2)$$

$$\text{Rule1 : If } y_2 \text{ is } P_2 \text{ and } y_2 \text{ is } Q_2, \text{ then } \gamma_2 = l_2 y_1 + m_2 y_2 + n_2, \quad (14.3)$$

where l, m, and n are the linear parameters determined by ANN in the model's consequent section. P_i and Q_i represent the linguistic labels obtained by a membership function (Fig. 14.3).

14.3.3 SVM

SVM is a two-layered supervised learning model based on an artificial intelligence system. In the first layer, the weights are nonlinear, whereas in the second, they are linear. SVM can be applied both to regression and classification problems. The mathematical function utilised in regression problems is represented as

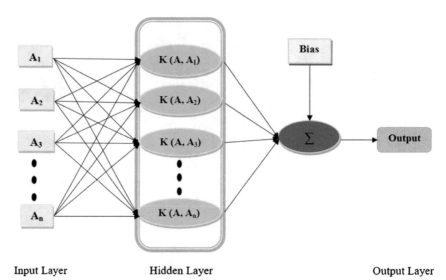

Fig. 14.4 Architecture of SVM

$$\gamma = f(x) = \left\{ \sum_{i=1}^{N} \alpha_i K(x_i, x) \right\} - \beta \qquad (14.4)$$

where K represents kernel function, number of training data points are represented by N, x is an independent vector, x_i denotes the vectors utilized in the training procedure, and the parameters α_i and β are the results of the objective function maximisation. Kernels that are often used are sigmoid kernels, RBF kernels, polynomial kernels, and linear kernels. For further information, refer Vapnik (1995) (Fig. 14.4).

14.3.4 Performance Criteria and Evaluation Methods

Here, 48 years [1974–2021] of rainfall, temperature and streamflow discharge data for monsoon season were collected from Central Water Commission, Bhubaneswar. From which the first 36 years of data were considered for training and the last 12 years of data were considered for testing purpose. For assessment of model three evaluating constraints such as mean squared error (MSE), Nash–Sutcliffe efficiency coefficient (NSE), and Willmott index (WI) were used.

$$\text{MSE} = \frac{1}{N} \sum_{k=1}^{N} (Q_P - Q_o)^2 \qquad (14.5)$$

$$WI = 1 - \left[\frac{\sum_{k=1}^{N}(Q_o - Q_P)^2}{\sum_{k=1}^{N}(|Q_P - \overline{Q_o}| + |Q_o - \overline{Q_o}|)^2} \right] \tag{14.6}$$

$$NSE = 1 \left[\frac{\sum_{k=1}^{N}(Q_p - Q_0)^2}{\sum_{k=1}^{N}(Q_0 - \overline{Q_o})^2} \right] \tag{14.7}$$

where Q_P = Predicted variables, Q_O = Observed variables, $\overline{Q_O}$ = Mean observed variables.

14.4   Results and Discussion

In this section, the performance of the algorithms is tested on the basis of various statistical indicators like MSE, WI, and NSE. The statistical results of the algorithms and the best solutions are expressed in Table 14.1. Table 14.1 reveals that SVM model outperforms ANFIS and ANN models for all the input combinations during training and testing phase. The best value of WI and NSE is 0.9936, 0.9889 and 0.95483, 0.95108 during training and testing phases, respectively, whereas the minimum value of MSE is 0.0067 and 4.089 in case of SVM.

Table 14.1 Results of ANN, SVM, and ANFIS

Technique	Model	Training			Testing		
		MSE	WI	NSE	MSE	WI	NSE
ANN	F*i	11.9043	0.948	0.93358	15.0837	0.9122	0.90827
	F*ii	11.678	0.9498	0.93481	14.7839	0.9139	0.90991
	F*iii	11.0205	0.94006	0.93592	13.9841	0.91468	0.91043
	F*iv	10.9032	0.94118	0.9366	12.564	0.9157	0.91138
	F*v	10.88	0.9426	0.93714	12.0083	0.91692	0.91236
ANFIS	F*i	3.1185	0.9749	0.97005	8.2008	0.94661	0.9436
	F*ii	2.8007	0.97518	0.97042	7.874	0.9474	0.94488
	F*iii	2.486	0.976	0.97166	7.2503	0.94862	0.94501
	F*iv	2.0431	0.97773	0.9728	6.743	0.94917	0.9469
	F*v	1.896	0.9781	0.97393	6.2896	0.9499	0.94772
SVM	F*i	1.0742	0.9872	0.98114	5.5004	0.95184	0.9493
	F*ii	0.924	0.988	0.9837	5.3685	0.95299	0.95005
	F*iii	0.1159	0.98917	0.98481	5.1987	0.9537	0.95032
	F*iv	0.0183	0.99182	0.98603	4.6008	0.95392	0.95067
	F*v	0.0067	0.9936	0.9889	4.089	0.95483	0.95108

In Fig. 14.5, a scatter plot depicts the observed and anticipated monthly streamflow for the best input combinations (f*v) data using the SVM, ANFIS, and ANN models. It can be observed from the scatter plot for the SVM model; the fit line coefficients a and b (assuming the fit line equation as $y = ax + b$) are nearer to 1 and 0, respectively, having a greater R^2 value compared to that of the ANN and ANFIS models.

Additionally, the histogram plot graphs of the algorithms during testing phases are depicted in Fig. 14.6. Result shows that SVM model found better than the ANFIS and ANN model.

According to the graphical and statistical results, it can be found that SVM algorithms obtained optimal solution of the problem than the ANFIS and ANN.

Fig. 14.5 Scatter plot between actual and predicted streamflow

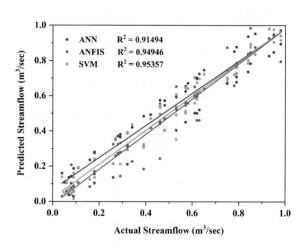

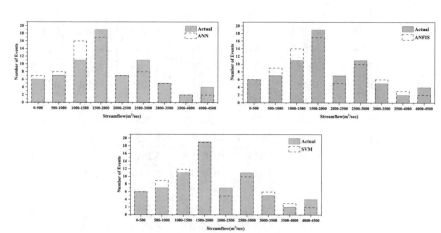

Fig. 14.6 Histogram plot between actual and predicted streamflow

14.5 Conclusion

In this paper, ANN, ANFIS, and SVM were used to model development for monthly streamflow data of the Rushikulya River Basin, Odisha, India. For the SVM Model, the kernel functions used were sigmoid and Gaussian. The results obtained from all the models were compared and found that SVM model showed the highest efficacy than rest of the models. Although the ANFIS and ANN produced fairly good results, ANFIS was able to produce better prediction in terms of variability and efficiency due to fuzzy inference system. The MSE values for the three models were 10.88, 1.896, and 0.0067 while the NSE values were 0.93714, 0.97393, and 0.9889, and WI values were 0.9426, 0.9781, and 0.9936 for ANN, ANFIS, and SVM model, respectively, which clearly indicate that SVM model is most accurate. Thus, it can be concluded that SVM is the most suitable to conduct data estimation of streamflow and can also be extended for other hydrological problems.

References

1. Besaw, L.E., Rizzo, D.M., Bierman, P.R., Hackett, W.R.: Advances in ungauged streamflow prediction using artificial neural networks. J. Hydrol. **386**, 27–37 (2010)
2. Jeong, D.I, Kim, Y.O.: Rainfall-runoff models using artificial neural networks for ensemble streamflow prediction. J. Hydrol. Process. **19**, 3819–3835 (2005)
3. Kagoda, P.A., Ndiritu, J., Ntuli, C., Mwaka, B.: Application of radial basis function neural networks to short-term streamflow forecasting. J. Phys. Chem. Earth **35**, 571–581 (2010)
4. Makwana, J.J., Tiwari, M.K.: Intermittent streamflow forecasting and extreme event modelling using wavelet based artificial neural networks. J. Water Resour. Manag. **28**, 4857–4873 (2014)
5. Danandeh Mehr, A., Kahya, E., Şahin, A. and Nazemosadat, M.J.: Successive-station monthly streamflow prediction using different artificial neural network algorithms. Int. J. Environ. Sci. Technol. **12**, 2191–2200 (2015)
6. Ali, S., Shahbaz, M.: Streamflow forecasting by modeling the rainfall–streamflow relationship using artificial neural networks. J. Model. Earth Syst. Environ. **6**, 1645–1656 (2020)
7. Wagena, M.B., Goering, D., Collick, A.S., Bock, E., Fuka, D.R., Buda, A., Easton, Z.M.: Comparison of short-term streamflow forecasting using stochastic time series, neural networks, process-based, and Bayesian models. J. Environ. Model. Softw. **126**, 104669 (2020)
8. Dalkiliç, H.Y., Hashimi, S.A.: Prediction of daily streamflow using artificial neural networks (ANNs), wavelet neural networks (WNNs), and adaptive neuro-fuzzy inference system (ANFIS) models. J. Water Sci. Technol. Water Supply **20**, 1396–1408 (2020)
9. Yaseen, Z.M., Ebtehaj, I., Bonakdari, H., Deo, R.C., Mehr, A.D., Mohtar, W.H.M.W., Diop, L., Singh, V.P.: Novel approach for streamflow forecasting using a hybrid ANFIS-FFA model. J. Hydrol. **554**, 263–276 (2017)
10. Dariane, A.B., Azimi, S.: Forecasting streamflow by combination of a genetic input selection algorithm and wavelet transforms using ANFIS models. J. Hydrol. Sci. J. **61**, 585–600 (2016)
11. Adnan, R.M., Liang, Z., Kuriqi, A., Kisi, A., Malik, A., Li, B.: Streamflow forecasting using heuristic machine learning methods. In: 2020 2nd International Conference on Computer and Information Sciences ICCIS 2020, pp. 5–10 (2020)
12. Anusree, K., Varghese, K.O.: Streamflow prediction of karuvannur river basin using ANFIS, ANN and MNLR models. J. Procedia Technol. **24**, 101–108 (2016)
13. Dariane, A.B., Azimi, S.: Streamflow forecasting by combining neural networks and fuzzy models using advanced methods of input variable selection. J. Hydroinformatics **20**, 520–532 (2018)

14. Khadangi, E., Madvar, H.R., Ebadzadeh, M.M.: Comparison of ANFIS and RBF models in daily stream flow forecasting. In: 2009 2nd International Conference on Computer Control Communication IC4 2009 (2009)
15. Hadi, S.J., Tombul, M.: Forecasting daily streamflow for basins with different physical characteristics through data-driven methods. J. Water Resour. Manag. **32**, 3405–3422 (2018)
16. Khazaee Poul, A., Shourian, M., Ebrahimi, H.: A comparative study of MLR, KNN, ANN and ANFIS models with wavelet transform in monthly stream flow prediction. J. Water Resour. Manag. **33**, 2907–2923 (2019)
17. Zamani Sabzi, H., King, J.P., Abudu, S.: Developing an intelligent expert system for streamflow prediction, integrated in a dynamic decision support system for managing multiple reservoirs: a case study. J. Expert Syst. Appl. **83**, 145–163 (2017)
18. Meng, E., Huang, S., Huang, Q., Fang, W., Wu, L., Wang, L.: A robust method for non-stationary streamflow prediction based on improved EMD-SVM model. J. Hydrol. **568**, 462–478 (2019)
19. Maity, R., Bhagwat, P.P., Bhatnagar, A.: Potential of support vector regression for prediction of monthly streamflow using endogenous property. J. Hydrol. Process. **24**, 917–923 (2010)

Chapter 15
A Survey on Medical Image Analysis Using Deep Learning

Trishaani Acharjee, Roshni Pradhan, Amiya Kumar Dash, Suresh Chandra Satapathy, and Milan Simic

Abstract Alzheimer's disease can cause permanent damage to the memory cells and thus its early detection and diagnosis is essential. Over the years the complexity of medical imaging has increased considerably and thus made it difficult for the physicians or radiologists to provide accurate diagnosis all the time. We require highly discriminative characteristics extracted from MRI scans to accurately classify dementia stages. Although in the recent years the deep learning based models especially CNN models have helped a lot in enhancing the accuracy, but the number of image samples available is very less, which may lead to overfitting. An inaccurate or delayed diagnosis can have a great impact on the patient's health as well as the expert's reputation. Thus computer technologies are extensively used to help give out a proper diagnosis. In this paper, we are providing a vivid discussion of three such research works where deep learning, especially convolutional neural network (CNN) is used for accurate detection of Alzheimer's Disease.

T. Acharjee (✉) · R. Pradhan · A. K. Dash · S. C. Satapathy
Kalinga Institute of Industrial Technology University, Bhubaneswar 751024, India
e-mail: ani.sha.tri.98@gmail.com

R. Pradhan
e-mail: roshni.pradhanfcs@kiit.ac.in

A. K. Dash
e-mail: amiya.dashfcs@kiit.ac.in

S. C. Satapathy
e-mail: suresh.satapathyfcs@kiit.ac.in

M. Simic
STEM, School of Engineering, RMIT University, Melbourne, Australia
e-mail: milan.simic@rmit.edu.au

15.1 Introduction

Medical imaging has been conducted by physicians and radiologists all through the ages in clinics or hospitals. However, over the years the medical image complexity has increased so much that it has become increasingly difficult for physicians or radiologists to provide accurate diagnoses all of the time [1]. A wrong or delayed diagnosis can harm the patient and also the expert's reputation. Thus, computer technologies are extensively used to help give out proper diagnosis [2]. Artificial intelligence (AI), machine learning (ML), the Internet of Things, data mining, and cloud computing have all grown in popularity as a result of the advancement of cutting-edge innovation and a massive change has been noticed in the medical imaging field. This has led to the growth of smart medicine which offers personalized medical services to all. Some of the medical imaging techniques have been used over the time of some decades to detect and diagnose and provide proper treatments "Computerized Tomography (CT), mammography, Magnetic Resonance Imaging (MRI), ultrasound, Positron Emission Tomography (PET) and X-ray" are only a few examples [3, 4].

Learning algorithms based on machine learning have power to profoundly involve in all aspects of medicine, such as researching on certain drugs, decision making, completely transforming the way medicine is carried out [2]. The later victory of machine learning algorithms for computer vision assignments comes at a basic crossroads since medical records are continuously getting to be digitalized [2]. Artificial Intelligence innovations will upgrade medical results and raise survival rates.

There are several forms of medical imaging, including magnetic resonance imaging, X-ray, PET or positron emission tomography, ultrasound (US), and CT scans or computed tomography, histology slides, retinal photography, and dermoscopy images are all examples of digital medical images [2].

An example of medical image in retinopathy is that, for over last few years the automated analysis of retinal colour images for DR has been studied by many groups, and although finding widely accepted, well characterized dataset is hard to find, studies shows that algorithms such as Lowa Detection Program (LDP), DR screening can be achieved almost accurately and safely [5].

The most often utilized deep learning model is convolutional neural networks, having the most success in medical picture processing to date. The fact that CNNs do not require feature engineering makes them a popular alternative to traditional machine learning methods like SVM, KNN, and logistic regression [2]. Due to their remarkable performance in medical imaging and the fact that they can be parallelized with GPUs, CNNs have recently acquired appeal amongst medical imaging researchers. Until present, no survey articles on the application of CNNs in medical image processing had been published. As a result, we've prepared this survey paper to give you an overview of today's state-of-the-art CNNs in medical image processing. In this paper, we are going to discuss about the different techniques that these three groups of researcher have undertaken and also the results that they have obtained after implementing their algorithms.

15.2 Literature Survey

There were a lot of studies and research work conducted on this field. Out of which some resources that were highly helpful and recommended are: A comprehensive list of works published in the field is provided by Litjens et al. [6]. Justin Ker et al. [2] and Guang-Di Liu [7] correctly summarize many topics and advancements in the field. Michael David Abramoff et al. [5] provides a thorough study of detec- tion of diabetic retinopathy with integration of deep learning. Dinggang Shen et al. [3] and Justin Ker [2] provides a detailed analysis of medical imaging using deep learning.

We did some thorough research and observed significant papers using terms like "deep learning," "Convolution Neural Network," "Medical Imaging," and "Medical Image Analysis" on Google Scholar after a thorough search. We also came across a website for Biomedical Image Analysis Grand Challenges (https://grand-challe nge.org/all challenges [8]) which is a website that collects and links to a variety of competitions and image datasets [3].

Some of the authors are practicing surgeons and radiologists and this paper aims to give a general summary of the current situation of machine learning algorithms in medical imaging, with an emphasis on which features are most valuable to the clinician [4] which might be a topic of interest for the practicing professionals in the field. This frame of view is meant to help researchers move away from the local minima of speculative research and towards practical answers that will have an impact on medical research and patient treatment.

15.3 Discussion

As found during the survey the different methodology that has been used over the years in this field of medical imaging has brought about a revolution in this field.

In the research paper by Atif Mehmood et al. [9] "A deep Siamese Convolu- tion Neural Network for multi-class classification of Alzheimer Disease" [9], they found out a suitable technique that gives almost accurate results for the detection of Alzheimer disease. Alzheimer's disease has the potential to permanently destroy memory cells, ultimately leading to dementia [9]. Overfitting occurs as a result of the lesser amount of picture samples available in the datasets, lowering performance. They developed a Siamese convolution neural network (SCNN) model based on VGG16 to diagnose dementia phases [9].

In this approach, the algorithm is based on three steps-data preprocessing and augmentation, feature extraction from the input images, and classification of dementia classes. For the classification of dementia phases, the approach is mostly CNN-based, but it was inspired by VGG16 [10]. They added an extra convolution layer to the model, which helped them obtain the most features out of a tiny dataset. Two adjusted VGG16 layers worked in couple with 14 Convolution layers, three

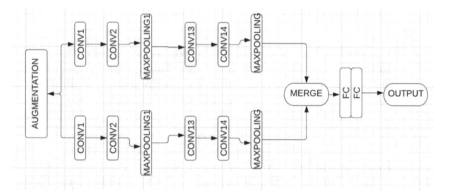

Fig. 15.1 Proposed Siamese convolutional neural network (SCNN) model for the classification of Alzheimer's

batch normalization, five max-pooling and three Gaussian noise layers in the algorithm. The parallel work to extract the more significant information is the reason for the model. The classification accuracy was increased by using multiple parallel layers [9] (Fig. 15.1).

15.3.1 Data Selection

In their study, they analysed the OASIS open-access dataset [11]. They created four classes (Table 2) based on dementia ratings (CDR) which are: No Dementia (CDR-0), Very Mild Dementia (CDR-0.5), Mild Dementia (CDR-1), and Severe De-mentia (CDR-2) [9]. The patients ranged in age from 20 to 88 years old. They also used the augmentation technique, which helped the model learn faster. For all the methods that are constructed with accordance to the CNN model [9], data preprocessing is a critical step in obtaining efficient and accurate results [12].

15.3.2 Image Preprocessing

The MRI images generally goes through a lot of deterioration due to low brightness obtained from visual devices, inappropriate setting of the lens slit in the scanning device, and also uneven light distribution. To solve this issue, picture enhancement techniques were used to improve the pixel distribution across a wide range of intensities. Unwanted information (noise) was introduced into the image during the acquisition process. Contrast stretching was used to expand the dynamic range of light intensity since the output images after this operation had better contrast and proper light distribution [9, 13]. Public OASIS collection was used to collect the MRI images and segmented using K-mean clustering to regain the white matter (WM), grey matter (GM), and cerebrospinal fluid (CSF) concentrations [9].

15.3.3   Data Augmentation

As privacy is a main concern, access to substantial amounts of data is a big problem in medical research. The tiny, unbalanced dataset causes overfitting issues during model training, lowering model efficiency. To solve this problem, they needed increased amount of data to improve the precision of their suggested model. As a result, they employed the augmentation approach to add 10 extra pictures to each MRI scan [9, 14].

15.3.4   Convolution Neural Network

In CNN, the preceding layer's output is combined with a trainable kernel, as well as weight sharing is used to reduce the number of weights in training. Generally convolution layer can be calculated using a formula which is as follows:

$$\text{Height} = \frac{\text{Image height} - \text{kernel Height} + 2(\text{Padding})}{\text{strides}} + 1, \qquad (15.1)$$

$$\text{Width} = \frac{\text{Image width} - \text{kernel width} + 2(\text{Padding})}{\text{strides}} + 1, \qquad (15.2)$$

$$C = \frac{W - K + 2P}{S} + 1, \qquad (15.3)$$

The image height and width are denoted by W, filter size is represented by K, padding is represented by P, and strides are represented as S. The pooling layer, usually max-pooling layer, is introduced between the convolution layers. This is meant to down sample which then minimizes the level of computational difficulty. The convolutional layer's the feature map's result was further separated into tiny areas, with each of the region describing the region's value [9]. The backpropagation approach was utilised to minimise the cost function when training the CNN, and the weights of each node are updated on a regular basis. They used the terms "random uniform" and "zeros" to describe the kernel and bias initializers in the model [9]. Using the sequential strategy, they developed a design concept with the rectified linear unit activation function (ReLU) on a layer-by-layer basis [9].

15.3.5 Regularization and Increased Learning Rate

Because the input of each layer altered as the parameters of the preceding layer changed, the CNN model's training procedure was difficult. Activation functions like sigmoid and ReLU, on the other hand, lose their gradient quickly, making deep neural network learning challenging and lowering the model's learning rate. To get over this, a proposed model batch normalization was used [9, 15].

$$y_i = BN_{\gamma,\beta}(x_i), \tag{15.4}$$

$$\mu_b = \frac{1}{n}\sum_{i=1}^{n} x_i, \tag{15.5}$$

$$\sigma_b^2 = \frac{1}{n}\sum_{i=1}^{n}(x_i - \mu_b)^2, \tag{15.6}$$

$$\overline{x_i} = \frac{x_i - \mu_b}{\sqrt{\sigma_b^2 + \epsilon}}, \tag{15.7}$$

$$y_i = \gamma * \overline{x_i} + \beta, \tag{15.8}$$

where the number of batches is indicated by the letter n, μ denotes mean and σ^2 denotes variance, x_i denotes each row. Equations (15.4)–(15.6) are used to calculate each activation's mean and variance over a little volume [9]. The two hyperparameters (γ and β), present in Equation (15.7), for each of the input dimensions, generates learnable scale parameters [9]. To improve the model's robustness and regularisation, they added Gaussian noise [16]. Gaussian noise provides excellent results while teaching the deep model also aids in the reduction of training loss [9].

15.3.6 Results Obtained for the SCNN Model

After preprocessing, they retrieved the characteristic of roughly 3820 samples. For validation of the proposed model's efficacy, they divided the samples into training and testing datasets. The testing dataset included 20% of the samples from each of the four classes, whereas the training dataset included 80% of the samples from each of the four classes [9]. They used the entire dataset as validation data, therefore the validation accuracy's final epoch result can be referred to as test accuracy or validation accuracy, and they terminated early to minimise overfitting. If the model would not have been early stopped, it would have lead to overfitting as the validation error starts to increase after a certain epoch which is harmful for the training process.

The suggested model's classification results were evaluated using several evaluation criteria, and they achieved a test accuracy of 99.05% [9].

The technology uses convolutional neural networks (CNN), with the input and output being two volumetric modalities, according to Rongjian Li et al. [17] research paper "Deep Learning Based Imaging Data Completion for Improved Brain Disease Diagnosis" [17]. They evaluated their proposed method on the Alzheimer's Disease Neuroimaging Initiative (ADNI) database [18], where the input modality is MRI images and output modality is PET images. They detected missing PET signals from MRI data using their 3D CNN model [17]. To train the model on the subject, they merged MRI and PET data, with MRI data's serving as the input and the PET data's serving as the outcome. A great number of variables in the trained network represent the MRI and PET data have a nonlinear relationship. They attempted to predict PET patterns using the trained network via a 3D CNN when the input MRI pattern was provided. And as a result, their model outperformed the prior methods of disease diagnosis.

15.3.7   Methodology Used

They used data from the Alzheimer's Disease Neuroimaging Initiative (ADNI) database [18] in their research. T1-weighted MRI was obtained for each patient by changing the intensity and then stripping the skull and removing the cerebellum. The white matter, grey matter, and cerebrospinal fluid regions of each magnetic resonance image were splitted and after that the data was geometrically normalised into a template space [17, 19–21]. In their research, they used the grey matter tissue density map. In order to further enhance the signal-to-noise ratio, PET pictures and density maps of grey matter tissue were normalized with a Gaussian kernel (with standard deviation of one unit). They reduced the processing cost by downsampling both the PET pictures and the grey matter tissue density maps to $60 \times 60 \times 60$ voxals [17].

The trainable filters are applied by the convolutional layer to the preceding layer's feature maps, while the subsampling layer reduces feature map resolution [17]. To achieve nonlinear mapping, 3D-CNNs, like 2D-CNNs, compute convolutions using 3D filters.

Assume the value on the jth layer of the jth feature map be $v_{ij} xyz$ in the occasion (x, y, z). Then the formula for the 3D convolution can be stated as:

$$V_{ij}^{xyz} = \sigma \left(b_{ij} + \sum_m \sum_{p=0}^{P_{i-1}} \sum_{q=0}^{Q_{i-1}} \sum_{r=0}^{R_{i-1}} w_{ijm}^{pqr} v_{(i-1)m}^{(x+p)(y+q)(z+r)} \right), \tag{15.9}$$

where the sigmoid function is denoted as $\sigma(\cdot)$, bias as b_{ij}, and the $(i-1)$th layer's set of feature maps that are related to the existing map of features is indexed by m, In three spatial dimensions, the 3D kernel's sizes are P_i, Q_i, and R_i, respectively, the filter's (p, q, r) th value implemented to the preceding layer's mth feature map is Wijm pqr. It's worth noting that Eq 15.9 [17] is a general three-dimensional convolution method that may be, with any number of feature mappings, can be employed on any layer of a three-dimensional CNN model.

Various sorts of CNN architectures can be created using this 3D convolution. The CNN model was trained using both MRI images and PET images. Because each procedure of convolution shrinks by a factor the size of the feature map along each axis connected to the filter size, the size of the input patch was regulated by the output patch's size [17]. To obtain the right PET imaging patches, they picked a massive amount of regions at random from each 3D MRI chunk.

They used 3D convolution with a 7 * 7 * 7 filter size on the patch of input on its first hidden layer, to make ten feature maps [17]. On the first concealed layer, they constructed ten feature maps using 3D convolution with a 7 * 7 * 7 filter size on the input patch [17]. To lower the computational cost, the size of the filter for transferring the last hidden layer's feature maps was similarly lowered in the output to 1 [17] (Fig. 15.2).

This CNN design has a good blend of representational capacity and computational expense. The richer the training data, the more layers and feature maps there are, but the computational cost for sophisticated networks is prohibitive. With error back-propagation, the weights of this network were changed using a stochastic gradient descent method. In all of the experiments, the learning rate was set to 10^{-2}, with the other variables kept at their default CNS package values [17, 22]. The network was trained over a number of epochs, with each epoch requiring each network to train once for each example. As the performance started to converge after almost 10 epochs, and as this training process was time consuming, they stopped the training after those 10 epochs. In total, the number of training patches they had were 398 * 50, 000 = 19.9 million [17].

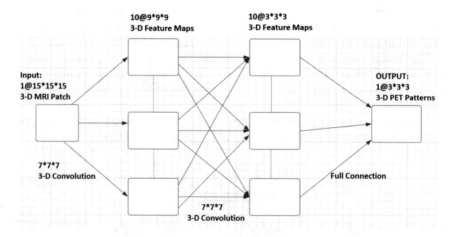

Fig. 15.2 The 3-D CNN architecture for imaging data completion used in this work. There are 2 hidden layers between the input and output layers. Each of the hidden layers contains 10 feature maps. The total number of trainable parameters in this network is 37,761

15.3.8   Results Obtained

The goal of the experiment was to see how well their 3D CNN model completed the missing PET data. They compared the predicted and real PET image data using a series of controlled trials. For the comparison, they didn't utilise any complex feature extraction or classification methods; instead, they employed basic standard methods to keep things simple and easy. They focus on the three binary-class classification tasks in their study (AD vs. NC, NC vs. MCI, and pMCI vs. sMCI), where MCI comprises both sMCI and pMCI. They compared this method to others such as the K-Nearest Neighbour and Zero Methods. The initial phase in their experiment is to fill in the gaps in the PET data with the help of either KNN, CNN, or Zero techniques [23]. The L 2-norm regularized logistic regression classifiers are used in the second stage to evaluate classification performance for all strategies using reconstructed data [17]. They performed the random partition 30 times to boost the performance estimate and presented the statistics computer throughout those 30 trials. As a result, they developed a single CNN model that they applied to all randomly select trials. They did the feature selection by deleting all of the voxals with a value of zero for all of the subjects. They visually matched the expected PET signals of each subject to the ground truth details. The figure illustrates the anticipated and ground truth results for two participants slice by slice [17]. According to their findings, the projected PET patterns matched the ground reality. This demonstrated that their strategy worked in locating the missing PET patterns. They objectively evaluated the suggested data completeness approach by categorizing the results using true and anticipated PET scans. They also provide classification results using the zero and KNN algorithms [17] (Fig. 15.3).

Based on their findings, they found that their model outperformed the KNN and Zero techniques by a wide margin. The substantial disparity in performance demonstrated that their method was able to uncover extremely nonlinear correlations between PET and MRI images. They also discovered that the 3D CNN model's results are equivalent to real PET images. This suggests that their anticipated PET scans could be utilized to improve disease diagnosis accuracy [17].

In the research paper by Ehsan Hosseini-Asl et al. [24] "ALZHEIMER'S DISEASE DIAGNOSTICS BY ADAPTATION OF 3D CONVOLUTIONAL NETWORK" [24], they proposed to predict Alzheimer Disease using deep 3D Convolution Neural Network [24]. Experiments were carried out on the CADDementia MRI dataset without the use of any preprocessing techniques, such as skull stripping. A 3D convo- lutional autoencoder was used to develop this 3D CNN model, which had been well trained formally to recognise structural shape variations in fundamental brain MRI data.

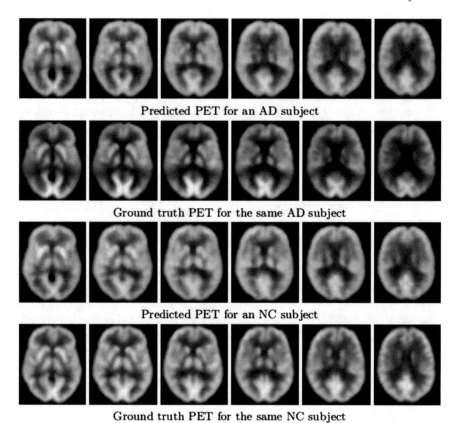

Predicted PET for an AD subject

Ground truth PET for the same AD subject

Predicted PET for an NC subject

Ground truth PET for the same NC subject

Fig. 15.3 Comparison of the predicted and the ground truth PET images on two subject. Each row corresponds to the data (either ground truth or predicted) of one subject, and each column corresponds to slice with the same drain position

15.3.9 Methodology for the 3D-CNN Model

The suggested approach for detecting Alzheimer's disease retrieves characteristics and attributes from a MRI of the brain by using a 3D-CAE which is trained by source domain and conducts classification which is specific to the task using a 3D-CNN that is target-domain-adaptable [24]. A conventional unsupervised autoencoder builds some scalar feature maps that are aligned by the combination of data encoding and decoding from a series of 3D photos given as input containing scalar or vectorial voxel-wise signals. The input image is encoded in the hidden layer by converting each fixed voxel region into a vectorial feature space, which is subsequently reconstructed to the original image space in the layer of output. To lower the reconstruction error, the auto encoder is programmed to extract characteristics that encapsulate various patterns of input data oscillations using backpropagation and limitations on feature space attributes. Because the number of features to be checked for the layers of input (encoding) and output (decoding) [25] is always growing. Using vectorial voxelwise signals to extract global characteristics from 3D images is computationally demanding and demands huge training data sets. [24]. Local features, rather than global features, are effective at pattern retricval from high-dimensional photos as learning is attempted using autoencoders with complete interconnections among all the nodes in the layers. They extracted local attributes from 3D photos utilising voxel-wise signal vectors that might be rather lengthy utilising unsupervised CAE with shared convolutional weights and local connected nodes [26–28]. Each CAE's hidden feature (activation) map is used to train the next-layer CAE, resulting in a hierarchical reduction of each input image [24].

To record the unique changes of a three-dimensional visual input, x, each voxel-by- voxel characteristic, $h_{i:j:k}$, correlated to the i-th node of the 3D lattice, the jth element of the signal vector input voxel-by-voxel and the k-th feature map ($k = [1, ...,$ $K]$) is retrieved by a convolution termed as moving-window (indicated underneath $*$) of the specified $n \times n \times n$ area, $x_{i: \text{neib}}$. This node has a linear encoding filtration system at- tached to it, which is controlled based on its weights, $W_k = [Wj: k: j = 1,$ $... , J]$ fol- lowed by feature-specific biases for every voxel-wise signal component j and each relative close position with reference to the node l, $b_k = [b_j, k: j = 1, ... ,$ $J]$ and also modifications that are nonlinear but have a specified activation function, $f(\cdot)$: [24].

$$h_{i:j:k} = f(W_k * X_{i:\text{neib}} + b_{j:k}) \tag{15.10}$$

Suppose $h_k = T (x: W_k, b_k, f())$ signify the whole of the encoding of the initial 3D picture and J-vectorial voxel-by-voxal signals with the k-th 3D feature map, h_k, in such a way that its scalar components are generated using Eq. (15.10) for a particular voxel area with the weights W_k and bias vectors b_k. $T_{\text{inv}}(...)$, Decodes or recon- structs the initial 3D picture using the original voxel region but various convolutional weights, P_k, biases, $b_{\text{inv}: k}$, and, perhaps, activation function, $g()$: [24].

$$\hat{x} = \sum_{k=1}^{K} \underbrace{T_{\text{inv}}(h_k : P_k, b_{\text{inv}:k}g(\cdot))}_{a_k} \tag{15.11}$$

Layers of L encoding are given, each layer l produces a picture of the output feature, $h(l) = [h(l): k: k = 1, ..., Kl]$, using voxel-wise Kl-vectorial characteristics and obtains the previous output, $h(l1) = [h(l1): k : k = 1, ..., Kl1]$, as the source image (i.e., $h(0) = x$. Eqs. (15.10) and (15.11) train the 3D-CAE by lowering given training input photos, the mean - square reconstruction error for T; $T 1$ (x [t]; $t = 1,$ ..., T) [24].

$$E(\theta) = \frac{1}{T} \sum_{t=1}^{T} \left\| \hat{x}^{[t]} - x^{[t]} \right\|_2^2 \tag{15.12}$$

where $\theta = [W_k; P_k; b_k; \text{bin}_{v:k:k=1}, ..., K]$, and $k ... k\ 2\ 2$ denotes across the T training photos, denotes the mean vectorial '2-norm and all unrestricted parameters, respectively [24]. To reduce the amount of unrestricted parameters, the decoding weight, Pk, and encoding weight, Wk by rearranging all of their proportions, they were able to come together, as detailed in [24]. A stochastic gradient descent approach coupled with backpropagation was used in order to bring down the cost of vector space of Eq. (15.12) [24].

The proposed classifier captures broad information using only a pile of locally connected bottom convolutional layers, with the fully connected top layers' parameters fine-tuned for job-specific fine-tuning [24]. The training procedure includes prior training, fundamental convolution layer training, and job-specific fine tuning.

The general feature extraction convolutional layers are created like a pile of 3D-CAEs that in the image recipient or source-domain had been trained previously during the preliminary step [24]. Deep supervision fine-tunes each sMRI classification's task specific binary or multi-class classification top fully connected layers once the layers have been initialised by encoding the weights of the 3D-CAE [24]. On the domain's data of the target, the weight values of the higher 3D-CNN layers that are fully coupled are fine-tuned by reducing a specified loss function to assign a task-specific classification to the collected characteristics. Given the properties of the input acquired from the target domain for training pictures by the network's bottom element that has been pre-trained, the loss is similar to the genuine output classes' negated log likelihood [24].

Each internal layer of the 3D-CNN employs ReLU activation functions, as well as totally connected higher tiers with a softmax layer at the top, to evaluate whether a sMRI scan of the brain as an input belongs to the AD, MCI, or NC groups (Fig. 15.4) [24].

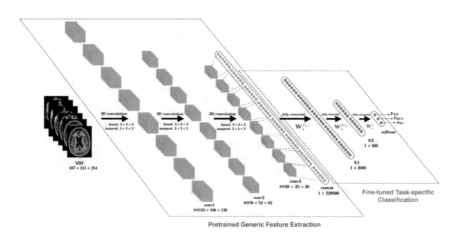

Fig. 15.4 The 3D adaptable CNN (3D*ACNN) for Alzheimer's disease diagnosis that they used

15.3.10 Results Obtained for the 3D-CNN Model

Utilizing CADDementia 1 as the input space, the implementation of the developed 3D-CAES for generalised feature extraction is assessed [24]. The database provides T1-weighted MRI (T1w) images of patients with possible AD, MCI, or NC without a dementia syndrome [24]. sMRI is spatially standardized using a strong registration approach to pretrain 3DCAES on CAD Dementia [24]. The cranium is then eliminated, and the image intensity levels are adjusted to [0, 1], yielding a sMRI with proportions of (200 150 150) [24]. In terms of accuracy, the proposed 3D-ACNN classifier was assessed and compared to competing systems for each task-specific classification (ACC) [24]. These trials show that the suggested in all 5 job-specific tests, 3D-ACNN beats the other techniques. Circumstances despite sMRI was the sole imaging modali ty used, and there was no previous skull-stripping [24].

15.4 Conclusion

We conducted a survey of convolution neural networks (CNN) for medical image processing in this article. According to the report, convolution neural networks play a significant role in computer-assisted diagnosis via evaluating medical images. Due to the lack labelled datasets in machine learning, it becomes difficult to train the models accurately which in turn reduces the performance and the accuracy of the models [2]. Attempts have been made to avoid limited data in computer vision tasks on deeper levels, by employing smaller filters, innovative combinations of CNN architectures, or hyperparameter adjusting [3]. Most of the dataset in this survey had very few patients involved. Despite the small dataset, the researchers have managed to get a relatively satisfactory result for their models. Although their approaches had some success, still there exists a challenge to find a relevant dataset for the purpose of training the models. Incorporating transfer learning into CNNs is a good potential area of research for tackling the problem of limited medical training data sets in future [2].

References

1. Alzheimer's Association.: 2019 Alzheimer's disease facts and fifigures. Alzheimer's Dement **15**, 321–387 (2019). [CrossRef]
2. Ker, J., Wang, L., Rao, J., Lim, T.: Deep learning applications in medical image analysis
3. Shen, D., Wu, G., Suk, H.-I.: Deep learning in medical image analysis
4. Weiner, M.W., et al.: The Alzheimers disease neuroimaging initiative: a review of papers published since its inception. Alzheimer's & Dementia **8**(suppl. 1), S1–S68 (2012)
5. Abramoff, M.D., Lou, Y., Ergnay, A., Claridia, W., Amelon, R., Folk, J.C., Niemeijer, M.: Improved automated detection of diabetic retinopathy on a public available dataset through integration of deep learning

6. Khatami, A., Khosravi, A., Nguyen, T., Lim, C.P., Nahavandi, S.: Medical image analysis using wavelet transform and deep belief networks
7. Liu, G.-D., Li, Y.-C., Zhang, W., Zhang, L.: A brief review of artificial intelligence applications and algorithms for psychiatric disorders
8. https://grand-challenge.org/allchallenges
9. Mehmood, A., Maqsood, M., Bashir, M., Shuyuan, Y.: A deep siamese convolution neural network for multi-class classification of Alzheimer disease
10. Arge, F.O.R.L.; Mage, C.I. V d c n l -s i r. 1–14 (2015)
11. OASIS Open- Access Dataset
12. Chincarini, A., Bosco, P., Calvini, P., Gemme, G., Esposito, M., Olivieri, C., Rei, L., Squarcia, S., Rodriguez, G., Bellotti, R.: Local MRI analysis approach in the diagnosis of early and prodromal Alzheimer's disease. Neuroimage **58**, 469–480 (2011). [CrossRef] [PubMed]
13. Ateeq, T., Majeed, M.N., Anwar, S.M., Maqsood, M., Rehman, Z.-U., Lee, J.W., Muhammad, K., Wang, S., Baik, S.W., Mehmood, I.: Ensemble-classififiers-assisted detection of cerebral microbleeds in brain MRI. Comput. Electr. Eng. **69**, 768–781 (2018). [CrossRef]
14. Bjerrum, E.J.: SMILES enumeration as data augmentation for neural network modeling of molecules. arXiv 2015, arXiv:1703.07076
15. Wiesler, S., Richard, A., Schl, R.: Mean-Normalized Stochastic Gradient for large-scale deep learning. In: Proceedings of the International Conference on Acoustics, Speech and Signal Processing (ICASSP), Florence, Italy, 4–9 May 2014, pp. 180–184
16. Boyat, A.K., Joshi, B.K.: A review paper: noise models in digital image processing. arXiv 2015, **6**, 63–75. [CrossRef]
17. Li, R., Zhang, W., Suk, H.-I., Wang, L., Li, J., Shen, D., Ji, S.: Deep learning based imaging data completion for improved brain disease diagnosis
18. ADNI Database
19. Turaga, S.C., Murray, J.F., Jain, V., Roth, F., Helmstaedter, M., Briggman, K., Denk, W., Seung, H.S.: Convolutional networks can learn to generate affinity graphs for image segmentation. Neural Comput. **22**(2), 511–538 (2010)
20. Jain, V., Seung, S.: Natural image denoising with convolutional networks. In: Koller, D., Schuurmans, D., Bengio, Y., Bottou, L. (eds.) Advances in Neural Information Processing Systems 21, pp. 769–776. Curran Associates, Inc. (2009)
21. Cire̦san, D.C., Giusti, A., Gambardella, L.M., Schmidhuber, J.: Mitosis detection in breast cancer histology images with deep neural networks. In: Mori, K., Sakuma, I., Sato, Y., Barillot, C., Navab, N. (eds.) MICCAI 2013, Part II. LNCS, vol. 8150, pp. 411–418. Springer, Heidelberg (2013)
22. Mutch, J., Knoblich, U., Poggio, T.: CNS: a GPU-based framework for simulating cortically organized networks. Technical Report MIT-CSAIL-TR-2010-013 / CBCL- 286, Massachusetts Institute of Technology, Cambridge, MA (February 2010)
23. Yuan, L., Wang, Y., Thompson, P.M., Narayan, V.A., Ye, J.: Multi-source feature learning for joint analysis of incomplete multiple heterogeneous neuroimaging data. Neuroimage **61**(3), 622–632 (2012)
24. Hosseini-Asl, E., Keynton, R., El-Baz, A.: Alzheimer's disease diagnostics by adaptation of 3D convolutional network
25. LeCun, Y., et al.: Gradient-based learning applied to document recognition. Proc. IEEE **86**(11), 2278–2324 (1998)
26. Masci, J. et al.: Stacked convolutional auto-encoders for hierarchical feature extraction. In: ICANN 2011, pp. 52–59. Springer (2011)
27. Makhzani, A., Frey, B.: A winner-take-all method for training sparse convolutional autoencoders. arXiv preprint arXiv:1409.2752 (2014)
28. Leng, B., et al.: 3D object retrieval with stacked local convolutional autoencoder. Signal Process. **112**, 119–128 (2015)

Chapter 16
A Descriptive Analysis of Reddit Comments Using Data Analytics Approach

Souvick Palit, Chittaranjan Pradhan, and Ahmed A. Elngar

Abstract We can use data from social media to examine and uncover links between positive and negative emotions. The findings of this article can be applied to a variety of subjects, including psychology and sociology. The goal was to provide insight into various observations made during our analysis. For this, we used Reddit comments as a reference point and data analytics to obtain various results. Our contributions to this paper are to grasp the concept of emotions and appropriately evaluate the data and analyzing Reddit comments and attempting to answer questions regarding the data acquired and discussing potential improvements to our analysis for future works in the field. We selected Reddit since it includes roughly 50 default subreddit themes that are accessible on the home page, including news, gaming, movies, music, and many others. Furthermore, redditors have the option of creating their own subreddit. Moreover, Reddit is one of the world's most prominent social networking websites. This makes it appropriate for our needs because it has a large number of user comments, which we can simply retrieve using the easy to use Pushshift API. The acquired comments data was then cleaned and analyzed using R in Rstudio and results were generated. Based on the results, conclusions were generated like correlations and mean values.

16.1 Introduction

Emotions fascinate us. Facial, verbal, or gestural expressions of our cognitive state are how we convey emotions as humans. The topic of emotional computing is concerned with the research and development of technologies that can identify these emotions. A

S. Palit (✉) · C. Pradhan
Kalinga Institute of Industrial Technology, Bhubaneswar, India
e-mail: souvickpalit@gmail.com

A. A. Elngar
Beni-Suef University, Beni-Suef City, Egypt

College of Computer Information Technology, American University in the Emirates, Dubai, United Arab Emirates

great deal of emotional expressiveness may be found in written language. A novelist's mastery of their profession may often be apparent in the way they use emotions in their words to fascinate their audience. With the recent evolution of social media, there has been a tremendous increase in the number of online interactions that are expressive in terms of emotion. Today's Internet users have a variety of ways to express themselves through text, including acronyms, emoticons, and hashtags [1].

Paul Eckman, a psychologist, identified six basic emotions that he claimed were common in all human civilizations in the 1970s. Disgust, surprise, happiness, fear, anger, and sadness were among the feelings he identified. Later, he added embarrassment, pride, excitement, and shame to his list of primary feelings.

Psychologists differ over whether the initial six basic emotions are genuinely fundamental, meaning that they underpin all other emotions. They have also claimed that fundamental emotions are not actually emotions, but rather building blocks for more complex psychological states, and that it's these more complex states that get the moniker "emotions" [2].

There are five components of emotions, according to Scherer [3]:

- Neurophysiological
- Motor expression
- Subjective feeling
- Motivational
- Cognitive.

Emotions are responsible for new action inclinations, new kinds of motivation, and a wide range of behaviors that have evolved throughout the history of human development to cope with or manage the environment and life's demands [4].

16.2 Background

16.2.1 Previous Work

A lot of work has been previously conducted in this field. A report was delivered on the performance of three publicly accessible word-emotion lexicons (EmoSenticNet, NRC, and DepecheMood) on a set of Facebook and Twitter communications [5]. They also created and developed an algorithm that incorporates natural language processing methods as well as a set of heuristics that mimic how humans intuitively analyze emotions in written texts. A research conducted by *Gunilla Widen, Johanna Lindstrom, Malin Brannback, Isto Huvila, Anna-Greta Nystrom* [6] states that they conducted an exploratory media diary study among undergraduate students to identify media usage patterns. Nicholas Botzer, Shawn Gu, and Tim Weninger's research [7] examines moral judgments made on social media by recording moral judgments made in a specific Reddit forum.

16.2.2   Online Disinhibition Effect

Individuals say and do things online that they would not ordinarily say or do in person, according to everyday Internet users, therapists, and studies. They loosen up and express themselves more freely as they grow less confined. The phenomenon has been dubbed the "online disinhibition effect" since it is so ubiquitous. This disinhibition might work in two seemingly opposing ways. People may provide a lot of personal information about themselves. They reveal secret emotions, fears, and wants. They show extraordinary compassion and generosity by going out of their way to help others. This condition is known as benign disinhibition.

Disinhibition, on the other hand, is not necessarily a good thing. Insults, harsh judgments, fury, hatred, and even threats are all experienced. People also visit the dark side of the Internet, where they may encounter pornography, crime, and violence in places they would never visit in the real world. Toxic disinhibition is the term for this [8].

16.2.3   Pearson Product-Moment Correlation

A monotonic link between two variables is a sort of linear relationship. The Pearson product-moment correlation, which is a linear relationship between two continuous, random variables and is frequently abbreviated as "r", is most commonly connected with the phrase "correlation". The covariance of the variables may be used to show how related one continuous variable is to another continuous variable. In the same way that variance displays the variability of two variables when they fluctuate together, covariance depicts the variability of a single variable. The absolute value of covariance, on the other hand, is difficult to analyze or compare between studies since it is dependent on the measurement scale of the variable. A Pearson correlation coefficient is commonly used to make understanding easier. This coefficient is a scaled, dimensionless measure of covariance with a range of -1 to $+1$ [9].

$$r = \frac{\sum (x_i - \overline{x})(y_i - \overline{y})}{\sqrt{\sum (x_i - \overline{x})^2 \sum (y_i - \overline{y})^2}} \tag{16.1}$$

where $r \longrightarrow$ correlation coefficient, $x_i \longrightarrow$ values of x-variable in a sample, $\overline{x} \longrightarrow$ mean of the values of the x-variable, $y_i \longrightarrow$ values of y-variable in a sample, and $\overline{y} \longrightarrow$ mean of the values of the y-variable.

16.2.4 Spearman's Rank-Order Correlation

It is a Pearson correlation coefficient which is calculated using the rankings of the values of the variables rather than the actual values of the variables. ρ is an abbreviation for a Spearman coefficient. Spearman coefficient is not limited to continuous variables because ordinal data can also be ranked. The coefficient quantifies monotonic interactions between two variables by employing rankings (data ranking translates a nonlinear strictly monotonic relationship into a linear relationship). Furthermore, because of this trait, a Spearman coefficient is generally resistant to outliers.

Spearman coefficient goes from -1 to $+1$. It can be used to describe anything from a perfect monotonic connection ($\rho = -1$ or $+1$) to no association ($\rho = 0$). The same rules apply to interpreting ρ values and confidence intervals for a Pearson correlation as they do for a Spearman coefficient [9].

$$\rho = 1 - \frac{6 \sum d_i^2}{n(n^2 - 1)} \tag{16.2}$$

where $\rho \longrightarrow$ Spearman's rank correlation coefficient, $d_i \longrightarrow$ Difference between the two ranks of each observation, and $n \longrightarrow$ number of observations.

16.2.5 Gamma Coefficient

The Goodman and Kruskal Gamma γ is an ordinal measure of association between two variables developed by Goodman and Kruskal [10]. It assesses the level of agreement or correlation between two ordinal-level datasets. There are two ways to look at Gamma γ. When there are no ties in the rankings, the first method is used, and when there are ties in the rankings, the second method is used [10].

$$\gamma = \frac{\sum f_a - \sum f_i}{\sum f_a + \sum f_i} \tag{16.3}$$

where $\gamma \longrightarrow$ Gamma, $f_a \longrightarrow$ frequency of agreements, and $f_i \longrightarrow$ frequency of inversions.

16.2.6 NRC Emotion Lexicon

The NRC Emotion Lexicon (EmoLex) was the first crowdsourced emotion lexicon. It has 14 k entries for English words. Each item has ten binary scores (0 or 1) showing whether or not there is a link between the eight fundamental emotions and positive and negative mood [11]. Annotators were initially given a basic word choice question for each term, with one option (the right answer) coming from a thesaurus and the rest possibilities consisting of randomly picked words. This question serves as a check to see if the annotator understood the word and was responding questions correctly. The question is also used to direct the annotator's response to a specific sense of the term. (Different meanings of the same word can convey various feelings.) The vocabulary that resulted has entries for around 25 K different word senses. The union of the emotions linked with each of a word's senses was used to generate a word-level lexicon (for 14 K words) [11].

16.3   Methodology

The overall data cleaning process is shown in Fig. 16.1.

The steps for the entire data collection and cleaning process are:

1. The Reddit comments data was acquired using the Pushshift API. The API requests' code was written in Python, and the data was kept locally. For the years 2017 through 2021, we gathered data. Rstudio was used to clean and analyze the datasets. The data is shown in Table 16.1.
2. The original datasets had three attributes: "author" (character), "comment" (character), and "created" (unknown). When importing the data into *R*, any rows with NULL values were excluded. All numbers and special characters in the "comment" are replaced with "". All of the comments were turned into lowercase. Because the "created" column did not have a datatype, they were

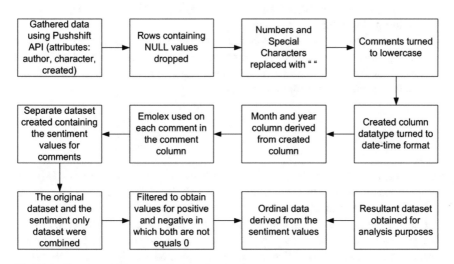

Fig. 16.1 Flowchart depicting entire data cleaning process

Table 16.1 Comments data for the year 2017 (first three entries)

Authors	Comments	Created
NaturalThe1	What a reference	2017–09–10, 10:16:51
Wagzzy	Rat it s pretty fun but you ve gotta feel the right w.	2017–01–01, 11:30:01
Immortal_Bisshoppe	I think part of it has to do with safety older cars ten…	2017–01–01, 11:30:01

    changed to date-time format. From the attribute "created", two derived attributes were produced: "month" and "year". The cleaned dataset (Table 16.1) now included five attributes: "author" (character), "comment" (character), "created" (data-time), "month" (ordered factor), and "year" (double class).

3. We used Saif Mohammad's NRC Word-Emotion Association Lexicon (EmoLex) to get a sentiment score for each comment in the "comment" attribute.
4. We created a separate dataset containing only the negative and positive values foe each year and then combined those values with the original dataset. We filtered from the obtained dataset so that only those values will be available in which both the positive and negative score are not equal to zero.

   - If negative/positive score is greater than or equal to 0 and lesser than 5, then the value will be 1 [12].
   - If negative/positive score is greater than or equal to 5 and lesser than 10, then the value will be 2 [12].
   - If negative/positive score is greater than or equal to 10, then the value will be 3 [12].

5. For the purpose of analysis, we also derive ordinal data from the positive and negative attributes and create another separate dataset. We consider the following conditions for both:
6. The ordinal values are also now combined with the original dataset. After that all respective year datasets were combined to form one final dataset which was then again filtered to remove values containing 0 in both negState and posState and hence obtained the dataset in Table 16.2.

16.4   Results

To examine the data from Table 16.2, we will now run some computations. We will compute the mean and standard deviation for both positive and negative scores, as well as Pearson, Spearman, and Gamma correlation coefficients.

    For calculating Pearson and Spearman correlation, we used cor() method in R's stats package and for Gamma correlation we used rcorr.cens() method in Hmisc package. For mean, we used R's inbuilt mean() method and for standard deviation

Table 16.2 Resultant dataset combined with data of all years (first three entries)

Author	Comment	Created	Month	Year	Negative	Positive	NegState	PosState
Wagzzy	Rat it s pretty …	2017–01–01, 11:30:01	January	2017	1	3	1	1
Immortal_Bis…	I think part of …	2017–01–01, 11:30:001	January	2017	2	0	1	1
FanimeGamer…	Good to hear…	2017–01–01, 11:30:001	January	2017	0	1	1	1

Table 16.3 Mean and standard deviation values

	Positive comments	Negative comments
Mean	1.865377	1.272493
Standard deviation	2.469874	1.908448

Table 16.4 Coefficient values

Coefficients	Values
Pearson	0.5912396
Spearman	0.4174389
Gamma	0.9150657

we used sd() method from stats package. The mean and standard deviation are shown in Table 16.3.

The Pearson, Spearman, and Gamma coefficients are calculated as in Table 16.4.

Pearson, Spearman, and Gamma are all terms that are used to describe the relationship between two variables. There is a positive correlation if the value is closer to 1, and there is no correlation if it is closer to 0. Pearson's coefficient was 0.59 and Spearman's coefficient was 0.41, which cannot be called significant correlations. However, we observed a Gamma correlation of 0.91, indicating a substantial link between positive and negative ordinal values.

We may make a lot of conclusions based on these findings. In comparison with the mean negative score, the mean positive score is greater. It is possible that this is because users on social media are prone to making sarcastic remarks. The lexicons will be unable to distinguish between sarcasm and genuine negative/positive comments.

Emotion ratings are extracted from brief sentences using the word-emotion lexicons like DepecheMood, EmoSenticNet, and NRC [5]. Here are some of the issues they discovered while using simply word-emotion lexicons for sentiment analysis.

- "I am very sad!" resulted in sadness + 1
- "I broke my favourite toy." resulted in [toy] + 0, [favourite] joy + 1, [broke] surprise + 1
- "I am not happy" resulted in joy + 1
- "Snakes!!!" resulted in fear + 0.

That being said in [6] the results of their research showed that 24% of the total population were positive 35% of the total population were neutral 17% of the total population were negative which are in accordance with our results showing low negativity and high positivity.

The Gamma correlation coefficient indicates a positive association between positive and negative remarks, although the Pearson and Spearman values do not. For the time being, we may assume that strong correlations suggest that good comments lead to bad remarks and vice versa. This may be observed on social media when a user publishes a positive comment and another user responds with either a positive

Fig. 16.2 Example of comment thread

or negative comment. A positive comment is followed by a negative comment, and vice versa, according to the correlation coefficients.

Let's focus on the negative comments. Negative comments on social media can be caused by a variety of factors. Sometimes a person is really upset by another user's comment and expresses his displeasure in his response; other times, a user is unconcerned with the comment and posts a negative comment for no apparent reason, which may be traced back to the online disinhibition effect. As a result, users can disseminate negativity without fear of penalties. Figure 16.2 shows an example of one such comment thread.

According to [7], there are three sorts of unfavorable users. Explainer who claim that what they did was not bad, Stubborn opinion who refuse to accept the responders' majority opinion, and returner who again publish the same issue in the hopes of eliciting more positive comments.

Now for the positive comments. Although some positive comments are real, word-emotion lexicons have difficulties discerning between sarcasm and a true positive comment, as previously said. A user may express unhappiness with another user's comment by making a sarcastic comment, similar to how negative comments are expressed as shown in Fig. 16.3.

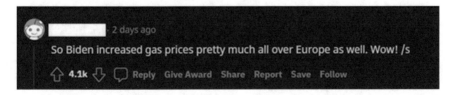

Fig. 16.3 Example of sarcastic comment

16.5 Conclusion

Our findings reveal that there are clear connections between positive and negative remarks, but only when the high correlation provided by Gamma correlation is taken into account. The mean positive comments are higher than the mean negative comments which means that there are more positive comments on Reddit rather than negative comments but, as we mentioned there are multiple possibilities for this in our results section. There are other flaws in our investigation, such as the fact that we did not analyze the impact of emoticons or hashtags. Another issue with our study is that the comments we obtained were chosen at random from all subreddits. If all of the comments were pulled from a single subreddit, the results could have been different and more accurate.

As discussed earlier, there are multiple improvements that can be made to our analysis to obtain a different and maybe more accurate result. So future studies could try and incorporate theories of emotions into their research which could lead to other fascinating results.

References

1. Narwekar, A.: Affective analysis of text in tweets. University of Illinois (2018). https://www.ideals.illinois.edu/handle/2142/101090
2. Griffiths, P.E.: III. Basic emotions, complex emotions, machiavellian emotions. Royal Inst. Philos. Suppl. **52**, pp 39–67 (2003)
3. Scherer, K.R.: What are emotions? And how can they be measured? Soc. Sci. Inf. **44**(4), 695–729 (2005)
4. Madhala, P.: Detecting consumer emotions on social networking websites. Tampere University (2019). https://www.researchgate.net/publication/336891523_Detecting_Consumer_Emotions_On_Social_Networking_Websites
5. Kusen, E., Cascavilla, G., Figl, K., Conti, M., Strembeck, M.: Identifying emotions in social media: comparison of word-emotion lexicons. In: IEEE International Conference on Future Internet of Things and Cloud Workshops, Czech Republic (2017)
6. Widen, G., Lindstrom, J., Brannback, M., Huvila, I, Nystrom, A.G.: Mixed emotions in active social media use—fun and convenient or shameful and embarrassing? iConference (2015)
7. Botzer, N., Gu, S., Weninger, T.: Analysis of moral judgement on reddit. Soc. Inf. Netw. (2021)
8. Suler, J.: The online disinhibition effect. Cyber Psychol. Behav. **7**(3), (2004)
9. Schober, P., Boer, C., Schwarte, L.A.: Correlation coefficients: appropriate use and interpretation. Anesth. Analg. **126**(5), 1763–1768 (2018)
10. Adeyemi, O.: Measures of association for research in educational planning and administration. Res. J. Math. Stat. **3**(3), 82–90 (2011)
11. Mohammad, S.M.: Sentiment analysis: automatically detecting valence, emotions and other affectual states from text. Comput. Lang. (2021)
12. Al-Amin, M., Islam, M.S., Uzzal, SD: A comprehensive study on sentiment of bengali text. In: International Conference on Electrical, Computer and Communication Engineering, IEEE, Bangladesh, pp 267–272 (2017)

Chapter 17
Intra-Domain Text Classification: A Hybrid Approach

Soumak Chakraborty, Himadri Mukherjee, and Alo Ghosh

Abstract The amount of textual information has been increasing at an enormous rate in the digital world. This has led to the development of efficient indexing mechanisms for easier retrieval. One of the primal attributes for categorizing texts is based on their domain. This is a challenging affair due to the commonality of vocabulary. The challenge further aggravates during deeper sub-domain classification. This very important as the rustics (especially students) often need information which concerns a particular subject. Systems capable of organizing information based on subjects can tremendously aid towards efficient retrieval in these scenarios. In this paper, a system is presented to classify educational documents amidst three subjects: computer science, physics and mathematics. Experiments were performed with over 13K research papers, and the highest accuracy of 93.35% was obtained for intra-domain classification using a hybrid technique comprising both handcrafted features and deep learning.

17.1 Introduction

There has been a large increase in the amount of textual information in the digital world. This has been accompanied by a tremendous number of accesses as well. To ensure efficient retrieval of such information, proper indexing is critical. One of the most common approaches of indexing textual information is based on domain. However, this is not adequate in disparate scenarios, especially for educational

The authors contributed equally

S. Chakraborty (✉) · H. Mukherjee · A. Ghosh
AILabs, Kolkata, West Bengal, India
e-mail: soumak.chakraborty@ailabs.academy

H. Mukherjee
e-mail: himadri.mukherjee@ailabs.academy

A. Ghosh
e-mail: alo.ghosh@ailabs.academy

documents. The education domain comprises of multiple subjects and students often search information concerning a particular subject. This sets the need for systems which can categorize educational information with respect to subjects. This will enable easier and more efficient access of information. Performing intra-domain classification is a challenging task. This is mostly due to the overlap of not only words but phrases as well.

Dhar et al. [1] have discussed disparate techniques of text categorization. Parida et al. [2] performed text categorization on the Reuters-21758 dataset. The texts were parameterized using TF-IDF features, and thereafter Chi-square test was used to select the best 1000 features. Random forest and naive Bayes were used for classification, and better performance was reported for naïve Bayes. Dhar et al. [3] performed text categorization of Bangla news text. The dataset consisted of nine domains, namely Business, Entertainment, Food, Literature, Medical, State Affairs, Sports, Science and Technology, and Travel. The texts were parameterized using graph-based feature which was followed by LSTM-based classification and the highest accuracy of 99.21% was reported. Hao et al. [4] distinguished eight different domains from 400 pages of web text data. The considered domains were Finance, IT, Health, Sport, Travel, Education, Culture and Military. The texts were modelled using TF-IDF features along with naïve Bayes and SVM, wherein better performance was obtained for naïve Bayes. Xue and Li [5] presented a system to categorize texts from the Sogou laboratory text categorization corpus. They used the bigram method for word segmentation followed by random forest Classifier. They performed experiments for different tree and feature sizes, wherein the best result was reported for tree and feature size of 250 and 900, respectively.

Kibriya et al. [6] performed text categorization using disparate datasets including 20 newsgroups WebKB and Reuters-21578 Standard and normalized TF-IDF features were extracted post-bag-of-words technique. Different varieties of naïve Bayes classifier were explored along with SVM, wherein the best result was obtained using SVM. Vaissnave and Deepalakshmi [7] attempted text categorization from Indian legal documents. The classes included Fact, Issue, Arguments of petitioner, Arguments of responder, Reasoning, Decision, Majority concurring and Minority dissenting. The texts were vectorized using Word2Vec technique, and an accuracy of 88% was reported using bidirectional LSTM-based classification. Lade and Dhore [8] attempted to classify Marathi texts amidst three categories, namely sports, entertainment and economy. The texts were parameterized using TF-IDF features and then fed to a KNN-based classifier which yielded the highest accuracy of 91.27%. Ahmed et al. [9] categorized Bangla texts into ten categories. They experimented with attention-based RNNs and BiLSTM which fetched acccuracies of 97.72% and 86.56%, respectively. Bahassine et al. [10] presented an improved version of Chi-square feature selection technique for categorizing Arabic texts. Experiments were performed with 5070 documents spanning over 6 domains, and the best F-score of 90.50% was reported with 900-dimensional features.

It is observed that most of the works concentrate on broad domain categorization and developments for intra-domain categorization has been on the lower side. This is an important aspect for faster retrieval of documents considering the fact that every domain encompasses a vast variety of information.

17.2 Proposed Method

In the current experiment, two different approaches were employed. In the former, a deep learning-based technique was used on the raw data. In the other part, the articles were parameterized using handcrafted features which were then supplied to a deep learning-based classifier. The details are presented in the subsequent paragraphs.

17.2.1 Deep Learning-Based Approach

In this technique, the words were vectorized to a numerical format which were then fed to a LSTM-based classifier which is detailed hereafter.

17.2.1.1 Word Embedding

The words for each of the articles were embedded/vectorized using Word2Vec [11] technique. These produced vector representations for every word which was constructed by considering the part of speech, disambiguated sense, syntax and semantics of the text. For every word, a 300-dimensional vector was obtained. Each of these vectors was amalgamated to form the entire text block. As different instances, had disparate word counts, so feature vectors of multifarious dimensions were obtained. In order to vectors of constant dimension, the amalgamated vectors were zero-padded to the length of the largest vector. The vectorized representation of 30 different words is presented in Fig. 17.1.

17.2.1.2 Long Short-Term Memory Network

LSTM or long short-term memory network [12] is an improvement over the standard recurrent neural network which solves the problem of remembering long context in text due to its memory capability and the problem of vanishing gradient that is evident in recurrent neural networks. The LSTM network is composed of multiple cells along with forget, input and output gates. The forget gate is responsible for discarding information and takes previous hidden cell state as the forget gate which works with the previous hidden state and the present input. This is followed by the input gate which adds information to the cell state. It uses a sigmoidal function to

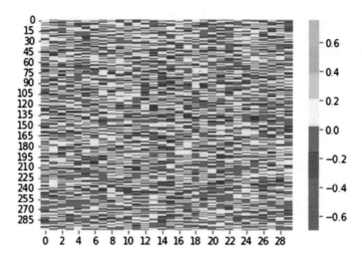

Fig. 17.1 Embedded representation of 30 words from the dataset

regulate the added values. This is finally followed by the output gate which selects meaningful information from the current cell state. This information is provided as output. The LSTM layer is followed by dense layers which are fully connected layers. They perform classification based on the input from the previous layers. In this experiment, the network comprised of a single LSTM layer with a default dimension of 50. Thereafter, a dropout layer was introduced with a 50% parameter discarding scheme (default) to prevent overfitting. This was followed by two dense layers. The 1st dense layer had a default dimension of 50 and the final output layer had a dimension of 3. The LSTM layer had tanh activation while the dense layers had ReLU and softmax activations which are presented in Eqs. (17.1) and (17.2), respectively. The number of generated parameters for the best network is presented in Table 17.1.

$$f(x) = \max(0, x), \tag{17.1}$$

here, x is the input to a neuron.

Table 17.1 Number of generated parameters for the proposed network

Layer	Parameters
Embedding	1500000
LSTM	70200
Dense 1	6375
Dense 2 (output)	378
Total	1576953

$$\sigma(z)_j = \frac{e^{z_j}}{\sum_{k=1}^{K} e^{z_k}},$$

(17.2)

where z is an input vector of length K.

17.2.2 Hybrid Approach

In this technique, term frequency–inverse document frequency (TF-IDF) features [13] were extracted for the words at the outset. The TF-IDF feature is composed of two aspects which are discussed as follows:

- Term Frequency (TF): Number of times a word is present in a sentence/total number of words in the sentence.
- Inverse Document Frequency (IDF): log(Total number of sentences/Number of sentences the specific word appears)

Finally, the TF-IDF feature is obtained by multiplying TF and IDF matrices. The maximal feature dimension was set to 5000 after trial. The words were converted to lowercase for avoiding duplicate entry due to capitalization, and only unigrams were considered. Stop words were removed using a standard list of English stopwords comprising of 318 entries. The extracted features were thereafter fed to the LSTM network.

17.3 Results and Discussion

The experiments were performed in two phases whose results are presented hereafter. The first phase presents the results for the deep learning-based approach. This phase was used to tune the LSTM network as well, and a 80:20 train test split was used. The second phase presents the outcome of the hybrid approach, and to test the robustness of the system, fivefold cross-validation was used.

17.3.1 Dataset

Data is a very important aspect for any experiment. The quality of data plays an important role in judging the robustness of a proposed system. It is important for the dataset to uphold real-world characteristics. In this experiment, a dataset of abstracts from research articles was used.[1] There are five different subjects, namely physics,

---

[1] https://www.kaggle.com/datasets/vetrirah/janatahack-independence-day-2020-ml-hackathon.

mathematics, computer science, statistics, quantitative finance and quantitative biology. Out of them the 1st were considered to class imbalance which was observed for the other subjects. Moreover, these three subjects are very much related to one another and at times have same keywords as well. There were 4910, 5120 and 3610 instances for physics, computer science and maths, respectively, totalling to 50412 unique tokens.

17.3.2 Deep Learning-Based Technique

Initially, the dimension of the LSTM layer was varied from 25 to 125 with a step of 25 whose results are presented in Table 17.2. The best performance was obtained for 50-dimensional LSTM. The dimension of the dense layer was set to 50 (default).

The dimension of the dense layer was varied from 25 to 150 with a step of 25. The results are presented in Table 17.3. The LSTM dimension was set to the default value of 75. It is noted that the best performance was obtained for 100-dimensional dense layer.

Finally, the best LSTM network comprising of 50-dimensional LSTM layer accompanied by 50% dropout followed by 100 and three-dimensional dense layers was obtained which was trained using disparate batch sizes whose performances are tabulated in Table 17.4. It is noted that the best performance was obtained for a batch size of 64 instances.

The best-performing setup was thereafter used to evaluate the system on the entire dataset using fivefold cross-validation where an accuracy of 90.32% was obtained. This was done to test the robustness of the proposed system for disparate train–test combinations and also to ensure that every instance in the dataset was subjected to test set atleast once.

Table 17.2 Performance for different LSTM dimensions

LSTM dimension	25	50	75	100	125
Accuracy	90.61	91.42	88.56	89.52	90.76

Table 17.3 Performance for different dimensions of the intermediate dense layer

Dense dimension	25	50	75	100	125	150
Accuracy	89.59	90.14	90.18	90.47	91.50	91.02

Table 17.4 Performance for different batch sizes during training

Batch size	8	16	32	64	128
Accuracy (%)	89.66	89.88	91.61	91.24	90.54

Table 17.5 Confusion matrix for the hybrid system

	Physics	Computer	Maths
Physics	4597	115	198
Computer	159	4853	108
Maths	206	121	3283

Table 17.6 Performance of TF-IDF features

Classifier	SVM	RF	MLP	NB	LSTM
Accuracy (%)	93.27	90.32	92.61	91.77	93.35

17.3.3 Hybrid Technique

The handcrafted TF-IDF features were fed to LSTM network, and an accuracy of 93.36% was obtained using cross-validation whose interclass confusions are presented in Table 17.5. It is noted that the highest confused pair was physics and maths. One of the probable reasons for this is the similarity of terminology amidst the two subjects. The confusions for computer science with the other two subjects were almost 43% lower than that of physics and maths. One of the primal reasons for this was the marginally higher number of data for computer science. Further analysis also revealed lower similarity of computer science texts with the other two subjects in disparate cases which possibly led to lower confusion.

The TF-IDF features were tested with other popular classifiers in the thick of SVM [14], random forest (RF) [15], multilayered perceptron (MLP) [16] and naive Bayes (NB) [17]-based classifiers. The results are presented in Table 17.6. It is noted that LSTM produced the best average performance amidst all the classifiers which was followed by SVM. The lowest performance was obtained for random forest. The foldwise performance is presented in Fig. 17.2. It is seen that LSTM consistently

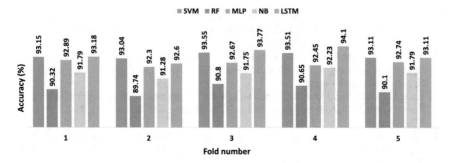

Fig. 17.2 Performance of the classifiers in every fold

produced the best performance for every fold except for fold 2 where SVM performed better. In the last fold, the performance of SVM and LSTM was the same. In all the folds, MLP ranked 3rd followed by naive Bayes and random forest.

17.4 Conclusion

In this paper, a hybrid technique of intra-domain text classification is presented. The system works with TF-IDF features and LSTM-based classification. The system was tested with abstracts of research articles from three different subjects, wherein a weighted precision of 0.93 was obtained. In future, we will test the system with more subjects from single and multiple domains. We also plan to parameterize the texts with other handcrafted features. Usage of other embedding techniques as well as deeper networks is also in our future plans. Finally, the system will be trained using data augmentation for better performance and will be deployed in the web to test its performance for real-time text categorization.

References

1. Dhar, A., Mukherjee, H., Dash, N.S., Roy, K.: Text categorization: past and present. Artif. Intell. Rev. **54**(4), 3007–3054 (2021)
2. Parida, U., Nayak, M., Nayak, A.K.: News Text Categorization using random forest and Naïve Bayes. In: 2021 1st Odisha International Conference on Electrical Power Engineering, Communication and Computing Technology (ODICON), pp. 1–4. IEEE (2021)
3. Dhar, A., Mukherjee, H., Obaidullah, S., Santosh, K.C., Dash, N.S., Roy, K.: Web text categorization: a LSTM-RNN approach. In: International Conference on Intelligent Computing and Communication, pp. 281–290. Springer, Singapore (2019)
4. Hao, P., Ying, D., Longyuan, T.: Application for web text categorization based on support vector machine. In: 2009 International Forum on Computer Science-Technology and Applications, vol. 2, pp. 42–45. IEEE (2009)
5. Xue, D., Li, F.: Research of text categorization model based on random forests. In: 2015 IEEE international conference on computational intelligence and communication technology, pp. 173–176. IEEE (2015)
6. Kibriya, A.M., Frank, E., Pfahringer, B., Holmes, G.: Multinomial naive bayes for text categorization revisited. In: Australasian Joint Conference on Artificial Intelligence, pp. 488–499. Springer, Berlin, Heidelberg (2004)
7. Vaissnave, V., Deepalakshmi, P.: A keyword-based multi-label text categorization in the Indian legal domain using Bi-LSTM. In: Soft Computing: Theories and Applications, pp. 213–227. Springer, Singapore (2022)
8. Lade, S., Dhore, M.L.: Text categorization of Marathi news articles using machine learning. In: Proceeding of First Doctoral Symposium on Natural Computing Research, pp. 63–72. Springer, Singapore (2021)
9. Ahmed, M., Chakraborty, P., Choudhury, T.: Bangla document categorization using deep RNN model with attention mechanism. In: Cyber Intelligence and Information Retrieval, pp. 137–147. Springer, Singapore (2022)

10. Bahassine, S., Madani, A., Al-Sarem, M., Kissi, M.: Feature selection using an improved Chi-square for Arabic text classification. J. King Saud Univ. Comput. Inf. Sci. **32**(2), 225–231 (2020)
11. Church, K.W.: Word2Vec. Natural Lang. Eng. **23**(1), 155–162 (2017)
12. Yu, Y., Si, X., Hu, C., Zhang, J.: A review of recurrent neural networks: LSTM cells and network architectures. Neural Comput. **31**(7), 1235–1270 (2019)
13. Aizawa, A.: An information-theoretic perspective of TF-IDF measures. Inf. Process. Manage. **39**(1), 45–65 (2003)
14. Pisner, D.A., Schnyer, D.M.: Support vector machine. In: Machine learning, pp. 101–121. Academic Press (2020)
15. Breiman, L.: Random forests. Mach. Learn. **45**(1), 5–32 (2001)
16. Gardner, M.W., Dorling, S.R.: Artificial neural networks (the multilayer perceptron)-a review of applications in the atmospheric sciences. Atmos. Environ. **32**(14–15), 2627–2636 (1998)
17. Webb, G.I., Keogh, E., Miikkulainen, R.: Naïve Bayes. Encyclop. Mach. Learn. **15**, 713–714 (2010)

Chapter 18
Prediction of Air Quality Using Machine Learning

Arun Kumar⊙ **and Anupam Jamatia**⊙

Abstract In recent years, air quality has become a significant environment and health issue, this impact people's everyday life. How to predict air quality accurately and precisely in urban cities has become one of the hot research issues. The majority of research papers talked about one, two, or three methods, and there rises a question as to select which method is superior to others. To resolve this issue, there is a need for a comprehensive study of various work done. This paper provides a comparative analysis of the most relevant studies related to air quality prediction. After studying comprehensively various experiments have been conducted using machine learning methods like linear regression (Linear R), lasso regression (Lasso R), ridge regression (RR), decision tree regression (DTR), random forest regression (RFR), extreme gradient boosting (XGBoost), and artificial neural network (ANN). These experiments successfully resolve limitations like data instability, overfitting, and multicollinearity. RFR, XGBoost, and ANN perform better and help to resolve air prediction issues, and specifically, ANN outperforms all. Results and discussion of this paper provide a holistic view of methods to researchers. Compared with other various models, the precision of prediction data has been greatly improved.

18.1 Introduction

In recent years, India has achieved rapid economic growth and become the world's sixth-largest economy in nominal terms. However, rapid industrialization, economic development, urbanization, and globalization associated with its fast economic growth but leading to severe air pollution [1]. Air pollution cost Indian economy 95 billion dollar if India achieves safe air quality, it improves the Indian economy and businesses [2]. According to IQRA (2021), World Air Quality Report: India ranked 5th in the world for the worst air quality in 2021.[1] Air pollution is causing massive

---

[1] https://www.iqair.com/in-en/india.

A. Kumar (✉) · A. Jamatia
Department of Computer Science and Engineering, National Institute of Technology, Agartala, Tripura, India
e-mail: arunraghav384@gmail.com

© The Author(s), under exclusive license to Springer Nature Singapore Pte Ltd. 2023 199
V. Bhateja et al. (eds.), *Evolution in Computational Intelligence*, Smart Innovation, Systems and Technologies 326, https://doi.org/10.1007/978-981-19-7513-4_18

public health diseases and environmental problems. Air pollutants like particulate matter has the potential of causing visibility impairment and can reduce visibility, causing environmental damage by depleting the nutrients in the soil, making lakes and streams acidic and material damage such as statues and monuments. According to research work [3], particulate matter and household air pollution claimed 1.24 million lives in 2017. Air quality has a great impact on people's daily life wherein PM2.5 (fine particles with a diameter of 2.5 µm or fewer in the atmosphere) is the main factor of air pollution and the increasing PM2.5 concentration will also directly affect human health. There is a positive correlation between PM2.5 concentration and lung cancer mortality ($r = 0.0052$, $P = 0.036$); this was found in research [4]. Technologies in machine learning can help in development of a smart environment for example prediction of air quality plays a key role in health and daily activities.

This paper provides a comparative analysis of the most relevant studies related to air quality prediction in brief, methods, evaluation criteria, results and limitations were discussed. Statistical analysis and feature engineering approaches were considered in the dataset evaluation. After studying comprehensively various methodologies, we have conducted experiments using machine learning methods like linear R, lasso R, RR, DTR, RFR, XGBoost, and ANN. These experiments successfully resolve limitations like data instability, overfitting, and multicollinearity. To evaluate the model prediction error rates and model performance of the proposed methods, four evaluation functions including the R-squared (R^2), mean square error (MSE), root mean square error (RMSE), and mean absolute error (MAE) were used in this study.

Motivation to work in this area rises because of the severe impact of PM2.5 on the human body, the economy, and the environment. How to effectively manage PM2.5 is an urgent problem to be solved. In the smart cities, we need air quality predictors because if the concentration of air pollutants like PM2.5 can be predicted, effective safety measures can be taken in advance to manage the generation of atmospheric pollutants and prevent atmospheric pollution. This can be helpful to all living creatures to maintain their ecosystem.

The related work with a comparative analysis of the most relevant studies will be discussed in Sect. 18.2. Section 18.3, describes the data collection and preprocessing. The detailed description of our proposed methods will be described in Sect. 18.4. Section 18.5, will display the evaluation function, experimental results of the model, as well as the comparisons and analysis between different methods. Section 18.6, presents the conclusion and the direction of further research.

18.2 Related Work

According to National Air Quality Index, eight major pollutants include PM10, PM2.5, nitrogen dioxide, sulfur dioxide, ozone, ammonia, and lead. Today, a variety of research work has been used by scholars at a national and global levels to predict the air pollutants, and various good result have been attained. These work includes chemical transmission method, time series method, and machine learning methods.

The time series methods to predict air pollutant concentration include: autoregressive method [5], autoregressive moving average, and the notion of integration. Chemical transport methods are also used to evaluate air pollutants on the scientific basis. These methods use both numerical methodologies with meteorological principles [6]. Input factors for the model include a source of emission, progress of the area, and meteorological factors. But, these models include complex numerical problems which need more time and accuracy and are still also not up to the mark.

In [7], authors have used RFR to predict Air Quality Index (AQI) in Beijing and the nitrogen oxides concentration in city of the Italy. This method performed better in the prediction of the nitrogen oxide concentration (RMSE = 83.6, $R^2 = 0.84$). Authors in [8] predict air quality using RFR in a total error framework. In their analysis, they have found RFR results close to deep neural network and give better results than multi-layer perceptron (MLP). They also discuss air quality prediction suffering from outliers (measurements were inconsistent with the dataset); this is due to a harsh environment and human error. In work [9], authors performed different kinds of regression techniques, namely linear R, support vector regression (SVR), DTR, and Lasso R. In this work, the regression analysis technique is used to evaluate the relationship between these factors and predict carbon monoxide based on other parameters. In a research paper [10], RR was used to find quantitative relation between different parameters, so to optimize time complexity. They form the RR-XGBoost model where RR helps in the optimization of input variable and XGBoost gives the final output result. Authors in [11] evaluate the experiment using SVR, RR, XGBoost, and integrated model. In an integrated model, they used sequence-to-sequence technology to establish single factor prediction model, and then, long short-term memory (LSTM) used as multifactor prediction, and finally, XGBoost tree is used for integration of two. Results show that the dual integrated LSTM model improves the accuracy. In work [12], authors proposed a prediction method by integrating the forecasting data and high-dimensional statistical analysis. They have used AdaBoost, XGBoost, and light gradient boosting method (LGBM) methodology. They have found that LGBM improved the air quality prediction. However, papers suffer from challenges such as air pollutant concentration variation and instability of data.

ANN is a deep learning model that is flexible and powerful. Traditional techniques need more computational power, to tackle this deep learning method to produce satisfactory results. Authors in [13] study for forecasting air pollution to predict next-day hourly ozone concentrations. They have used MLR and feed-forward artificial neural network (FANN). The result shows that linear models suffer from non-linearity, and the model is sensitive to the training dataset and not produce desired results for lower concentration. Another study [14] has presented a hybrid approach combining ANN and clustering; it was found that it is not suitable for high concentration. But, the results show hybrid model outperforms classical MLP for high concentration. In the field of mining engineering, where prediction of air quality is a must for well being of people, ANN shows the best result in the harsh environment of coal mines. In a

Table 18.1 Statistical characteristics of experimental data

Variable	Unit	Mean	St. dev
Average temperature (T)	°C	24.90	7.20
Maximum temperature (T_M)	°C	32.23	6.85
Minimum temperature (T_m)	°C	19.27	7.45
Atmospheric pressure at sea level (SLP)	hPa	1008.10	7.56
Humidity (H)	%	64.13	15.46
Average visibility (VV)	km	1.89	0.68
Average wind speed (V)	km/h	6.63	3.96
Maximum sustained wind speed (VM)	km/h	15.69	7.65
Particulate matter (PM2.5)	$\mu g/m^3$	107.9	82.57

study [15], principal component analysis (PCA) identifies gases that affect mine air quality. Result of PCA fed into ANN model. The results showed PCA-based ANN has better performance. Rybarczyk and Zalakeviciute [16] studied 46 articles on machine learning to predict air pollutants; their systematic review shows there are two kinds of studies that first focus on estimation of air quality pollutant which use regression algorithms. Second, focus on forecasting problems that uses neural networks and SVM. From the thorough analysis of research papers, some observations were made like, PM2.5 was major prediction target of various studies, and a majority of research papers carried out the prediction of air quality for the next-day.

18.3 Data Collection

According to report by Greenpeace, Bangalore had the third-worst air pollution among Indian cities.[2] Bangalore city was taken as our study area; it is located in the southeast of the South Indian state of Karnataka. The meteorological dataset derived from the Web site[3] it covers all standard parameters as compared to bench mark data. Our study collected two-year data due to free availability from January 1, 2017, to December 31, 2018. Dataset has 9 columns and 731 rows. The distribution of the column is as follows; it has eight independent features T, T_M, T_m, SLP, H, VV, V, VM, and one dependent feature, i.e., PM2.5 as mentioned in Table 18.1. Each row contains the meteorological according to the respective column name.

To have a statistical analysis of the dataset mean and standard deviation (st.dev) has been calculated, this is represented in Table 18.1. Potential relationship between attribute with the help of correlation coefficient [17] has been calculated. Pearson correlation coefficient has been calculated, it measures the linear correlation between

---

[2] https://www.greenpeace.org/india/en/press/.

[3] https://en.tutiempo.net/.

Table 18.2 Pearson correlation coefficients

	T	T_M	T_m	SLP	H	VV	V	VM	PM2.5
T	1.0	0.97	0.95	−0.88	−0.50	0.64	0.32	0.28	−0.62
T_M	0.96	1.0	0.89	−0.82	−0.58	0.60	0.32	0.29	−0.56
T_m	0.95	0.89	1.00	−0.91	−0.28	0.57	0.31	0.25	−0.66
SLP	−0.88	−0.82	−0.90	1.0	0.24	−0.51	−0.34	−0.29	0.61
H	−0.51	−0.59	−0.31	0.24	1.0	−0.46	−0.39	−0.37	0.17
VV	0.62	0.59	0.56	−0.51	−0.46	1.0	0.40	0.34	−0.62
V	0.32	0.32	0.31	−0.32	−0.38	0.37	1.0	0.76	−0.35
VM	0.28	0.29	0.25	−0.31	−0.36	0.34	0.76	1.0	−0.28
PM2.5	−0.62	−0.56	**−0.66**	**0.62**	0.13	−0.57	−0.35	−0.28	1.0

two attributes ranging from (0, 1), where 0 represents two-variable completely unrelated and 1 indicates related. Table 18.2 shows the correlation matrix results of PM2.5 with other meteorological variables. It can be seen temperature is negatively correlated with PM2.5 which means if the temperature is low concentration of PM2.5 is more, similarly visibility decreases as the concentration of PM2.5 increases.

In the data preprocessing stage, the imputation technique was used for replacing the missing data. This helps to prevent deletion of the dataset because every time deletion of the dataset is not feasible can lead to a reduction in the dataset and raises concern for biasing. Handling missing data is important during preprocessing as many machine learning algorithms do not support missing values. In this study, the missing values are treated by imputation to recover the corresponding value, and missing values are imputed with mean values. This helps to prevent data loss and works well with small a dataset. The experiment was conducted on both with-imputation and without-imputation dataset.

18.4   Experiment

In this paper, seven machine learning methods have been performed. The methods have been implemented using the Python programming language for this Anaconda3 has been used. Libraries used were scikit-learn, plotly, sklearn metrics, XGBoost, keras-tuner, etc. The workflow diagram is represented in Fig. 18.1.

Linear R is probably the method where most of the academicians started their first machine learning method is used to examine two things first it checks whether a set of predictor variables is doing a good job in predicting an outcome (dependent) variable, and second, it checks which variables are important predictors of the outcome (dependent) variable. Lasso R is referred to as a type of regularized linear R. It can shrink the coefficients of input variable that does not contribute to the prediction. This regression is well suited for models showing high levels of multicollinearity

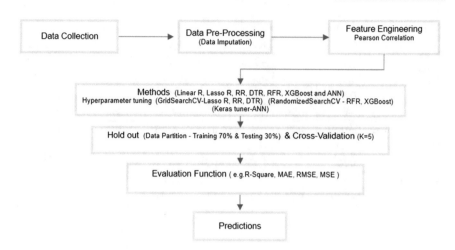

Fig. 18.1 Workflow diagram

or if you want to automate certain parts of model selection, like a variable selection and a parameter elimination. RR built a parsimonious model when the number of predictor variables in a set exceeds the number of observations. It helps to reduce the complexity of the model by coefficient shrinkage. Lasso R which uses the L1 regularization technique RR uses L1. In terms of interpretation and accuracy, lasso R is better. DTR can predict the value of a target variable by learning from decision tree rules inferred from dataset features [18]. This method allows visualizing the model with the help of a flow chart in which each node represents a condition for non-leaf nodes. A major drawback of this model is the overfitting problem. To overcome the drawbacks of the decision tree RFR is being used. RFR is a meta-estimator that fits several classifying decision trees on various sub-samples of the dataset and uses averaging to improve the predictive accuracy and control overfitting [19]. This method can be used for weather normalization techniques that can predict the concentration of an air pollutant at a specific measured time point. XGBoost is faster and accurate than RFR because the gradient of data is considered for each tree. In work, [20] gives the reasons for better performance of XGBoost as weights of newly built tree can be scaled down by a given constant, which reduces an influence of a single tree on the final score. ANN displays a complex relationship between outputs and inputs variables to discover a new pattern. It has three layers input layer, hidden layer (mathematical computations on input information) and the output layer (results after computation). Prediction of air quality can be easily done with the help of ANN models like MLP, convolutional neural network.

18.5 Result Analysis and Discussions

To evaluate the model prediction error rates and model performance of the proposed methods, four evaluation functions including R^2, MSE, RMSE, and MAE were used in this study. The calculation method is shown in Eqs. (18.1)–(18.4), respectively,

$$R^2 = \frac{\sum_{i=1}^{n} (y_i - \hat{y}_i)^2}{\sum_{i=1}^{n} (y_i - \bar{y})^2} \tag{18.1}$$

$$\text{MSE} = \frac{1}{n} \sum_{i=1}^{n} (y_i - \hat{y}_i)^2 \tag{18.2}$$

$$\text{RMSE} = \sqrt{\frac{1}{n} \sum_{i=1}^{n} (y_i - \hat{y}_i)^2} \tag{18.3}$$

$$\text{MAE} = \frac{1}{n} \sum_{i=1}^{n} |y_i - \hat{y}_i| \tag{18.4}$$

where, y_i represents the true value, $\hat{y}_i$ represents the predicted value, and n represents the number of true values.

The coefficient of determination known as R^2 value was chosen to see the goodness of fit regression models. R^2 value does not consider the overfitting issue as it shows to what extent the variance of one variable explains the variance of the other variable, and a value near one indicates a better model. MSE is most suitable for the dataset that contains a lot of noise such as outliers or unexpected values. A lower MSE value indicates a better model. RMSE can penalize large errors in the dataset by assigning a higher weight to larger errors. It gives a higher error metric for large outliers, which are desirable in the case of air quality concentration forecasting, and as such, it is well suited for the comparison of different models' performance. MAE does not penalize large errors and reflects possible outliers in the dataset in the same way as RMSE does.

Results of all methods on dataset with and without imputation have been presented in Table 18.3. To be able to evaluate the ability of each model, 70% of data points will be fed into training process while the remaining 30% are for testing a purpose; this process is called hold out mentioned in the table. To better utilize our data cross-validation is used K-fold cross-validation with $k = 5$.

Hyperparameter optimization technique has been performed to find optimal parameter values for a method while applying it to a dataset. GridSearchCV is applied to lasso R, ridge R, and DTR. This technique helps to produce good results, but this technique, itself, has some limitations like the best parameters results are limited, and this increases the time complexity because it is time-consuming. Randomized-

SearchCV is being applied in RFR and XGBoost. This technique helps to produce good results as it allows one to determine the range in which hyperparameters falls; grid search takes the assumption that the researcher has already determined this range. We can analyze the performance of RFR is better if compare to linear R, lasso R, RR, and DTR. This can be seen from Table 18.3 that R-square and RMSE result improved significantly. For ANN hyperparameters used were 'adam' optimizer, '100' epochs, and 'relu' as an activation function with 5 layers (one input and one output layer rest 3 as hidden layers).

In the hold out, XGBoost was shown to be the second best method, with MSE results 34.44 (without imputation) and 31.90 (with imputation), while ANN outperforms all other methods with MSE results 30.90 (without imputation) and 24.65 (with imputation). It can be analyzed in most cases random forest performs well than linear, ridge, lasso, and decision tree in various evaluation criteria. Imputation allows to produce improvement in metrics results like XGBoost MAE results 50.33 (without imputation) and 46.02 (with imputation). Similarly, random forest MAE results 51.26 (without imputation) and 48.63 (with imputation). In cross-validation, the results do not have significant differences with imputation and without imputation like random forest RMSE 3172.44 (without imputation) and 3160.07 (with imputation); it shows slight improvement while in many cases it did not show improvement like XGBoost MAE 58.29 (without imputation) and 59.02 (with imputation).

The results can also be visualized from the scatter map of Fig. 18.2 gives the correlation between a test and prediction values. If we compare the scatter plots of Fig. 18.2a–c and d–g, it is visible that from Fig. 18.2d–g, scatter points get tighter and form a straight line which shows a better correlation.

The decision tree is easy to interpret and even work when there is a nonlinear relationship between variable. This generally causes an overfitting problem. Similarly, in prediction work that depends on the independent variable, decision tree in this case can not yield the best result as it assumes all independent variables interact with each other. This can be seen in Table 18.3 if the results of DTR and RFR were compared. RFR overcome the drawback of DTR as it is an ensemble method. It also solves low bias and high variance problems. In terms of efficiency and accuracy, it outperforms. Observation made from Table 18.3, in hold out case with-imputation MAE error ranges from 55.29 to 40.95; random forest result is 48.63 which is the third best. XGBoost in each step generates a week learner and combines it into the model. Simply, in GBM, a weak learner is based on the gradient direction of a loss function. GBM adds a new tree to complement already built one tree; this makes it different from RFR method. The performance of this method can be analyzed through its result. ANN outperforms all other methods. ANN is the first choice of researchers where the conventional model fails. Well-trained ANN model gives good results in large amounts data; it also can accommodate multiple input parameters.

Table 18.3 Comparison of the prediction error based on various models

Models	Without imputation								With imputation							
	Hold out				Cross validation				Hold out				Cross validation			
	R^2	MSE	RMSE	MAE	R^2	MSE	RMSE	MAE	R^2	MSE	RMSE	MAE	R^2	MSE	RMSE	MAE
Linear R	0.56	41.98	3433.44	58.95	0.49	42.80	3425.85	58.53	0.53	40.28	3057.66	55.29	0.46	43.27	3499.35	59.15
RR	0.56	42.02	3392.63	58.24	0.49	42.76	3419.73	58.47	0.531	39.69	3026.42	55.01	0.468	43.23	3492.49	59.09
Lasso R	0.55	43.00	3490.78	59.08	0.50	42.47	3370.98	58.06	0.53	40.05	3103.26	55.70	0.47	42.98	3447.93	58.71
DTR	0.49	41.59	3337.49	57.77	0.29	50.19	5224.11	68.63	0.30	37.99	2852.28	53.40	0.37	54.13	5203.12	76.39
RFR	0.65	35.84	2628.04	51.26	0.53	39.70	3172.44	55.79	0.63	34.55	2365.84	48.63	0.50	40.08	3160.07	56.48
XGBoost	0.63	34.44	2533.34	50.33	0.49	42.27	3398.29	58.29	0.61	31.90	2118.72	46.02	0.45	42.78	3483.76	59.02
ANN	0.69	30.90	2421.76	48.95	0.45	38.65	3233.89	56.91	0.65	24.65	2111.72	40.95	0.49	39.65	3172.89	58.90

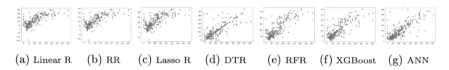

(a) Linear R   (b) RR   (c) Lasso R   (d) DTR   (e) RFR   (f) XGBoost   (g) ANN

Fig. 18.2 Scatter graphs of various model

18.6 Conclusion

In this paper, seven existing methods for air quality prediction have been evaluated and compared. The methods used are linear R, lasso R, RR, DTR, RFR, XGBoost, and ANN. Results have been calculated for without imputation and with imputation on hold out case and cross-validation case. According to the results RFR, XGBoost and ANN perform well. Their error rates are much lower than other models which reflects their accuracy and superiority over another methods. It was found that ANN can deal with a large amount of training data than random forest or the XGBoost. In addition, the RFR method performance was close to that of XGBoost. One of the limitation in the various studies was found that a prediction of air quality in short-term prediction, i.e., for one hour, it provides good results, but in the long-term prediction for 24 h or more, it doesn't yield good results because of the time interval between prediction and training result will become very long. The conceptual and methodological enhancement of this work could help in smart city pollution management.

References

1. Gautam, D., Bolia, N.: Air pollution: impact and interventions. Air Qual. Atmos. Health **13** (2020). https://doi.org/10.1007/s11869-019-00784-8
2. Mele, M., Nieddu, L., Abbafati, C., et al.: An ANN experiment on the Indian economy: can the change in pollution generate an increase or decrease in GDP acceleration? Environ. Sci. Pollut. Res. **28**, 35777–35789 (2021). https://doi.org/10.1007/s11356-021-13182-4
3. Chatterjee, P.: Indian air pollution: loaded dice. Lancet Planet. Health **3**, e500–e501 (2019). https://doi.org/10.1016/S2542-5196(19)30247-5
4. Cao, Q., Rui, G., Liang, Y.: Study on PM2.5 pollution and the mortality due to lung cancer in China based on geographic weighted regression model. BMC Public Health **18**, 925 (2018)
5. Bhalgat, P., Pitale, S., Bhoite, S.: Air quality prediction using machine learning algorithms. Int. J. Comput. Appl. Technol. Res. (IJCATR) **8**, 367–370 (2019)
6. Wang, M., Sampson, P.D., Hu, J., et al.: Combining land-use regression and chemical transport modeling in a spatiotemporal geostatistical model for ozone and PM2.5. Environ. Sci. Technol. **50**(10), 5111–5118 (2016). https://doi.org/10.1021/acs.est.5b06001.
7. Liu, H., Li, Q., Yu, D., Gu, Y.: Air quality index and air pollutant concentration prediction based on machine learning algorithms. Appl. Sci. **9**, 4069 (2019). https://doi.org/10.3390/app9194069
8. Lepioufle, J.-M., Marsteen, L., Johnsrud, M.: Error prediction of air quality at monitoring stations using random forest in a total error framework. Sensors **21**, 2160 (2021). https://doi.org/10.3390/s21062160

9. Aarthi, A., Gayathri, P., Gomathi, N.R., Kalaiselvi, S., Gomathi, V.: Air quality prediction through regression model. Int. J. Sci. Technol. Res. **9**(3), 923–928 (2020)
10. Liu, B., Tan, X., Jin, Y., Yu, W., Li, C.: Application of RR-XGBoost combined model in data calibration of micro air quality detector. Sci. Rep. **11** (2021). https://doi.org/10.1038/s41598-021-95027-1
11. Chen, H., Guan, M., Li, H.: Air quality prediction based on integrated dual LSTM model. IEEE Access **9**, 93285–93297 (2021). https://doi.org/10.1109/ACCESS.2021.3093430
12. Zhang, Y., et al.: A predictive data feature exploration-based air quality prediction approach. IEEE Access **7**, 30732–30743 (2019). https://doi.org/10.1109/ACCESS.2019.2897754
13. Sousa, S., Martins, F., Alvim-Ferraz, M.C.M., Pereira, M.C.: Multiple linear regression and artificial neural networks based on principal components to predict ozone concentrations. Environ. Model. Softw. **22**, 97–103 (2007). https://doi.org/10.1016/j.envsoft.2005.12.002
14. Tamas, W., Notton, G., Paoli, C., Nivet, M.-L., Voyant, C.: Hybridization of air quality forecasting models using machine learning and clustering: an original approach to detect pollutant peaks. Aerosol Air Qual. Res. **16**(2), 405–416 (2016). https://doi.org/10.4209/aaqr.2015.03.0193
15. Jo, B., Khan, R.: An Internet of Things system for underground mine air quality pollutant prediction based on azure machine learning. Sensors (Basel, Switzerland) **18** (2018). https://doi.org/10.3390/s18040930
16. Rybarczyk, Y., Zalakeviciute, R.: Machine learning approaches for outdoor air quality modelling: a systematic review. Appl. Sci. **8**(12), 2570 (2018). https://doi.org/10.3390/app8122570
17. Liu, B., Yu, W., Wang, Y., Lv, Q., Li, C.: Research on data correction method of micro air quality detector based on combination of partial least squares and random forest regression. IEEE Access (2021). https://doi.org/10.21203/rs.3.rs-241776/v1
18. Kaur, G., Gao, J., Chiao, S., Lu, S., Xie, G.: Air quality prediction: big data and machine learning approaches. Int. J. Environ. Sci. Dev. **9**, 8–16 (2018). https://doi.org/10.18178/ijesd.2018.9.1.1066
19. Rubal, Kumar, D.: Evolving differential evolution method with random forest for prediction of air pollution. Procedia Comput. Sci. **132**, 824–833 (2018). https://doi.org/10.1016/j.procs.2018.05.094
20. Pan, B.: Application of XGBoost algorithm in hourly PM2.5 concentration prediction. IOP Conf. Ser.: Earth Environ. Sci. **113**, 012127 (2018). https://doi.org/10.1088/1755-1315/113/1/012127

Chapter 19
A Novel COVID-19-Based Optimization Algorithm (C-19BOA) for Multimodal Optimization Processes

Sheikh Safiullah, Asadur Rahman, and Shameem Ahmad Lone

Abstract The advent of ongoing pandemic due to novel coronavirus disease-2019 (COVID-19) has rapidly unsettled the health sector with a considerable fatality rate. The main factors which help in minimizing the spread of this deadly virus are the proper use of masks, social distancing, and antibody growth rate in a person. Based on these factors, the authors propose a new nature-inspired meta-heuristic algorithm, named COVID-19-based optimization algorithm (C-19BOA). The proposed C-19BOA mimics the spread and control behavior of coronavirus disease centered on three containment factors: (1) social distancing, (2) use of masks, (3) antibody rate. Initially, the mathematical models of containment factors are presented, and further, the proposed C-19BOA is developed. To ascertain the effectiveness of developed C-19BOA, its performance is verified on standard mathematical benchmark functions. These performances are compared with established bio-inspired optimization algorithms available in literature. Performance analyzes reveal the developed C-19BOA which is at par with the established optimization algorithms in terms of minimization of benchmark functions and convergence to optimal values.

19.1 Introduction

The maiden instance of ongoing coronavirus disease-2019 (COVID-19) pandemic is diagnosed at China's Wuhan city, the epicenter of COVID-19 epidemic, during Dec 2019. The disease has globally infected around 492 million people and taken life of more than 6 million so far [1]. The virus responsible for COVID-19 spreads primarily from one infected person to another person in contact with each other. Minute drops and aerosols comprising the virus spreads from an infected person to another person through nose and mouth during breathing, coughing, sneezing,

S. Safiullah · A. Rahman (✉) · S. A. Lone
Electrical Engineering Department, National Institute of Technology Srinagar, Srinagar, J&K, India
e-mail: asadur2003@yahoo.co.in

S. A. Lone
e-mail: salone@nitsri.ac.in

or speaking. Another factor of transmission, though not the main mode, is through contaminated surfaces.

Bio-inspired models are nature-inspired replicas that are aimed for successful application in hybrid methodologies. These are designed to evaluate the parameters in artificial-intelligence-based machine-learning optimizations. Meta-heuristics deal through vast search spaces and discover sub-optimal solutions in fairly desired execution times [2]. Numerous such meta-heuristic algorithms based on natural genetics [3], biogeography [4], particle swarms [5], bee colony [6], cuckoo [7], firefly [8], gray wolf [9], arithmetic-optimization-algorithm (AOA) [10], archimedes-optimization-algorithm [11] are available in literature.

19.1.1 Related Works

One of the most protracted meta-heuristic algorithms implemented to advance deep learning problems is genetic algorithm (GA) [3]. GA is constructed on the perception Darwin's well-known evolution and natural selection theories, which are expressed in terms of mathematical operators for biological features. Inspired by the mathematics of GA and biological neurons, author in [4] has formulated biogeography-based optimization (BBO). Another substantial advancement in this field is the particle-swarm-optimization (PSO) by [5]. PSO algorithm is inspired by fish and bird swarm intelligence. Authors in [6] proposed artificial-bee-colony (ABC) algorithm built on the smart behavior of honey-bee swarm. Cuckoo-search-algorithm (CSA) which is motivated based on the coercive brood-parasitic conduct of cuckoo breeds is reported in [7]. Firefly algorithm (FA) reported in [8, 12, 13] is inspired by the flashing light behavior of fireflies. MBO technique is reported in [14–17]. Authors in [9] have proposed an interesting algorithm called gray-wolf-optimizer the concept of which is driven by the supervision capabilities for hunting behavior of gray wolves persistent in a wildlife. Likewise, other bio-inspired algorithms available in literature are salp-swarm-algorithm (SSA) [18], whale-optimization-algorithm (WOA) [19], laying-chicken-algorithm [20], big bang algorithm [21], and swine-influenza-inspired-optimization (SIIO) [22].

Hence, it can be inferred from the above literature that meta-heuristic models are rising areas of interest in research. However, very few studies based on virus proliferation models are available in literature. The virus-optimization-algorithm (VOA) was suggested [23] in 2016 and further enhanced for continuous optimization problems in [24]. Authors in [25] suggested a bio-inspired meta-heuristic mimicking the coronavirus spread pattern and its infection nature. Authors in [26] have proposed COVID-19 models based on infection spreads. Similarly, authors in [27, 28] have modeled the disease propagation of COVID-19 natured processes. This motivates the authors of the present paper to propose an optimization algorithm based on COVID-19 behavior. It needs further testing for its applicability in various standard benchmark functions and modern engineering problem.

19.1.2 Objectives of the Present Study

The objectives defined for application of the proposed C-19BOA are as follows:

1. To develop population-based, nature-inspired COVID-19-based optimization algorithm (C-19BOA) based on the behavior of present-day coronavirus disease propagation. The proposed algorithm mimics the virus infection propagation and decimation phenomenon in nature.
2. To compare and authenticate the performance of C-19BOA with established optimization algorithms available in literature, based on the convergence for standard mathematical benchmark functions.

19.2 COVID-19-Based Optimization Algorithm (C-19BOA) Methodology

This section describes the methodology of the proposed C-19BOA. Initially, the containment factors are introduced, and their mathematical modeling is presented. Further, the algorithm C-19BOA is framed based on these containment factors. Attributes to be estimated for C-19BOA are explained in the following subsections.

19.2.1 Initial Population

An initial population with 'n' number of individuals (row) and 'p' parameters (column) is considered. The primary population involves one infected individual, referred as patient-zero (PZ). It is assumed that PZ infects some of the population. Primarily, a random initialization is done to infect certain individuals of the population. An initial matrix $x(n, p)$ is formed using (19.1).

$$x(n, p) = L_l + \text{random}(n, p) \times \{U_l - L_l\} \tag{19.1}$$

where L_l and U_l are the upper and lower limits of the solution. The values of upper and lower limits vary according to the problem definition. Hence, it becomes necessary to normalize the initial matrix for further calculations. The normalization of initial matrix $x(n, p)$ is done using (19.2) and represented as $x_{\text{norm}}(n, p)$ matrix.

$$x_{\text{norm}}(n, p) = \frac{x(n, p) - \min(x)}{\max(x) - \min(x)} \tag{19.2}$$

Table 19.1 Description of different elements with their respective values

Element	Description	Value
$T_{\text{infection}}$	Time taken by PZ to infect new individual	3 (user-defined)
R_0	Basic virus reproduction rate (no. of newly infected individuals produced by an infected individual)	2.4 (mean value for different states or provinces reported in [29])
W	Virus level proliferation	0.35 (reported in [29])
K	Maximum carrying capacity of virus replication	0.31 (reported in [29])
c	Virus clearance rate	2.4 (reported in [29])

19.2.2 Containment Factors

The containment factors are based on disease propagating nature of COVID-19. Containment factors majorly include (a) social distancing, (b) use of mask, and (c) antibody rate calculation of each individual after initial infection. After patient-zero (PZ) infects some of the population, containment factors for each individual are evaluated. Table 19.1 provides the description of different elements used for the containment factors' evaluation as reported in [29]. The mathematical modeling of each containment factors is discussed below.

(a) *Social Distancing (SD)*: As in case of coronavirus infection, an infected person (individual) has to be isolated from rest of the population in order to minimize the infection rate. Similar approach is dealt in the present study after initial infection. Every individual is checked for SD factor calculation. An SD matrix is formulated representing the distance among parameters in a population. In the simplest form, the distance (D) between any two normalized parameters, i and j, can be calculated as shown in (19.3).

$$\text{Distance}(D) = i - j; i \neq j \tag{19.3}$$

The normalized matrix $x_{\text{norm}}(n, p)$ represented in (19.2) is considered for SD factor calculation. The matrix $x_{\text{norm}}(n, p)$ can be elaborated as shown in (19.4). An SD matrix is formulated (19.5) with the help of $x_{\text{norm}}(n, p)$ by considering the distances among parameters having dimensions equal to $n \times j \times p$. The SD matrix formed consists of p sub-matrices each having dimensions equal to $n \times j$.

$$x_{\text{norm}}(n, p) = \begin{bmatrix} x_{\text{norm}11} & x_{\text{norm}12} & \cdots & x_{\text{norm}1p} \\ x_{\text{norm}21} & x_{\text{norm}22} & \cdots & x_{\text{norm}2p} \\ | & | & | & | \\ x_{\text{norm}n1} & x_{\text{norm}n2} & \cdots & x_{\text{norm}np} \end{bmatrix} \tag{19.4}$$

$$\text{SD}(n, j, p) = \{x_{\text{norm}}(n, p) - x_{\text{norm}}(j, p)\}; j = 1 : n, n \neq j \tag{19.5}$$

If any individual violates SD norms and comes close to an infected one, it has high chances of getting infected. IR is the infection rate due to violation of SD factor. So, higher the value of SD factor lesser is the IR. Let the distance below which infection can spread due to violation of SD factor be named as threshold distance (TD). The practical value of threshold distance (TD) according to guidelines of World Health Organization (WHO) is considered as 6 ft [30]. For the present work, the practical distance is normalized between a range of 0–1 such that 0 indicates the least value of TD, while 1 indicates the maximum value (6 ft). Let ρSD be the social distancing probability in the present work denoting the normalized value of practical distance; $0 \leq \rho SD \leq 1$.

(b) **Use of Masks**: Medical/surgical face masks block the spread of breathing drops, thereby, guarantee a higher protection against the disease propagation. As soon as the person get exposed to the infection on effective contacts (when the SD is less than TD), the individual develop infection based on rate μR. μR is the rate at which susceptible individuals of a population are exposed to infection. The value of μR is calculated using R_0 (refer Table 19.1) and given by (19.6).

$$\mu R = \frac{R_0}{T_{\text{infection}}} \tag{19.6}$$

Thus, the disease propagation rate (DPR) based on mask use for '$n \times p$' population is given by (19.7).

$$DPR(p, j, n) = \mu R\{1 - SD(n, j, p)\} \tag{19.7}$$

Consider a proportion of the population correctly and constantly wearing face masks. Let η_m be the efficiency of each mask to block virus proliferation. Based on the above assumptions, the DPR model equation given in (19.7) is further modified to mask infection rate MIR (19.8), as stated in [28].

$$MIR(p, j, n) = \mu \times R\{1 - SD(n, j, p)\} - \{SD(n, j, p) + \eta_m x_{\text{norm}}(n, p)\} \tag{19.8}$$

where $SD(n, j, p)$ is the SD matrix formed using (19.5).

Consider ρM be the mask use probability, denoting the normalized value of proper mask usage; $0\ 0 \leq \rho M \leq 1$. It is considered that any individual having SD and MIR more than social distancing probability (ρSD) and mask use probability (ρM) respectively is reinfected. Finally, the individuals are checked for their antibody rates using the following antibody rate (AR) factor.

(c) **Antibody Rate (AR)**: The previous studies reported in [31, 32] have used the response of T-immune cells to mitigate influenza virus. The present model adopts the ability of T-immune cells for killing the virus propagation (W) reported in [29]. For the present work, the antibody rate of population $x_{\text{norm}}(n, p)$ (19.2) is calculated using (19.9):

$$AR(n, j, p) = x_{\text{norm}}(n, p) \left\{ r \left(1 - \frac{x_{\text{norm}}(j, p)}{K} \right) - (c + 1) \right\} \qquad (19.9)$$

where

- AR symbolizes the infection killing rate of cells by immune response due to evolved antibody.
- K is the maximum carrying capacity of virus replication.
- r is the replication rate.
- c is the rate at which virus is cleared.

On a scale of 0–1, let ρAR be the antibody rate probability of an individual. The minimum infection killing rate of cells by immune response of an individual be denoted by 0. Whereas 1 denoting maximum infection killing rate of cells by immune response. Thus, individuals having AR more than ρAR value are considered as recovered and are treated as healthy individuals of the population. The individual with highest AR is treated as healthiest individual out of the recovered population.

19.2.3 Procedure and Flowchart for C-19BOA

The step-by-step procedure of the developed optimization algorithm C-19BOA is described as mentioned below:

1. Generate initial population with PZ as infected.
2. Normalize population.
3. For (time < iteration limit)
4. Calculate SD and MIR of individuals in the population using (19.5) and (19.8), respectively.
5. Check violations for SD and MIR.
   Individuals with SD < ρSD and MIR < ρM are reinfected and discarded. Others go for AR check.
6. Calculate AR of individuals in the population using (19.9).
7. Individuals having AR > ρAR are treated as recovered. However, individuals with AR < ρAR are unhealthy and discarded.
8. The recovered population are sorted according to their recovery rate. Store the *Best individual* from the sorted population having maximum recovery rate.
9. Continue, until point no. 3 is terminated.
10. The latest *Best individual* is the final optimum solution.

The flowchart for the developed algorithm C-19BOA is described in Fig. 19.1.

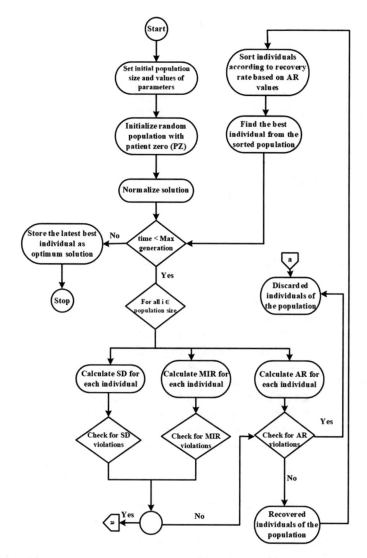

Fig. 19.1 Flowchart for developed algorithm C-19BOA

19.3   Performance Evaluation of Proposed C-19BOA on Standard Benchmark Functions

The performance of C-19BOA is evaluated on some of the standard mathematical benchmark functions based on the minimization and convergence characteristics. Performance of C-19BOA is compared with few of the established population-based optimization algorithms, viz. GA [3], BBO [4], PSO [5], FA [8], and AOA [10].

Table 19.2 Benchmark functions and their characteristics

Function	Multimodal (MM) or unimodal (UM)	Separable (S) or non-separable (NS)	Regular (R) or irregular (IR)	Dimension range
Ackley	MM	NS	R	±30
Quartic	UM	S	R	±1.28
Rastrigin	MM	S	R	±5.12
Rosenbrock	UM	NS	R	±2.048
Schwefel 2.21	MM	NS	IR	±100
Schwefel 2.22	MM	NS	IR	±10
Sphere	UM	S	R	±5.12
Schubert	MM	S	R	±10

19.3.1 Standard Mathematical Benchmark Functions

In order to discover the diverseness of C-19BOA and depict the nature of problems upon which the proposed algorithm meets the required characteristics, several standard mathematical benchmark functions are considered. To inspect the performance of C-19BOA in contrast to other optimization algorithms reported in [3–5, 8, 10], the performance comparison on these standard benchmark functions is done. This set of benchmark functions used for comparison is sufficient enough to account for all kinds of problems. Some benchmark functions are unimodal, and some are multimodal. Some of them are differentiable (regular), while some are irregular in their domains. Similarly, some functions are separable and some non-separable. Table 19.2 shows the different benchmark functions used for comparison in the present work.

19.3.2 Results Analysis of Benchmark Functions

The result comparison of proposed C-19BOA with GA, PSO, FA, BBO, and AOA on different benchmark functions is listed in Table 19.3. Each experiment is executed 50 times with diverse random seeds. The best *mean* and standard deviation (*Std. Dev.*) values are noted for comparison in Table 19.3.

It can be observed from Table 19.3, that both C-19BOA and PSO execute the average best on most of the benchmark functions. PSO is efficient enough in finding the minimum function value for majority of the cases. However, C-19BOA is the next best followed by FA. For a more clear comparison, values less than 10^{-8} are set to 0. Critical observation of the results in Table 19.3 highlights that, except for Ackley and Rastrigin function, proposed C-19BOA performs well ahead of all other optimization algorithms. C-19BOA is most dominant for Schubert benchmark function.

Table 19.3 Result of proposed C-19BOA versus GA, PSO, FA, BBO, AOA on Benchmark functions

Function	Study	GA [3]	PSO [5]	FA [8]	BBO [4]	Proposed C-19BOA	AOA [10]
Ackley	Mean	7.5498	0	0.0147	3.4287	0.2516	2.5609
	Std	5.2027	0	0.0727	1.5843	0.2663	1.4702
Quartic	Mean	−0.3417	−0.3442	−0.0129	−0.1952	−0.3442	−0.3441
	Std	0.0054	0	0.0636	0.2148	0	0.0049
Rastrigin	Mean	5.1106	1.1569	0.1328	22.1216	5.1795	2.1736
	Std	3.3795	1.0676	0.6571	14.1858	2.0060	1.5278
Rosenbrock	Mean	0.0275	0	0.0024	0.1275	0	0.0014
	Std	0.1017	0	0.0121	0.1506	0	0.0075
Schwefel 2.21	Mean	0.4267	0	0.0017	0.0539	0	0.0107
	Std	0.1044	0	0.0087	0.0550	0	0.0109
Schwefel 2.22	Mean	0.3311	0.2926	0.0117	0.2936	0.2926	0.2927
	Std	0.0294	0	0.0580	0.0011	0	0.00040
Sphere	Mean	0.7841	0	0.0020	3.6441	0	0.0893
	Std	0.7393	0	0.0099	4.4599	0	0.1350
Schubert	Mean	−242.1090	−271.2091	−8.2987	−157.0483	−299.6329	−222.8429
	Std	16.2289	0	41.0681	52.4311	0	24.7525

It should be noted that no special effort was taken to adjust the algorithm. Some optimization algorithm may take certain significant alterations for their performance evaluation, which may affect the following. Firstly, the benchmark functions may or may not have any dependency with the real-world applications. Secondly, if any alteration is done in tuning the algorithms, the benchmark functions can yield different results altogether. Hence, an effort was made to examine the best possible results by running the algorithms 50 times in order to reduce the possibility of any error on a certain population size with large iterations. This may yield the results as to how efficient a particular technique is by converging from a certain initial random value to optimal value.

Fig. 19.2 depicts the convergence characteristics of proposed C-19BOA versus GA, PSO, FA, BBO, and AOA on different benchmark functions. It is clearly seen in Fig. 19.2 that the convergence characteristics of C-19BOA are at par with the other algorithms. C-19BOA is thus effective enough in converging the function to its optimal value. Figure 19.2 shows that the proposed C-19BOA has converged to a particular optimal value like most of the other established optimization algorithms. To conclude this section, the benchmark function results indicate that the proposed C-19BOA has promising outcomes, and in the plethora of population-based optimization algorithms, this new approach might be able to find a niche.

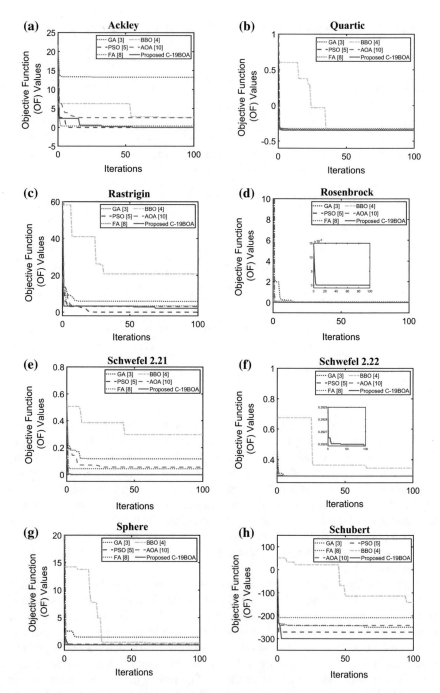

Fig. 19.2 Convergence characteristics for proposed C-19BOA versus GA, PSO, FA, BBO on benchmark functions **a** Ackley, **b** Quartic, **c** Rastrigin, **d** Rosenbrock, **e** Schwefel 2.21, **f** Schwefel 2.22, **g** Sphere, **h** Schubert

19.4 Conclusion

This paper introduces a novel bio-inspired meta-heuristic optimization algorithm, based on the behavior of present-day coronavirus disease (COVID-19). The spread of COVID-19 infection from one person to another through social contacts is modeled mathematically. Similarly, the infection spread control through containment factors is also modeled mathematically, and COVID-19-based optimization algorithm (C-19BOA) is developed. The proposed C-19BOA is successfully tested on standard mathematical benchmark functions evaluating its performance. The optimized benchmark function values for C-19BOA are compared with already established bio-inspired optimization algorithms, viz. GA, PSO, FA, BBO, and AOA. Performance analyzes reveal the developed C-19BOA which is at par with the established optimization algorithms in terms of minimization of benchmark functions and convergence to optimal values. Future research can be carried out on post vaccination effects, with possible decline in virus proliferation.

References

1. Worldometer. https://www.worldometers.info/coronavirus
2. Del Ser, J., Osaba, E., Molina, D., Yang, X.S., Salcedo-Sanz, S., Camacho, D., et al.: Bio-inspired computation: where we stand and what's next. Swarm Evol. Comput. **48**, 220–250 (2019)
3. Holland, J.H., et al.: Adaptation in Natural and Artificial Systems: An Introductory Analysis with Applications to Biology, Control, and Artificial Intelligence. MIT Press (1992)
4. Simon, D.: Biogeography-based optimization. IEEE Trans. Evol. Comput. **12**(6), 702–713 (2008)
5. Kennedy, J., Eberhart, R.: Particle swarm optimization. In: Proceedings of ICNN'95—International Conference on Neural Networks, vol. 4, pp. 1942–1948. IEEE (1995)
6. Karaboga, D., Basturk, B.: A powerful and efficient algorithm for numerical function optimization: artificial bee colony (ABC) algorithm. J. Glob. Optim. **39**(3), 459–471 (2007)
7. Yang, X.S., Deb, S.: Cuckoo search via Lévy flights. In: World Congress on Nature & Biologically Inspired Computing (NaBIC), pp. 210–214. IEEE (2009)
8. Yang, X.S.: Nature-Inspired Metaheuristic Algorithms. Luniver Press (2010)
9. Mirjalili, S., Mirjalili, S.M., Lewis, A.: Grey wolf optimizer. Adv. Eng. Softw. **69**, 46–61 (2014)
10. Archimedes optimization algorithm: a new metaheuristic algorithm for solving optimization problems. Appl. Intell. **51** (2021)
11. The arithmetic optimization algorithm. Comput. Methods Appl. Mech. Eng. **376**, 113609 (2021)
12. Sheikh, A., Bakhsh, F.: Optimal power flow through variable frequency transformer using different optimization techniques. Distrib. Gener. Alternative Energy J. (2022). https://doi.org/10.13052/dgaej2156-3306.37410
13. Safiullah, S., Rahman, A.: Performance evaluation of hybrid power system incorporating electric-vehicles. Distrib. Gener. Alternative Energy J. **37**(4), 1055–1082 (2022). https://doi.org/10.13052/dgaej2156-3306.3748
14. Safiullah, S., Rahman, A., Ahmad Lone, S.: Optimal control of electrical vehicle incorporated hybrid power system with second order fractional-active disturbance rejection controller. Opt. Control Appl. Methods (2021). https://doi.org/10.1002/oca.2826

15. Safiullah, S., Rahman, A., Lone, S.A.: State-observer based IDD controller for concurrent frequency-voltage control of a hybrid power system with electric vehicle uncertainties. Int. Trans. Electr. Energy Syst. **31**(11), e13083 (2021). https://doi.org/10.1002/2050-7038.13083
16. Farooq, Z., Rahman, A., Lone, S.A.: Load frequency control of multi-source electrical power system integrated with solar-thermal and electric vehicle. Int. Trans. Electr. Energy Syst. **31**(7), e12918 (2021)
17. Farooq, Z., Rahman, A., Lone, S.A.: System dynamics and control of EV incorporated deregulated power system using MBO-optimized cascaded ID-PD controller. Int. Trans. Electr. Energy Syst. **31**(11), e13100 (2021)
18. Mirjalili, S., Gandomi, A.H., Mirjalili, S.Z., Saremi, S., Faris, H., Mirjalili, S.M.: Salp Swarm Algorithm: a bio-inspired optimizer for engineering design problems. Adv. Eng. Softw. **114**, 163–191 (2017)
19. Mirjalili, S., Lewis, A.: The whale optimization algorithm. Adv. Eng. Softw. **95**, 51–67 (2016)
20. Hosseini, E.: Laying chicken algorithm: a new meta-heuristic approach to solve continuous programming problems. J. Appl. Comput. Math. **6**(1) (2017)
21. Eghbal, H.: Big bang algorithm: a new meta-heuristic approach for solving optimization problems. Asian J. Appl. Sci. **10**(3), 134–144 (2017)
22. Pattnaik, S.S., Jadhav, D.G., Devi, S., Ratho, R.K.: Swine influenza inspired optimization algorithm and its application to multimodal function optimization and noise removal. Artif. Intell. Res. **1**(1), 18–30 (2012)
23. Liang, Y.C., Cuevas Juarez, J.R.: A novel metaheuristic for continuous optimization problems: virus optimization algorithm. Eng. Optim. **48**(1), 73–93 (2016)
24. Liang, Y.C., Juarez, J.R.C.: A self-adaptive virus optimization algorithm for continuous optimization problems. Soft Comput. 1–20 (2020)
25. Martínez-Álvarez, F., Asencio-Cortés, G., Torres, J., Gutiérrez-Avilés, D., Melgar-García, L., Pérez-Chacón, R., et al.: Coronavirus Optimization Algorithm: a bioinspired metaheuristic based on the COVID-19 propagation model. Big Data **8**(4), 308–322 (2020)
26. Hosseini, E., Ghafoor, K.Z., Sadiq, A.S., Guizani, M., Emrouznejad, A.: Covid-19 optimizer algorithm, modeling and controlling of coronavirus distribution process. IEEE J. Biomed. Health Inform. **24**(10), 2765–2775 (2020)
27. Samui, P., Mondal, J., Khajanchi, S.: A mathematical model for COVID-19 transmission dynamics with a case study of India. Chaos Solitons Fractals **140**, 110173 (2020)
28. Srivastav, A.K., Tiwari, P.K., Srivastava, P.K., Ghosh, M., Kang, Y.: A mathematical model for the impacts of face mask, hospitalization and quarantine on the dynamics of COVID-19 in India: deterministic vs. stochastic. Math. Biosci. Eng. **18**(1), 182–213 (2021)
29. Hernandez-Vargas, E.A., Velasco-Hernandez, J.X.: In-host mathematical modelling of covid-19 in humans. Annu. Rev. Control (2020)
30. COVID-19: Keep on Keeping Your Distance. https://www.healthychildren.org/English/health-issues/conditions/COVID-19/Pages/Social-Distancing-Why-Keeping-Your-Distance-Helps-Keep-Others-Safe.aspx
31. Hernandez-Vargas, E.A., Wilk, E., Canini, L., Toapanta, F.R., Binder, S.C., Uvarovskii, A., et al.: Effects of aging on influenza virus infection dynamics. J. Virol. **88**(8), 4123–4131 (2014)
32. Hancioglu, B., Swigon, D., Clermont, G.: A dynamical model of human immune response to influenza A virus infection. J. Theor. Biol. **246**(1), 70–86 (2007)

Chapter 20
Improving English-Assamese Neural Machine Translation Using Transliteration-Based Approach

Sahinur Rahman Laskar, Bishwaraj Paul, Partha Pakray, and Sivaji Bandyopadhyay

Abstract Natural language translation is a well-defined task of linguistic technology that minimizes communication gap among people of diverse linguistic backgrounds. Although neural machine translation attains remarkable translational performance, it requires adequate amount of train data, which is a challenging task for low-resource language pair translation. Also, neural machine translation handles rare word problems, i.e., low-frequency words translation at the subword level, but it shows weakness for highly inflected language translation. In this work, we have explored neural machine translation on low-resource English-Assamese language pair with a proposed transliteration approach in the data preprocessing step. In the transliteration approach, the source language is transliterated into target language script that leverages a smaller subword vocabulary for the source-target languages. Moreover, the pre-trained embeddings on the monolingual data of transliterated source and target languages are used in the training process. With our approach, the neural machine translation significantly improves translational performance for English-to-Assamese and Assamese-to-English translation and obtain state-of-the-art results.

20.1 Introduction

Human spoken or natural languages are grouped into low, medium and high-resource languages depending on their resource availability. Resources include texts written by native speakers, online data [1, 2] and computational data resources [3]. A language refers to the low-resource category which have few online resources or computational data and in contrast to neural machine translation (MT), minimal data require for the model training [4]. Determining which pair is a low-resource possesses a research question. Because of the different styles of inflected words causes, the presence of sparsity issues further complicates definition of "low-resource". Therefore, more parallel data is required to attain reasonable results for the languages having less inflected

S. R. Laskar (✉) · B. Paul · P. Pakray · S. Bandyopadhyay
Department of Computer Science and Engineering, National Institute of Technology, Silchar, India
e-mail: sahinurlaskar.nits@gmail.com

© The Author(s), under exclusive license to Springer Nature Singapore Pte Ltd. 2023
V. Bhateja et al. (eds.), *Evolution in Computational Intelligence*, Smart Innovation, Systems and Technologies 326, https://doi.org/10.1007/978-981-19-7513-4_20

words [5]. The language pairs in a train set, containing parallel data less than one million, are typically considered as a low-resource [6]. Due to insufficient resources, English-Assamese (En-As) is considered as a low-resource pair. The Assamese language belongs to the Indo-Aryan language family and official/regional language of the Indian north-eastern (NE) state of Assam. Unlike English, the Assamese script is derived from Bengali-Assamese script[1] and it is a morphological rich language [7]. The word order of English is subject-verb-object (SVO), whereas Assamese follows subject-object-verb (SOV). Because of the impediment of the adequate data, the En-As MT task is in the beginning stage [8–11]. In this paper, we have investigated to improve En-As pair translation in both directions using neural machine translation (NMT) techniques. The contributions of this paper are as follows:

- Proposed a source language transliteration-based approach to enhance the translational performance of low-resource NMT for En-As pair.
- Achieved a state-of-the-art MT results by exploring different NMT models for both forward and backward directions of En-As pair.

The structure of the paper is organized as follows: Sect. 20.2 briefly discusses the background concept of NMT and review of related works. The proposed approach is described in Sect. 20.3. Sections 20.4 and 20.5 present experimental results and conclusion of paper with future scope.

20.2 Background Concept and Related Work

In this section, the background of NMT is discussed briefly and review the prior works on En-As MT task.

20.2.1 NMT

Prior to NMT, statistical machine translation (SMT) [12, 13] is the state-of-the-art approach in the MT task. However, long-term dependency issue, inability of contextual analysis and complexity of system lead to the development of NMT [14–17]. The encoder-decoder architecture of RNN-based NMT uses long short-term memory (LSTM) to encounter long-term dependency issue. Moreover, the attention mechanism is introduced [14, 16], for very long sentences. The concept behind the attention mechanism is that the decoder focuses on different parts of the source sentence globally at various decoding steps. Here, input sequences $s_1, s_2 \ldots s_n$ pass to the encoder and generate a context vector X. Based on conditional probability, as given in Eq. (20.1), the decoder generates the output sequences $t_1, t_2 \ldots t_m$.

---

[1] https://rb.gy/owt9mq.

$$P(t \mid s) = \sum_{i=1}^{m} P\left(t_i \mid t_{<1}, X\right) \tag{20.1}$$

Then, an attention vector a is computed by estimating hidden states of source side $\left(h_{ip}\right)$ and target side $\left(h_{op}\right)$. Equation (20.2) presents the computation of an attention vector a.

$$a_v = \frac{\exp\left(\text{score}\left(h_{op}, h'_{ip}\right)\right)}{\sum_{i'} \exp\left(\text{score}\left(h_{op}, h_{i'}\right)\right)} \tag{20.2}$$

In current work, we have employed the score function using Eq. (20.3).

$$\text{score}\left(h_{op}, h'_{ip}\right) = h_{op} W_a h'_{ip} \tag{20.3}$$

By considering the average weights of hidden states (input) and the attention vector, context vector c_l is calculated. An attentional hidden vector generates by the concatenation of h_{op} and c_t as shown in Eq. (20.4).

$$h'_{op} = \tanh\left(W_c\left[c_t, h_{op}\right]\right) \tag{20.4}$$

Lastly, softmax layer incorporated to the vector h'_{op} using Eq.(20.5) to produce the predicted target sentence.

$$P\left(t_j \mid t_{<1}, X\right) = \text{softmax}\left(W_s h'_{op}\right) \tag{20.5}$$

The bidirectional RNN (BRNN) considers two independent RNNs for forward and backward directions [18]. The drawback of RNN-based NMT is lack of parallelization, which introduces transformer-based NMT [19]. Here, the self-attention mechanism computes attention multiple times, called as multi-head attention. It helps the model to represent different words through multiple positions. In this work, NMT models, namely, NMT (BRNN) with global attention [16] and NMT (transformer) [19] models are used for both directions of translation, i.e., As-to-En and En-to-As.

20.2.2 Related Work

In the literature survey, we focused the existing MT work on, particularly the En-As language pair. It is found that all the existing works are focused on the preparation of dataset [8–11]. In [8, 9], the authors build SMT systems on a very limited dataset. A parallel corpus contributed, namely, EnAsCorp1.0 [10] and En-As baseline systems were developed using SMT (phrase-based) and NMT (RNN). Moreover, [11] contributed parallel corpora, namely, Samanantar which consists of 11 Indian languages including Assamese, and developed transformer-based multilingual NMT model for En-to-Indic and Indic-to-En translations. In [20], the authors investigated En-As pair translations by utilizing EnAsCorp1.0 [10] via data augmentation approach. In this work, we have utilized parallel dataset of both EnAsCorp1.0 and Samanantar and investigates NMT models using transliteration-based approach with pre-trained embeddings for En-to-As and vice versa translations.

20.3 Transliteration Approach

In this work, we have proposed a transliteration-based approach to improve the translational performance of NMT. The objective of the transliteration approach is to allow subword-level lexical sharing between source and target languages and provide a smaller subword vocabulary that will be shared during the training process. Figure 20.1 presents the pictorial diagram of the proposed transliteration-based approach. Our approach consists of the following steps:

- The source language text is transliterated into the target language script (TSL). For instance, En-to-As translation, the source language English is transliterated into the Assamese script. And, for As-to-En translation, the source language Assamese is transliterated into the English script. We have used Indic-trans[2] [21] for forward and backward directions (En-to-As/As-to-En) of transliteration.
- We have applied jointly learn byte pair encoding (subword level) [22] on the transliterated source and target sentences in the data preprocessing step.
- We have pre-trained embeddings (PE) using GloVe [23] on the monolingual data of transliterated source and target languages, and obtained GloVe vectors at subword level are used in the training process.

20.4 Experiment and Result and Analysis

In this work, parallel corpora, namely, EnAsCorp1.0 [10] and Samanantar [11] are used, and the data statistics are presented in Table 20.1. To increase the training amount of data, we have merged both dataset, EnAsCorp1.0 and Samanantar, as shown in Table 20.1. We have used test set from [10] only because the source of test data [11] is PMIndia[3] which is already present in the train data of [10].

For experiments, NMT setup is utilized. For NMT setup, BRNN (BiLSTM) and transformer models are build by utilizing the OpenNMT-py [24] toolkit. The OpenNMT-py toolkit is employed to perform data preprocessing, training and translation steps. The two layer LSTM-based encoder-decoder architecture is used for the

---

[2] https://github.com/libindic/indic-trans.

[3] http://data.statmt.org/pmindia/v1/parallel/.

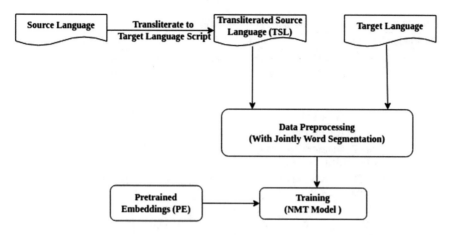

Fig. 20.1 Transliteration-based approach

Table 20.1 Data statistics of train, validation and test set

Type	Sentences	Tokens	
		En	As
Train set-1 [10]	203,315	2,414,172	1,986,270
Train set-2 [11]	138,353	1,715,435	1,377,336
Total	341,668	4,129,607	3,363,606
Validation set-1 [10]	4500	74,561	59,677
Validation set-2 [11]	1000	19,922	16,824
Total	5500	94,483	76,501
Test set [10]	2500	41,985	34,643

BRNN models with global attention mechanism [16] and 0.3 drop-out. However, six layers with drop-out of 0.1 are used in the transformer model. And, Adam optimizer with learning rate 0.001 is used in the NMT models. The word segmentation technique, byte pair encoding (BPE) [22] with 32k merge operations, is employed in the data preprocessing step. A single NVIDIA Quadro P2000 GPU is used to train the models.

For the evaluation of quantitative results of generated translation obtained from different MT models, automatic evaluation metrics, namely, BLEU [25], RIBES [26], TER [27], METEOR [28] and F-measure scores are used. Except TER, in all other metrics, more the score, more is the prediction accuracy while in case of TER it is exactly opposite. Table 20.2 presents the automatic evaluation scores of different NMT models. It is observed that the transformer model with transliterated source language (TSL) and pre-trained embeddings (PE) achieves better automatic evaluation scores for both direction of translations. Figure 20.2 presents comparative analysis of BLEU scores on the same test data which is used in this work. It is observed that

Table 20.2 Experimental results of automatic evaluation scores

Translation	Model	BLEU	TER	RIBES	METEOR	F-measure
En-to-As	NMT (BiLSTM)	6.32	93.5	0.281621	0.0994	0.2014
	NMT (BiLSTM) (With TSL)	8.12	88.9	0.394610	0.1512	0.3671
	NMT (BiLSTM) (With TSL + PE)	8.90	87.0	0.422308	0.1571	0.3790
	NMT (Transformer)	6.74	93.2	0.293234	0.1021	0.2099
	NMT (Transformer) (With TSL)	8.58	86.8	0.419994	0.1537	0.3731
	NMT (Transformer) (With TSL + PE)	9.14	85.9	0.425550	0.1608	0.3844
As-to-En	NMT (BiLSTM)	12.33	88.7	0.408232	0.1321	0.2771
	NMT (BiLSTM) (With TSL)	14.24	78.1	0.539332	0.2409	0.6003
	NMT (BiLSTM) (With TSL + PE)	14.96	77.5	0.546596	0.2436	0.6025
	NMT (Transformer)	12.54	88.2	0.405376	0.1354	0.2827
	NMT (Transformer) (With TSL)	14.38	77.4	0.544878	0.2768	0.6094
	NMT (Transformer) (With TSL + PE)	15.60	77.2	0.551847	0.2858	0.6127

TSL Transliterated source language, *PE* Pre-trained embeddings

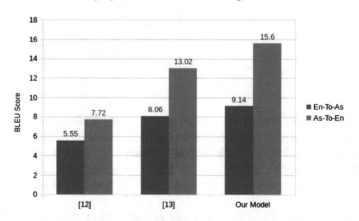

Fig. 20.2 BLEU score comparison among [10, 20] and our best model for English-Assamese pair translation

our best NMT model (Transformer with TSL and PE) attains higher scores than that of [10, 20]. In the predicted translations, it is manually observed the fluency (well-formed sentence) is better in case of As-to-En than that of En-to-As translation. Moreover, the NMT models show weakness in the adequacy (contextual meaning) factor of translation for both direction of translation.

20.5 Conclusion and Future Work

In this work, we have explored NMT using BiLSTM and transformer models on low-resource English-Assamese pair for both forward and backward directions of translation. We have utilized transliteration-based approach with pre-trained embeddings in NMT models, and the transformer-based NMT model achieves the best results in terms of automatic evaluation scores. In the future work, the parallel data will be increased. And, a multilingual transfer learning approach will be investigated for further research.

Acknowledgements We want to thank the Center for Natural Language Processing (CNLP), the Artificial Intelligence (AI) Lab and the Department of Computer Science and Engineering at the National Institute of Technology, Silchar, India, for providing the requisite support and infrastructure to execute this work.

References

1. Megerdoomian, K., Parvaz, D.: Low-density language bootstrapping: the case of Tajiki Persian. In: Proceedings of the Sixth International Conference on Language Resources and Evaluation (LREC'08), pp. 3293–3298. European Language Resources Association (ELRA), Marrakech, Morocco (2008)
2. Probst, K., Brown, R., Carbonell, J., Lavie, A., Levin, L.S., Peterson, E.: Design and implementation of controlled elicitation for machine translation of low-density languages, pp. 3293–3298 (2001)
3. Hogan, C.: OCR for minority languages. In: Symposium on Document Image Understanding Technology (1999)
4. Gu, J., Hassan, H., Devlin, J., Li, V.O.K.: Universal neural machine translation for extremely low resource languages. In: Walker, M.A., Ji, H., Stent, A. (eds.) Proceedings of the 2018 Conference of the North American Chapter of the Association for Computational Linguistics: Human Language Technologies, NAACL-HLT USA, June 1–6, vol. 1 (Long Papers), pp. 344–354. Association for Computational Linguistics, New Orleans, Louisiana (2018)
5. Denkowski, M., Neubig, G.: Stronger baselines for trustable results in neural machine translation. In: Proceedings of the First Workshop on Neural Machine Translation, pp. 18–27. Association for Computational Linguistics, Vancouver (2017)
6. Kocmi, T.: Exploring benefits of transfer learning in neural machine translation (2020)
7. Saharia, N., Das, D., Sharma, U., Kalita, J.: Part of speech tagger for Assamese text. In: Proceedings of the ACL-IJCNLP 2009 Conference Short Papers, pp. 33–36. Association for Computational Linguistics, Suntec, Singapore (2009)
8. Barman, A., Sarmah, J., Sarma, S.: Assamese WordNet based quality enhancement of bilingual machine translation system. In: Proceedings of the Seventh Global Wordnet Conference, pp. 256–261. University of Tartu Press, Tartu, Estonia (2014)
9. Baruah K.K., Das P., Hannan A., Sarma, S.K.: Assamese-English bilingual machine translation. Int. J. Nat. Lang. Comput. (IJNLC) **3** (2014)
10. Laskar, S.R., Khilji, A.F.U.R., Pakray, P., Bandyopadhyay, S.: EnAsCorp1.0: English-Assamese corpus. In: Proceedings of the 3rd Workshop on Technologies for MT of Low Resource Languages, pp. 62–68. Association for Computational Linguistics, Suzhou, China (2020)

11. Ramesh, G., Doddapaneni, S., Bheemaraj, A., Jobanputra, M., AK, R., Sharma, A., Sahoo, S., Diddee, H., Kakwani, D., Kumar, N., et al.: Samanantar: the largest publicly available parallel corpora collection for 11 Indic languages. Trans. Assoc. Comput. Linguist. **10**, 145–162 (2022)

12. Koehn, P.: Statistical Machine Translation, 1st edn. Cambridge University Press, USA (2010)

13. Tarakeswara Rao, B., Patibandla, R., Murty, M.R.: A comparative study on effective approaches for unsupervised statistical machine translation. In: Embedded Systems and Artificial Intelligence, pp. 895–905. Springer (2020)

14. Bahdanau, D., Cho, K., Bengio, Y.: Neural machine translation by jointly learning to align and translate. In: Bengio, Y., LeCun, Y. (eds.) 3rd International Conference on Learning Representations, ICLR 2015, May 7–9, Conference Track Proceedings, pp. 1–15. arXiv, San Diego, CA, USA (2015)

15. Cho, K., van Merriënboer, B., Gulcehre, C., Bahdanau, D., Bougares, F., Schwenk, H., Bengio, Y.: Learning phrase representations using RNN encoder–decoder for statistical machine translation. In: Proceedings of the 2014 Conference on Empirical Methods in Natural Language Processing (EMNLP), pp. 1724–1734. Association for Computational Linguistics, Doha, Qatar (2014)

16. Luong, T., Pham, H., Manning, C.D.: Effective approaches to attention-based neural machine translation. In: Proceedings of the 2015 Conference on Empirical Methods in Natural Language Processing, pp. 1412–1421. Association for Computational Linguistics, Lisbon, Portugal (2015)

17. Sutskever, I., Vinyals, O., Le, Q.V.: Sequence to sequence learning with neural networks. In: Proceedings of the 27th International Conference on Neural Information Processing Systems, vol. 2, pp. 3104–3112. NIPS'14, MIT Press, Cambridge, MA, USA (2014)

18. Ramesh, S.H., Sankaranarayanan, K.P.: Neural machine translation for low resource languages using bilingual lexicon induced from comparable corpora. In: Proceedings of the 2018 Conference of the North American Chapter of the Association for Computational Linguistics: Student Research Workshop, pp. 112–119. Association for Computational Linguistics, New Orleans, Louisiana, USA (2018)

19. Vaswani, A., Shazeer, N., Parmar, N., Uszkoreit, J., Jones, L., Gomez, A.N., Kaiser, L.u., Polosukhin, I.: Attention is all you need. In: Guyon, I., Luxburg, U.V., Bengio, S., Wallach, H., Fergus, R., Vishwanathan, S., Garnett, R. (eds.) Advances in Neural Information Processing Systems 30, pp. 5998–6008. Curran Associates, Inc. (2017)

20. Laskar, S.R., Ur Rahman Khilji, A.F., Pakray, P., Bandyopadhyay, S.: Improved neural machine translation for low-resource English–Assamese pair. J. Intell. Fuzzy Syst. **42**(5), 4727–4738 (2022)

21. Bhat, I.A., Mujadia, V., Tammewar, A., Bhat, R.A., Shrivastava, M.: IIIT-H system submission for FIRE2014 shared task on transliterated search. In: Proceedings of the Forum for Information Retrieval Evaluation, pp. 48–53. FIRE '14, Association for Computing Machinery, New York, NY, USA (2014)

22. Sennrich, R., Haddow, B., Birch, A.: Neural machine translation of rare words with subword units. In: Proceedings of the 54th Annual Meeting of the Association for Computational Linguistics (vol. 1: Long Papers), pp. 1715–1725. Association for Computational Linguistics, Berlin, Germany (2016)

23. Pennington, J., Socher, R., Manning, C.D.: Glove: global vectors for word representation. In: Moschitti, A., Pang, B., Daelemans, W. (eds.) Proceedings of the 2014 Conference on Empirical Methods in Natural Language Processing, EMNLP 2014, October 25–29, A meeting of SIGDAT, a Special Interest Group of the ACL, pp. 1532–1543. ACL, Doha, Qatar (2014)

24. Klein, G., Kim, Y., Deng, Y., Senellart, J., Rush, A.: OpenNMT: open-source toolkit for neural machine translation. In: Proceedings of ACL 2017, System Demonstrations, pp. 67–72. Association for Computational Linguistics, Vancouver, Canada (2017)

25. Papineni, K., Roukos, S., Ward, T., Zhu, W.J.: Bleu: a method for automatic evaluation of machine translation. In: Proceedings of the 40th Annual Meeting on Association for Computational Linguistics, pp. 311–318. ACL '02, Association for Computational Linguistics, Stroudsburg, PA, USA (2002)

26. Isozaki, H., Hirao, T., Duh, K., Sudoh, K., Tsukada, H.: Automatic evaluation of translation
    quality for distant language pairs. In: Proceedings of the 2010 Conference on Empirical Meth-
    ods in Natural Language Processing, pp. 944–952. Association for Computational Linguistics,
    Cambridge, MA (2010)
27. Snover, M., Dorr, B., Schwartz, R., Micciulla, L., Makhoul, J.: A study of translation edit rate
    with targeted human annotation. In: Proceedings of Association for Machine Translation in the
    Americas, pp. 223–231 (2006)
28. Lavie, A., Denkowski, M.J.: The METEOR metric for automatic evaluation of machine trans-
    lation. Mach. Transl. **23**(2–3), 105–115 (2009)

Chapter 21
Sentimental Analysis of Twitter Users' Text Using Machine Learning

Debashri Debnath and **Nikhil Debbarma**

Abstract Globally, most people are suffering from various mental health issues because of workload, relationship loss, death of close people, failure, etc. Social media has become a great place to share opinions related to particular issues. Among the different microblogging sites, Twitter is the best platform to analyze the mental status of its users as tweets support 280 characters. This paper aims to build a model which can predict Twitter users' emotions based on their tweets. For preparing the dataset, Twitter scrapper has been used over the publicly available data with the help of Python Programming and to remove the noise from the dataset used nltk library and also used emotion-related hashtags and their relatable hashtags to scrap the tweets. Machine learning is used for text analysis. The results of different models were tested through measuring parameters like Precision, Recall, and $F1$ score, and the k-fold cross-validation has been done.

21.1 Introduction

Social networks have evolved into a great place for their users to communicate with their friends and become a very important part of everyone's life. People express and share their thoughts, interests, and hobbies on various microblogging sites such as Facebook, Twitter, and Instagram, and the mental state of a person can be assessed by examining little facts posted on social media. The mental state defines the perspective of a person and gives signs about his/her general behavior. Emotions are complicated neurological expression which has three sections: Sentimental occurrence, Physical and Psychological reaction, and Sociological reaction. Mental illness is one of the most well-known issues that is affecting the everyday life of a person. Most people are suffering from mental health issues because of workload, relationship loss, death of close people, failure, abuse, drug addiction, etc. Sentiment analysis, also known as opinion mining, is employed in this study to evaluate the quality of information derived from text data. It is used to analyze the opinions of the customer, product

D. Debnath (✉) · N. Debbarma
National Institute of Technology Agartala, Jirania 799046, Tripura, India
e-mail: debashri50@gmail.com

© The Author(s), under exclusive license to Springer Nature Singapore Pte Ltd. 2023
V. Bhateja et al. (eds.), *Evolution in Computational Intelligence*, Smart Innovation, Systems and Technologies 326, https://doi.org/10.1007/978-981-19-7513-4_21

review, feedback, and unstructured data. Unstructured data like Twitter, Facebook, and E-mail is very difficult to maintain because of various problems such as virtual noise effect and unspecific data. Here, total 202 hashtags are used to collect data from Twitter through scrapping. A total of 16,658 tweets have been collected using scrapping. This study aims to determine the mental health state of Twitter users by analyzing their tweets. The dataset is segmented into test and training data before the machine learning algorithm is applied, and the model is trained using several classifiers such as Naive Bayes, Random Forest, and Support Vector Machine. Precision, Recall, and $F1$ score are used to compare the accuracy of each model.

21.2   Related Work

Chhinder et al. [1] predicted sentiment related to the social issues of women, data collected through Twitter scrapping and Twitter API and used Machine Learning tools and Python programming. Rakshitha et al. [8] analyzed tweets to know about mental health and used supervised machine learning approaches and data collected using Twitter API and used two classifiers. Sudha Tushara et al. [11] used R language and word frequency method, and Singular Value Decomposition method. Mehak et al. [4] predicted mental health disease and collected data using Twitter API. Machine learning algorithms and function extraction received accuracy of 92%. Rahul et al. [7] collected data from mental health survey 2019 which contains data of technical nontechnical employees. Received highest accuracy of 84% with the Decision tree classifier. Iwan et al. [2] build a corpus model and collected data using Twitter API and data crawling and used the features of SenticNet's four dimensions. Siddharth et al. [10] explored how text analysis can be used to analyze some issues, used Twitter API, MongoDB, and No SQL to store data from Twitter, and used Naive Bayes and Neural Network. Kausar et al. [3] predicted sentiment during the COVID-19 and Twitter API, RTweet package, and different hashtags were used to collect data. Rehab et al. [9] predicted Arabic tweets and Facebook comments for sentimental analysis. Munmun et al. [5] predicted depression before the beginning of it is reported and received 70% accuracy. Srinivasulu et al. [13] analyzed stress and mental health conditions and received 75.3% accuracy. Tripti et al. [12] received the highest 88.51% accuracy using Decision tree and measured accuracy, Recall, Precision, and F1 score. Murty et al. [6] used cross-domain techniques for text categorization. Reddy et al. [13] applied machine learning algorithms to predict stress and mental health conditions, and the results were 75.13% accurate.

21.3   Methodology

This research work focuses on mining the text written in "English" language to predict the mental status of Twitter users through their tweets by following the sentiment analysis steps. Tweets were extracted using the Twitter scraper, 202 hashtags were

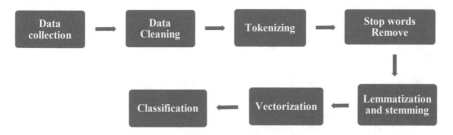

Fig. 21.1 Steps involved in methodology

1.46E+18	AlumasiR;Alumasi R	0	0	3	TRUE	Recovery is hard but regret is harder #MentalHealth #Mental	['MentalHealth', []	https://twitter.com/AlumasiRehab/st
1.47E+18	jesswithie;Messy	1	0	1	TRUE	That time someone told me drinking orange peel will cure m	['depression'] []	https://twitter.com/jesswithie/status
1.47E+18	ASadGuy5;A Sad Guy	0	0	1	TRUE	#depression has the upper hand this morning. Just I was was	['depression'] []	https://twitter.com/ASadGuy52/statu
1.47E+18	Rivka_m_;Rivka M. S	0	0	1	TRUE	Disassociation is one of the brains defense mechanisms you	['anxietyrelief', '[]	https://twitter.com/Rivka_m_stieh/st
1.46E+18	drunkbutt Drunking l	0	0	0	TRUE	I feel so alone and empty â€¦ #edtwt #depression #sad	['edtwt', 'depress []	https://twitter.com/drunkbutter/stat
1.46E+18	TrizzyJam;Pat Triano	2	8	11	TRUE	To anyone who has ever dealt with #Anxiety #Depression or	['Anxiety', 'Depre []	https://tw https://t.co/lcpyOcG9nC?ar
1.47E+18	Cherishec;Cherishec	0	1	1	TRUE	SEASONAL MOODS... â â Can you identify with this? â â â #Ch	['Cherishedvoice []	https://twitter.com/CherishedVoices,
1.47E+18	Muad_dib;Muad'dib	0	0	1	TRUE	Being on the razor edge of despair is survival, not living,#me	['mentalhealthav []	https://twitter.com/Muad_dib_UK/st:
1.47E+18	Irie_Well;Irie Welln	0	2	2	TRUE	These CBD oils to help prepare going back to work/school.-m	['mentalhealth', '[]	https://twitter.com/Irie_Wellness/st:
1.47E+18	kimba715;kimba7	0	0	0	TRUE	Welcome back #winter #depression :D long time no see :)	['winter', 'depres []	https://twitter.com/kimba715/status/
1.46E+18	pruthuviy;Pruthuviy	2	3	8	TRUE	People around the world are getting #SemicolonTattoos to s	['SemicolonTatto []	https://twitter.com/pruthuviya/statu:
1.46E+18	tristimani stu alexar	1	0	4	TRUE	Hypomanic and Anxious - not a nice mix. #manic #hypomani	['manic', 'hypom: []	https://twitter.com/tristimania101/st
1.47E+18	mindyour Skye	0	0	1	TRUE	Stop wasting time with people who canâ€™t even give you t	['depression', 'an []	https://twitter.com/mindyourbusiess
1.46E+18	Salomen_Salomen	0	0	0	TRUE	I know itâ€™s bad but atm this helps to calm me. #cigarettes	['cigarettes', 'sm[]	https://twitter.com/Salomen_Enigma
1.46E+18	Saachi515 Rayburn	1	7	11	TRUE	The mental and physical health problems this batch is facing	['depression', 'Ar []	https://twitter.com/Saachi51531049/s
1.47E+18	JoS475848 Jo	0	0	0	TRUE	Today I took the first steps in looking after my mental health.	['anxiety', 'panic', ['vitaminc https://twitter.com/JoS475848/statu
1.47E+18	UkHealing Natural H	0	0	0	TRUE	Banish the Blues Now by @CSCarrigan: https://amazon.com/	['healing', 'intuiti ['CSCarrig https://tw https://t.co/bOOFblqtv3?ar
1.47E+18	Journal_C Journal_C	1	3	7	TRUE	Here's a look inside ourChasing Peace JournalThis is our Anx	['MentalHealthM []	https://twitter.com/Journal_Cue/stat
1.47E+18	iamjenma;AnakNiHe	0	0	1	TRUE	Things aren't that good and smooth but LIFE GOES ON, so yes	['loveyourself', 'å []	https://twitter.com/iamjenmariano/s

Fig. 21.2 Tweets collected for #Depression

used for data collection, and then all were stored in standard CSV format. Here, Plutchik's wheel of emotion and different mental health disease-related keywords were used to collect the 16,658 data. Our methodology includes the following steps shown in Fig. 21.1.

21.3.1 Data Collection

Twitter Scraping is used here because the Twitter API has some limits when it comes to get data. Figure 21.2 shows the raw data scraped for #Depression using Python libraries. There were thirteen columns in the scraped data, but we need only Content and Hashtag column. Any unwanted and uninformative tweets present on scraped data must be removed as such uninformative things may train the model wrongly or may give misleading results.

Now, we attempted to locate the word frequency graph, which is a well-known method for determining the frequency of each word in NLP (Natural Language Processing). In Fig. 21.3, word frequency graph is generated for all the hashtags which we used to collect data from the Twitter. It shows the number of times a word is tweeted on Twitter by its users, where X line represents "Maximum keywords" and the Y line represents "Number of keywords" used.

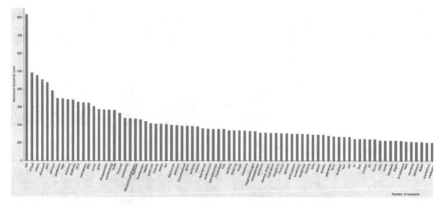

Fig. 21.3 Word frequency graph

Fig. 21.4 Word cloud created for different hashtags

The generated WordClouds of Fig. 21.4 shows the emotion-related hashtags which were used by Twitter users in their tweets. "Wordcloud" library is used in Python for creating this WordClouds. Here, color and size of each word show the frequency of it, that is, it gives more sense to the words used frequently in tweets about a particular topic.

21.3.2 Data Pre-processing

Text pre-processing is a crucial step in the text mining process. The high dimensionality and large size of textual data are the most significant challenges in text mining. Text data can be reduced in dimensionality and size using natural language processing and morphological methods. Data pre-processing will be performed on the merged dataset that contains all the raw data collected through individual hashtags. It is nothing but filtering the data to remove the incomplete, noisy, and inconsistent data is what data pre-processing is all about. Different steps of data pre-processing are mentioned below:

- Convert to lowercase
- Remove all the punctuation
- Remove URLs from tweets
- Tokenization
- Remove stop words
- Stemming and lemmatization.

Read "content" and the "hashtag" column data before performing the cleaning operation. Since machine learning cannot handle categorical variables, we have to convert it into numerical values. Encoding is applied to read data with the help of scikit-learn library. Here, Labeled encoding is used which provides values based on the alphabets. Since the presence of punctuation adds noise to the text and it brings ambiguity, punctuations are removed and applied tokenization. Stop words like "a, an, the, in, for, our" do not provide much information, so they were removed from the text. In this case, "not" has not been removed from the list because if we remove this stopword then it may change the meaning of a text. For example: "I am not happy" here the word "not" should not be removed from the text to hold the exact meaning of the text. We used Feature engineering method. The main purpose of using the feature engineering method is to create new variables that are currently not available in the training set. It helps to simplify and speed up the data transformation to enhance model accuracy. The cleaned data has been shown in below Fig. 21.5:

21.3.3 Model Building

After cleaning and pre-processing, the dataset was ready for further analysis. Each tweet was labeled systematically using a Level encoder in Python with the help of the Sklearn library. For tweets classification, the machine learning approach was used and TF-IDF vectorizer is used here to extract the feature from the text as it carries all the necessary information about the data in a detailed manner.

	content	hashtag	body_len	punct%	cleaned_data
0	The hustle is against #poverty not competition...	19	56	3.6	[hustl, poverti, competit, happi, decemb]
1	There are far too many starving kids globally ...	19	73	6.8	[far, mani, starv, kid, global, saveourkid, po...
2	Asking your kids to have children without esta...	19	224	0.9	[ask, kid, children, without, establish, gener...
3	Powerful testimonials #period #poverty \n@New...	19	220	6.8	[power, testimoni, period, poverti, newheritag...
4	Visit to outskirts of Kasur\n#Poverty #Attachm...	19	51	5.9	[visit, outskirt, kasur, poverti, attach, 49th...

Fig. 21.5 Tweets after cleaning

	body_len	punct%	0	1	2	3	4	5	6	7	...	35339	35340	35341	35342	35343	35344	35345	35346	35347	35348
0	53	3.8	0.0	0.0	0.0	0.0	0.0	0.0	0.0	0.0	...	0.0	0.0	0.0	0.0	0.0	0.0	0.0	0.0	0.0	0.0
1	38	5.3	0.0	0.0	0.0	0.0	0.0	0.0	0.0	0.0	...	0.0	0.0	0.0	0.0	0.0	0.0	0.0	0.0	0.0	0.0
2	195	8.2	0.0	0.0	0.0	0.0	0.0	0.0	0.0	0.0	...	0.0	0.0	0.0	0.0	0.0	0.0	0.0	0.0	0.0	0.0
3	242	5.4	0.0	0.0	0.0	0.0	0.0	0.0	0.0	0.0	...	0.0	0.0	0.0	0.0	0.0	0.0	0.0	0.0	0.0	0.0
4	97	4.1	0.0	0.0	0.0	0.0	0.0	0.0	0.0	0.0	...	0.0	0.0	0.0	0.0	0.0	0.0	0.0	0.0	0.0	0.0

5 rows × 35351 columns

Fig. 21.6 After vectorization

In Fig. 21.6 when Vectorization applied on data, dataset contains 5 rows and 35,351 columns. The word is treated as a set of documents. Depending on how frequently the documents appear, some words are labeled zeroes and some non-zeroes. Once the dataset has been split into training and test sets (80:20 ratio), different machine learning algorithms have been applied to train the model. This research uses three different classifiers to train the model: Naive Bayes, Support Vector Machine (SVM), and Random Forest.

21.4 Classification Algorithms

1. Naive Bayes: The Naive Bayes classifier is extremely scalable and it predicts based on the probability which can be defined as

$$P(S/A) = P(A/S) * P(S)/P(A) \tag{21.1}$$

where A is the text to be classified and S denotes a single class. $P(S)$ and $P(A)$ are probabilities of text and class identity, respectively; $P(A/S)$ is the chance of text A occurring in class S. It works effectively with a big dataset where this condition holds.
2. Support Vector Machine (SVM): It is used to find the best splits available after the training dataset has been split. New data is placed on one side of the split with a score, i.e., the likelihood, to determine whether the data is on the correct side of the split or not.
3. Random Forest classifier: It uses both Classification and regression. Based on the dataset, it generates a Decision tree and predicts with each of the possible paths. Then, it chooses the best path or the solution from all of them.

21.5 Result and Discussion

Through Python programming, tweets were collected directly using a Twitter scraper by utilizing different emotion-related keywords associated with mental health and commonly used by psychiatrists. The data was then cleaned and pre-processed, using label encoding, and data splitting has done into two sets (training and testing) and used TF-IDF vectorizer. Various machine learning algorithms were employed to train the model for classification in order to find the best match for the data. In the result part, text analysis is done by providing a text line as an input and then it returns an emotion of that text as the output.

21.5.1 Measurements

Precision: It indicates the positive prediction made by the model which can be defined as

$$TP/(TP + FP) \tag{21.2}$$

Recall: It is the measure of identifying the True Positives (TP) which can be defined as

$$TP/(TP + FN) \tag{21.3}$$

$F1$ score: It is a measure of test accuracy which is calculated using Precision and Recall.

$$2 * ((Precision * Recall)/(Precision + Recall)) \tag{21.4}$$

Accuracy: It is an evaluating metric that shows the model predictions and is used for comparing the different model predictions and finding the label of the dataset during the validation process. Here, TP stands for True Positives, TN is True Negative, FP stands for False Positives, and FN stands for False Negatives. Confusion matrix generated for the model is shown in Fig. 21.7.

The classification models are evaluated using the Confusion Matrix, and we categorize the Twitter data using a machine learning approach. The Scikit module in Python was used to construct this confusion matrix. A person can be seen tweeting either favorable, bad, or neutral messages. The diagonal is visible.

After applying ML tools and techniques, the performance of each classifier has been obtained and shown in Table 21.1.

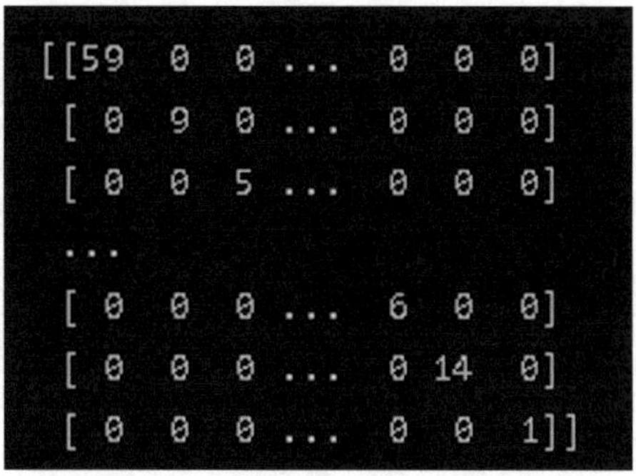

Fig. 21.7 Confusion matrix

Table 21.1 Prediction of different algorithms

S. No.	Algorithm	Precision (%)	Recall (%)	F1	Accuracy (%)
1	Naive Bayes	89	68	77	80.32
2	SVM	84	77	80	83.23
3	Random Forest	87	88	87	86.94

21.5.2 K-Fold Cross-Validation

To assess the model's correctness, tenfold cross-validation was used. Data is randomly partitioned into ten parts, nine of which are utilized for training the model and one of which is used for testing. Second, this process is repeated ten times, with the inaccuracy determined each time. The mean of the errors created in each iteration will be the model's overall error. To assess the model's correctness, tenfold cross-validation was used, yielding an 86.70% result. Standard deviation is vital to know because it indicates how widely the values are spread out. As a result, this model has a standard deviation of 0.27%. Table 21.1 displays the accuracy of the various classifiers. The Random Forest classifier has an accuracy of 86.94%, a Precision of 88%, and an $F1$ score of 87%. As a result, the Random Forest classifier has a better performance than the other two classifiers. Here, the text sentence will be utilized as the input and the output will be based on the sentiment of the text. Figures 21.8, 21.9, 21.10 , and 21.11 depict the resultant output where it shows the result with its relatable suggestive words.

```
You: i am very happy
Your emotion: happy
some related sentiment:  {'glad', 'felicitous', 'well-chosen', 'happy'}
```

Fig. 21.8 Text analysis using machine learning

```
You: i am not happy
Your emotion: unhappy
some related sentiment:  {'unhappy', 'dysphoric', 'infelicitous', 'distressed'}
```

Fig. 21.9 Text analysis using machine learning

```
You: i am in pain
Your emotion: pain
some related sentiment:  {'hurting', 'pain', 'painful_sensation', 'hurt'
```

Fig. 21.10 Text analysis using machine learning

```
You: i am working hard to achieve my goals
Your emotion: Focused
some related sentiment:  {'focused', 'rivet', 'centre', 'center', 'concentre',
'focalize', 'pore', 'concentrate', 'focalise', 'concenter', 'focus',
'sharpen', 'focussed'}
```

Fig. 21.11 Text analysis using machine learning

21.6 Conclusion

Here, sentiment analysis is used, and for text vectorization, TF-IDF vectorizer is used. For classification, machine learning algorithms like Naive Bayes, Support Vector Machine, and Random Forest classifiers are used. As a result, accuracy of the Random Forest classifier is best (86.94%) than the other classifiers. Though the accuracy can be improved more and more in the future, it is not cent percent accurate.

21.7 Future Work

In the future, a recommendation-based system can be implemented to help the users who are dealing with mental health problems. So that they can be more conscious of their mental health state and perform appropriate exercises.

References

1. Chhinder, K., Anand, S.: Social issues sentiment analysis using Python. In: 2020 5th International Conference on Computing, Communication and Security (ICCCS), pp. 1–6. IEEE (2020)
2. Iwan, S., Nadia, N., Tessy, B.: Study on mental disorder detection via social media mining. In: 2019 4th International Conference on Computing, Communications and Security (ICCCS), pp. 1–6. IEEE (2019)
3. Kausar, M.A., Soosaimanickam, A., Nasar, M.: Public sentiment analysis on Twitter data during covid-19 outbreak. Int. J. Adv. Comput. Sci. Appl **12**(2), 415–422 (2021)
4. Mehak, F., Munwar, I., Hifsa, T.: Mental health diseases analysis on Twitter using machine learning. iKSP J. Comput. Sci. Eng. **1**(2), 16–25 (2021)

5. Munmun, D.C., Michael, G., Scott, C., Horvitz, E.: Predicting depression via social media. In: Seventh International AAAI Conference on Weblogs and Social Media (2013)
6. Murty, M.R., Murthy, J., Reddy, P.P., Satapathy, S.C.: A survey of cross-domain text categorization techniques. In: 2012 1st International Conference on Recent Advances in Information Technology (RAIT), pp. 499–504. IEEE (2012)
7. Rahul, K., Saurav, M.: Predicting mental health disorders using machine learning for employees in technical and non-technical companies. In: 2020 IEEE International Conference on Advances and Developments in Electrical and Electronics Engineering (ICADEE), pp. 1–5. IEEE (2020)
8. Rakshitha, C., Gowrishankar, S.: Machine learning based analysis of Twitter data to determine a person's mental health intuitive wellbeing. Int. J. Appl. Eng. Res. **13**(21), 14956–14963 (2018)
9. Rehab M, D., Raed, M., Narmeen, S., Rushaidat, S.: Sentiment analysis in Arabic tweets. In: 2014 5th International Conference on Information and Communication Systems (ICICS), pp. 1–6. IEEE (2014)
10. Siddharth, S., Darsini, R., Sujithra, M.: Sentiment analysis on Twitter data using machine learning algorithms in Python. Int. J. Eng. Res. Comput. Sci. Eng. **5**(2), 285–290 (2018)
11. Sudha Tushara, S., Zhang, Y.: Analyzing tweets to discover Twitter users' mental health status by a word-frequency method. In: 2019 IEEE International Conference on Intelligent Systems and Green Technology (ICISGT), pp. 5–53. IEEE (2019)
12. Tripti, A., Archana, S.: An efficient knowledge-based text pre-processing approach for Twitter and Google+. In: International Conference on Advances in Computing and Data Sciences, pp. 379–389. Springer (2019)
13. Srinivasulu Reddy, U., Thota, A.V., Dharun, A.: Machine learning techniques for stress prediction in working employees. In: 2018 IEEE International Conference on Computational Intelligence and Computing Research (ICCIC), pp. 1–4. IEEE (2018)

Chapter 22
Influence of Spectral Bands on Satellite Image Classification Using Vision Transformers

Adithyan Sukumar, Arjun Anil, V. V. Variyar Sajith, V. Sowmya, Moez Krichen, and Vinayakumar Ravi

Abstract Neural networks play an important role in satellite image classification. We know that the most common neural networks used in image classification tasks are convolutional neural networks (CNNs) (Boesch in Vision transformers (VIT) in image recognition—2022 guide, 2022). In this paper, we explored the influence of the spectral bands in image classification using the Vision Transformer (ViT). Convolution is a local operation, and a convolution layer typically models only the relationships between neighborhood pixels. Transformer is a global operation, and a transformer layer can model the relationships between all pixels. This motivated us to use ViT for satellite image classification. Sentinel-2 EuroSAT image dataset, which consists of 27,000 images in ten classes, is used for the experiment. ViT model is trained with three band dataset, Red-Green-Blue (RGB) and compared with ViT model trained with RGB along with Near InfraRed (NIR) and with multispectral satellite image dataset (13 bands). Experimental results shows that NIR band combined with RGB was able to produce more accurate results comparing to RGB alone, whereas 13 band dataset outperformed both RGB and RGB and NIR datasets.

22.1 Introduction

Land cover refers to the surface on the ground that covers vegetation, infrastructure, water bodies, soil, etc. Land cover classification is important to monitor changes on the surface. Presently, image classification has become an important task of remote

A. Sukumar · A. Anil · V. V. V. Sajith · V. Sowmya (✉)
Amrita School of Engineering, Center for Computational Engineering and Networking (CEN),
Amrita Vishwa Vidyapeetham, Coimbatore, India
e-mail: v_sowmya@cb.amrita.edu

M. Krichen
FCSIT, Al-Baha University, Al Baha, Kingdom of Saudi Arabia

ReDCAD, University of Sfax, Sfax, Tunisia

V. Ravi
Center for Artificial Intelligence, Prince Mohammad Bin Fahd University, Khobar, Saudi Arabia

© The Author(s), under exclusive license to Springer Nature Singapore Pte Ltd. 2023   243
V. Bhateja et al. (eds.), *Evolution in Computational Intelligence*, Smart Innovation,
Systems and Technologies 326, https://doi.org/10.1007/978-981-19-7513-4_22

sensing for pattern analysis, and image analysis. Various researchers have developed and used different image classification methods. The classification of satellite images helps us to monitor various aspects such as pollution, forest cover mapping, wetland mapping, and land cover analysis, etc [2].

ViT was proposed by Dosovitskiy et al. [3] as a replacement for CNN in image classification tasks [4, 5]. ViT is used to classify datasets such as ImageNet, CIFAR-100, VTAB, etc., and was able to yield excellent results compared to CNN. ViT is now commonly used for remote sensing scene classification tasks for datasets such as Merced, AID, Optimal31, and NWPU [6]. Pritt and Chern [7] integrated metadata and image features from the IARPA fMoW dataset and used CNN as a set to produce good precision when classifying 63 different classes [8]. Leveraging the feature extraction capability of pretrained Resnet50 delivered promising results when trained on SAT4 dataset. Jiang et al. [9] used two CNN trained in RGB and NIR data, as normal CNN cannot fully exploit the information due to the high correlation between the bands. The normalized difference vegetation index can be calculated using NIR along with the red band. Using this as input to VGG, Alexnet, and Convnet helps the models to be efficient by giving high accuracy with fewer parameters that can be trained [10, 11]. Color to grayscale image conversion is one of the most important steps in image pre-processing. Sowmya et al. [12] used singular value decomposition-based image conversion techniques to convert SIFT features combined with structure similarity index. The results show that the accuracy of the model increases with the conversion of images. In this paper, we use ViT to compare the RGB dataset with RGB and NIR and with all the 13 spectral band dataset. The objective is to show the influence of the addition of more bands into the dataset helps the ViT classify images more accurately. RGB bands along with the NIR bands were taken from the 13-band dataset to compare the RGB and NIR dataset with the RGB datasets. 13-band dataset is taken to compare the multispectral dataset with the fewer-band data set. This paper is organized into sections as follows. Section 22.2 discusses the materials and methodology. Section 22.3 shows the results obtained provides information on the metrics used to evaluate the model and the values obtained followed by conclusion.

22.2 Methodology

ViT was trained and compared using three datasets (RGB, RGB and NIR, Multi-spectral, or 13 bands). For this paper, we used Sentinel-2 satellite images, which are freely accessible and provided by the Earth observation program Copernicus [13]. Dataset consists of a total of 13 bands, of which we require four bands for the study Red, Green, Blue, and Near-Infrared to make comparison between RGB and RGB and NIR. We use the 13 band dataset to compare both 3 band and 4 band dataset. Position of Red, Green, Blue, and NIR channels are one, two, three, and seven (considering the indices from zero), respectively. Helber et al. [13] as shown in Fig. 22.2. The data used for the study are Sentinel-2 EuroSAT images which is divided into ten class labels, Annual Crop, Forest, Herbaceous Vegetation, Highway, Industrial,

Fig. 22.1 Representation of classes in the sentinel-2 EuroSAT image dataset

Table 22.1 A detailed description of the dataset

Class No.	Class labels	Number of images
1	Annual crop	3000
2	Forest	3000
3	Herbaceous vegetation	3000
4	Highway	2500
5	Industrial	2500
6	Pasture	2000
7	Permanent crop	2500
8	Residential	3000
9	River	2500
10	Sea Lake	3000

Pasture, Permanent crop, Residential, River, SeaLake. The dataset consist of 27,000 .tif files which have all the bands of the spectrum as collected from the Sentinel-2 satellite. Each class label contains 2000–3000 images. A detailed breakdown of this dataset is given in Table 22.1 (Fig. 22.1).

Data augmentation is frequently used to help neural networks generalize better to unseen data and to increase the richness of the training data utilized. Various data augmentation techniques are applied to the dataset like resizing, normalization, random horizontal flip, random rotation, and random zoom. In order to maintain equality in comparison, augmentation is applied equally in all three datasets.

The deep learning architecture used for image classification is Vision Transformer. ViT is a self-attention-based architecture, similar to the transformers we use in natural language processing (NLP) tasks. By the introduction of transformers in NLP was able to create a large success; here ViT is also such a transformer that was developed by making very few possible modifications. Usually, tokens are given as input for the

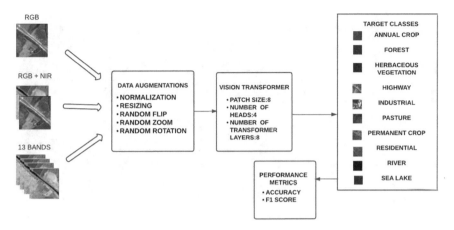

Fig. 22.2 Workflow for the satellite image classification using ViT

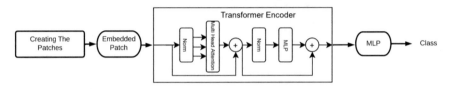

Fig. 22.3 Vision transformer architecture

NLP transformer, whereas in ViT, we split large images into patches along with linear embedding as the input. We train this model for image classification in a supervised method.

The architecture consists of a patch and position embedding unit that splits input images into patches and gives positional embedding to those patches. These patches along with positional details are fed as input to the transformer encoder. The transformer encoder has alternating layers of multiheaded self-attention and MLP blocks. Layernorm (LN) is applied before every block and residual connections after every block. The last layer is a softmax layer as shown in Fig. 22.3 [3].

To perform the experiment without bias, we used the same hyper-parameters for training ViT on these datasets. To make the model generalize, certain data augmentations like resizing, normalization, flipping, rotation, etc. have been applied. The efficiency and accuracy of the classification of satellite images depend on the model selected for classification. ViT splits the images into a series of position-embedded patches, which are sent to the transformer encoder; By doing this, ViT is able to understand the local and global features of the image. The model was evaluated using accuracy, precision, recall, and the $F1$ score. These metrics serve as the basis to identify the errors during the classification process. To determine these metrics, testing data from each class are used.

22.3 Results and Discussions

The data augmentation applied were resizing, normalization, random horizontal flipping, random rotation, and random zoom. The original dataset had 27,000 images, but the experiment was conducted only for 20,000 images in order to maintain class imbalance problems. Out of the 20,000 images, 70% of the data was taken for training, 10% of the data for validation and 20% for testing. Experimental setup was kept the same for all three experiments (RGB, RGB and NIR, Multispectral). The experiment was carried out under the following circumstances in tensorflow, the learning rate was set at 0.001 with a batch size of 256. The image was broken down to 8 patches since it produced the best results via trial and error method. The number of transformer layers for ViT is taken as eight.

From Fig. 22.3, it can be inferred that 100 epochs give better validation accuracy for RGB data. In order to study the effect of addition of NIR and multispectral bands experimental setup was kept same.

The metrics used to evaluate the performance of the models are accuracy, recall, precision, and $f1$ score. Accuracy, precision and $F1$ Score is computed using the equation.

$$\text{Accuracy} = \frac{\text{TP} + \text{TN}}{\text{TP} + \text{FN} + \text{FP} + \text{TN}}$$

$$\text{Precision} = \frac{\text{TP}}{\text{TP} + \text{FP}}$$

$$F_1 = 2 \cdot \frac{\text{precision} \cdot \text{recall}}{\text{precision} + \text{recall}} = \frac{\text{TP}}{\text{TP} + \frac{1}{2}(\text{FP} + \text{FN})}$$

The ViT trained with the RGB dataset gave an accuracy of 0.929. The $f1$ scores for classes such as Sealake, Forest, and Residential Area were 0.97,0.98 ,0.96, respectively. From the classification report, it is understood that very few classes were classified with a precision, recall and $F1$ score greater than 0.95.

Additionally, using the NIR band along with the RGB band for ViT gave an accuracy of 0.939. Classes like Forest, River, SeaLake, Industrial and Residential area was classified correctly in appreciable manner. In addition to this, the precision for annual crops, herbaceous vegetation, highway and pasture was high, and the recall for industrial was also high (Figs. 22.4 and 22.5).

By using all the 13 bands provided in the dataset, the ViT gave an accuracy of 0.95. All classes except highway was predicted correctly with high precision, recall and $F1$ score. Most of the classes gave an $F1$ score greater than 0.96 and all SeaLake class test images were correctly classified ($F1$ score = 1). Comparing the results in Table 22.2, the RGB dataset was able to classify only very few classes efficiently. The addition of the NIR band to the RGB dataset resulted in the increase of true positive values. Annual Crop, Herbaceous Vegetation, Highway, Industrial, Pasture, River and SeaLake are the classes were 4 band dataset outperformed 3 band dataset.

EPOCHS-RGB
Validation Accuracy
0.95
0.90
0.85
0.80
0.75
0.70
0.65
0.60
0
20
40
60
80
100
Epochs

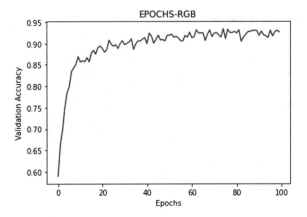

Fig. 22.4 Epochs versus validation accuracy for RGB dataset

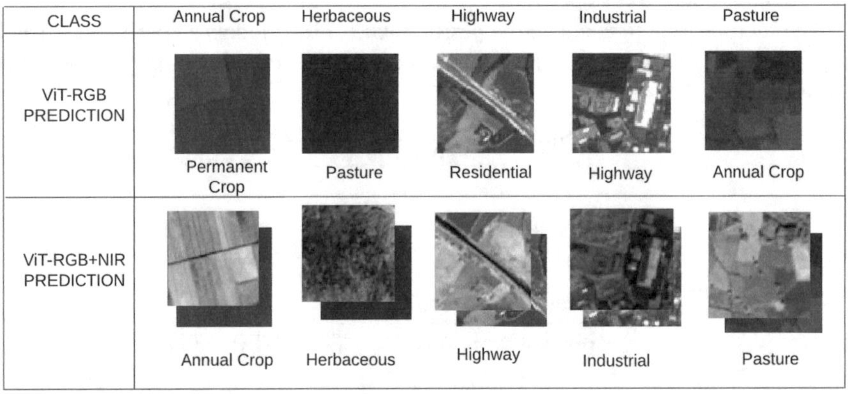

CLASS
Annual Crop
Herbaceous
Highway
Industrial
Pasture
ViT-RGB PREDICTION
Permanent Crop
Pasture
Residential
Highway
Annual Crop
ViT-RGB+NIR PREDICTION
Annual Crop
Herbaceous
Highway
Industrial
Pasture

Fig. 22.5 Images predicted via model trained on RGB and NIR dataset which were misclassified by the model trained on the RGB dataset

Whereas Multispectral dataset on Vision transformer predicted most of the classes better than both 4 bands and 3 bands.

The RGB bands contain information about rich color features of the image, and the NIR band contains details regarding sharp edges in the image.This shows that adding more bands to the dataset can provide more information, thus improving performance of machine learning models. Table 22.3 shows which bands perform better for each class, and for most classes RGB and NIR produced improved results. Capturing multispectral images requires more sensors and is expensive, compared to the requirement of RGB and NIR. Since RGB and NIR images are able to give better classification accuracy compared to RGB images and was also close to the accuracy of 13 band images, it is more cost efficient and computationally inexpensive.

Table 22.2 Performance analysis of the ViT model for three datasets

| | ViT | | | | | | | | |
| | RGB and NIR | | | RGB | | | Multispectral | | |
Classes	Precision	Recall	F1	Precision	Recall	F1	Precision	Recall	F1
Annual crop	0.99	0.83	0.9	0.96	0.84	0.89	0.97	0.96	0.97
Forest	0.97	0.98	0.97	0.98	0.98	0.98	0.98	0.99	0.99
Herbaceous vegetation	0.93	0.93	0.93	0.9	0.93	0.92	0.98	0.96	0.97
Highway	0.92	0.89	0.9	0.89	0.89	0.89	0.93	0.9	0.92
Industrial	0.96	30.95	0.96	0.97	0.93	0.95	0.95	0.98	0.96
Pasture	0.92	0.9	0.91	0.89	0.92	0.9	0.96	0.96	0.96
Permanent crop	0.84	0.91	0.8	0.84	0.92	0.88	0.97	0.94	0.96
Residential area	0.89	0.99	0.94	0.93	0.98	0.96	0.95	0.97	0.96
River	0.98	0.98	0.98	0.91	0.91	0.91	0.98	0.98	0.98
Sealake	0.99	0.99	0.99	0.98	0.96	0.97	1	1	1
Accuracy	0.939			0.929			0.955		

Table 22.3 Classwise comparison of RGB and RGB and NIR

Classes	Bands
Annual crop	RGB and NIR
Forest	RGB
Herbaceous vegetation	RGB and NIR
Highway	RGB and NIR
Industrial	RGB and NIR
Pasture	RGB and NIR
Permenant crop	RGB
Residential area	RGB
River	RGB and NIR
Sealake	RGB and NIR

22.4 Conclusions

In this paper, we have compared the performance of the vision transformer trained on RGB, RGB and NIR, and multispectral data. Results show that addition of more bands increases ViT performance. ViT trained with RGB and NIR images outperformed model trained with RGB images, whereas multispectral (13 bands) outperformed both. This shows the influence of addition of spectral bands on ViT. In the future works, we can improve the performance of the model using datasets with more bands and using the right combinations of bands. Apart from increasing the number of bands, the model performance can be further improved by increasing transformer layers and multilayer perceptron heads.

References

1. Boesch, G.: Vision transformers (VIT) in image recognition—2022 guide (2022). https://viso. ai/deep-learning/vision-transformer-vit
2. Kaplan, G., Avdan, U.: Mapping and monitoring wetlands using sentinel-2 satellite imagery (2017)
3. Dosovitskiy, A., Beyer, L., Kolesnikov, A., Weissenborn, D., Zhai, X., Unterthiner, T., Dehghani, M., Minderer, M., Heigold, G., Gelly, S., Uszkoreit, J., Houlsby, N.: An image is worth 16×16 words: transformers for image recognition at scale (2020)
4. Himabindu, G., Murty, M.R., et al.: Extraction of texture features and classification of renal masses from kidney images. Int. J. Eng. Technol. **7**(2.33), 1057–1063 (2018)
5. Yarlagadda, A., Malla, R.M., Janapana, H.: Seizure prediction based on convolution neural network with subnet loss function
6. Bazi, Y., Bashmal, L., Al Rahhal, M.M., Al Dayil, R., Al Ajlan, N.: Vision transformers for remote sensing image classification. Remote Sens. **13**(3), 516 (2021)
7. Pritt, M., Chern, G.: Satellite image classification with deep learning. In: 2017 IEEE Applied Imagery Pattern Recognition Workshop (AIPR) (2017)

8. Deepika, J., Sowmya, V., Soman, K.P.: Image classification using convolutional neural networks. Int. J. Sci. Eng. Res. **5**(6), 1661–1668 (2014)
9. Jiang, J., Feng, X., Liu, F., Xu, Y., Huang, H.: Multi-spectral RGB-NIR image classification using double-channel CNN. IEEE Access **7**, 20607–20613 (2019)
10. Sasidhar, T.T., Sreelakshmi, K., Vyshnav, M.T., Sowmya, V., Soman, K.P.: Land cover satellite image classification using NDVI and simpleCNN. In: 2019 10th International Conference on Computing, Communication and Networking Technologies (ICCCNT), pp. 1–5. IEEE (2019)
11. Unnikrishnan, A., Sowmya, V., Soman, K.P.: Deep learning architectures for land cover classification using red and near-infrared satellite images. Multimedia Tools Appl. **78**(13), 18379–18394 (2019)
12. Sowmya, V., Govind, D., Soman, K.P.: Significance of contrast and structure features for an improved color image classification system. In: 2017 IEEE International Conference on Signal and Image Processing Applications (ICSIPA) (2017)
13. Helber, P., Bischke, B., Dengel, A., Borth, D.: Eurosat: a novel dataset and deep learning benchmark for land use and land cover classification. IEEE J. Sel. Top. Appl. Earth Observations Remote Sens. **12**(7), 2217–2226 (2019)

Chapter 23
Smart Garbage Collection System for Indian Municipal Solid Waste

Hritesh Ghosh, Chittaranjan Pradhan, and P. H. Alex Khang

Abstract With the extension in the human populace, the summary of neatness as for trash the executives is disparaging massively. The flood of trash in open regions makes an unhygienic condition in the close by encompassing. It might incite a few genuine illnesses among the close by individuals. To stay away from this and to improve the cleaning, savvy and effective trash the executives framework is proposed in this paper. In the proposed framework, people can send pictures and videos of a garbage station overflowing with its particular location with the help of a flutter-based Smartphone app. With the help of cloud computing and machine learning, the system will store and analyze the locations and send cleaners to the spots to pick up the garbage by notifying them with another Smartphone app for cleaners. Authorities can also monitor every ongoing process and check the feedback. This will assist with dealing with the trash assortment productively.

23.1 Introduction

India's populace is quickly developing and there is a huge development of waste. The reason is due to the shifting of rural population to the urban area [1]. As of now, we don't have a concentrated arrangement or appropriate chain framework to track and screen the waste [2, 3]. So, we are attempting to address the waste administration issues for certain arrangements in this paper. The society member can record the issue through the mobile application and also can give the feedback to the authority. Next, as the requests go to the cloud, we'll have a garbage pickup engine, which will give us the locations for a particular area that should be cleaned within 24 h. And we'll have an admin application for the admins and the authority that'll be monitoring over the process. Lastly, an application we'll have for cleaners who will be assigned

H. Ghosh (✉) · C. Pradhan
Kalinga Institute of Industrial Technology, Bhubaneswar, India
e-mail: ghoshhritesh@gmail.com

P. H. Alex Khang
VIUST Vietnam, VUST.SEFIX.EDXOPS, Global Research Institute of Technology and Engineering, North Carolina, USA

V. Bhateja et al. (eds.), *Evolution in Computational Intelligence*, Smart Innovation, Systems and Technologies 326, https://doi.org/10.1007/978-981-19-7513-4_23

to clean a particular location at a time by the garbage pickup engine. The thought process of making this model is to place one stage into the arrangement of waste administration [4].

23.2 Related Work

In paper [5], the current framework is designed as 'preseparated squander' for separating the data set of waste gathered from the sensor concerning its class for example natural, plastic, paper, bottle, metal, and so forth. This empowers a proficient waste administration framework and has been taken on in Korea. In the portrayal of such a framework, a nonexclusive work process has been given, wherein, on receipt of ready message, assortment is to be organized and when the undertaking is done the status in the framework is refreshed likewise [6]. While the kinds of savvy containers and cloud engineering are arranged, the genuine execution technique which deals with the various factors of the framework has not been portrayed.

In paper [7], the component to focus on the assortment in light of the area for example schools or clinics has been incorporated together and, in like that, a powerful waste administration framework has been proposed. Further, the comparative needs have been recognized for the perilous waste (for example, causing the speedy well-being effect on individuals living regions). The referenced objective is accomplished through original calculations which enhance the need and related cost. In current technique, information is assessed with ongoing and manufactured information is recovered by the district of Saint Petersburg, Russia, and for this they have planned and created models like devoted trucks model, diversion model, least distance model, and reassignment model [8]. In paper [9], an optimized framework has been proposed for the waste collection in the development of smart sustainable cities.

23.3 Proposed System

A garbage collection system requires the interaction of the municipality, people, or the users and cleaners with the admin [10]. So some of the key requirements of this system are listed down below:

- Municipalities can register themselves.
- Users can create, update, delete, and get their profiles.
- Users can search for the municipality using an area name or city name or zip code.
- Municipalities can add and update the list of garbage stations.
- Users can place or cancel a request for cleaning a garbage station.
- Cleaning teams can get the locations to clean made by the admin using his/her Id.
- Users can get the status of their placed requests, success, in process, not possible right now, etc.

To satisfy above mentioned points some core design principles we need to have for this system are:

- User Centricity—We need to guarantee that each cooperation our clients have with this framework is a positive one and settles for something.
- Inclusivity—We need to plan and fabricate our foundation which is practical by, however, many individuals as could be allowed, paying little mind to capacity, age, and geography.
- Simplicity—By this system, we need to convey fulfillment through a perfect and centered insight with the help of flutter-based UI of our system. So user friendliness is a must for this system.
- Consistency—This system needs constant interaction between users and admin and then admin to cleaners.

For the interaction with our different individuals of this system, three user-interfaces (Fig. 23.1) needed first for the users, an user application. Then an admin application for the municipality. Lastly, an application for the cleaners.

Core features of user app are:

- Spot Locator—Allows users to find the location of a particular uncleaned garbage station, using integrated Google Map.
- Request Placement—The users can place a request of a selected location. They just need to cross-verify their selected location, and can also send a picture or 30 s long video of the location.
- Verification of Locations—With the help of the user app, users from a particular municipality will get the notifications of garbage spots to be cleaned in their area. If they think it's a spot which needs to be cleaned, they will confirm the location.
- Tracking The Process—With a tracking feature, it becomes easy for users to track their requests, until the garbage is collected from the locations in the integrated map it'll show processing after the garbage collection it'll show done but if for some reason a certain request got canceled it'll show canceled you can enter the location again.
- Feedback—Users can send direct feedback to the municipality if anything goes wrong with the application or the ceanears are not performing well.

Fig. 23.1 List of UIs for the system

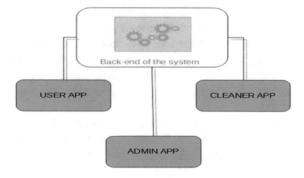

Core features of admin app are:

- Management of Users and Cleaners—Being on the administrator board, one can straightforwardly deal with all the User Ids and Cleaner Ids by adding, updating, and removing from the database. He can likewise actually take a look at dynamic Users and Cleaners status.
- Delivering Pickup Locations—After finding the confirmed locations admin makes sure, cleaners will get the pickup locations and when they will pick up, etc.
- Analytics and Report Generation—Utilizing investigation and report highlights, admin can get constant experiences of reports and other bookkeeping data which assists with distinguishing the development and chances to extend reach.
- Monitoring Every Action—Screen every one of the mentioned areas, cleaners, pickups, feedback of users, canceled locations, and other significant information related to the system's performance.

Core features of cleaner app are:

- Cleaner's Profile—Through this feature, a team leader of a cleaning team updated his profile. It contains his complete name, address, email, contact number, photograph, and other individual data, and so forth.
- Notification For Pickups—Through this the cleaning team gets the listed locations to cleanup under a certain limit of time.
- Map For The Delivery Route—Integrate Google Map or different suppliers and permit cleaners to pick the most limited and quickest courses to come to the locations.
- Sending Response—After picking up the garbage cleaners, send a response to the admin and a picture of the spot after cleaning.

23.3.1 System Design

The overall system design is shown in Fig. 23.2.

All solicitations produced using a ripple-based versatile applications or UI will go to various administrations by means of the API gateway. Programming interface entryway will deal with load adjusting and directing solicitations to administrations. This will confirm and approve the client and send back a requestID. This requestID is utilized for additional correspondence.

Various administrations like, client enrollment and the board administration, cleaning administration, and so on will utilize conditional information bases. We will utilize the Amazon Aurora relational database. This is a profoundly adaptable data set assistance to oversee clients and simultaneous solicitations for cleaning and so forth.

Information about different municipalities, their cleaners, garbage collection sites, etc., will be stored in JSON document storage in ElasticSearch. Multi-node clusters can be used here. Whenever a user searches for a spot registered under a certain

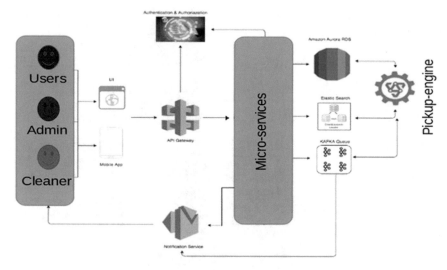

Fig. 23.2 Proposed system's sequence diagram

municipality, it will be gotten from elastic search. Elastic search gives quick versatile hunt choices.

Once the user selects a particular location from the Integrate Google Map in user app. He will go to the send location with photos and videos option and then the request will move forward. From the location we'll get the municipality of that particular location, next with the help of notification service that location and photos and videos of that particular location will be sent to 50 random users of that municipality if five users confirm that fact. The location is selected for further process. The user who sends the request will get one notification, location selected for cleaning.

Once the request for cleaning is placed, all the information is sent to the central message Queue like Kafka. The pickup handling unit peruses the solicitation data and afterward advises the chosen municipality admin about the cleaning spot. Simultaneously, it looks for accessible cleaning groups to local areas to get the trash. It also gets the information like how many more requests are placed for the same location in 24 h from this estimated urgency, of the location and other details. It will select the best available cleaning team then.

The users get push notification about the selected team and the pickup. Tracking service will work then and the users can track their placed request's status, live location of the cleaning team, etc.

Cleaning team picks up the garbage and sends a response to the admin with a photo of the location after cleaning.

Admin gets the response from the cleaning team, based on the response he/she will give further instructions to the cleaning team to clean properly or something else otherwise he/she can end the process after the response.

Users get the notification of garbage picked up from the requested location.

Fig. 23.3 Lifecycle of a
garbage pickup

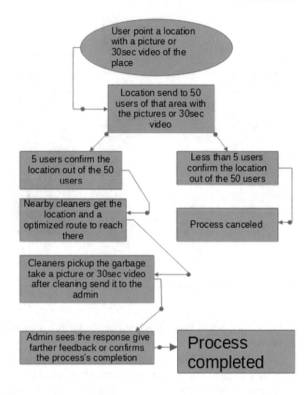

23.3.2 Garbage Pickup Engine

This will read information like request of a location's time, same request for the
location within 24 h, time to reach to the location from the near municipality or
any authority office, available cleaning teams, historic data about last pickup of the
location, cleaning time and predict the pickup time and urgency for the location in
light of chronicled information and ongoing information like guides information,
current cleaning group information, etc. [11]. This will likewise deal with relegating
a cleaning group for trash pickups. This continues to refresh the pool of cleaning
groups for given locations. The details are shown in Fig. 23.3.

23.4 Expected Results

There are 4378 urban areas and towns in India. Class I urban areas alone add to
in excess of 72% of the absolute metropolitan strong waste (MSW) produced in
metropolitan regions. For 2015, a worth of 32.2% is anticipated. For 2015, a worth
of 32.2% is anticipated in spite of the fact that there is no far reaching information on
squander age rates, assortment inclusion, capacity, transport, and removal volumes

and practices; the Central Public Health and Environmental Engineering Organization (CPHEEO) assessed a for each capita squander age in Indian cities and towns in the scope of 0.2–0.6 kg each day.

The projected waste generation of an Indian urban city is shown in Fig. 23.4.

If the proposed system is implemented impeccably, like in India, there are around 3700 metropolitan neighborhood bodies with 100 enterprises, 1500 civil chambers, and 2100 nagar panchayats. If these bodies connect them with the system using the admin app, we can track 1683.23 million tons of municipal solid waste (MSW) generated yearly. The top MSW generating states are shown in Fig. 23.5.

The per capita of MSW created everyday in India goes from around 100 g/individual/day in humble communities to 500 g/individual/day in enormous towns/urban areas. We should consider the normal per capita of MSW created everyday is 300 g/individual/day. Toward the year's end, we have 1683.23 million tons of city strong waste (MSW). Here, we are thinking about the numbers in light of India's absolute populace. The optimized route of the proposed system is shown in Fig. 23.6.

Integrated Google Map is utilizing the Geohash which encodes a geographic area into a short series of letters and digits, here's hash ezs42 for instance; this is the way it is decoded into a decimal scope and longitude. The initial step is deciphering it from text-based base 32 ghs to get the parallel portrayal:

$$[e]_{32\ ghs} = [13]_{10} = [01101]_2$$

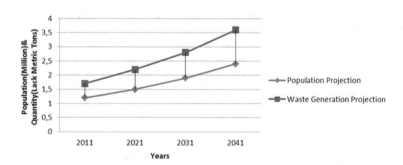

Fig. 23.4 Waste generation projection of an urban city in India

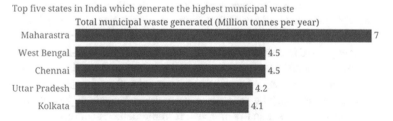

Fig. 23.5 Top MSW generating states

Fig. 23.6 Optimized route on Integrate Google Map feature proposed by the system

$$[z]_{32\ ghs} = [31]_{10} = [11111]_2$$
$$[s]_{32\ ghs} = [24]_{10} = [11000]_2$$
$$[4]_{32\ ghs} = [4]_{10} = [00100]_2$$
$$[2]_{32\ ghs} = [2]_{10} = [00010]_2.$$

This activity brings about the pieces 01101, 11111, 11000, 00100, and 00010. Beginning to count from the left side with the digit 0 in the principal position, the digits in the odd positions structure the longitude code (0111110000000), while the digits in the even positions structure the scope code (101111001001). Every double code is then utilized in a progression of divisions, thinking about the slightest bit at a time, again from the left to the right side. For the scope esteem, the stretch − 90 to + 90 is separated by 2, delivering two spans: − 90 to 0, and 0 to + 90. Since the principal bit is 1, the higher stretch is picked, and turns into the ongoing span. The system is rehashed for all pieces in the code. At last, the scope esteem is the focal point of the subsequent span. Longitudes are handled in a comparable manner, remembering that the underlying span is − 180 to + 180.

In developing countries like India, we need a cost effective but efficient system to solve the problem of garbage disposal. Previously, there were some smart garbage collection systems like deploying smart garbage bins in the municipal solid waste stations, etc. [12]. Our proposed system can be a fruitful solution for this problem:

- Sejong City, South Korea's new managerial capital has taken on this framework, to deal with the misuse of over 1.2 million individuals through a few high-profile establishments. The establishment of the framework, which is estimated at nearly £23 m, is supposed to start during the principal half of 2019 and become functional toward the start of 2022. Just to mechanize the framework, we'll require more than 1,000,000 shrewd canisters that will cost us around 400 million USD which is near 40 crore rupees (INR). The proposed system doesn't need any heavy investment like this but still can be equally effective if implemented accurately.

- The older systems can only point out the locations that need to be cleaned, for identifying the locations they use a weight measurement system, if the weight of the garbage exceeds a certain level then the sensors sends the feedback to the servers, this system's limitation is that it can't identify the locations that need urgent cleaning. But the system we proposed can also predict the urgency for cleaning of a location before cleaning and make such a schedule before sending cleaners for pickup. The pickup engine of this system can calculate the urgency by analyzing the total number of user confirmation out of 50 users that receive the request and time duration between every confirmation sent. This way our system can perform the task more efficiently.

23.5 Conclusion

With the increasing population of India, the problem of irregular garbage collection will increase further. In this era of digitization and automation, we need a centralized solution for this issue particularly. This paper shows the brilliant waste administration framework by utilizing vacillate-based advanced cell applications fit for sending areas of trash stations to the municipality. By carrying out this framework, it diminishes time, cost, and advances courses.

References

1. Devi, P., Ravindra, W.S., Sai Prakash, S.K.L.V.: An IoT enabled smart waste management system in concern with Indian smart cities. In: International Conference on Trends in Electronics and Informatics, IEEE, India, 2018, pp. 64–69
2. Opara, J.A.: Urban waste control and management: issues and challenges. J. Environ. Manage. Educ. 1(1), 1–61 (2008)
3. Murugesan, S., Ramalingam, S., Kanimozhi, P.: Theoretical modeling and fabrication of smart waste management system for clean environment using WSN and IOT, materialstoday. Elsevier, vol. 45, pp. 1908–1913 (2021)
4. Sivaraman, A.K., Alex Khang, P.H., Rani, S.: AI-Centric smart city ecosystem: technologies, design implementation. ISBN: 978-1-032-17079 (2022). https://doi.org/10.1201/978103 217079
5. Folianto, F., Low, Y.S., Yeow, W.L.: Smartbin: smart waste management system. In: International Conference on Intelligent Sensors, Sensor Networks and Information Processing. IEEE, Singapore (2015)
6. Garach, P.V., Thakkar, R.: A survey on FOG computing for smart waste management system. In: International Conference on Intelligent Communication and Computational Techniques. IEEE, Jaipur (2018)
7. Anagnostopoulos, T.V., Zaslavsky, A.: Effective waste collection with shortest path semi-static and dynamic routing. In: International Conference on Next Generation Wired/Wireless Networking, pp. 95–105. Springer (2014)
8. Gutierrez, J.M., Jensen, M., Henius, M., Riaz, T.: Smart waste collection system based on location intelligence. Procedia Comput. Sci. 61, 120–127 (2015)

9. Shah, P.J., Anagnostopoulos, T., Zaslavsky, A., Behdad, A.: A stochastic optimization framework for planning of waste collection and value recovery operations in smart and sustainable cities. Waste Manage. **78**, 104–114 (2018)
10. Vinoth Kumar, S., Senthil Kumaran, T., Krishna Kumar, A., Mathapati, M.: Smart garbage monitoring and clearance system using internet of things. In: International Conference on Smart Technologies and Management for Computing, Communication, Controls, Energy and Materials, IEEE, Chennai (2017)
11. Abdullayev V.H., Alex Khang, P.H., Hahanov, V., Litvinova, E, Chumachenko, S., Abuzarova V.A.: Autonomous robots for smart city: closer to augmented humanity. In: AI-Centric Smart City Ecosystem: Technologies, Design and Implementation, pp. 161–176 (2022). https://doi.org/10.1201/978103217079
12. Hahanov, V., Alex Khang, P.H., Litvinova, E., Chumachenko, S., Vugar Abdullayev, V.H., Abuzarova, V.A.: The key assistant of smart city—sensors and tools. In: AI-Centric Smart City Ecosystem: Technologies, Design and Implementation, pp. 361–373 (2022). https://doi.org/10.1201/978103217079

Chapter 24
Heart Disease Prediction Using Machine Learning

Jyoti Kiran, Nikhil Debbarma, and Sushanth Ganjala

Abstract Over the last few decades, heart disease has become the leading cause of death and the most life-threatening condition on the planet. Early detection of heart disease will aid in lowering the death rate (Dinesh et al. in Prediction of cardiovascular disease using machine learning algorithms [1]). Heart disease has become one of the most difficult problems in the medical field. Machine learning is a rapidly emerging branch of research and technology that can assist people in detecting heart disease before it causes significant damage. It can determine if a person has heart disease or not by taking into account criteria such as the person's age, cholesterol level, chest pain, and other characteristics. Our main goal is to identify the best trustworthy machine learning method that is also computationally efficient. Our main goal is to identify the best robust machine learning technique for heart disease diagnosis that is both computationally efficient and accurate. To predict and categorize patients with heart disease, we used different machine learning methods such as decision tree classifier, random forest, Naive Bayes, K-nearest neighbor, logistic regression, and support vector machine. The given model is helpful in relieving a lot of strain from determining the probability of the classifier correctly and accurately identifying heart disease. It increases medical care while lowering costs.

24.1 Introduction

The most important part of human body is heart. Nowadays, cardiovascular diseases are one of the major cause for global death. It is calculated that 17.9 million die annually [2]. Adults are also affected from this disease. It has become most life-threatening disease in the world currently. As a result, preventing heart disease has become more important now. The main challenges in today's healthcare services are providing of best quality services and effective treatment. Machine learning helps in the prognosis of heart illnesses, and the results are precise. Machine Learning can handle massive amount of data efficiently. Nowadays in the medical sector, machine

J. Kiran (✉) · N. Debbarma · S. Ganjala
National Institute of Technology Agartala, Agartala, Tripura, India
e-mail: kiranjyoti858@gmail.com

© The Author(s), under exclusive license to Springer Nature Singapore Pte Ltd. 2023
V. Bhateja et al. (eds.), *Evolution in Computational Intelligence*, Smart Innovation, Systems and Technologies 326, https://doi.org/10.1007/978-981-19-7513-4_24

learning can be used for diagnosis, detection, and prediction of various diseases. It is also known for plays a very important role to detect the crucial discrete patterns and thereby analyze the given data. After analysis of data, various machine learning techniques or algorithms help in heart disease prediction and early diagnosis [3].

Our project can help the patients who are willing to diagnose with a heart disease by help of their medical history information. It finds who all are having any symptoms of heart disease such as chest pain or high blood pressure and can help in diagnosing disease with their medical information and effective treatments, so that they can be cured from this life-threatening disease [4]. The proposed work is trying to find out heart diseases at early stage to avoid disastrous consequences. A dataset is taken from the Kaggle Website with patient's medical history and attributes. This dataset contains information of 5 popular datasets for heart disease. Based on information in this dataset, we predict whether the patient can have a heart disease or not. For achieving objective of this project, we use 12 medical attributes of a patient and classify if the patient is likely to have a heart disease. These medical attributes are trained under different machine learning classification algorithms: support vector machine, K-nearest neighbor, Naive and random forest classifier, decision tree, and logistic regression. And, eventually, we classify cases that are at threat of getting a heart disease or not, and also this system is completely cost effective. Different machine learning algorithms will give different accuracy, so after comparing among them, we can achieve the best algorithm which predicts heart disease with accurate result. The main objective of our project is to increase efficiency for predicting heart disease rate.

24.2 Literature Survey

In [5], authors used various features related to heart disease and implemented different supervised machine learning algorithms such as random forest and K-nearest neighbor. The accuracy achieved by K-nearest neighbor (KNN) is 86.885%, and the accuracy achieved by random forest algorithm is 81.967%. The dataset was taken from the Cleveland database of UCI repository of heart disease patients. This research paper aims to predict the probability of developing heart disease in the patients.

In [6], authors predict the probability of heart disease and classify patient's risk level by implementing different machine learning algorithms such as decision tree, Naive Bayes, logistic regression, and random forest. This paper presents a comparative study by analyzing the performance of different machine learning algorithms by use of heart disease dataset available in UCI machine learning repository. This experiment shows that random forest algorithm has obtained the highest accuracy of 90.16% as compared to other implemented ML algorithms.

In [2], author used various machine learning algorithms such as support vector machine (SVM), random forest, decision tree, and Naive Bayes to predict whether a person is suffering from a heart disease by considering certain symptoms like age of the person, cholesterol level, chest pain, and some other factors. Classification algorithms have been applied for the development of model. The accuracy obtained by result shows that random forest gives more accuracy.

In [7], the authors used various supervised ML classifiers like logistic regression, gradient boosting, decision tree, random forest and that have been used to deploy a model for cardiovascular disease prediction. It uses the popular datasets from the Framingham database and others from the database of the UCI Heart repository.

In [8], authors compared the results of different machine learning algorithms and deep learning algorithms. For this work, it uses very known dataset from the Cleveland database of UCI repository of heart disease patients. Using deep learning approach, 94.2% accuracy was achieved and classifies patient's risk.

In [9], the authors applied the four machine learning algorithms. Dataset was trained for all the different algorithms and finally tested with testing data. The most accurate algorithm was to be selected based on various criteria. They found out that logistic regression algorithm has the highest efficient algorithm out of the four with an accuracy of 82.89%.

24.3   Proposed Model

The proposed work uses multiple classification algorithms to determine cardiac disease and perform analysis. The goal of this proposed project is to obtain accurate prediction of whether or not a patient has cardiac disease. The input values come from the patient's health report information, which the medical expert enters. The information is fed into a model that forecasts the likelihood of heart disease. Figure 24.1 depicts the full process. The proposed structure includes various steps. The heart disease dataset in .csv format was obtained from Kaggle Website as a first step. The data was then entered into the software program, where we looked at the properties, kinds, value ranges, and other statistical data. The next stage was to pre-process the data, to ensure that machine learning classifiers work better, actions including detecting missing values in the dataset and replacing them with the user constant or mean value, depending on the kind of attribute, and were performed. Data pre-processing refers to deal with the missing values, noise in data and normalization based on algorithms used. The machine learning classifier used in the proposed model is K-nearest neighbor, support vector machine, Naive Bayes, decision tree classifier, random forest classifier, and logistic regression. After that, we evaluated our model on the basis of accuracy, and performance is model using various performance metrics. Here in this model, an effective heart disease prediction system has been developed using different machine learning classifiers. This model uses 12 medical attributes such as blood pressure, age, sex, chest pain, fasting sugar, and cholesterol for prediction of heart disease.

24.3.1 Data Collection

Gathering data on the characteristics of persons with and without heart disease, as well as the outcome of whether or not they have the disease, is the initial step in the process. This heart disease dataset is taken from the Kaggle Website. Kaggle is one of the finest places for data professionals and machine learners to find datasets. Data provided by Kaggle is authentic, and it is recommended by various researchers. On the basis of information of this dataset, the pattern which leads to the detection of patient prone to getting is a heart disease which is extracted. These datasets are split into two parts: training data and testing data. This dataset contains 1190 rows and 12 columns, where each row corresponds to a single record. The working of the model proposed is pictorially described in Fig. 24.1. Basically, this dataset is a combination of 5 most famous datasets for heart disease which are Cleveland, Switzerland, Hungarian, Long Beach, and Statlog heart dataset. The dataset contains a total of 1190 records with 11 features. This dataset has shape of (1190, 12). Total no. of attributes in the dataset are 12 out of which are 11 features and 1 target variable. All 12 attributes are as shown in Table 24.1.

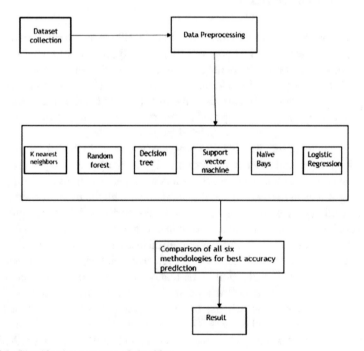

Fig. 24.1 Classification accuracy of classifier

Table 24.1 Details of selected attribute

S. No.	Attribute	Details
1	Age	Patient's age in years
2	Sex	Gender of patient: 0 = female; 1 = male
3	Chest pain type	1 = typical angina 2 = atypical angina 3 = non-anginal pain 4 = asymptomatic
4	Resting bps	Resting blood pressure (mm Hg)
5	Cholesterol	Serum cholesterol in mg/dl
6	Fasting blood sugar	0 = less than 120 mg/dl 1 = more than 120 mg/dl
7	Resting ecg	Resting electrographic results 0 = normal 1 = having ST-T wave abnormality
8	Max heart rate	Maximum heart rate achieved
9	Exercise angina	Exercise induced angina 0 = no 1 = yes
10	Oldpeak	Exercise induced ST depression in comparison with rest state
11	ST slope	Slope of exercise ST segment 0 = normal 1 = unsloping 2 = flat 3 = downsloping
12	Target	Has heart disease or not 0 = no 1 = yes

24.3.2   Data Pre-processing

Data pre-processing is an important step in machine learning as the quality of data and the useful information that can be derived from it directly affects the ability of our model to learn; therefore, it is extremely important that we pre-process our data before applying it into our model [8].

1. Processing missing values. The collected data usually has missing values. These data must be cleaned for noise, and missing data must be filled in to provide an efficient and precise result. Here are five methods for dealing with missing data. Deleting rows, substituting mean/median/mode, designating a distinct category, predicting the values that are missing using methods that accommodate missing data.
2. Evaluating the Distribution of the Data. The dataset should be nearly balanced. A dataset that is very unbalanced can render the entire model training ineffective and produce erroneous results.
3. Transformation of Data. Data transformation transforms the format of data from one form to another in order to make it more understandable. It involves activities including as normalization, aggregation, and smoothing.
4. Checking Duplicate Values in the Data. Duplicates should be removed as soon as possible, else the model's generalization and accuracy prediction will suffer.
5. Splitting the Dataset. Training part has a size of 80%, and test part has a size of 20%.

24.3.3 Proposed Methods

1. K-nearest neighbors—The supervised algorithm approach includes the K-nearest neighbors algorithm. This algorithm identifies items based on their immediate neighbors. It takes into account the similarity between the new case and the existing cases and places the new example in the category that is closest to the existing categories. As a result, the data is grouped based on their similarities, and KNN can be used to fill in missing values. After the missing values have been filled in, the dataset is subjected to several prediction approaches. The exact location of the K-nearest neighbor should be decided with the help of the training dataset. Nearest neighbors are those data points that have less distance in feature space from new data point. Euclidean distance is used for calculation of distance of an attribute from its neighbors. It is chosen over other algorithms due to its simplicity and ease of implementation, as well as its high speed. The algorithm determines if the patient has disease or not using the cardiac disease dataset.

2. Support vector classifier—One of the most common supervised learning algorithms is support vector machine [10]. The purpose of the SVM algorithm is to find the optimal line or decision boundary for dividing n-dimensional space into categories so that future data points can be easily classified. A hyperplane is the most optimal decision boundary. The extreme locations that assist create the hyperplane are selected by the support vector machine. Support vectors are the extreme examples, and the algorithm is named after them. The support vector machine algorithm's goal is to find a hyperplane in an N-dimensional space (N—the number of features) that distinguishes between data points.

3. Decision Tree Classifier—A decision tree is a machine learning approach for categorical and numerical data classification. It uses a flowchart like a tree-structure classifier, and it can be concluded by two entities, namely decision nodes and leaves. Internal nodes show dataset attributes, branching show decision rules, and each leaf node represents the overall result. A decision tree is a basic and commonly used method for dealing with medical data. It is a graphical depiction for obtaining all viable alternatives to a problem/decision depending on certain parameters. It is termed a decision tree because, like a tree, it starts with a root node and grows into a tree-like structure with additional branches [9]. The decision tree model completes analysis based on some terminologies.

    (a) Root Node: Decision tree starts from root node. It represents the entire dataset, which further gets divided according to various features.
    (b) Decision Node: It is obtained by after splitting the root nodes.
    (c) Leaf Node: Leaf nodes are the final output, and tree cannot be expanded further after getting a leaf node.
    (d) Sub-tree: Sub-tree is a subsection of decision tree.
    (e) Pruning: Pruning is nothing but removing some nodes to stop overfitting.
    (f) Interior Node: Handles various attributes.

4. Random forest—andom forest is a well-known machine learning algorithm that uses supervised learning techniques. In machine learning, it can be applied for both classification and regression problems. It is based on ensemble learning, which is a technique of integrating numerous classifiers to solve a complicated problem and increase the performance of the system. Random forest adds extra randomness to the model while growing the trees. Rather than sorting out the most important characteristic when rendering a node, it looks for the best. A feature from a collection of features chosen at random. As a result, there is a lot of variety, which is usually a good thing and results in a more stable model. As a result, only a random subset of the characteristics is used in random forest. Random forest is an ensemble model that resembles a forest and consists of many regression trees [5].

5. Naive Bayes Classifier—Naive Bayes algorithm is basically based on the Bayes rule. It is mainly used in classification problems. The basic and most important assumption in constructing a classification is that the attributes of the dataset are independent. It considers that the presence of a particular feature in a class is unrelated to presence of others. It is simple to anticipate and operates best when the premise of attribute independence is true. As illustrated in Eq. 24.1, Bayes' theorem calculates the posterior likelihood of an outcome (Q) given a probability value of event R.

$$P(Q/R) = (P(R/Q)P(Q))/P(R) \tag{24.1}$$

The posterior probability is $P(Q/R)$, the prior probability is $P(Q)$, the predictor prior probability is $P(R)$, and the likelihood is $P(R/Q)$ [6].

6. Logistic Regression—Logistic regression is a statistical analysis method to predict a binary outcome, such as yes or no, based on prior observations of a dataset. A logistic regression model predicts a dependent variable after analyzing the relationship between one or more existing independent variables [11]. An example of logistic regression could be applying machine learning to predict whether a political candidate will win or lose an election. These binary results allow straightforward decisions between two alternatives. In machine learning, logistic regression is frequently used to estimate the likelihood of occurrence. When a collection of explanatory independent characteristics is specified, response attributes are used. It is employed when the A dependent variable with categorical values such as yes/no or true/false, and so on.

24.4 Experimental Implementation

Some of the implementation specifics were already mentioned in the methodology section. Python is the programming language utilized in this project. In Anaconda Navigator's Jupyter Notebook, we are running Python code. Jupyter Notebook is significantly faster than Python IDEs such as PyCharm or Visual Studio for ML algorithm implementation. The benefit of using a Jupyter Notebook is that, while writing code is quite useful for data visualization and displaying graphs such as histograms and bar charts.

24.5   Result and Analysis

The overall objective of this project is to find out the suitable machine learning technique that is computationally efficient as well as most accurate for heart disease prediction. This work is based on the application of machine learning algorithms, of which we have selected the some most used algorithms, named random forest, support vector machine, Naive Bayes, decision tree classifier, K-nearest neighbor, and logistic regression. From the result, it is been estimated that the random forest gives highest accuracy as compared to other five machine learning algorithms. The dataset used is split into a training set and testing set. Here, 80% of the dataset is chosen for training, and the remaining is considered for testing. At first, the six algorithms were implemented. Dataset was trained for all the algorithms individually. After this, all of them were tested. The most efficient algorithm was to be selected based on various factors. We found out that random algorithm has the most efficient with an accuracy of 94.96%. Decision tree and Naive Bayes had accuracy of 88.24% and 85.29% respectively, and SVM, K-nearest neighbors, and logistic regression had accuracy of 80.25%, 68.49% and 80.67%, respectively. The classification accuracy of different machine learning algorithms is graphically represented in Fig. 24.2.

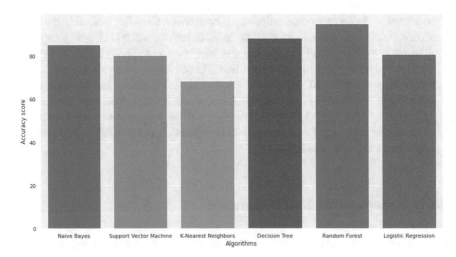

Fig. 24.2 Classification accuracy of classifier

Table 24.2 Result of machine learning classifiers

Classifiers	Accuracy (%)	Precision	Recall
K-neighbors	68.49	0.79	0.79
Random forest	94.96	0.89	0.90
Decision tree	88.24	0.75	0.90
Naive Bayes	85.29	0.79	0.79
Support vector machine	80.25	0.79	0.79
Logistic regression	80.67	0.79	0.78

24.5.1 Performance Metrics

Different performance metrics are applied to evaluate how well certain algorithms perform in terms of various factors including accuracy, recall, and precision. The accuracy, precision, sensitivity (recall), and F-measure performance metrics used in this study are detailed here. The confusion matrix shows many performance criteria for evaluating classifiers (Table 24.2).

24.6 Conclusion

Heart disease is the main source of death [12]. Therefore, predicting the disease before becoming more infected decreases the risk of death, and it will great advantage of heart disease patients. The project included data processing and analysis of dataset. This dataset contains information regarding heart disease patients. Also, correlation matrix is used to detect the dependencies between the attributes. Then, we used six different machine learning algorithm such as K-nearest neighbor, Naive Bayes, support vector machine, logistic regression, random forest, and decision tree; on a real dataset of people, we collected dataset from Kaggle Website. This is a classification problem with a range of input features as parameters and a binary target attribute for predicting whether heart disease is present or not. We achieved highest accuracy of 94.96% by using random forest classifier.

24.7 Future Work

This approach can be enhanced in several aspects, for example

1. Applying more deep learning and machine learning algorithms.
2. Using other methods for attribute selection.
3. Increasing the size of the dataset.

References

1. Dinesh, K.G., Arumugaraj, K., Santhosh, K.D., Mareeswari, V.: Prediction of cardiovascular disease using machine learning algorithms. In: 2018 International Conference on Current Trends towards Converging Technologies (ICCTCT) (2018). https://doi.org/10.1109/ICCTCT.2018.8550857
2. Sharma, V., Yadav, S., Gupta, M.: Heart disease prediction using machine learning. In: 2nd International Conference on Advances in Computing, Communication Control and Networking (ICACCCN) (2020)
3. Rindhe, B.U., Ahire, N., Patil, R., Gagare, S., Darade, M.: Heart disease prediction using machine learning algorithms. Int. J. Adv. Res. Sci. Commun. Technol. (IJARSCT) 5(1) (2021)
4. Jindal, H., Agarwal, S., Khera, R., Jain, R., Nagrath, P.: Heart disease prediction using machine learning. In: IOP Conference Series: Materials Science and Engineering, Volume 1022, 1st International Conference on Computational Research and Data Analytics (ICCRDA 2020), Rajpura, India, 24 Oct 2020
5. Garg, A., Sharma, B., Khan, R.: Heart disease prediction using machine learning techniques. In: IOP Conference Series Materials science and Engineering (2021). https://doi.org/10.1088/1757-899X/1022/1/012046
6. Rajdhan, A., Agarwal, A., Sai, M., Ravi, D., Ghuli, P.: Heart disease prediction using machine learning. Int. J. Eng. Res. Technol. (IJERT) 09(04) (2020). https://doi.org/10.17577/IJERTV9IS040614
7. Gupta, S.K., Shrivastava, A., Upadhyay, S.P., Chaurasia, P.K.: A machine learning approach for heart attack prediction. Int. J. Emerg. Technol. Innov. Res. (2020)
8. Bharti, R., Khamparia, A., Shabaz, M., Dhiman, G., Pande, S., Singh, P.: Prediction of heart disease using a combination of machine learning and deep learning. Comput. Intell. Neurosci. (2021). https://doi.org/10.1155/2021/8387680
9. Raut, S., Magar, R., Memane, R., Rupna, V.S.: Heart disease prediction using machine learning. J. Emerg. Technol. Innov. Res. (JETIR) 7(6) (2020)
10. https://www.javatpoint.com/machine-learning-support-vector-machine-algorithm
11. https://www.techtarget.com/searchbusinessanalytics/definition/logisticregression
12. Ramalingam, V.V., Dandapath, A., Raja, M.K.: Heart disease prediction using machine learning techniques. Int. J. Eng. Technol. 7, 684–687 (2018). https://doi.org/10.14419/ijet.v7i2.8.10557

Chapter 25
An Efficient Approach for Credit Card Fraud Identification with the Oversampling Method

Mitul Biswas and Swapan Debbarma

Abstract The financial business is rapidly expanding, and as a result, banking online transactions are increasing as the government pushes digital transactions. The bulk of financial transactions has been made with debit or credit cards. As a result, the amount of fraud linked with it is increasing. However, because existing fraud detection machine learning algorithms are trained and subsequently assessed on severely uneven datasets, their performance in real-world situations suffers. We presented a method in this study that may operate better after turning these unbalanced datasets into balanced datasets using the oversampling approach, ensuring that the system is not biased when the algorithm is used. The results show that oversampling method with the Random Forest algorithm performs better than the Logistic Regression algorithm with the oversampling method with 80% of accuracy.

25.1 Introduction

Cybersecurity is becoming increasingly critical as we get closer to a digital world. Detecting anomalous activity is the most challenging challenge when it comes to digital security. A big percentage of consumers prefer to utilize credit cards while doing any type of online transaction. Credit card credit limits might occasionally let us make purchases even if we don't have the dollars on hand right now. Cybercriminals, on the other hand, make use of these characteristics. We need a system that can block a transaction if it detects something suspect to handle this problem. This involves the creation of a system that can track the pattern of all transactions and terminate them if any of the patterns are abnormal.

M. Biswas (✉)
NIT Agartala, Krishnagar, West Bengal, India
e-mail: mitulbiswaschottu@gmail.com

S. Debbarma
NIT Agartala, Jirania Agartala, Tripura, India

25.1.1 Classification of Credit Card Fraud

Application fraud: When a fraudster gains control of a transactional application, it is known as application fraud. Card not present: This type of fraud occurs when a transaction is completed without the use of a physical card. The card that has been lost or stolen: If a card has been lost or stolen, anybody can use it. Fake website: If we made a payment without knowing if the website was genuine. The information on the card may be stolen and misused by the website. Merchant issue: This fraud occurs when a merchant misappropriates and shares card information with a third-party organization. Credit card fraud occurs when a card holder's or card's credentials are shared with any other fraudulent group without the card owner's authorization, and certain transactions occur. Because the government is pushing digital transactions like Digital India, the financial industry is growing and expanding at a rapid pace. As a result, banking online transactions are also increasing. Debit or credit cards are used for the majority of banking transactions. As a result, fraud is also increasing day by day. However, because present fraud detection machine learning systems are trained on severely unbalanced datasets, they are unable to correctly detect fraudulent transactions. As a result, the accuracy of these strategies in real-life situations varies greatly. So, using pattern recognition and machine learning, my objective is to turn these unbalanced datasets into balanced datasets and determine which of the available methods or methodologies gives the best outcomes in all fraudulent instances.

25.2   Problem Statement

Machine learning techniques had previously been employed to work on unbalanced credit card fraud datasets. No authors have converted imbalanced datasets to balanced datasets and used oversampled data for machine learning algorithms' training and testing phase to determine which machine learning algorithm provides the best results in a real-world scenario for detecting fraudulent transactions.

25.3   Related Works

Asha [1] uses support vector machine (SVM), k-nearest neighbor (KNN), and artificial neural network (ANN) methodologies to discover credit card fraud, as well as machine learning and deep learning techniques to find non-fraud transactions.

Another paper by VENGATESAN and KUMAR [2] suggests credit card fraud detection in the banking industry using a machine learning algorithm and compares the accuracy of logistical regression and KNN techniques. The banking industry provides a variety of services to its consumers, including ATM cards, Internet banking, gold loans, education loans, debit cards, and credit cards, in order to entice people to create bank accounts. Customers often use credit cards 24 h a day, seven days a week, allowing the bank server to keep track of all transactions using machine learning algorithms. It must locate or anticipate fraud detections. The data collection contains all of the characteristics of each transaction, and we must categorize each one and find out whether transactions are happening legally or not.

A credit card fraud detection model was created by Bahnsen, Aouada, Stojanovic, and Ottersten [3], and extracting the proper features from transactional data is crucial. In order to uncover customer purchasing trends, this is frequently achieved by aggregating transactions. We propose to develop a new set of characteristics based on the von Mises distribution, which they used to examine the periodic behavior of transaction time in their research.

Krishna Modi [4] conducted a comparison study of numerous strategies for detecting fraud, including decision trees, rule-based mining, neural networks, fuzzy clustering approaches, hidden Markov models, and hybrid approaches of these methods.

The Random Forest algorithm was used by Suresh Kumar, Soundarya, Kavitha, Keerthika, and Aswini [5] to design a system for identifying fraudulent transactions and assessing their accuracy (RFA). This approach is based on supervised learning and uses decision trees to categorize the dataset. After the dataset has been categorized, a confusion matrix is generated. The Random Forest algorithm's performance is measured using the confusion matrix.

The performance of Nave Bayes, k-nearest neighbor, and Logistic Regression on highly skewed credit card fraud data was evaluated in comparative research by Awoyemi, Adewunmi, and Oluwadare [6].

Varmedja, Karanovic, Sladojevic, Arsenovic and Anderla [7] conducted research that supports many approaches for determining if a transaction is fraudulent or not. Their suggested model might be utilized to spot more abnormalities.

The transaction aggregation approach is extended by Bahnsen, Aouada, Stojanovic, and Ottersten [8], who propose a new set of characteristics based on analyzing the periodic behavior of transaction time using the von Mises distribution.

Goel, Abhilasha, and Agarwal [9] developed a system that analyzes fraud in online buying using the Random Forest algorithm (RFA), detecting fraud and preventing odd actions.

Malini and Pushpa [10] have done an analysis-based approach to credit card fraud identification. They have made the fraud detection analysis approach with the help of KNN and outlier detection.

Devi, Janani, Gayathri, Indira [11] devised an approach that uses a Random Forest algorithm to improve the accuracy of fraud detection. The classification phase of the Random Forest algorithm is used to assess the dataset and the user's current dataset. Finally, increase the output data accuracy. The accuracy, sensitivity, specificity, and precision of the methodologies are used to assess their effectiveness.

Zheng and Liu [12] have detected an efficient way to detect transactional fraud by utilizing relation and behavior diversity or behavior profiles of users of their past transactional behavior.

25.4   Existing Systems

To distinguish between fraudulent and authentic credit card transactions, machine learning techniques such as Random Forest, *k*-nearest neighbor, SVM, and Logistic Regression approaches have been applied. In the training and testing phase, however, all of the authors used extremely skewed datasets of credit card frauds. As a result, the accuracy of all available machine learning algorithms varies depending on the experiments of different authors.

25.5   Proposed System

The performance of a machine learning algorithm is determined by the dataset presented to the system throughout its learning period. The algorithm learned from the test data and made a judgment based on that learning whenever the program received input during real-time application. As we all know, according to a Business Standard study [14], the overall number of financial transactions is 41 million, and the total number of scams reported by the New India Express [13] is 1194. So, if we look at the percentage of fraudulent transactions compared to the total number of transactions. If we study the dataset created by real-time transactions, we can see that it is very skewed. As a predictive modeling assignment, unbalanced classification is particularly tough due to the severely skewed class distribution. This is why traditional machine learning algorithms and assessment methods that assume a balanced class distribution fail miserably. Machine learning systems have a significant barrier incorrectly predicting scams. Although there are a few alternative algorithms that can operate on an unbalanced data collection with excellent accuracy, there are few approaches. One-class SVM (support vector machine), for example, is a classification technique that aids in the detection of outliers in data. This technique may be used to deal with challenges with unbalanced data, such as fraud detection [15]. Another method for dealing with uneven datasets is the Random Forest algorithm [16]. As a result, we must transform the existing unbalanced dataset into a balanced dataset since, as far as we know, without a properly balanced dataset, we will be unable to effectively assess any type of fraudulent transaction because of the machine learning algorithm will not be properly trained.

25.6   Methodology

This section is intended to discuss implementation, which covers the algorithms and other components required for the proposed system's implementation. The implementation in this article will begin with importing the dataset and then cleaning and normalizing it. The dataset is separated into two parts, one for training and the other for testing. Finally, the system could be able to distinguish between fraudulent and legitimate transactions. Python is the programming language to be utilized in the proposed system. Python is a scripting language that is simple to use, interpreted, object-oriented, and high level. Python is an excellent programming language for implementing machine learning algorithms. It comes with several important packages and library functions for machine learning.

25.6.1 The Following Are the Packages and Libraries that Were Used

Numby: It's one of the Python libraries' members. Multidimensional arrays and linear algebraic procedures are the most common applications.

Pandas: One of the Python libraries is Pandas. Pandas are mostly used as a tool for data analysis and manipulation. It is mostly used to read data from a dataset and load it.

Scikit-learn: This Python package may be used to create statistical and machine, learning models. It is the most appropriate Python module for machine learning modeling.

Keras: Keras is a high-level API. It's a neural network API, to be precise. It is possible to execute it on top of the tensor flow. Its primary purpose is to build deep learning algorithms. It can run on both the CPU and the GPU. Keras and backend running tensor flow were employed in this work. Keras with the tensor flow as a backend aid in the training of neural network design.

MySQL: The MySQL database is used to store data. For storing user information, we utilized MySQL. The user must register by providing credentials, which are then saved in the MySQL database.

Tkinter: Tkinter is a graphical user interface (GUI) library in Python. It's compatible with both Unix and Windows. Importing the Tkinter module is the first step, followed by the creation of a GUI and the addition of one or more widgets, which are then called in the loop.

The unbalanced datasets are first balanced using SMOTE, and then machine learning techniques are employed to determine which algorithm provides the most precision, recall, and accuracy, as well as the best $F1$-score. The oversampled and undersampled datasets to run for the training and testing phases have never been tested before by previous authors. The majority classes are heavily favored by the unbalanced datasets. That's why we need to create a dataset that's 50/50 fraudulent and non-fraudulent, and then put it through the algorithms' training and testing phases. SMOTE is used to generate new synthetic points, ensuring that the classes are evenly distributed. This is used to correct problems with unbalance. It is used mostly to generate synthetic points from the minority class to achieve a balance between the minority and the majority. SMOTE additionally selects the distance between the minority class's nearest neighbors and makes synthetic points in the space between them. We don't have to remove any rows, unlike with random sampling, which lets us keep more information. After all, SMOTE may be more accurate than random undersampling because no rows are deleted.

25.6.2 Dataset Collection

The credit card fraud detection dataset, which can be downloaded from Kaggle, was used in this study. This dataset covers transactions done by European cardholders in September 2013 over the course of two days. There are 31 numerical features in the dataset. Because some of the input variables contain financial information, the PCA transformation of these variables was used to ensure that the data remained anonymous. Three of the given characteristics were not altered. The feature "Time" displays the time between the first and subsequent transactions in the dataset. The amount of credit card transactions are displayed in the "Amount" feature. The label is represented by the feature "Class."

25.6.3 The Mainly Used Algorithms for Our Experiments Are

Logistic Regression. Random Forest classifier

The working processes of those above methods are described below:

Logistic Regression process

Step 1: Data pre-processing and cleansing is the first step. The dataset was re-sampled as fraudulent and non-fraudulent. The data has been standardized and scaled. The dataset is split into two parts: training and testing. The unbalanced dataset is converted to a balanced dataset using SMOTE.
Step 2: Fitting Logistic Regression to the training set is the second step.
Step 3: Predict the exam's outcome.
Step 4: Double check that the outcome is right (making of confusion matrix using it).
Step 5: Visualize the test set's findings. Logistic regression has several advantages.
Step 6: Come to a halt.

Random Forest classifier process

Step 1: Data pre-processing and cleaning. The dataset was re-sampled as fraudulent and non-fraudulent. The data has been standardized and scaled. The dataset is split into two parts: training and testing. The unbalanced dataset is converted to a balanced dataset using SMOTE.

Step 2: Select K data points from the training set at random.

Step 3: For the data points you've picked, make decision trees (subsets).

Step 4: Choose a N for the number of decision trees you want to make.

Step 5: Go through Steps 2 and 3 again.

Step 6: Locate each decision tree's forecasts for new data points, and assign the new data points to the most popular category.

Step 7: Come to a halt.

25.7   Results and Discussion

The dataset used in the proposed technique may be acquired from the website www. kaggle.com. The transactions done by clients of a European bank in 2013–14 are utilized as a dataset. It is used to train and test our system in order to detect fraudulent and legitimate transactions (Fig. 25.1).

25.7.1   Evaluation Measure

We will discuss the multiple ways to check the performance of the machine learning algorithms. We can check the efficiency of the algorithms based on the following parameters.

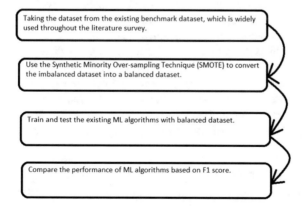

Fig. 25.1 It shows the workflow model of the proposed approach

25.7.2 Confusion Matrix

The confusion matrix is basically the best representation of the following parameters.

True positives (TP): Predicted positive and they are actually positive.
False positives (FP): Predicted positive and they are actually negative.
True negatives (TN): Predicted negative and they are actually negative.
False negatives (FN): Predicted negative and they are positive.

$$\text{Accuracy Measurement} = (TP + TN)/(TP + FP + TN + FN)$$

$$\text{Precision measurement} = TP/(TP + FP)$$

$$\text{Recall measurement} = TP/(TP + FN)$$

$$F1 - \text{score measurement} = (2 * \text{Precision} * \text{Recall})/(\text{Precision} + \text{Recall})$$

25.7.3 Results

Table 25.1 [17] is used as a guide for comparing the performance of various machine learning algorithms that provide results based on unbalanced training and testing datasets. The $F1$-score is used to determine how robust and exact our classifier is. It is the harmonic mean of recall and accuracy. The $F1$-score is being used to assess performance. Based on the $F1$-score in the table above, we can conclude that the Random Forest method outperforms the other three machine learning algorithms in detecting fraud transactions in an unbalanced dataset.

Figure 25.2 depicts the dataset's unbalanced nature. The dataset contains 99.83% legitimate transactions and 0.17% fraudulent ones. So, we'll utilize oversampling to balance the dataset, and then the oversampled dataset will be used for training and testing. To determine how skewed certain characteristics in the dataset are, we will utilize transaction amount and transaction time distributions. We can't access the names or other details of the privacy policies. All other characteristics have

Table 25.1 Shows the performance results of different ML methods

Evaluation parameters in (%)	Logistic regression	Random forest
Precision	92.8956	95.9887
Recall	93.112	95.1234
Accuracy	90.448	94.9991
$F1$ Score	92.112	95.1102

undergone PCA transformation, which implies they have already been scaled. The transaction amount and transaction duration will be scaled using the feature scaling mechanism. Figure 25.3 shows that the distribution of time and amount is scaled.

Figure 25.4 displays the outcomes of balancing the unbalanced dataset. According to the findings, the dataset now contains 50% real transactions and 50% fraudulent ones. After cross-validation, oversampling (SMOTE) is employed to prevent over-fitting and data leaking. The machine learning algorithms will now be trained and tested using this dataset.

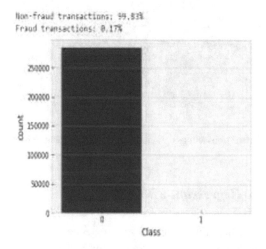

Fig. 25.2 This shows that the taken dataset is consist of fraud and non-fraud transactions

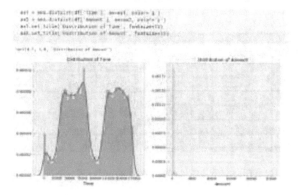

Fig. 25.3 It shows the distribution of transaction time and amount

```
Before OverSampling, counts of label '1': [391]
Before OverSampling, counts of label '0': [227454]

After OverSampling, the shape of train_x: (454908, 30)
After OverSampling, the shape of train_y: (454908,)

After OverSampling, counts of label '1', %: 50.0
After OverSampling, counts of label '0', %: 50.0
<matplotlib.axes._subplots.AxesSubplot at 0x7fcEaaaf4b38>
```

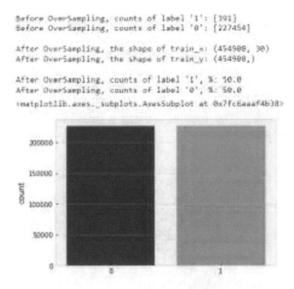

Fig. 25.4 It shows that the transactions dataset of fraud and non-fraud transactions is now balanced after oversampling

25.7.4 Logistic Regression with SMOTE

It has a high recall value, indicating that the model can detect the greatest number of fraud transactions; however, the precision is quite poor, indicating that the model incorrectly labels many legitimate transactions as fraud. It's critical to have a high accuracy number (Figs. 25.5, 25.6 and 25.7).

	precision	recall	f1-score	support
0	1.00	0.98	0.99	56861
1	0.06	0.94	0.12	101
accuracy			0.98	56962

Fig. 25.5 It shows the results of different measurement matrices after logistic regression are run with SMOTE

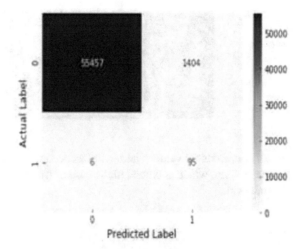

Fig. 25.6 It shows the confusion matrix OF LOGISTIC REGRESSION with SMOTE results

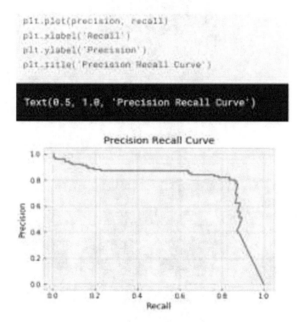

Fig. 25.7 It shows the precision–recall curve of logistic regression with SMOTE

25.7.5   *Random Forest with SMOTE*

In Figs. 25.8, 25.9 and 25.10, we see that Random Forest with oversampling method performs better than Logistic Regression with oversampling method. It provides us

Fig. 25.8 It shows the
results of different
measurement matrices after
random forest is run with
SMOTE

	precision	recall	f1-score	support
0	1.00	1.00	1.00	56861
1	0.88	0.83	0.86	101
accuracy			1.00	56962

with a superior recall and accuracy value. The recall has been reduced slightly, but
the precision has greatly risen, which is crucial in the case of fraud detection, and as
we all know, it's a trade-off.

Fig. 25.9 It shows the
confusion matrix of random
forest with SMOTE results

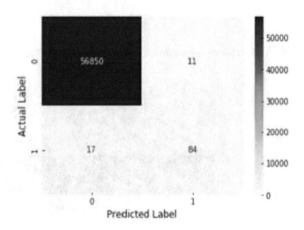

Fig. 25.10 It shows the
precision–recall curve of
random forest with SMOTE

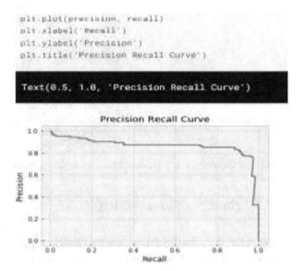

25.8   Conclusion

When employing balanced datasets, good prediction results can be obtained. With balanced datasets, the Random Forest classifier provides the best results in this article. It assists us in detecting over 80% of fraud transaction situations while also not labeling numerous non-fraud transaction cases as fraud. In this study, we exclusively utilize SMOTE with the Random Forest classifier and Logistic Regression, and we find that Random Forest with SMOTE produces the best results. With both balanced and unbalanced datasets, Random Forest yields the best outcomes. There is always a trade-off between recall and accuracy, and the appropriate strategy in each case is determined by the situation objectives. So, the best approach is to use oversampled method to balance the available imbalanced dataset and then use the dataset to train and test the Random Forest classifier and then implement that approach in real-life scenarios to find out or identify fraudulent transactions, and we will be able to get the best results.

References

1. Asha, R.B., Kumar, S.: Credit card fraud detection using artificial neural network. In: Global Transactional Proceeding, 23rd Jan, vol 2. Publishing services by Elsevier B.V., pp. 35–41 (2021). https://doi.org/10.1016/j.gltp.2021.01.006. (Elsevier, 2666-285X/2021)
2. Vengatesan, K., Kumar, A., Yuvraj, S., Ambeth, K.V.D, and Sabnis, S.S.: Adv. Math: Sci. J. **9**(3), 1185–1196 ISSN: 1857-8365 (printed); 1857-8438 (electronic). https://doi.org/10.37418/amsj.9.3.43. Spec. Issue on RDESTM (2020)
3. Bahnsen, A.C., Aouada, D., Stojanovic, A., Ottersten, B.: Detecting credit card fraud using periodic features. In: IEEE 14th International Conference on Machine Learning and Applications, 978-1-5090-0287-0/15 (2015).
4. Modi, K., Dayma, R.: Review on fraud detection methods in credit card transactions. In: International Conference on Intelligent Computing and Control (I2C2'17) (2017)
5. Kumar, M.S., Soundarya, V., Kavitha, S., Keerthika, E.S., Aswini, E.: Card fraud detection using random forest algorithm. In: IEEE 3rd International Conference on Computing and Communication Technologies ICCCT 2019, 978-1-5386-9371-1/19 (2019)
6. Awoyemi, J.O., Adewunmi, A.O., Oluwadare, S.A.: Credit card fraud detection using machine learning techniques. In: IEEE Conference 978-1-5090-4642-3/17 (2017)
7. Varmedja, D., Karanovic, M., Sladojevic, S., Arsenovic, M., Anderla, A.: Credit card fraud detection—machine learning methods. In: IEEE Conference 978-1-5386-7073-6/19/ (2019)
8. Bahnsen, A.C., Aouada, D., Stojanovic, A., Ottersten, B.: Feature engineering strategies for credit card fraud detection. Expert Syst. Appl. **51**, 134–142 (2016). https://doi.org/10.1016/j.eswa.2015.12.030 0957-4174
9. Goel, E., Abhilasha, Agarwal, A.: Fraud detection using random forest algorithm. Int. J. Comput. Sci. Eng. (IJCSE) **5**(05) (2016)
10. Malini, N., Pushpa, M.: Analysis on credit card fraud identification techniques based on KNN and outlier detection. In: 3rd International Conference on Advances in Electrical, Electronics, Information, Communication and Bio-Informatics (AEEEICB17) (2017)
11. Meenakshi, D., Janani, B., Gayathri., Indira, N.: Credit card fraud detection using random forest. Int. Res. J. Eng. Technol. (IRJET) **6** (2019)
12. Zheng, L., Liu, G.: Transaction fraud detection based on total order relation and behaviour diversity. IEEE Trans. Comput. Soc. Syst. (2018)

13. The New India Express NCRB report 2020: Debit credit card fraud climbs steeply, more cases compared to 2019. In: https://www.edexlive.com/news/2021/sep/16/ncrb-report-2020-debit-credit-card-fraud-online-climbs-steeply-225-more-cases-when-compared-to-2-24053. html (2020)
14. Business Standard India leads the world with approximately 41M transactions in a day. In: https://www.edexlive.com/news/2021/sep/16/ncrb-report-2020-debit-credit-card-fraud-online-climbs-steeply-225-more-cases-when-compared-to-2-24053.html (2020)
15. Chuprina, R., Kovalenko, O.: Credit card fraud detection: Top ML solutions 2021. In: Card Fraud Detection: Top ML Solutions in 2021 (spd.group) (2021)
16. Frei, L.: Detecting credit card fraud using machine learning. In: Detecting Credit Card Fraud using Machine Learning—by Lukas Frei—Towards Data Science
17. Trivedi, N.K., Simaiya, S.,Lilhore U.K., Sharma, S.K.: An efficient credit card fraud detection model based on machine learning methods. Int. J. Adv. Sci. Technol **29**(5), 3414 3424 (2020)

Chapter 26
Content-Based Secure Image Retrieval in an Untrusted Third-Party Environment

Sandeep Singh Sengar and Sumit Kumar

Abstract In this digital world, where availability of the image-generating tools is quite common and owing to the rapid growth of Internet knowledge, people use to exchange massive volume of images every day which results in creating large image repositories. So, retrieving appropriate image available on these repositories is one of the vital tasks. This problem leads to evolving content-based image retrieval (CBIR). As the generation of image increases, people start transferring these images to a remote third-party server, but these images may have personal information. This leads to adding privacy concerns toward the system as transferring personal data to some other place might be a cause of leakage of information or transfer to an unauthorized person. So, to keep this in mind, sensitive images like medical and personal images require encryption before being a contracted out for the privacy-preserving resolutions. In this work, we have deployed ACM for image encryption as well as asymmetric scalar product preserving encryption (ASPE) for feature vector encryption and similarity matching. We have demonstrated our results based on various benchmark databases.

26.1 Introduction

In this section, background, aim, and motivation of this work have been elaborated.

S. S. Sengar
Department of Computer Science, Cardiff Metropolitan University, Cardiff, UK
e-mail: SSSengar@cardiffmet.ac.uk

S. Kumar (✉)
Department of Informatics Cluster, School of Computer Science, University of Petroleum and Energy Studies, Dehradun, Uttarakhand 248007, India
e-mail: sumitvarshney68@gmail.com

© The Author(s), under exclusive license to Springer Nature Singapore Pte Ltd. 2023
V. Bhateja et al. (eds.), *Evolution in Computational Intelligence*, Smart Innovation, Systems and Technologies 326, https://doi.org/10.1007/978-981-19-7513-4_26

26.1.1 Background

Nowadays, knowledge has developed progressively advanced. This leads to cheap and cutting-edge multimedia devices and has given rise to vast data volumes of multimedia content. Vast data volumes of multimedia consist of audio, video, images, etc. Today, billions of images are uploaded and downloaded which is creating large amount of data [1–3]. This large amount of data needs to be stored in the database for several requests such as medical, crime prevention, and security.

These solicitations have generated a necessity for effective, secure, and efficient ways of storage, search, and image retrieval via similar processing procedures. We share images with each other and also publish them for instance on the Internet. Those image collections are important contributors to the public domain of the Internet which is of billions of images. Private image collections or images on the Internet might be the most obvious example, but the use of digital imaging has spread too many application areas.

Modern hospitals are one of the best examples, in which large collections of medical images are succeeded, stored, and used every day. Newspapers, image providers, and other firms in the graphic design trade use digital images in their workflow and databases. One more example is of security industry in which surveillance cameras produce large amount of images. Suppose an institute has advanced a novel face recognition algorithm. The institute will wish that those input images and images present in their database are not revealed publicly or advertise by chance. Another example, we can take of clinicians who can use CBIR [4, 5] to find the similar cases of patients and facilitate the clinical decision-making processes. The patients may or may not want to reveal their medical images to any other except a particular doctor in medical CBIR applications whether physically or electronically. Therefore, we effort to find a solution to secure storage and image retrieval byway that such goal can be achieved.

Previously, the images were kept with linked labels or thread, and search was accomplished on the ground of these labels. Since applying thread to an individual image is a very time-consuming task and also it incorporates human perception about each image. It has been seen very frequently that different human being can perceive different images differently which could result in either getting the wrongly matched image or leave correctly associated images. Another aspect which makes this process unrealistic is today every process is time-dependent, i.e., we need to design a system which could produce the result within a particular time-bound, and this is not possible which earlier process with such huge volumes of multimedia repositories. Nowadays, with extensive reach of social networks and multimedia tools like digital camera and mobile phones, and users became a lot of involved concerning their secrecy and their information kept on servers. Some users even choose to hide their details from database administrator. If stored information on the server is seen by database admin, a user's privacy may be on the urge of breaching if the database or server administrator is faulty and the chances of misusing of this information increase. If an organization's employee details are stored on a server along with their photographs for face recognition, the organization prefers to keep this information out of reach of any other user or database admin. If this information is not hidden perfectly, a compromised database admin may be able to access them and use it for his/her own benefits, such a situation needs to be avoided. Therefore, going with this direction, it is very much clear that multimedia data security is one of the main concern of today's retrieval system.

26.1.2 Aim and Motivation

The first part of the paper will focus on image retrieval technique based on feature vector extraction method using both texture and shape feature. Second part will show how an image is securely encrypted at owner's end and decrypted at user side. The main idea is that the owner extracts feature of an image, creates an image feature vector database, and creates an encrypted image database and sends it to centralize database we assume the database admin to be honest-but-curious in nature. On the other hand, user also extracts the desired features to form query image feature vector and sends it to the centralized database. In the centralized database, the query processor will find similar encrypted images with the query image given by users and sends the result to user. At user end, user decrypts images with the owner keys. The output is images with the most similar image in the database. The motivation for creating such a secure system is that the authorized image users can only retrieve the images he needs and database admin also can't breached the privacy of the owner image.

26.2 Literature Review

In the last couple day decades, multimedia data have been flourished very radipdly. Due to this, keeping all the data with oneself is next to impossible. Hence, people start storing this data to any third-party storage. But, this transfer of data must be secure. Not only the data must be encrypted but the relevant searching should also be protected. To the best of our knowledge, the first ever searchable encryption valid for the large volume of images was proposed by Song et al. [6]. Another work presented by Weng et al. [7] which preserve the confidentiality. This work is based on hash values with neglecting some values for enhancing the security measures. Zhihua et al. [1] proposed a work which form the feature vector using color descriptor, EDH, and scalable color histogram. But, his scheme suffers from high computational overhead. Recently, Majhi et al. [8], proposed a secure image retrieval work in cloud environment for the health care. But, in this work, cloud service provider or the cloud administrator has been considered honest-but-curious. This assumption may not be true for all real-world applications.

26.3 Preliminaries

In this section, we will demonstrate the various techniques that we have incorporated during the development of this work. They will be described under the various subsections as follows.

26.3.1 Local Binary Pattern (LBP)

LBP [9] is a kind of visual descriptor employed for grouping in PC vision. LBP is the specific case of the surface range illustration proposed in 1990. Since 1990, LBP has been observed to be an intense module for texture ordering. It has additionally been resolved that when LBP is joined to histogram of oriented gradients (HOG) [10] descriptor, it add to the recognition performance remarkably on few datasets. A correlation of a few deviations of the first LBP in the arena of foundation subtraction was brought on 2015 by Silva et al. A complete study of the various executions of LBP can be found in Bouwmans et al. The LBP feature vector, in its most straightforward shape, is done in different way:

- Separate the analyzed window into cells (e.g., 16 × 16 pixels for every cell).
- For every pixel in a cell, contrast the pixel with every one of its 8 neighbors (to its left side best, left-center, left-base, right-top, and so forth.). Take after the pixels along a circle, i.e., clockwise or counter-clockwise.
- Where the center pixel's value is taken if the center value is greater than the neighbor's value, express '0'. Else, state '1'. This results in 8-digit binary number (which is typically changed over to decimal gradually).
- Histogram is computed, over each cell value, of the recurrence of each 'number' happening (i.e., every blend of which pixels are smaller and greater than the center). This histogram can be viewed as a 256-dimensional feature vector.
- Alternatively standardize the histogram.
- histograms of all cells which results in the form of feature vector for the whole window (Fig. 26.1).

26.3.2 Arnold Cat Map

In this section, Arnold Cat Map [11] is described in briefly which is used in the image encryption process. The ACM has wide application on image encryption process which is capable to confuse any square image significantly. According to the ACM, when an image undergoes chaotic transformation, pixels of the original image get scrambled to some other random position. The ACM works on square image, so in our proposed scheme, we have used ACM on 128 × 128 blocks. The

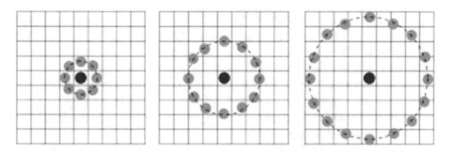

Fig. 26.1 Three neighborhood cases used to characterize a texture and LBP are calculated

ACM process is performed as follows: $\begin{bmatrix} u' \\ v' \end{bmatrix} = A \begin{bmatrix} u \\ v \end{bmatrix}$ where $A = \begin{bmatrix} 1 & p \\ q & pq+1 \end{bmatrix}$ The ACM moves the intensity value presence at the co-ordinate (u, v) position to the co-ordinate (u', v') position. The parameters p and q may be considered as a secret key. The user can perform the inverse operation with knowing $\begin{bmatrix} u \\ v \end{bmatrix} = A^{-1} \begin{bmatrix} u' \\ v' \end{bmatrix}$ where $A^{-1} = \begin{bmatrix} 1+pq & -p \\ -q & 1 \end{bmatrix}$.

26.3.3 Asymmetric Scalar Product Preserving Encryption (ASPE)

ASPE [12] was recommended in deprive of any data structures use over encrypted data kNN activities were executed. Many techniques are being used by ASPE for encrypting database points and query point. Invertible matrix is used to encrypt the database points, and similarly, its inverted matrix to encrypt query points, it avoids attack centered on distance preservation among encrypted data and unencrypted data, henceforth escaping distance recovery. Two feature vector given in DB, i.e., f_{v1} and f_{v2}, their distance $d(f_{v1}, f_{v2})$ could be known from their encrypted values $E_T(f_{v1}, K)$ and $E_T(f_{v2}, K)$. These distance permits the attacker to figure signature, and in this way, signature connecting attack is done. To oppose these attacks, we require an encryption work that does not uncover distance information. For kNN seek, we watch that correct distance computation isn't fundamental. Or maybe, we just need a distance examination activity. Given two feature vector f_{v1} and f_{v2} in DB, we must choose which of the two feature vector of image is closer to the feature vector of query image, i.e., f_{vq}.

- **Key**: a $(d + 1) \times (d + 1)$ invertible matrix M.
- **Feature vector encryption function** $E_T(\cdot)$: Considering a database feature vector f_v. Firstly, creating a feature vector $f_v = (f_v^T, -0.5\|f_v\|^2)^T$ of dimension $(d + 1)$. Secondly, compute an encrypted feature vector $f_v' = M^T f_v$.

- **Query Image feature vector encryption function** $E_q(\cdot)$: Considering a query image feature vector f_{vq}. Firstly, generating a random number $r > 0$. Secondly, creating a feature vector $f_{vq} = r(f_{vq}^T, 1)^T$ of dimension $(d + 1)$. Thirdly, the compute the encrypted query feature vector as $f_{vq}' = M^{-1} f_{vq}$.
- **Distance comparison operator**: Let f_{v1}' and f_{v2}' be the encrypted feature vectors of f_{v1} and f_{v2}, respectively. To know whether the feature vector f_{v1} is nearer to query image feature vector f_{vq} than f_{v2}, we calculate $(f_{v1}' - f_{v2}') \cdot f_{vq}' > 0$, where f_{vq}' is the encrypted feature vector f_{vq}.
- **Decryption Function**: Considering an encrypted feature vector f_v'. The feature vector $f_v = \pi_d M^{T^{-1}} f_v'$, where π_d is a $d \times (d + 1)$ matrix and $\pi_d = (I_d, 0)$, where I_d is $d \times d$ identity matrix.

26.4 Proposed Work

CBIR is the efficient image retrieval technique, but it lacks some security feature. Now, as the generation of image increases, one is not always keen to contain each and every image with oneself, and image may contain some personal stuff, or those images might be needful to someone at the various point of time, so people use to transfer them to some centralized repository. Now, this leads to add privacy concern toward the system as transferring personal data to some other place might be a cause of leakage of information or transfer to not authorized person. Therefore, we have proposed secure and efficient technique for image retrieval.

26.4.1 Content-Based Image Retrieval

Traditionally, searching of the images is utilizing content, labels, or watchwords or comment allotted to the image while keeping it into the databases. While if the image which is kept into the databases are not remarkably or particularly labeled or wrongly depicted at that point, it is deficient, relentless, and to a great degree of tedious work on the vast volume of pictures. This issue prompts the development of efficient CBIR [13, 14] process which is utilized for recovering the correct images from the database. In a CBIR system, user selects a query image and extracts its visual image features and combined them together to form the query image feature vector. Now, the same feature extraction process has been employed to each image of the respective database to form feature image database. Afterward, using a similarity measurement technique, query image feature vector is matched with every single feature vector of the feature database to receive some most similar images. This process has been depicted in Fig. 26.2.

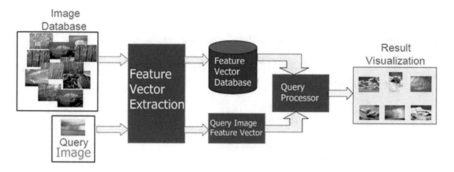

Fig. 26.2 Basic of content-based image retrieval

26.4.2 Image Encryption

As per Arnold's change, a picture is hit with the change that evidently randomizes the first association of its pixels. Be that as it may, if iterated enough circumstances, in the end, the first picture returns. The quantity of considered cycles is known as the Arnold's time frame. The period depends on the picture estimate, i.e., for various size pictures, Arnold's period will be diverse. The various steps involved in the image encryption process are given below.

26.4.3 Feature Extraction

Feature extraction is done on all images of respective database in a unique way so that the result of image retrieval is accurate. Block diagram for the feature extraction is given in Fig. 26.3.

26.5 Experimental Results

In this section, we have eleborated the various databases and their experimental results.

26.5.1 Experimental Implementation

All the proposed techniques and the derived results are obtained in MATLAB environment. The experiment has been performed on MATLAB environment installed over a system having, Intel(R) Core(TM) i7-4770 CPU@3.4-GHz,4 GB RAM, Windows 8 operating system, configuration.

Step 1: Input a RGB image

Step 2: Use numb as variable which is denoted as the No. of Iterations

Step 3: Find the No. of rows and columns.

Step 4: for incr = 1 to numb

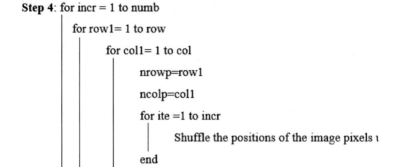

    for row1= 1 to row

        for col1= 1 to col

            nrowp=row1

            ncolp=col1

            for ite =1 to incr

                Shuffle the positions of the image pixels ı

            end

            Result the new encryption image

        end

    end

end

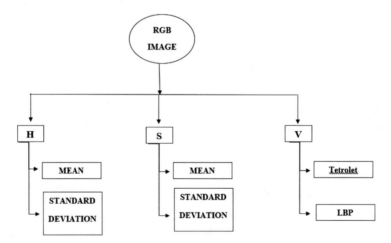

Fig. 26.3 Image feature extraction

26.5.1.1 Benchmark Databases

We have used four standard image database, namely:

1. GHIM-10,000 [15]: It contains 10,000 different images which is of 20 categories like firecracker, building, cars, horse, and insects.
2. Sorted Produce-1400 [4]: It contains 1400 different images which is of 14 categories of food items like potato and apple.

Table 26.1 Demonstrating the average precision, recall, F-score for various databases

Database name	Average precision	Average recall	Average F-score
Corel-1000	64.82	12.96	19.42
GIIIM-10,000	62.50	2.50	4.80
Produce-1400	67.80	13.56	20.36
Olivia-2688	68.20	4.67	8.72

3. Oliva and Torralba Scene (OT-Scene) [16]: It contains 2688 different images which is of 8 categories of scenic beauty.

4. Coral 1000 [17]: It contains 1000 different images which is of 10 categories like dinosaur, beach, bus, and roses.

26.5.1.2 Results and Analysis

All classifications contain categories of images of different sizes. The experiment is checked by various number of returning pictures, which fluctuates from 3 to 20. Calculation of precision, recall, F-score is done on the image databases. Proposed strategies are superior to the older research work as better features of image are being extracted and security is provided to it at top level. Tetrolet as an image feature is never been used for content-based image retrieval process. Figure 26.4 will show first the secure content-based image retrieval results followed by the calculation of precision, recall, F-score is done on the image databases (Table 26.1).

26.6 Conclusion

In this work, the authors have presented a secure CBIR scheme that not only efficiently retrieved image based on the primitive visual image feature but simultaneously impose security aspects to its transmission. There are three main entities in this work, i.e., (i) owner who have the original image database: Its responsibility is to generate an encrypted image database, generate an encrypted image feature database, and transfer both databases to the centralized database (ii) user who has a query image and wants to receive the corresponding similar images securely: Its responsibility is to extract query image features, form encrypted query image feature vector, and send it to the centralized database (iii) centralized database: by which the main repository and searching process has been carried out at pseudo-encrypted domain. The owner has created an encrypted image database using modified ACM with a logistic map, created an encrypted image feature database using the ASPE technique, and transfer these databases to a centralized database. Now, the user has

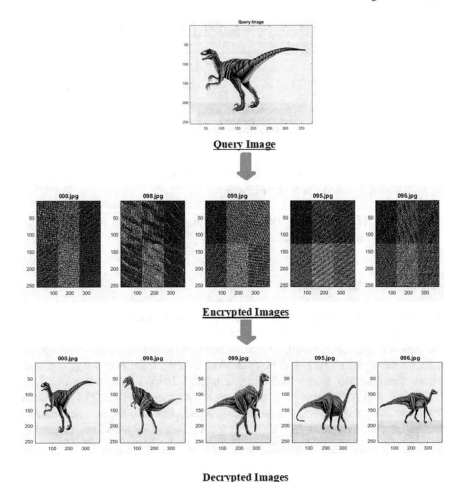

Fig. 26.4 Result for a category of corel-1000 image

to create the encrypted query image feature vector using the same ASPE technique and transfers it to a centralized database. Afterward, similarity measurement of this query image feature vector among all the encrypted feature vectors of the encrypted feature image database has been carried out in a centralized database to retrieve a few of most similar encrypted images which are to be transferred to the user side. Now, through the received key from the owner side, user will decrypt those received images to get the final output. Similarity measurement in a fully encrypted domain may lead us to the wrong outcomes. Therefore, we have incorporated similarity measurement in the pseudo-encrypted domain and got the desired result efficiently and securely.

References

1. Xia, Z., Wang, X., Zhang, L., Qin, Z., Sun, X., Ren, K.: A privacy-preserving and copy-deterrence content-based image retrieval scheme in cloud computing. IEEE Trans. Inf. Forensics Secur. **11**(11), 2594–2608 (2016)
2. Sengar, S.S., Mukhopadhyay, S.: Motion segmentation-based surveillance video compression using adaptive particle swarm optimization. Neural Comput. Appl. **32**(15), 11443–11457 (2020)
3. Sengar, S.S., Mukhopadhyay, S.: Moving object detection using statistical background subtraction in wavelet compressed domain. Multimedia Tools Appl. **79**(9), 5919–5940 (2020)
4. Kumar, S., Pradhan, J., Pal, A.K.: Adaptive tetrolet based color, texture and shape feature extraction for content based image retrieval application. Multimedia Tools Appl. **80**(19), 29017–29049 (2021)
5. Kumar, S., Pradhan, J., Pal, A.K.: A CBIR scheme using GLCM features in DCT domain. In: 2017 IEEE International Conference on Computational Intelligence and Computing Research (ICCIC), pp. 1–7. IEEE (2017)
6. Song, D.X., Wagner, D., Perrig, A.: Practical techniques for searches on encrypted data. In: Proceeding 2000 IEEE Symposium on Security and Privacy. S&P 2000, pp. 44–55. IEEE (2000)
7. Weng, L., Amsaleg, L., Morton, A., Marchand-Maillet, S.: A privacy-preserving framework for large-scale content-based information retrieval. IEEE Trans. Inf. For. Secur. **10**(1), 152–167 (2014)
8. Majhi, M., Pal, A.K., Pradhan, J., Islam, S.K., Khan, M.K.: Computational intelligence based secure three-party CBIR scheme for medical data for cloud-assisted healthcare applications. Multimedia Tools Appl. 1–33 (2021)
9. Guo, Z., Zhang, L., Zhang, D.: Rotation invariant texture classification using LBP variance (LBPV) with global matching. Pattern Recogn. **43**(3), 706–719 (2010)
10. Wang, X., Han, T.X., Yan, S.: An HOG-LBP human detector with partial occlusion handling. In: 2009 IEEE 12th International Conference on Computer Vision, pp. 32–39. IEEE (2009)
11. Bao, J., Yang, Q.: Period of the discrete Arnold cat map and general cat map. Nonlinear Dyn. **70**(2), 1365–1375 (2012)
12. Kumar, S., Pal, A.K., Islam, S.K., Hammoudeh, M.: Secure and efficient image retrieval through invariant features selection in insecure cloud environments. In: Neural Computing and Applications, pp. 1–26 (2021)
13. Kumar, S., Pal, A.K.: A CBIR scheme using active contour and edge histogram descriptor in YCbCr color space. IJCTA **9**(41), 889–898 (2016)
14. Kumar, S., Pradhan, J., Pal, A.K.: A CBIR technique based on the combination of shape and color features. In: Advanced Computational and Communication Paradigms, pp. 737–744. Springer, Singapore (2018)
15. Liu, G.H., Yang, J.Y., Li, Z.: Content-based image retrieval using computational visual attention model. Pattern Recogn. **48**(8), 2554–2566 (2015)
16. Oliva, A., Torralba, A.: Modeling the shape of the scene: a holistic representation of the spatial envelope. Int. J. Comput. Vis. **42**(3), 145–175 (2001)
17. Li, J., Wang, J.Z.: Real-time computerized annotation of pictures. IEEE Trans. Pattern Anal. Mach. Intell. **30**(6), 985–1002 (2008)

Chapter 27
Artificial Intelligence-Based Model for Detecting Inappropriate Content on the Fly

Awanit Ranjan, Pintu, Vivek Kumar, and Mahendra Pratap Singh

Abstract Social media made it convenient for users to express, communicate, discuss, and exchange their opinions on various issues in recent years. For example, Twitter, YouTube, Facebook, and News portals allow users to express themselves through comments. However, such platforms are being misused in the name of freedom of speech. Numerous improper messages towards specific persons or communities can be found in them that use abusive, vulgar, hostile, or harsh words. Moreover, bots are also involved in exchanging such messages nowadays. As a result, user experiences are sometimes ruined on social media. Therefore, automatic identification and filtering of such offensive messages is a significant issue for improving user experience. This paper proposes a heterogeneous ensemble-based machine learning (ML) model powered by artificial intelligence (AI) that can classify messages into Threat, Obscenity, Insult, Identity Hate, Toxic, and Severe Toxic categories. The experimental evaluation of the proposed model on a standard dataset demonstrates the accuracy and adaptability of the proposed model.

27.1 Introduction

With the increased digitization, billions of young users use social media to share personal views, pictures, and political opinions. This information revolution is good as long as there are no malpractices and destructive behaviours, which unfortunately exist and growing exponentially. Sometimes, users get to engage in tug of war over

A. Ranjan · Pintu · V. Kumar · M. P. Singh (✉)
Department of CSE, NIT Karnataka, P. O. Srinivasnagar, Suarthkal, Mangaluru, Karnataka 575025, India
e-mail: mp_singh@nitk.edu.in

A. Ranjan
e-mail: awanitranjan.181me214@nitk.edu.in

Pintu
e-mail: pintu.181co139@nitk.edu.in

V. Kumar
e-mail: vivekkumar.181co159@nitk.edu.in

© The Author(s), under exclusive license to Springer Nature Singapore Pte Ltd. 2023   299
V. Bhateja et al. (eds.), *Evolution in Computational Intelligence*, Smart Innovation, Systems and Technologies 326, https://doi.org/10.1007/978-981-19-7513-4_27

social platforms abusing and fighting each other, which is enjoyed by some, but that can hurt the sentiments of others and cause mental trauma or depression. This unwanted and unethical behaviour, directly misused by freedom of speech, is inappropriate content. Inappropriate content spreads across multiple platforms and is associated with the growing engagement trend.

In some cases, user's discussions rapidly crash and become improper, like reviling, passing impolite hateful or toxic, offensive or abusive comments, and inconsiderate remarks towards users. The users are spread over vast networks and anonymous to each other, thus often do not care about morality and misbehave. Therefore, inappropriate messages or remarks are transforming into an internet-based threat, gradually corrupting the viability of user encounters.

Consequently, programmed detection and separation of such improper language have turned into a significant issue from the perspective of the quality of conversations [1]. Several existing approaches deal with sub-classes of inappropriate content. In contrast, our focus is on the broad domain of improper content detection, which social media can use to avoid publishing such content on their respective platforms.

Users use foul language to benefit themselves by hurting the sentiments of others and believing that they are hiding behind a screen, they are away from anything. They are correct to a certain extent and have total control over how they interact with other users. Every social media platform that allows users to comment or send messages has an issue that mainly affects the mental health of children and young women. When individuals are free of any repercussions, it reveals the gloomy essence of our society. Various machine learning models exist for detecting inappropriate content such as toxic and abusive comments on different social platforms. These models suffer from significant drawbacks, such as being trained using a specifically labelled dataset which gives high training accuracy but fails to give outstanding results on the live contents.

In this paper, our primary focus is on designing and implementing a machine learning-based model backed by AI for detecting inappropriate content on the fly. The proposed model is intended to be more aggressive in reducing misclassification to ensure that inappropriate contents, such as toxic, hostile, rude, abusive, or bullying content, do not go live on social media. Additionally, the model periodically retrains itself on a continually growing dataset, making it autodidacticism, unique and ineffable.

The rest of the paper is structured as follows. Section 27.2 reviews literature related to the detection of inappropriate content on the fly. The proposed model, which includes machine learning and AI modules, is described in Sect. 27.3. Section 27.4 presents experimental results and analysis, including system specification, and dataset description. Section 27.5 concludes the paper and presents the future research directions.

27.2 Literature Survey

On various social media platforms, it has been observed that user conversations often derail and become inappropriate such as hurling abuses and passing rude and discourteous comments on individuals or particular groups/communities.

To address this, Yenala et al. [2] have proposed a deep learning (DL)-based approach for detecting inappropriate content in text. Their approach considers two scenarios: Query completion suggestions in search engines, and user conservation's in messengers. The first scenario, i.e. query suggestion, uses CNN and Bi-LSTM, whereas the second, i.e. conversation messages, uses LSTM and Bi-LSTM. They have also evaluated various proposed techniques and assessed their performance. On the other side, Anitigoni et al. [3] have presented a detailed view of large-scale crowdsourcing and characterization of Twitter abusive behaviours. However, the above approaches cannot adapt over time because they consider the static data and limited inappropriate content class.

Modha et al. [4] have presented an approach for detecting and visualizing online aggression, a particular hate speech case over social media platforms. They considered three types of aggression which are: Overtly aggressive (OAG), covertly aggressive (CAG), and non-aggressive (NAG). The authors have used SVM, CNN, attention-based, and BERT models and standard databases like TRAC for aggression detection, SemEval for offensive content in English, HASOC hate, and offensiveness. Their approach focuses on hate speech, a subcategory of inappropriate content.

Monirah et al. [5] have proposed a novel CNN-CB deep learning approach for cyberbullying detection that consists of four layers: embedding, convolution, pooling, and dense. Their approach eliminates the need for feature engineering and gives an accuracy of 95%. In [6], Rajesh et al. have developed an Automater called Blockshame for public shamming detection on Twitter. They have categorized the shamming tweets into six types: abusive, comparison, passing judgement, religion, ethnicity, sarcasm, joke, and whataboutery, and have used six SVM classifiers to classify each tweet into one of these types or as non-shamming. In contrast, Salawu et al. [7] have performed an extensive literature survey and classified its review into four main classes: Supervised learning, lexicon-based, rule-based, and mixed-initiative. They have used SVM, Naive Bayes, a rule-based approach, and matched text to predefined rules to identify bullying. The paper is limited to cyberbullying, which is only a minor part of the problem of inappropriate content. They collected datasets from YouTube, Myspace, Twitter, and emails. On the other hand, Fortunatus et al. [8] explained combining textual features to detect cyberbullying in social media posts. They have applied a lexicon enhanced rule-based method to detect cyberbullying on Facebook comments. They have reported an F1 score of 86.673% and a recall of 95.981%. Although with an excellent F1 score and recall value, the algorithm's accuracy is just 71.782%.

Shah et al. [9] have proposed a dynamic and flexible AIaaS, i.e. artificial intelligence as a service, to identify and prevent the propagation of inappropriate, immoral information like rude, cyberbullying, and detest content. The authors extracted the content expression using semantic and sentiment analysis, then utilized SVM, Naive Bayes, and decision tree to classify the content into immoral classes of rude, cyber-bullying, and dislike comments.

Gonzalo et al. [10] have addressed the issue of detecting erotic and sexual content on text documents using natural language processing (NLP). They analysed twelve models using three styles of word embeddings with four machine learning models. Their model is focused on erotic and sexual content, which is a subcategory of inappropriate content. In contrast, Shivakumar et al. [11] surveyed several deep learning approaches to detect inappropriateness in text. They have described various cutting-edge ways to capture inappropriateness. Finally, the survey is concluded by Siamese convolutional neural networks for detecting semantic similarities between given questions and input replies.

Yousaf et al. [12] carried out the work of filtering inappropriate content and classification of YouTube videos. The authors have proposed an EfficientNet-B7 pre-trained CNN and BiLSTM network to learn video descriptors effectively and classify video to one of the multiclass, which is ultimately tested on a significantly large dataset of 1.1 lakh video clips taken from YouTube. The proposed model outperforms with an accuracy of 95.66%.

It can be observed from the above description that most of the approaches deal with a specific category of inappropriate content and do not consider on the fly content. Therefore, we need a model that can deal with various categories of inappropriate content on the fly, adapting with time and giving perfect accuracy. In the following section, we describe the proposed approach for detecting inappropriate content on the fly.

27.3 Proposed Approach

Natural language processing (NLP) is a broad and multidisciplinary discipline that deals with the automatic translation of human languages. It is a candidate for every practical application that uses text and breakthroughs in the machine learning discipline primarily drive its success. Text classification is one of the candidate tasks for NLP. We have proposed a new end-to-end ML architecture backed by AI to classify the comment's appropriateness. Our architecture is robust, unique, and free from tedious and biased feature engineering processes; on the other hand, the AI system provides autodidacticism to the machine learning architecture.

Figure 27.1 shows a holistic view of the proposed architecture for detecting inappropriate content on the fly. The two main components of the architecture are machine learning module and artificial intelligence module. The description of these modules and the other architectural components is as follows.

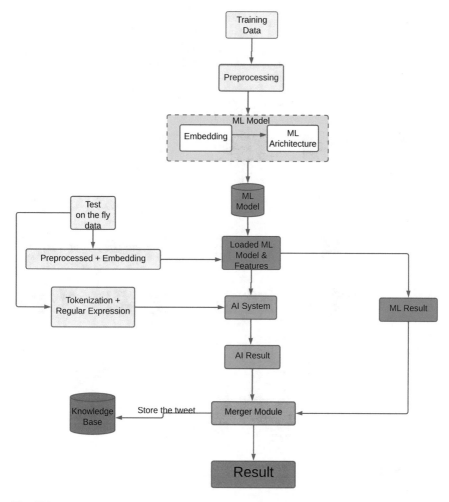

Fig. 27.1 Architecture for inappropriate content detection on the fly

27.3.1 Machine Learning Module

This subsection describes the proposed heterogeneous stacking-based ensemble learning module shown in Fig. 27.2 for inappropriate content detection on the fly. The existing empirical results have shown that ensemble-based learning models tend to perform better when there are significant differences among the ensemble models. The stacked model of several learning stages is the most popular ensemble learning approach. Thus, to solve the stated problem statement, we propose a novel machine learning module backed by an artificial intelligence module described in Subsect. 27.3.2.

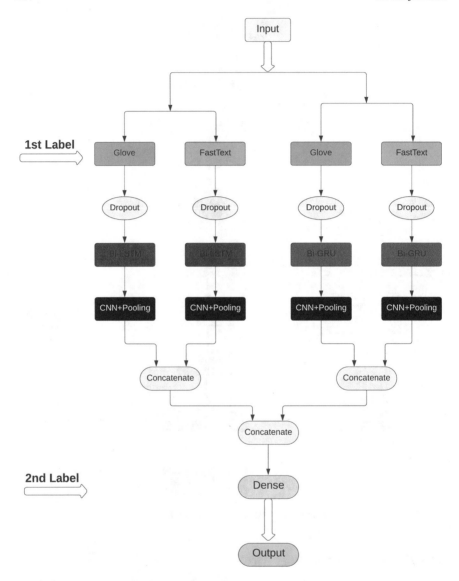

Fig. 27.2 The proposed machine learning module

We used two types of word embeddings, Glove and FastText, at the first level of ML module as shown in Fig. 27.2. Moreover, two popular RNN-based machine learning algorithms, Bi-LSTM and Bi-GRU, followed by convolution layer (CNN) and average and max pooling as shown in Fig. 27.2 are also used. The input tweet is preprocessed and entered into the proposed ML model by passing through both stated word embeddings. These word embeddings turn the real-time tweets into a 300 dimension vector, and this is then regularized using a dropout layer. Subsequently, after dropout, the input passes through the stated RNN models. The outputs from these models are convoluted and pooled using average and max pooling. Then all the results from the pooling layers are concatenated, as illustrated in Fig. 27.2. This concatenated output enters the second level, which is the ensemble network's meta-classifier, i.e. the dense layer (fully connected neural network), and the final output layer, which has sigmoid as an activation function for predicting 1 and 0 for each respective inappropriate classes: Toxic, Severe-Toxic, Obscene, Insult, Threat, and Identity Hate.

27.3.2 Artificial Intelligence Module

Humans are the most capable of comprehending, thinking, and interpreting data. We have information and use it to execute various actions in the real world. With the evolution of technology, machines can also learn and analyse data and things. Machine learning has risen to prominence as a potent tool for recognizing complicated patterns. Despite its numerous advantages and enormous popularity, machine learning is not without flaws. Because machine learning models learn from static data, it lacks adaptability with time. ML models are autonomous but algorithm-specific, and they require a large, unbiased, and high-quality dataset to train. The machine learning model works well on the training dataset. However, it fails to learn in real time, either gradually or interactively. We propose an AI-powered machine learning approach that makes the entire system autodidacticism in nature.

AI has recently grown into an emerging technology that closely resembles human brain capabilities. AI is studying how the human brain thinks, learns, makes decisions, and works to solve problems. Finally, it generates intelligent software systems. It still has a long way to go. The complexity of the human brain makes it challenging to imbue computers with human-like intelligence. Finally, we propose an AI module, keeping these restrictions in mind. Since it is responsible for making the entire system more intelligent, this module acts as the brain of the proposed ML architecture. Knowledge representation is an essential and crucial aspect of artificial intelligence that deals with how AI agents think and how thinking leads to intelligent behaviour. In our work, we used procedural language to represent knowledge responsible for understanding how to accomplish things. The proposed AI module is a predicate logic-based simple model which can infer the input comments as inappropriate or appropriate. The description of the AI Module is as follows.

The proposed inference-based predicate logic AI module is composed of two phases to tackle the challenges of detecting inappropriate content on the fly. In the first phase, AI module extracts words from sentences and matches them into standard dictionary class, i.e. Identity Hate, Obscene, Toxic, Severe-Toxic, Insult, or Threat. Thus, this phase aim is to map sentences to different words to get the sentence's nature or meaning.

Table 27.1 Predicate logic-based inference stating Inapp. (inappropriate) and app. (appropriate) classes examples

S. No.	Sentence	Infer-P	Infer-Q	Inapp	Class name
1.	God is fake as there is no proof of his presence	God	Fake	Inapp.	Identity Hate
2.	Many people believe that Christianity is not a misleading practice	Christianity	Not, misleading	App.	–
3.	The race of Muslims is not full of terrorists	Muslims	Not, terrorists	Inapp.	Identity Hate

Fig. 27.3 Identity Hate class inference

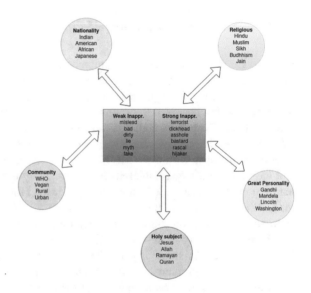

The second phase aims to find the relation between words, classify it as appropriate or inappropriate, and explain the user's intent to write the message. To do so, after inferring words from a sentence, the predicate logic is used to classify whether the sentence is inappropriate or appropriate.

The AI module also creates an internal knowledge base (KB) for every new message. When the KB has 500 MB to 1 GB in size, the proposed ML model gets trained on KB. As a result, the model becomes more intelligent in recognizing inappropriate content. For example, let us consider the sentences shown in Table 27.1. To explain the proposed AI module, we considered inappropriate class Identity Hate shown in Fig. 27.3 for the following three cases.

- Weak inappropriate words without negation, i.e. Case 1.
- Weak inappropriate words with negation, i.e. Case 2.
- Strong inappropriate words with negation, i.e. Case 3.

Case 1: Dealing with weak inappropriate word without presence of negation.

Example 1: God is fake as there is no proof of his presence.

PHASE-I

In the above sentence, the AI module captures the term *God* and *fake* and infer them in the following ways:

God ⇒ *Holy Subject*.

The AI module searches for the word *God* in the standard dictionary shown in Fig. 27.3. If the word is present in any cluster, its inference, *Holy Subject*, is marked as the Subject (*P*).

Fake ⇒ *Weakly Inappropriate*.

Similarly, the word *Fake*, which is inferred as *Weakly Inappropriate* from the following logic, is marked as Predicate (**Q**).

PHASE-II

The Subject (**P**) and Predicate (**Q**) are passed as input to following logic.

$P \wedge Q \Rightarrow$ *Identity Hate*.

Holy Subject $\wedge$ *Weakly Inappropriate* $\Rightarrow$ *Identity Hate*.

The above logic infers *Identity Hate* for the *Holy Subject* and *Weakly Inappropriate* that refers to inappropriate class. Therefore, it can be concluded that the user's intention is inappropriate.

Case 2: Dealing with weak inappropriate word with negation.

Example 2: Many people believe that Christianity is not a misleading practice.

PHASE-I

In the above sentence, the AI captures the term *Christianity*, *misleading* and the negation *not* and infers them in the following ways:

Christianity ⇒ *Religious*.

Thus, the word *Christianity* which is a religious is marked as Subject (**P**).

Misleading ⇒ *Weakly Inappropriate*

Similarly, the word *Misleading*, which is inferred as *Weakly Inappropriate* from the following logic, is marked as Predicate (**Q**).

PHASE-II

The Subject (*P*) and Predicate (*Q*) are passed to the following logic. Due to the presence of negation in the sentence, we use negated *Q* in the logic.

$P \wedge \neg Q \Rightarrow$ *Appropriate*

Religious $\wedge$ ¬*Weakly Inappropriate* $\Rightarrow$ *Appropriate*

The above logic does not infer *Identity Hate* for the *Subject (P)* and *Predicate (Q)*. Therefore, it can be concluded that the user's intention is appropriate.

Case 3: Dealing with strongly inappropriate word as shown in Figs. 27.3 and 27.4 with negation.

Example 3: The race of Muslims is not full of terrorists.

PHASE-I

The artificial intelligence model captures the term *Muslims*, *terrorists* and the negation *not* and infers them as following:

Fig. 27.4 Bag of highly (strongly) inappropriate words [13]

Muslims ⇒ *Religious.*

Thus, the word *Muslims*, which is a religious, is marked as Subject (**P**).

Terrorists ⇒ *Strongly Inappropriate.*

Similarly, the word *Terrorists*, which is inferred as strongly inappropriate from the following logic, is marked as Predicate (**Q**).

PHASE-II

The Subject (**P**) and Predicate (**Q**) are passed to the following logic.

$P \wedge \neg Q \Rightarrow$ *Inappropriate*

Religious $\wedge \neg$*Strongly Inappropriate* ⇒ *Inappropriate*

The above logic infers *Identity Hate* for the *Subject* (*P*) and *Predicate* (*Q*). Therefore, it can be concluded that the user's intention is inappropriate.

The experimental evaluation of the proposed model is presented in the next section.

27.4   Evaluation and Result

This section describes the dataset and evaluation metrics and analyses results obtained from the experimental study of the proposed model. The proposed model was trained and tested on Kaggle platform having a CPU of 4 cores with 16 GB RAM and a GPU of 2 cores with 13 GB RAM.

Table 27.2 Dataset showing six inappropriate categories

Category	Example sentence
Toxic	Beware of hounds, they will first f*#k you then bite your a$$
Severe toxic	The rapist inserted the beer bottle in her intestine
Obscene	Hi sexy, why are your b@@bs so small
Threat	How dare to break up with me? I will kill your family
Insult	Who the hell pissed on my shoes, the son of a bitch?
Identity hate	Nazis like you deserve worse than death

	id	comment_text	toxic	severe_toxic	obscene	threat	insult	identity_hate
92982	f89eba161bfcaa32	You will now note that CorporateM changed the ...	0	0	0	0	0	0
62623	a78fb12530c91a63	you have a message re your last change, go fuc...	1	0	1	0	1	0
142192	f88f88e4a56a6a35	John Sherry & Richard Sherry\nThe article need...	0	0	0	0	0	0

Fig. 27.5 Preview of Jigsaw toxic comment classification challenge dataset [14]

27.4.1 Dataset

We have used Jigsaw Toxic Comment Classification Challenge dataset hosted on Kaggle platform [14] containing 159,571 tweets. This dataset is developed by conversation AI team, a research community of Google and Jigsaw, to develop the model for identifying and classifying different types of toxic online comments. The dataset is a multi-headed, having six different inappropriateness categories, which are shown with examples in Table 27.2.

The dataset consists of training data and testing data CSV files of 68.8 and 60.35 MB, respectively. The training and validation data contain binary labelled comments for all six categories. This dataset is a public dataset under CC0 with the comments from Wikipedia's CC-SA-3.0 talk page edits. Figure 27.5 provides a glimpse of the dataset.

Although there are several other datasets for inappropriate content, we chose the dataset mentioned above. Because it is well-structured, has high availability, and, most importantly, provides multiple classes of inappropriateness, allowing the model to produce more than just a binary result of appropriate or inappropriate.

27.4.2 Evaluation Metrics

Inappropriate content detection is a candidate problem for the classification task. Therefore, we have considered classification accuracy to evaluate the proposed model performance. In the following formula, TP, TN, FP, and FN stand for true positive, true negative, false positive, and false negative, respectively.

$$\text{Accuracy} = \frac{\text{TP} + \text{TN}}{\text{TP} + \text{TN} + \text{FP} + \text{FN}} \qquad (27.1)$$

27.4.3 Results

We ran our model with the following parameters and found that the model performed admirably well.

- The training data was divided into training and validation data of 70 and 30% ratio.
- For better accuracy, we set the learning weight decay to 0.1 and the dropout to 0.2 shown in Table 27.4.

The model was trained for 50 epochs shown in Table 27.3 with batch size 128 (shown in Table 27.4) and took about 3 h to get trained on the Kaggle platform using the available GPU. On average, it took 0.05 s to forecast and classify the input into six categories.

Figures 27.6 and 27.7 show how accuracy and loss change with each progressing epoch for the inappropriate content detection tasks. The proposed model obtained 97.215 and 96.845% accuracy for training and testing data.

Table 27.3 Epochs tuning

Epochs	Training accuracy (%)	Validation accuracy (%)
10	96.630	96.420
20	96.831	96.422
30	96.910	96.562
50	**97.215**	**96.845**
60	97.160	96.721

The bold signifies the corresponding value of epochs, batch size, and dropout, giving the best accuracy

Table 27.4 Tuning of batch size and dropout rate

Batch size	Training accuracy (%)	Validation accuracy (%)	Dropout	Training accuracy (%)	Validation accuracy (%)
16	97.05	96.78	0.1	97.02	96.74
32	97.07	96.84	**0.2**	**97.147**	**96.905**
64	96.79	96.42	0.3	96.853	96.563
128	**96.83**	**96.67**	0.4	96.56	96.35
256	96.76	96.50	0.5	95.90	95.71

The bold signifies the corresponding value of epochs, batch size, and dropout, giving the best accuracy

Fig. 27.6 Training and
validation accuracy

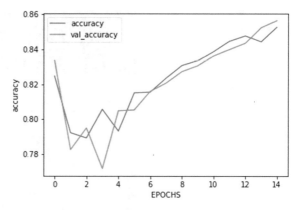

Fig. 27.7 Training and
validation loss

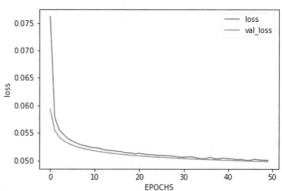

27.4.4   Performance Comparison

We compared the proposed model for inappropriate content detection with the model
presented in Hate speech detection with comment embedding [13], in which para-
graph2vec is used to model comments and words together. It contributes to low-
dimensional text embedding, in which similar words are clustered together. These
embeddings are used to train binary classifiers to identify hate speech. The researchers
claim that their model yields an accuracy of 80.07%. While the proposed model gives
an accuracy of 96.845%, thus outperforming in performance.

27.4.5   Demonstration of Result

To simulate results, we designed a flask web app. The screenshot in Fig. 27.8 shows
the user interface and prediction result for the given input as on the fly tweet.

Fig. 27.8 Example of on the fly tweet [4]

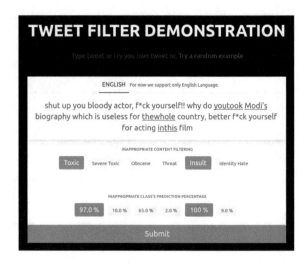

27.5 Conclusion

In this paper, we addressed the issue of automatically detecting inappropriate content on the fly. We proposed a novel ensemble-based machine learning model backed by artificial intelligence for automatically identifying inappropriate content. First, we used heterogeneous ensemble-based Bi-LSTM and Bi-GRU algorithms as the base model. Then, we applied an inference-based predicate logic-based artificial intelligence framework to improve the accuracy further and make our model more adaptive with time. Evaluating our model on Twitter comments data reveals that it significantly outperforms other existing models to the best of our knowledge. The current work has been limited to only the English language. It cannot handle data such as emojis, photos, videos, or any other data that is not text. The linear search approach identifies the subject and predicate from the standard dictionary. However, it can be further improved to reduce the time complexity of the search. Our ongoing and future works include working on these limitations.

References

1. Nakov, P., Nayak, V., Dent, K., Bhatawdekar, A., Sarwar, S.M., Hardalov, M., Zlatkova, Y.D.D., Bouchard, G., Augenstein, I.: Detecting abusive language on online platforms: a critical analysis. In: CoRR, pp. 1–9 (2021)
2. Yenala, H., Jhanwar, A., Chinnakotla, M.K., Goyal, J.: Deep learning for detecting inappropriate content in text. Int. J. Data Sci. Anal. **6**, 273–286 (2018)
3. Founta, A.-M., Djouvas, C., Chatzakou, D., Leontiadis, I., Blackburn, J., Stringhini, G., Vakali, A., Sirivianos, M., Kourtellis, N.: Large scale crowdsourcing and characterization of Twitter abusive behavior. In: Proceedings of the Twelfth International AAAI Conference on Web and Social Media, pp. 1–11 (2018)

4. Modha, S., Majumder, P., Mandl, T., Mandalia, C.: Detecting and visualizing hate speech in social media: a cyber Watchdog for surveillance. Exp. Syst. Appl. **13**(1), 591–603 (2020)
5. Al-Ajlan, M.A., Ykhlef, M.: Deep learning algorithm for cyberbullying detection. Int. J. Adv. Comput. Sci. Appl. **9**(9), 199 205 (2018)
6. Basak, R., Sural, S., Ganguly, N., Ghosh, S.K.: Online public shaming on Twitter: detection, analysis, and mitigation. IEEE Trans. Comput. Soc. Syst. **6**(2), 208–220 (2019)
7. Salawu, S., He, Y., Lumsden, J.: Approaches to automated detection of cyberbullying: a survey. IEEE Trans. Aff. Comput. **11**(1), 3–24 (2020)
8. Fortunatusa, M., Anthonya, P., Charters, S.: Combining textual features to detect cyberbullying in social media posts. Proc. Comput. Sci. **176**, 612–621 (2020)
9. Shah, F., Anwar, A., AlSalman, H., Hussain, S., Al-Hadhrami, S.: Artificial intelligence as a service for immoral content detection and eradication. Sci. Program. 1–9 (2022)
10. Barrientos, G.M., Alaiz-Rodríguez, R., González-Castro, V., Parnell, A.C.: Machine learning techniques for the detection of inappropriate erotic content in text. Int. J. Comput. Intell. Syst. (2020)
11. Teli, S.H., Kiran, V.: A survey on deep learning for the detection of inappropriate content present in text. Int. J. Eng. Appl. Sci. Technol. **6**, 354–360 (2021)
12. Yousaf, K., Nawaz, T.: A deep learning-based approach for inappropriate content detection and classification of youtube videos. IEEE Access **10**, 16283–16298 (2022)
13. Djuric, N., Zhou, J., Morris, R., Grbovic, M., Radosavljevic, V., Bhamidipati, N.: Hate speech detection with comment embeddings. In: Proceedings of the 24th International Conference on World Wide Web, pp. 29–30 (2015)
14. Jigsaw toxic comment classification challenge. https://www.kaggle.com/competitions/jigsaw-toxic-comment-classification-challenge/data

Chapter 28
Fuzzing REST APIs for Bugs: An Empirical Analysis

Sunil Kumar, Jyoti Gajrani, and Meenakshi Tripathi

Abstract Today every application needs to interact with many other applications to function. APIs are bringing applications together in order to perform a designed function built around exchanging data and executing pre-defined processes. With the increasing use of APIs, concern about API security is also increasing. Organizations use various testing techniques for security testing of their APIs including fuzzing. Fuzzing is an effective technique in security and recent research shows fuzzing will be an effective technique in API security as well. In this paper, we study various available open-source API fuzzers, their fuzzing methodologies, and issues security researchers face with these fuzzers, and empirical analysis of findings of these fuzzers on a vulnerable API. We present our experimental results and propose the concept of a better API fuzzer with better surface detection, fuzz input generation, and speedy fuzzing.

28.1 Introduction

Application programming interfaces (APIs) are primarily a way to implement the protocols and tools used for two software systems to talk to each other. Over the last decade, we have seen an explosion in adopting Representational state transfer (REST) APIs for designing a smooth way to make applications interact with others. Today almost every enterprise uses the cloud and the services that cloud service providers provide are programmatically accessed through REST APIs by third-party applications and other services [1]. Most APIs in enterprises deal with sensitive and customer personally identifiable information (PII) data. In today's scenario, regular Pentesting is not enough for API Security. Malicious users can find an exploitable endpoint which can lead to expensive data breaches. Organizations are using static

S. Kumar · J. Gajrani (✉)
Engineering College Ajmer, Ajmer, Rajasthan, India
e-mail: jyotigajrani@ecajmer.ac.in

M. Tripathi
Malaviya National Institute of Technology, Jaipur, Rajasthan, India
e-mail: mtripathi.cse@mnit.ac.in

code analysis, dynamic code analysis and some are testing against OWASP [2] API Security—Top 10 vulnerabilities [3]. Recent researches show Fuzzing is an effective way in API Security and has the ability to uncover undetected vulnerabilities [4]. API Fuzzing involves generating new fuzz or test inputs and feeding them to a target API which is continuously monitored for errors. Organizations are facing various challenges with adopting Fuzzing in their development cycle including the efficiency and knowledge of fuzzers. With the goal of proposing a better model of fuzzer, we compared available open-source API fuzzers. We tested these Fuzzers on open-source vulnerable API vAPI [5] and gave an empirical analysis of findings. With help of this analysis and understanding of the underlying architecture of these fuzzers [6], we proposed a model of a improved fuzzer. While understanding the architecture of fuzzers, we kept three points in consideration—1. Input generation, 2. Endpoint detection, 3. Content discovery. There are mainly two perspectives to categorize fuzz testing approaches: based on how they generate new test inputs and based on how they use feedback from the target under test. First, we categorize fuzzing into three types based on how feedback from the fuzzing target is being used—1. Black-box Fuzzing, 2. White-box Fuzzing and 3. Gray-box Fuzzing. Second, there is mutation-based fuzzing, grammar-based fuzzing, and random fuzzing based on how new inputs are being generated.

The rest of the paper is organized as follows: Sect. 28.2 briefly presents open source API fuzzers which are used in our analysis. Section 28.3 presents Related work. Section 28.4 describes working of vAPI. Our analysis findings are shown in Sect. 28.5. The model of proposed fuzzer is described in Sect. 28.6. Section 28.7 presents few open issues and Sect. 28.8 concludes the work.

28.2 Open Source API Fuzzers

For our analysis, we have chosen 5 popular open-source API fuzzers, RESTler [7], CATS [8], Burp Intruder (Community Version) [9], API-Fuzzer [10], TnT-Fuzzer [11] as shown in Table 28.1. Each fuzzer has a different architecture and different input generation, endpoint detection, and content discovery method.

Table 28.1 API fuzzers used in experiments

Name	Version	Language	Supported schemas
RESTler [7]	8.3.0	Python, F#	Open API 2 / 3
CATS [8]	7.0.1	Java	Open API 2 / 3
Burp Intruder [9]	Community 2021.10.3	Java	Open API 2 / 3
API-Fuzzer [10]	0.0.1	Ruby	Open API 2
TnT-Fuzzer [11]	2	Python	Open API 2

RESTler is the first automatic intelligent REST API fuzzer. RESTler analyzes OpenAPI specification and generates grammar-based tests. RESTer has its own automatic fuzz input generation technique. RESTler can automatically detect endpoints in API based on payload. CATS is a gray-box fuzzer with an automatic input generation process. CATS generate fuzz inputs randomly. Currently, CATS doesn't support automatically detecting endpoints in APIs and we need to specify all endpoints in CATS. We need OpenAPI specifications for API to run CATS fuzzer. Burp Intruder is an inbuilt fuzzer in popular Web Security [12] tool Burp Suite by Portswigger. It is a highly customizable fuzzer and we can extend its functionality by installing various plugins. Initially, Burp Intruder does not have the ability to generate a fuzz payload. API-Fuzzer is a simple fuzzer and it does not have the ability of endpoint detection and fuzz input generation. It can check for crashes based on random fuzz payloads we supply. TnT-Fuzzer is a schema-based web API fuzzer and it has the ability to generate an automatic fuzz payload based on schema rules but it doesn't have the ability to detect endpoints.

28.3 Related Work

To propose a better model of fuzzer, we need to have a deep understanding of recent research in REST API fuzzing. REST API is an influencing research area and many researchers published their studies related to API Fuzzing techniques. Most of these studies are based on black-box testing, automating fuzzing, Dynamic Symbolic Execution, etc. Researchers also published individual studies on capabilities and working of fuzzers, i.e., RESTler [7] and OSS-Fuzz [13]. However, to the best of our knowledge, no work is carried out to compare available REST API Fuzzers based on Empirical Analysis.

Comparison between RESTler and Pythia by authors of Pythia [14] developed a baseline for comparing two different technique based fuzzers, i.e., black box fuzzing and gray-box fuzzing in the case of RESTler and Pythia.

The authors of RESTTESTGEN [15] compared their work with EvoMaster [16] and shown RESTTESTGEN as robust and effective fuzzer. Both RESTTESTGEN and EvoMaster are black-box technique-based API Fuzzers.

28.4 Vulnerable API

vAPI [5] or Vulnerable API is a REST API that is used to illustrate common API vulnerabilities. vAPI is implemented using the Python Framework, it consists of a token database and a user database. We can install vAPI with docker images and containers from Docker Hub using *docker pull mkam/vulnerable-api-demo* command. vAPI has multiple endpoints and its working process is explained in Fig. 28.1.

vAPI Process Flow

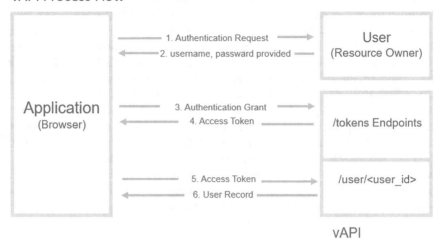

Fig. 28.1 Working of vAPI

Request

POST /tokens HTTP/1.1
Accept: application/json
Content-Length: 36
Content-Type: application/json
Host: 192.168.13.37:8081

{"auth":
    {"passwordCredentials":
        {"username": "USER_NAME",
         "password":"PASSWORD"}
    }
}

Response

HTTP/1.0 200 OK
Date: Tue, 07 Jul 2015 15:34:01 GMT
Server: WSGIServer/0.1 Python/2.7.6
Content-Type: text/html; charset=UTF-8

```
{
  "access":
    {
      "token":
        {
          "expires": "Tue Jul  7 15:39:01 2015",
          "id": "AUTH_TOKEN"
        },
      "user":
        {
          "id": 10,
          "name": "USER_NAME"
        }
    }
}
```

Fig. 28.2 Access token request and response in vAPI

In vAPI, /tokens and /user/USER_ID user endpoints play an important role. /token endpoint is used to request access tokens which will be used in further steps to get user data from the /user/USER_ID endpoint. Request and response samples for /tokens and /user/USER_ID endpoints are given in Figs. 28.2 and 28.3 respectively.

```
Request                                      Response
GET /user/1 HTTP/1.1                         HTTP/1.0 200 OK
Host: 192.168.13.37:8081                     Date: Mon, 06 Jul 2015 22:08:56 GMT
X-Auth-Token: AUTH_TOKEN                      Server: WSGIServer/0.1 Python/2.7.9
Content-type: application/json               Content-Length: 73
Accept: text/plain                           Content-Type: application/json
Accept-Language: en-US,en;q=0.5
Accept-Encoding: gzip, deflate               {
Connection: keep-alive                         "response":
Content-Length: 0                                {
                                                   "user":
                                                    {
                                                      "password": "PASSWORD",
                                                      "id": USER_ID,
                                                      "name": "USER_NAME"
                                                    }
                                                 }
                                               }
```

Fig. 28.3   User data request and response in vAPI

Table 28.2   Number Of unique defects, HTTP 500 server errors, endpoints detected and tests for each fuzzer on vAPI hosted on Azure

Fuzzer	Number of unique defects	Unique HTTP 500 server errors	Endpoints detected	Tests
RESTler	7	2	2	244
CATS	5	0	1	400
BurpIntruder	6	5	1	887
API-Fuzzer	3	0	1	No data
TnT-Fuzzer	4	1	1	180

28.5   Empirical Analysis of Findings

We tested vAPI with a total of five Fuzzers. We hosted Docker containerized vAPI on Azure Web Services [17] for testing purposes. Each fuzzer requires different dependencies and to fulfill them we installed each Fuzzer in different Azure Virtual Machines to avoid performance and dependencies conflict issues. We started fuzzing with each fuzzer on the root / endpoint and each fuzzer stopped within 5 min. Table 28.2 shows results from our experiments.

As shown in Table 28.2, RESTler found the highest number of defects in vAPI and we recorded two HTTP 500 errors as well. RESTler generated fewer tests and detected 2 endpoints without OpenAPI Specifications. API-Fuzzer didn't give any data about the number of test cases and was the least performer.

Through our experimentals, RESTler is found to be a robust fuzzer. However, with automated OpenAPI/Swagger schemas BurpIntruder performance is comparatively good in sense of time taken and well tool usability. In case of any error in OpenAPI/Swagger schemas, all fuzzers crash except BurpIntruder.

The results show different degrees of coverage for the different black-box, graybox and white-box testing tools, where RESTler and BurpIntruder seem the tool giving the best results in our experiment. However no Fuzzer is as efficient as manual testing when it comes to Endpoint Detection.

28.6 Model of Proposed Fuzzer

Based on empirical analysis of findings from 5 famous fuzzers on vAPI, we propose a hypothetical model of improved fuzzer. We see RESTler gives better results in comparison to other fuzzers. RESTler is a black-box learning-based fuzzer. It is a combination of black-box grammar-based fuzzing, coverage-guided fuzzing. From our experiment results, we learn by combining different fuzz input generation and its exception techniques we can get better results. There can be a single fuzzer for Black-box, Gray-box, and White-box fuzzing. In Black-box fuzzing, we can further have Random Fuzzing, Mutation-based Fuzzing, Grammar-based, and Learning-based fuzzing for better results. Similarly, for Gray-box fuzzing, we can combine Mutation-based, Learning-based fuzzing and Mutation-based, Grammar-based fuzzing for White-box fuzzing. For target API with OpenAPI/Swagger specifications, the proposed model analyzes APLs entire OpenAPI specifications and after that automatically generates and executes tests that exercise the service through its REST API.

It parallelly checks for various vulnerabilities and dynamically learns how the API behaves from prior test case responses. Its learning capability allows this model to explore API requests in deep and discover mode bugs that can be reachable only through specific request sequences. Further, after test case execution, it has a tool to detect and analyze responses from test cases for ease of understanding.

28.7 Open Research Issues

Our work aims at improving fuzz testing of REST APIs. At the initial phase, we found there are many limitations in available open-source API fuzzers. Building a complex fuzzer as proposed still requires a lot of research. Test input generation approaches and execution methods like symbolic execution, model-based testing, combinatorial test generation are major focus areas in fuzzing. Currently available commercial fuzzing tools like Postman, SoapUI which use fuzz based upon manual test cases generated by testers is also an area that needs researchers' attention. A major challenge in the development of the proposed fuzzer is integrating it in the development cycle for White-box and coverage-guided testing. Most enterprises currently adopting the DevSecOps model and integrating API fuzzing in the DevSecOps cycle required the researcher's attention.

Many general-purpose fuzzers adopted learning-based black-box fuzzing. The main idea of learning-based black-box fuzzing can be divided into two parts- first, approximate the structure of the target input domain using a model. This model can be based on formal grammars, such as CFGs, or statistical models, such as Recurrent Neural Networks. Second, sample the learned model to generate new fuzz inputs which are further alternated with mutations and then executed on the target API, hoping to uncover unknown errors. Simultaneously working on learning-based fuzzing and learning-based Black-box fuzzing in a fuzzer is still a challenge.

28.8 Conclusions

In this paper, we introduced a model on improved fuzzer based on our experiment on 5 popular fuzzers. Currently available open-source fuzzers have various limitations and are unable to perform automatic endpoint detection and generate intelligent fuzz payloads. Based on results from experiments and currently available technologies we prepared a model of an improved fuzzer. It is a learning-based fuzzer that has the ability to automate endpoint detection and its intelligent input generation will resolve current open issues. Integration of various fuzz input generation and execution along with a defect detection, analysis, and response tool will improve the output of fuzzers. Future research work includes working on the proposed model and building a fuzzer that resolves most of the current issues in API fuzzing.

References

1. Chen, Y., Yang, Y., Lei, Z., Xia, M., Qi, Z.: Bootstrapping automated testing for RESTful web services (2021). https://doi.org/10.1007/978-3-030-71500-7_3
2. OWASP: OWASP API Top 10 (2019). https://www.owasp.org/
3. Araujo, F., Medeiros, I., Neves, N.: Generating tests for the discovery of security flaws in product variants. In: Proceedings of the IEEE International Conference on Software Testing, Verification and Validation Workshops, pp. 133–142 (2020)
4. Sutton, M., Greene, A., Amini, P.: Fuzzing: Brute Force Vulnerability Discovery. Addison-Wesley Professional (2007)
5. vAPI. https://github.com/appknox/vapi. Last Accessed 20 Apr 2022
6. Seagle Jr, R.L.: A framework for file format fuzzing with genetic algorithms (2012)
7. Atlidakis, V., Godefroid, P., Polishchuk, M.: RESTler: stateful REST API fuzzing. In: 2019 IEEE/ACM 41st International Conference on Software Engineering (ICSE). IEEE (2019)
8. CATS. https://github.com/Endava/cats. Last Accessed 20 Apr 2022
9. Burp Intruder. https://portswigger.net/burp/documentation/desktop/tools/intruder. Last Accessed 20 Apr 2022
10. API-Fuzzer. https://github.com/Fuzzapi/API-fuzzer. Last Accessed 20 Apr 2022
11. TnT-Fuzzer. https://github.com/Teebytes/TnT-Fuzzer. Last Accessed 20 Apr 2022

12. Zhao, D.: Fuzzing technique in web applications and beyond. J. Phys.: Conf. Ser. **1678**(1) (2020) (IOP Publishing)
13. Ding, Z.Y., Le Goues, C.: An empirical study of OSS-Fuzz bugs. In: 2021 IEEE/ACM 18th International Conference on Mining Software Repositories (MSR). IEEE (2021)
14. Atlidakis, V., et al.: Pythia: grammar-based fuzzing of REST APIs with coverage-guided feedback and learning-based mutations. arXiv Preprint (2020). arXiv:2005.11498
15. Viglianisi, E., Dallago, M., Ceccato, M.: RESTTESTGEN: automated black-box testing of RESTful APIs. In: 2020 IEEE 13th International Conference on Software Testing, Validation and Verification (ICST). IEEE (2020)
16. Arcuri, A.: Evomaster: evolutionary multi-context automated system test generation. In: 2018 IEEE 11th International Conference on Software Testing, Verification and Validation (ICST). IEEE (2018)
17. Microsoft Azure. https://azure.microsoft.com/en-in/. Last Accessed 20 Apr 2022

Chapter 29
An Approach for Predicting Election Results with Trending Twitter Hashtag Information Using Graph Techniques and Sentiment Analysis

Chhandak Patra, D. Pushparaj Shetty, and Sonali Chakraborty

Abstract India is one of the largest democracies in the world where the Lok Sabha and the Rajya Sabha elections are held every five years. Nowadays, social media acts as an important and inexpensive platform for propagating messages of the political parties. In the present study, a methodology is proposed by combining sentiment analysis and graph techniques to look into the trending hashtag networks propagated by the political parties using Twitter. The demonstration of the proposed methodology is done on the trending hashtag's information collected from Twitter on the Uttar Pradesh (U.P) state elections, 2022.

29.1 Introduction

Advancement in low-cost Internet connectivity has ushered a surge in the use of social networking sites such as Facebook, Twitter and Instagram. Hundreds of millions of users communicate and exchange their thoughts and ideas with their loved ones and to the masses via social media. Twitter is a data hub where the users contribute large amounts of information and generate huge amount of data every day [1, 2]. Analysis of users' emotions expressed on social media platforms aids in understanding human behaviour and solving numerous challenges [3]. The current study attempts to frame a systematic method for analysing hashtag networks in order to include both the macro and the micro picture in these networks. Sentiment analysis and network analysis have been combined to provide a comprehensive picture of the performance of a party while propagating its hashtags and countering the infor-

C. Patra (✉) · D. Pushparaj Shetty · S. Chakraborty
National Institute of Technology Karnataka Surathkal, NH-66, Srinivasnagar, Surathkal, Mangalore, Karnataka 575025, India
e-mail: saikatpatra.21@gmail.com

D. Pushparaj Shetty
e-mail: prajshetty@nitk.edu.in

S. Chakraborty
e-mail: al.sonali@nitk.edu.in

© The Author(s), under exclusive license to Springer Nature Singapore Pte Ltd. 2023   323
V. Bhateja et al. (eds.), *Evolution in Computational Intelligence*, Smart Innovation,
Systems and Technologies 326, https://doi.org/10.1007/978-981-19-7513-4_29

mation flow of its rivals' hashtags. The various network parameters are weighted based on their importance in the network to generate a composite score for each hashtag, both for the message's propagator and its rivals. The proposed model's output is depicted using a sample case from the recently concluded U.P Vidhansabha Elections, 2022.

29.2 Related Works

Twitter is a microblogging platform where users create and share "tweets" [4–8]. There are two broad categories for analysing twitter space for election scenario:

29.2.1 Graph-Based Methods

Khan et al. [9] analyse the network graphs and the retweet graphs for the prominent leaders in Pakistan to analyse the information flow and the communities formed thus knowing who has more popularity on the social media space. Melo et al. [10] presented an exploratory study on the Twitter network emphasising on its dynamic aspects. They analysed the propagation of a message over time. Himelboim et al. [11] devised a method called Selective Exposure Clustering (SEC) to look into the interconnected networks and their Twitter conversation patterns. Networked communication has evolved into an essential link to consumers and partners [12].

29.2.2 Sentiment Analysis

Authors [13] collected 3000 tweets in Arabic and Saudi and determined that the hybrid technique outperforms both the supervised and unsupervised approaches. The study's main purpose was to use an opinion lexicon and an L.D.A scoring scheme to analyse the tweets of six hockey clubs. It is been discovered that the type of lexicon utilised has an impact on sentiment analysis results. The spectrum obtained is more positive when the lexicon is short and contains fewer relevant terms. The resulting spectrum, however, will encompass both positivity and negativity [14] if the vocabulary is sufficiently vast and contains more important words. In [15], the authors collected the tweets about 2019 Indian Loksabha elections from the prominent hashtags and performed a comparative analysis of the sentiment classes of the tweets using Long Short-term Memory (LSTM) network with machine learning approaches such as SVM, decision tree classifier, logistic regression and random forest classifier. In the study, the highest accuracy was achieved using LSTM networks and random forest classifiers.

29.3 Research Gaps

The literature study depicts that graphical methods and sentiment analysis are the two most widely used approaches for analysing the twitter space for election scenarios. Both methods are quiet promising but are pretty one-dimensional when considered individually. There is no study showing both the macro-level and the micro-level analyses.

Figure 29.1a, b depicts two different types of replies for a same tweet which contribute to the information propagation. The sentiment invoked from the tweets is entirely contrasting. This distinction is not made in the graphical representation since the sentiment associated with each tweet is not considered. The major drawback of this approach is that it fails to derive the sense of the coverage of the message. A message may have fully affirmative sentiments, but the network may seem to be a particular cluster having continuous affirmation of the message. Consider party A delivering a message. The party affiliates and sympathisers will always give an affirmation towards the message no matter how malicious it may have been originally. So, there is a need to take both the methods and utilise them in a manner where these approaches can be used judiciously and in a way that one becomes a compliment to the other. The macro and the micro-level scenarios are derived by graphical method and the sentiment analysis, respectively.

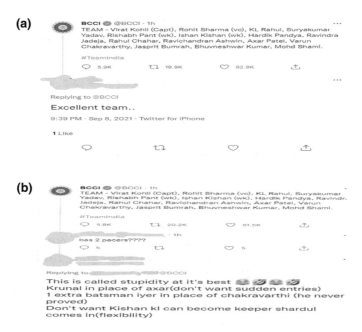

Fig. 29.1 a Positive sentiment, **b** negative sentiment

29.4   Approach and Methodology

29.4.1   Data Extraction and Graph Representation

Some of the popular tools for analysing the Twitter Networks are Gephi, SocioViz, Python (networkx package), NodeXL, etc. In the study, analysis of tweets in graphical form and computation of various metrics is done using NodeXL and Gephi.

NodeXL [16] a social network analysis tool that combines network analysis into the familiar Excel spreadsheet format. It helps to obtain network data, store it, analyse and visualise it and generates reports that communicate insights into connected structures.

Gephi [17] is a visualisation application based on Java. majorly used to visualise, alter and explore raw edge and node graph data networks and graphs. It is similar to Photoshop, but much more geared at graph data. To find hidden patterns, the user interacts with the representation and manipulates the structures, shapes and colours (Fig. 29.2).

Figure 29.3 shows the work flow of the proposed methodology.

29.4.1.1   Graph Formation from Tweets

The current study investigates the tweets with a hashtag in the form of a large network structure. The base account from which a tweet originates is represented by the graph's node. The person mentioned in the tweet becomes a new node, with a directed edge connecting the first and second. If user B responds to user A's tweets, a directed edge exists from B to A. Figure 29.4 depicts the creation of a graph from many cross-referencing tweets.

29.4.2   Proposed Methodology

- **Data Collection**: Approximately, 500K tweets were collected using NodeXL [16] from several trending hashtags related to the UP Vidhansabha Elections, 2022. Figure 29.5 represents the top trending hashtags collected for the month of January, 2022. In the background, NodeXL connects directly to the Twitter API to extract the given network. The tweets are extracted and internally formed into an edge list that can be visualised as a graph and from which other metrics can be derived, just like in a regular graph network. Only two opposing entities were considered: the ruling party (denoted as RULE) and the opposition party (denoted as OPP). Based on their agenda, the collected hashtags are classified as either RULE propagated or OPP propagated. The sentiment analyser was tuned to determine which sentiment (positive or negative) is preferred by both entities in order to fuel or counter the hashtag.

- **Data Pre-processing**: The tweets collected are cleaned before feeding into the sentiment analyser using the following steps:

  - *Conversion into lower case*: The tweet text is converted to lower case to avoid case specific understanding issues.
  - *Removing hashtags and extra spaces*: The hashtags and the extra spaces between the words were removed [18].
  - *Removal of repetitive letters from a word*: Twitter users post text in an unstructured form and so the possibility of writing words having multiples of same letter such as "hapyyyyyy", "lotttt", "awsomeeee" is very frequent. The consecutive occurences of each letter appearing more than two times are removed.
  - *Removal of URLs and user information*: All the URLs and the user names in the tweets are removed.

(a)

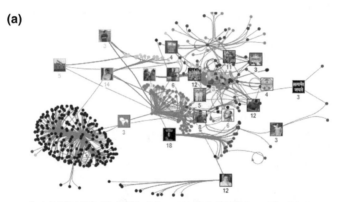

Created with NodeXL Pro (http://nodexl.codeplex.com) from the Social Media Research Foundation (http://www.smrfoundation.org)

(b)

Fig. 29.2 **a** NodeXL [16] visualisation of #indiawithmodi, **b** Gephi [17] visualisation clustered by groups with labels of highest page rank nodes of #LakhimpurKheriViolence

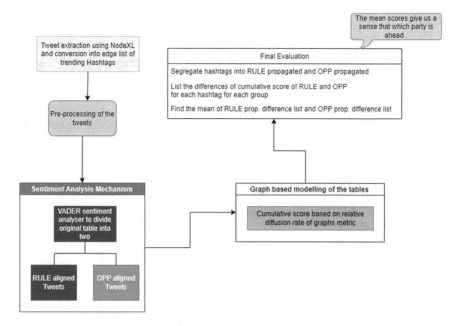

Fig. 29.3 Work flow of the proposed method

Fig. 29.4 Tweet network
formation mechanism [16]

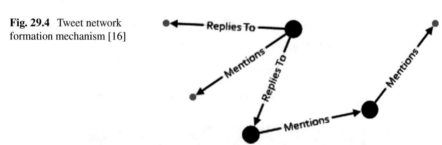

- **Lemmatisation**: Lemmatisation integrates several inflected word forms into a single one. It is the process of replacing words with their base such as, the words "caring" and "cars" can be reduced to "care" and "car", respectively [19].
- **Removal of stop words**: Words like "at", "the", "this" do not contribute to the sentiment of the tweet and so words called stop words are removed.

- **Sentiment Analysis**: In the present study, lexicon-based sentiment analyser are used. The pre-processed edge list of tweets are fed into Valence Aware Dictionary and sentiment Reasoner (VADER) sentiment analyser [20]. VADER is an unsupervised learning approach and there is no need of finding physical annotators having ideological biases. The analyser will segregate the output tweets as positive, negative and neutral and are assigned to the parties accordingly. Finally, two

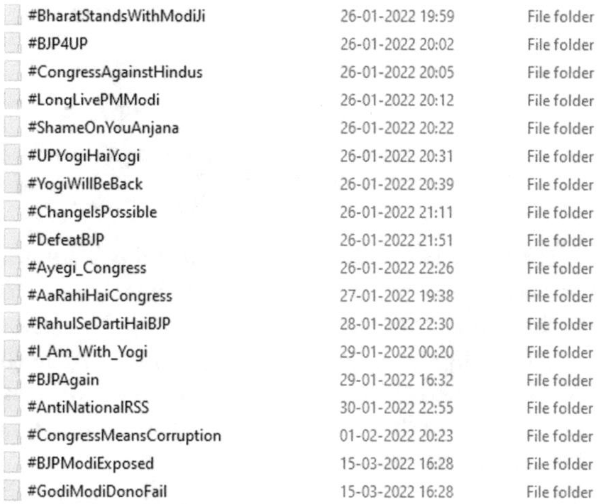

Fig. 29.5 Hashtags collected for January, 2022

#BharatStandsWithModiJi	26-01-2022 19:59	File folder
#BJP4UP	26-01-2022 20:02	File folder
#CongressAgainstHindus	26-01-2022 20:05	File folder
#LongLivePMModi	26-01-2022 20:12	File folder
#ShameOnYouAnjana	26-01-2022 20:22	File folder
#UPYogiHaiYogi	26-01-2022 20:31	File folder
#YogiWillBeBack	26-01-2022 20:39	File folder
#ChangeIsPossible	26-01-2022 21:11	File folder
#DefeatBJP	26-01-2022 21:51	File folder
#Ayegi_Congress	26-01-2022 22:26	File folder
#AaRahiHaiCongress	27-01-2022 19:38	File folder
#RahulSeDartiHaiBJP	28-01-2022 22:30	File folder
#I_Am_With_Yogi	29-01-2022 00:20	File folder
#BJPAgain	29-01-2022 16:32	File folder
#AntiNationalRSS	30-01-2022 22:55	File folder
#CongressMeansCorruption	01-02-2022 20:23	File folder
#BJPModiExposed	15-03-2022 16:28	File folder
#GodiModiDonoFail	15-03-2022 16:28	File folder

tables are created for each party. At the end of this phase, there will be three tables, namely the overall or original edge list, RULE's list and OPP list.

- **Cumulative Score Calculation for Each Network for Both Parties**: The proposed methodology calculates a score for each hashtag for both the contenders based on relative diffusion with respect to the original. The diffusion rate of information in social networks [21] is modified to fit the proposed model. The equation for the diffusion rate is given in Eq. (29.1):

$$\alpha(t) = \frac{\frac{d_i}{E[d]}}{1 + \alpha_o t} \tag{29.1}$$

where

$$E[d] = \frac{\sum_{i=1}^{i=n} d_i}{n} \tag{29.2}$$

where $\alpha_o = 0.5$ [21] denotes the decay rate of diffusion, t = time, n = number of nodes.

In this study, page rank is a better metric for gauging the influence of a node as compared to the degree of the node. Therefore, d_i is the page rank of a node in the network. The d_i in (29.1) is denoted as the node having the highest page rank since the most influential node will be responsible for wider diffusion in a network.

Three different α are generated for each of three networks which are: α_{FULL} depicting the diffusion of the original complete network, α_{RULE} for the diffusion of the RULE network and α_{OPP} for the diffusion of the OPP network. In order to find the relative diffusion of RULE and OPP with respect to FULL(χ) Eq. (29.3) used is as follows:

$$\chi_{RULE} = \frac{\alpha_{RULE}}{\alpha_{FULL}}, \quad \chi_{OPP} = \frac{\alpha_{OPP}}{\alpha_{FULL}} \tag{29.3}$$

Equation 29.3 when expanded cancels out the $1 + \alpha t$ factor and it comes to:

$$\chi_{\text{RULE}} = \frac{d_{i\text{RULE}}}{d_{i\text{FULL}}} \cdot \frac{E[d]_{\text{FULL}}}{E[d]_{\text{RULE}}}, \text{ or,}$$

$$\chi_{\text{RULE}} = \frac{n_{\text{RULE}}}{n_{\text{FULL}}} \cdot \frac{\sum d_{i\text{FULL}}}{d_{i\text{FULL}}} \cdot \frac{d_{i\text{RULE}}}{\sum d_{i\text{RULE}}} \tag{29.4}$$

The range of Eq. 29.4 is [0, 1] since the denominator has a factor of number of nodes of the original network which will weigh the denominator more and therefore the denominator of the fraction will always be greater than the numerator.

After computing (χ_{RULE}) and (χ_{OPP}) ranging in (0, 1), the values are multiplied with 100. The relative diffusion parameter contributes to understanding the strength and influence of that particular hashtag. The factor relative diffusion score out of 100 is calculated as follows:

$$\text{Total}_{\text{RULE}} = \chi_{\text{RULE}} \times 100 \tag{29.5}$$

This process after continuing for all hashtags will have two sets of values for each hashtag namely $\text{Total}_{\text{RULE}}$ and $\text{Total}_{\text{OPP}}$. Amongst these tweets, there will some propagated by RULE and others by OPP. So in order to know which party has done better in their own hashtags and also in countering, the narrative flow in their rival's hashtags is understood by the approach explained with Eq. (29.6).

Two sets of hashtags need to be separated, namely the RULE propagated hashtags and the OPP propagated hashtags. Thereafter, the difference (diff) of scores of RULE and OPP for each hashtag in RULE propagated and OPP propagated is found.

$$\text{diff}_{\text{RULE}i} = \text{Total}^i_{\text{RULE}^{\text{Rule Prop}}} - \text{Total}^i_{\text{OPP}^{\text{Rule Prop}}} \tag{29.6}$$

The mean score of both these list of values is found and whoever is leading the charge would have a higher value so as to depict that party has been successful in propagating its own agenda and reversing the agenda of the other.

29.4.3 Observations

The top trending hashtags leading up to the U.P Vidhan Sabha elections, 2022 were collected for the month of January, 2022 comprising 19 hashtags. Each hashtag had around 20k recent tweets from which the edge list was made ranging approximately 35K–40K tweets. A snippet of the trending hashtags collected and their cumulative scores for January, 2022 given in Fig. 29.6. The methodology mentioned in the above section is followed to get the following results:

Out of 19 hashtags, the OPP promoted ten and the BJP was promoted nine. The same method is used to compute the BJP's mean score of **20.20** and the OPP's mean score of **15.23** as depicted in Fig. 29.7. Apart from a couple of cases, there

1	Hashtag Name	relative diffusion of bjp	relative diffusion of opp	Propagated	Cumulative Score	Cumulative Score
2	#BharatStandsWithModiJi	0.196316432	0.247733189	BJP	19.6316432	24.77331888
3	#LongLivePMModi	0.254847519	0.103346452	BJP	25.48475188	10.33464525
4	#ModiRunsAway	0.24193754	0.345698059	OPP	24.19375401	34.56980591
5	#BJPModiExposed	0.527744897	0.616043069	OPP	52.77448974	61.60430686
6	#BJP4UP	0.159684181	0.110205065	BJP	15.9684181	11.02050653
7	#AaRahiHaiCongress	0.114472034	0.195858056	OPP	11.44720341	19.58580564
8	#Ayegi_Congress	0.262527646	0.824291562	OPP	26.25276465	82.42915621
9	#ShameOnYouAnjana	0.285710599	0.309873171	OPP	28.57105988	30.98731706
10	#GodiModiDonoFail	0.484764069	0.594397848	OPP	48.47640693	59.43978481
11	#ChangeIsPossible	0.14374397	0.439724275	OPP	14.374397	43.97242748
12	#DefeatBJP	0.323822565	0.383008674	OPP	32.38225648	38.30086739
13	#YogiWillBeBack	0.341738707	0.106722109	BJP	34.17387068	10.67221087
14	#UPYogiHaiYogi	0.283112201	0.059857609	BJP	28.31122009	5.985760949
15	#CongressMeansCorruption	0.265805599	0.192308682	BJP	26.58055994	19.23086825
16	#I_Am_With_Yogi	0.942709649	0.387651003	BJP	94.27096489	38.76510027
17	#CongressAgainstHindus	0.228759125	0.201437438	BJP	22.87591252	20.14374382
18	#RahulSeDartiHaiBJP	0.663723529	0.516914326	OPP	66.3723529	51.69143256
19	#BJPAgain	0.689393254	0.134936954	BJP	68.9393254	13.49369545
20	#AntiNationalRSS	0.268278856	0.613999406	OPP	26.82788556	61.39994065

Fig. 29.6 Cumulative scores of Jan. 2022 hashtags

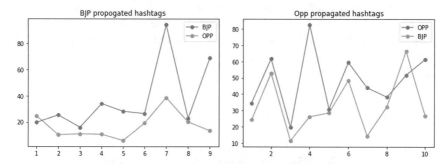

Fig. 29.7 BJP and OPP propagated hashtags' cumulative score for January 2022 plot

Fig. 29.8 ABP-C voter opinion poll [22]

is a significant difference between the BJP and OPP cumulative scores in the BJP propagated hashtag plot. In the OPP propagated hashtag plot, it is seen that the BJP is closing in on the cumulative scores overtaking the OPP in there own hashtags.

These results hold true against various opinion polls made by various psephology companies and news channels. One such opinion polls by ABP news and C voter [22] has been shown in Fig. 29.8 where the trend clearly shows the B.J.P leading the vote percentages in January, 2022 similar as the results in this study.

29.5   Conclusion

The current research contributes to understand the complexities that occur in the social media space prior to a major election. The study demonstrates the dynamic progression of an acceptance by a party or rejection on the popular social media platform Twitter. The sentiments associated with potential voters, as well as the changes in sentiments they experience over time, are reflected. These networks can be studied, and improvements for better information dissemination can be proposed in order to bring people's attention to situations more rapidly and in a more organised format.

References

1. Twitter. Retrieved 15 Mar 2022, from https://twitter.com/home
2. Neogi, A.S., Garg, K.A., Mishra, R.K., Dwivedi, Y.K.: Sentiment analysis and classification of Indian farmers' protest using twitter data. Int. J. Inf. Manag. Data Insights 1(2), 100019 (2021)
3. Chintalapudi, N., Battineni, G., Di Canio, M., Sagaro, G.G., Amenta, F.: Text mining with sentiment analysis on seafarers' medical documents. Int. J. Inf. Manag. Data Insights 1(1), 100005 (2021)
4. Twitter: Second quarter 2016 report (2016)
5. Twitter: Twitter IPO prospectus (2013)
6. DuVander, A.: Which APIs are handling billions of requests per day? Programmable Web (2012)
7. Alexa.com: Website traffic ranking (2017)
8. Satish, M., Srinivasa Rao, P., RamakrishnaMurty, M.: Identification of natural disaster affected area using Twitter. In: AISC Springer ICETC-2019, vol. 3, pp. 792–801 (2019)
9. Khan, A., Zhang, H., Shang, J., Boudjellal, N., Ahmad, A., Ali, A., Dai, L.: Predicting politician's supporters' network on Twitter using social network analysis and semantic analysis. Sci. Program. **2020** (2020)
10. Melo, C., Lechevallier, Y., Aufare, M.A.: Social Networks Analysis: A Case Study on the Twitter Network
11. Himelboim, I., Smith, M., Shneiderman, B.: Tweeting apart: applying network analysis to detect selective exposure clusters in Twitter. Commun. Methods Measures **7**(3–4), 195–223 (2013)
12. Hansen, D.L., Shneiderman, B., Smith, M.A.: Analyzing social media networks with NodeXL: insights from a connected world (2010). ISBN 978-0-12-382229-1
13. Oud, A., Sarah, T.A., Alohaideb, W.: Hybrid sentiment analyser for Arabic tweets using R. In: 7th International Joint Conference on Knowledge Discovery, Knowledge Engineering and Knowledge Management (IC3K), vol. 1, pp. 417–424 (2015)

14. Fiaidhi J et al (2012) Opinion mining over Twitter space: classifying tweets programmatically using the R approach. In: Seventh International Conference on Digital Information Management (ICDIM), pp. 313–319. IEEE

15. Ansari, M.Z., Aziz, M.B., Siddiqui, M.O., Mehra, H., Singh, K.P.: Analysis of political sentiment orientations on Twitter. Procedia Comput. Sci. **167**, 1821–1828 (2020)

16. Smith, M.A., Shneiderman, B., Milic-Frayling, N., Mendes Rodrigues, E., Barash, V., Dunne, C., Capone, T., Perer, A., Gleave, E.: Analyzing (social media) networks with nodexl. In: Proceedings of the Fourth International Conference on Communities and Technologies, pp. 255–264 (2009)

17. Bastian, M., Heymann, S., Jacomy, M.: Gephi: an open source software for exploring and manipulating networks. In: Proceedings of the International AAAI Conference on Web and Social Media, vol. 3, no. 1, pp. 361–362 (2009)

18. Waila, P., Singh, V.K., Singh, M.K.: Evaluating machine learning and unsupervised semantic orientation approaches for sentiment analysis of textual reviews. In: 2012 IEEE International Conference on Computational Intelligence and Computing Research, pp. 1–6. IEEE (2012)

19. Asghar, M.Z., Khan, A., Ahmad, S., Kundi, F.M.: A review of feature extraction in sentiment analysis. J. Basic Appl. Sci. Res. **4**(3), 181–186 (2014)

20. Hutto, C.J., Gilbert, E.E.: VADER: a parsimonious rule-based model for sentiment analysis of social media text. In: Eighth International Conference on Weblogs and Social Media (ICWSM-14), Ann Arbor, MI (2014)

21. Kumar, P., Sinha, A.: Information diffusion modeling and analysis for socially interacting networks. Soc. Netw. Anal. Min. **11**(1), 1–18 (2021)

22. ABP News: CVoter survey January UP assembly election 2022 opinion polls vote share seat sharing KBM BJP SP BSP Congress (2022 Jan 29). Retrieved 17 Apr 2022 from https://tinyurl.com/2p8swu64

Chapter 30
Open-World Machine Learning for Unknown Class Identification in Human-Machine Interaction Systems

Jitendra Parmar and Satyendra Singh Chouhan

Abstract Nowadays, different organizations widely use automated human–machine interaction systems for customer support and operational activities. The identification of appropriate queries is an essential part of these systems. The classical machine learning-based or rule-based systems cannot identify the unknown classes and may mislead the client by delivering incorrect responses. To identify these unknown classes, we proposed an open-world machine learning-based solution. The proposed model identifies the unknown classes and rejects them. After identifying the unknown classes, it also suggests the number of possible novel classes in rejected queries. The proposed model evaluated various performance matrix for both unknown class identification and the discovery of new possible classes. The result of the proposed models shows that it outperforms on state of the art.

30.1 Introduction

Many organizations rely on automated AI-based systems, including Financial Technology (Fin-Tech) organizations in the growing digital world. Apart from commercial organizations, customer-oriented platforms also use automation to serve their end-users smoothly. Social media users are overgrowing by the time; hence, these platforms also require techniques to deal with the unknown. Data is an integral part of the process of automation and machine intelligence. The modernity of machine learning has pushed the world towards intelligence applications employed on real-time data. Traditional machine learning approaches follow the closed-world assumptions (test instances appear during the training phase), which may not be suitable for artificial intelligence (AI)-based systems. These systems work in the real-world environment, where unknown classes may appear; it is called open or dynamic environment.

J. Parmar (✉) · S. S. Chouhan
Malaviya National Institute of Technology Jaipur, Jaipur, Rajasthan 302017, India
e-mail: 2019rcp9044@mnit.ac.in

S. S. Chouhan
e-mail: sschouhan.cse@mnit.ac.in

© The Author(s), under exclusive license to Springer Nature Singapore Pte Ltd. 2023   335
V. Bhateja et al. (eds.), *Evolution in Computational Intelligence*, Smart Innovation,
Systems and Technologies 326, https://doi.org/10.1007/978-981-19-7513-4_30

(a) HMI build with traditional machine learn-(b) HMI build with open-world machine
ing   learning

Fig. 30.1 a ChatBot using classical machine learning and **b** ChatBot using open-world machine learning

Open-world machine learning-based approaches require to deal with such conditions. Open-world machine learning-based models follow the open-world assumptions (test instances do not appear during the training).

Open-world machine learning is suitable for open environments where inputs change rapidly in size, occurrence rate, and nature. These unpredictable inputs can be handled if the system learns as a human does. To better understand this problem, let us consider the human–machine interaction (HMI) examples from real world.

We have taken a real-world human–machine interaction system ILA[1] to understand this problem. India's leading banking and finance organization, State Bank of India, SBI's subsidiary SBI cards, introduces ILA to resolve real-time human queries. This system is specially designed for credit cards-related queries and possible explanations. We can see that in Fig. 30.1a is exhibiting classical machine learning-based HMI that responds to the user's first query correctly. Still, it provides a piece of different/incorrect details for the second query which belongs to unknown class, instead reject it. Part (b) of Fig. 30.1, which is built with an open-world machine learning approach, shows ideal dialogue and indicates what should be the response of the system for queries belongs to unknown classes. The proposed framework aims to fulfil following objectives.

1. Prevent wrong responses of the system by rejecting queries belongs to unknown classes.
2. Find a possible number of new classes in rejected queries.

In this paper, we proposed a system to address the above objectives. The proposed system successfully identifies the unknown user queries and classifies them as unknown. Further, it finds the precise possible classes in data to be used in future to make the system more knowledgeable. The proposed model functions in two modules; the first module deals with unknown classes, while the second module finds the number of possible classes in unknown queries found by module one.

---

[1] https://www.sbicard.com/en/personal/benefits/easy-access-channels/chatbot.page.

30.2 Literature Reviews

Detection of unknown objects in classification is a common problem in image and natural language processing. Twenty four years ago, in [1], authors proposed open-world machine learning, a part of continual machine learning.

The discovery of an unknown instance is an essential part of the open-world environment. In [2], authors suggested the model with One-Vs-Set. The model is defined to recognise open sets. In [3], authors proposed a model for open space and nonlinear open set recolonisation for multi-class settings. The proposed model distinguishes the open space and known space for available data. The model drops the probability of cast as per the space, drops for open space, and identifies the unknown. In [4], authors proposed a convolutional neural network (CNN)-based architecture to identify the unknown instances of any class. Its layered architecture consists of embedding, convolution, pooling, and output layers. The proposed architecture, deep open classification (DOC), also uses the One-Vs-Rest concept to classify unknown data.

In [5], authors proposed cosine similarities based on the naive classical method to classify documents. It calculates vectors for the entire document and applies deep classification methods using the Weibull layer for output. The model also uses Word2Vec for data encodings [6]. There is some work also available to increase the knowledge-base for novel classes [7]. It increases the accuracy of classification for the models while encountering the unknown. In [8] and [9] suggested a meta learning-based framework to classify the unknown instances.

In [10], the authors proposed the framework for an automated dialogue-based system (ADS). The proposed model uses long short-term memory (LSTM) with the local outlier factor (LOF) [11]. The input layer uses the Glove embedding for sentence encoding [12]. In [13], authors proposed SoftMax and Deep Novelty (SMDN) end-to-end framework. It works with decision boundaries and confidence scores to classify instances with softmax as output [14]. In [13], authors proposed SoftMax and Deep Novelty (SMDN) end-to-end framework. It works with decision boundaries and confidence scores to classify instances with softmax as output [14].

30.3 Proposed Methodology

The proposed model is shown in Fig. 30.2, consisting of two major modules. First convolutional neural network (CNN)-based model to identify unknown classes. The second is the clustering model that finds the number of possible novel classes among unknown queries found by module one. The detailed description of the first and second modules is given in Sects. 30.3.1 and 30.3.2, respectively.

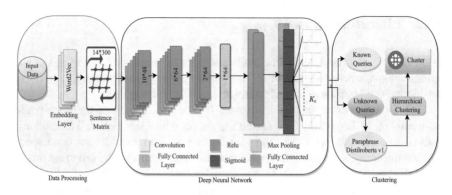

Fig. 30.2 Proposed model

30.3.1 Unknown Class Identification

The embedding layer of the proposed work uses Word2Vec encoding. The output of the embedding layer, sentence matrix, provides a 14×300 vector to the convolution layer. In the convolutional layer, let d_m is considered as the dimension of the word vector $w_{vi} \in \mathbb{R}^{d_m}$, it resembles vith word of the sentence. Let us assume with the padding l_e is the length of the sentence. Then the word concatenation is denoted as:

$$w_{vi} : l_e = w_{vi}1 \parallel w_{vi}2 \parallel w_{vi}3 \parallel, \ldots w_{vi}l_e \tag{30.1}$$

The $\parallel$ denotes the concatenation, whereas $w_{vi:vi+vj}$ is the general representation of concatenated words, that concatenate words: $w_{vi}, w_{vi+1}, \ldots w_{vi+vj}$. Now, we apply the filters for convolution $w\mathbb{R}^{w_s d_m}$ on w_s size of the window, and generate the feature map f_m from the $w_{vi:1+w_s-1}$, which can be denoted as:

$$f_m = f_n(w.w_{vi:1+w_s-1} + b_{\text{ias}}) \tag{30.2}$$

f_n is function which is nonlinear and $b_{\text{ias}} \in \mathbb{R}$ is represents a bias. Now, we can determine all the word windows possible in a sentence, that is $w_{1:w_s}, w_{2:w_s}, \ldots w_{l_e-w_s+1:l_e}$. Now, the filter f_n applies of all possible windows of the word in the sentence. It generates the feature map f_m, where $f_m \in \mathbb{R}^{l_e-w_s+1}$, and feature map f_m denotes as;

$$f_m = (f_{m1}, f_{m2}, \ldots, f_{ml_e-w_s+1}) \tag{30.3}$$

After every convolution, Relu activation is used to normalize convolution outputs. We have used three convolution layers with different filters in the proposed work, 2, 3, and 4, respectively. After the third convolution layer, a pooling layer used with the concept of max pool (1 pool). It reduces a feature vector by taking the maximum value from f_m that is $\hat{f}_m = \text{Max}(f_m)$ for each interrelated filter. The max pool layer

Table 30.1 Configuration of proposed model

Type	Sub-layer	Number of filters	Kernel size	Stride	Output shape	Parameters
Embeddings					(None, 14, 300)	706,200
Convolution (1D)	3.1	48	2	1	(None, 10, 48)	72,048
Convolution (1D)		64	3	1	(None, 6, 64)	15,424
Convolution (1D)		64	4	1	(None, 2, 64)	20,544
Max pooling (1D)	3.2	–	1 pool	–	(None, 1, 64)	0
Flatten		–	–	–	None, 64)	0
Dense	3.3	96 units	–	–	(None, 96)	6240
One-Vs-Rest	3.4	50 (= known classes)			(None, 50)	4850

provides d_m—dimension vectors of feature f_m, now fed into the fully connected layer.

Compared to classical models of machine learning that use the *SoftMax* function for output, the proposed model uses a *sigmoid* activation function with the concept of One-Vs-Rest for the output. One-Vs-Rest is a heuristic approach for using binary classification specific for multi-class classification. The proposed model uses the One-Vs-Rest concept as the output layer with K_c classes, where K_c is the number of known classes. Here, the ith *sigmoid* functions for K_{c_i} class. The proposed model defines all classes as positive if $y = K_{c_i}$, negative ($y \neq K_{c_i}$) otherwise.

The configuration of module one is given in Table 30.1.

30.3.2 Discovery of New Classes

The proposed model used the sentence-transformers model, paraphrase distilroberta-base-v1 [15], for encoding the queries sentences. After encoding, hierarchical clustering determines the number of possible classes in unknown data found by module one. We have calculated the distance matrix for all data points. Now, we consider all data points as a unique cluster. Merge the two most comparable data points (clusters) and update the distance matrix accordingly. It is repeated till we obtain the single cluster and update the distance matrix. The proposed module considered the minimum distance (Single linkage) for inter-cluster distance, which can be defined as:

$$\text{dis}_{\text{Min}}(Cl_i, Cl_i) = \text{Min}\left(d_p, d_q\right) \tag{30.4}$$

where Cl_i and Cl_j are two closet cluster, dis is distance between Cl_i and Cl_j. The $d_p \in Cl_i$ and $d_q \in Cl_j$.

30.4 Experiments and Result

The proposed model experimented with computers fitted with Intel(R) Core(TM) i5-2410M CPU @ 2.30 GHz and DDR3 8 GB RAM, running Windows 10. All experimentation is executed in python 3.10.1. The proposed model uses encoded data for each experiment; for unknown classes identification, it uses Google pre-trained Word2Vec, and for clustering data, it uses distilroberta-base-paraphrase-v1. The following performance matrices were used to validate and measure the performance analysis of the proposed model.

$F1$-Score: It is a harmonic mean of precision and recall that provides accuracy matrices for correctly classified queries (Eq. 30.5). It can be defined as:

$$F1\text{-score} = 2 \times \frac{P_{\text{re}} \times R_{\text{ec}}}{P_{\text{re}} + R_{\text{ec}}} \tag{30.5}$$

The proposed model uses the Normalized Mutual Information (NMI) as N_{MI}, Homogeneity (H_0), Completeness (C_m), and V-measure (V_m) to analyze the performance of module 2. Equations 30.6, 30.7, 30.8, and 30.9 define N_{MI}, H_0, C_m, and V_m, respectively.

$$N_{\text{MI}} = \frac{\sum_{c,k} S_{c,k} \log \frac{n_{c,k} \cdot N_a}{S_c \cdot S_k}}{\sqrt{\left(\sum_c S_c \log \frac{S_c}{N_a}\right)\left(\sum_c S_c \log \frac{S_c}{N_a}\right)}} \tag{30.6}$$

$$H_o = 1 - \frac{\sum_{c,k} S_{c,k} \log \frac{S_{c,k}}{S_k}}{\sum_c S_c \log \frac{S_c}{N_a}} \tag{30.7}$$

$$C_m = 1 - \frac{\sum_{c,k} S_{c,k} \log \frac{S_{c,k}}{S_c}}{\sum_c S_c \log \frac{S_k}{N_a}} \tag{30.8}$$

$$V_m = \frac{2 \times H_o \times C_m}{H_o + C_m} \tag{30.9}$$

where S_c = Number of sample in class c, S_k = Number of samples in cluster k, $S_{c,k}$ = Number of samples in class c as well as cluster k, N_a = Number of total samples in data.

30.4.1 Data Sets Used

Two publicly available benchmark data sets were used to evaluate the proposed model and obtain the $F1$-Score on three different thresholds that are 0.4, 0.5, and 0.6.

CLINC150 [16]²: A data set specially developed for out of scope intent detection. It has ten distinct domains, including 23,700 instances and 150 intents of different classes. Some of these classes are taken as unseen classes by excluding them from training data.

DBpedia [17]³: This data set contains more than 300k hierarchically labelled Wikipedia reports. A section of the data (after data processing, kernel contained) delivers taxonomic, hierarchical classifications for 342,782 Wikipedia reports. It has three levels with 9, 70, and 219 classes.

30.4.2 Baseline Methods Used for Comparison

The proposed model compared with the following benchmark works for open-query classification.

- DOC_CNN [4]: It uses the CNN network for open classification with kernel sizes of 3, 4, and 5. The threshold of rejection unknown is 0.5.
- DOC_LSTM [4]: It is a modified version of DOC-CNN with Bi-LSTM. It also uses a 0.5 threshold value for the rejection of unknown classes.
- L2AC [9]: It is a version of L2AC–Novote; it uses One-Vs-Many as the output layer to find similarities between instances.
- CBS-SVM [18]: It is equivalent to SVC with the 'linear' parameter but builds 'lib-linear' instead of 'libsvm' library, providing it better flexibility in terms of corrections and loss functions.

30.4.3 Results

Tables 30.2 and 30.3 give the proposed model's results to classify unknown classes and discovery of possible novel classes, respectively. The $F1$-score of the proposed methods is improved with both data sets. In the case of CLINC150 data set, for threshold value 0.4, the performance or model slightly decreasing, though the model outperforms with threshold values 0.5 and 0.6. The $F1$-score is improved by 1.48% to 1.85% for 0.5 and 0.6, respectively. The model outperforms on DBpedia data set in all threshold values. The model's performance improved by 5.4%, 7.4%, and 8.2% for threshold values 0.4, 0.5, and 0.6, respectively.

The high NMI values of 83.20% and 66.90% for both data sets, respectively, show no uncertainty in class while we validate ground truth. The proposed model produces 91.31% and 90.82% homogeneity for both data sets. The high homogeneity value shows when samples within a cluster belong to the identical class. Similarly, higher

---

² https://github.com/clinc/oos-eval.

³ https://www.kaggle.com/danofer/dbpedia-classes.

Table 30.2 Performance of the proposed model for unknown class identification

Threshold	CLINC150			DBPEDIA		
	0.4	0.5	0.6	0.4	0.5	0.6
DOC-CNN	0.6776	0.6935	0.7048	0.6894	0.7439	0.7713
DOC LSTM	0.5301	0.5747	0.6137	0.4353	0.4630	0.4924
L2AC	0.4179	0.6712	0.7135	0.0532	0.3294	0.4505
CBS-SVM	0.4926	0.6460	0.6967	0.5167	0.5971	0.6128
Proposed model	0.6679	0.7083	0.7320	0.7441	0.8182	0.8533

Table 30.3 Performance of the proposed model for discovering new classes

	NMI	Homogeneity	Completeness	V-measure
CLINC150	0.8320	0.9131	0.7642	0.8324
DBPEDIA	0.6690	0.9082	0.5293	0.6696

completeness, 76.42% and 52.93%, show that all the class samples belong to the same cluster. The harmonic mean of homogeneity and completeness (V-measure) was also high for both data sets. The V-measure is 83.24% and 66.96% for both data sets, respectively.

30.5 Conclusion

The world is rapidly changing in digitalising classical and conventional systems and adapting automation. The natural and dynamic environment encounters many open challenges that cannot be solved with classical machine learning. Hence, there is a need for technology that can incorporate solutions for open-world scenarios and dynamic environments. The proposed model is suitable for the rejection of anonymous and unknown classes that will improve the reliability of HMI or similar systems. The results show that it improves the performance of existing techniques in a similar field.

In future, we plan to improve the proposed model by assembling identification of novel classes with continual learning ability.

References

1. Ring, M.B.: Child: a first step towards continual learning. In: Learning to Learn, pp. 261–292. Springer, Berlin (1998)
2. Scheirer, W.J., de Rezende Rocha, A., Sapkota, A., Boult, T.E.: Toward open set recognition. IEEE Trans. Pattern Anal. Mach. Intell. **35**(7), 1757–1772 (2012)

3. Scheirer, W.J., Jain, L.P., Boult, T.E.: Probability models for open set recognition. IEEE Trans. Pattern Anal. Mach. Intell. **36**(11), 2317–2324 (2014)
4. Shu, L., Xu, H., Liu, B.: Doc: deep open classification of text documents. In: Proceedings of the 2017 Conference on Empirical Methods in Natural Language Processing, pp. 2911–2916 (2017)
5. Prakhya, S., Venkataram, V., Kalita, J.: Open set text classification using convolutional neural networks. In: International Conference on Natural Language Processing (2017)
6. Mikolov, T., Chen, K., Corrado, G., Dean, J.: Efficient estimation of word representations in vector space (2013). arXiv:1301.3781
7. Mazumder, S., Ma, N., Liu, B.: Towards a continuous knowledge learning engine for chatbots (2018). arXiv:1802.06024
8. Guo, X., Alipour-Fanid, A., Wu, L., Purohit, H., Chen, X., Zeng, K., Zhao, L.: Multi-stage deep classifier cascades for open world recognition. In: Proceedings of the 28th ACM International Conference on Information and Knowledge Management, pp. 179–188 (2019)
9. Xu, H., Liu, B., Shu, L., Yu, P.: Open-world learning and application to product classification. In: The World Wide Web Conference, pp. 3413–3419 (2019)
10. Lin, T.-E., Xu, H.: Deep unknown intent detection with margin loss. In: Proceedings of the 57th Annual Meeting of the Association for Computational Linguistics, pp. 5491–5496 (2019)
11. Breunig, M.M., Kriegel, H.-P., Ng, R.T., Sander, J.: Lof: identifying density-based local outliers. In: Proceedings of the 2000 ACM SIGMOD International Conference on Management of Data, pp. 93–104 (2000)
12. Pennington, J., Socher, R., Manning, C.D.: Glove: global vectors for word representation. In: Proceedings of the 2014 Conference on Empirical Methods in Natural Language Processing (EMNLP), pp. 1532–1543 (2014)
13. Lin, T.-E., Xu, H.: A post-processing method for detecting unknown intent of dialogue system via pre-trained deep neural network classifier. Knowl. Based Syst. **186**, 104979 (2019)
14. Goodfellow, I., Bengio, Y., Courville, A.: Deep learning (2016)
15. Reimers, N., Gurevych, I., Reimers, N., Gurevych, I., Thakur, N., Reimers, N., Daxenberger, J., Gurevych, I., Reimers, N., Gurevych, I., et al.: Sentence-bert: sentence embeddings using Siamese bert-networks. In: Proceedings of the 2019 Conference on Empirical Methods in Natural Language Processing. Association for Computational Linguistics
16. Larson, S., Mahendran, A., Peper, J.J., Clarke, C., Lee, A., Hill, P., Kummerfeld, J.K., Leach, K., Laurenzano, M.A., Tang, L., Mars, J.: An evaluation dataset for intent classification and out-of-scope prediction. In: International Joint Conference on Natural Language Processing (EMNLP-IJCNLP), pp. 1311–1316 (2019)
17. Auer, S., Bizer, C., Kobilarov, G., Lehmann, J., Cyganiak, R., Ives, Z.: Dbpedia: a nucleus for a web of open data. In: The Semantic Web, pp. 722–735. Springer, Berlin
18. Fei, G., Wang, S., Liu, B.: Learning cumulatively to become more knowledgeable. In: Proceedings of the 22nd ACM SIGKDD International Conference on Knowledge Discovery and Data Mining, pp. 1565–1574 (2016)

Chapter 31
Detection of Requirements Discordances Among Stakeholders Under Fuzzy Environment

Faiz Akram, Tanvir Ahmad, and Mohd Sadiq

Abstract Software requirements (SRs) elicitation is one of the key sub-processes of software development that is used in identifying the stakeholders' need. Several stakeholders participate in the requirements elicitation process. During the analysis, it has been observed that there are requirements discordances among stakeholders because each stakeholder has a different opinion on the same requirement. Based on our review, we found that one of the reasons for the failure of software is the disagreement among the stakeholders. Few methods have been developed for the detection of discordances among the stakeholders in which crisp values were used to record the stakeholders' preferences on SRs, and small-scale software systems were used during the analysis. Linguistic variables find their relevance in real-life applications to identify the preferences of requirements over other requirements rather than crisp values. Therefore, to deal with this issue, a method is presented to detect the requirements discordances among stakeholders in which fuzzy logic is used to address imprecision and vagueness. An example is shown to discuss the applicability of the proposed method.

F. Akram (✉) · T. Ahmad
Department of Computer Engineering, Faculty of Engineering and Technology, Jamia Millia
Islamia, New Delhi 110025, India
e-mail: faizakram1988@jmi.ac.in

T. Ahmad
e-mail: tahmad2@jmi.ac.in

M. Sadiq
Software Engineering Laboratory, Computer Engineering Section, UPFET, Jamia Millia Islamia,
New Delhi 110025, India
e-mail: msadiq@jmi.ac.in

© The Author(s), under exclusive license to Springer Nature Singapore Pte Ltd. 2023   345
V. Bhateja et al. (eds.), *Evolution in Computational Intelligence*, Smart Innovation,
Systems and Technologies 326, https://doi.org/10.1007/978-981-19-7513-4_31

31.1 Introduction

The term software can be defined as a set of programs, documentation, and operating procedure [1]. Among these components of the software, documentation includes different types of requirements, elicited by various requirements elicitation techniques, viz. traditional methods, package-oriented method, goal-oriented method, etc. After completing the requirements elicitation process, a system might have a very large number of requirements [2]. Depending on the number of requirements and the participation of stakeholders during the elicitation process, a system may be classified as a small-scale, medium-scale, and large-scale system. Lim and Finkelstein [3] define the large-scale software project as one in which tens of thousands of users participate during the elicitation process, and these users are categorized in a number of groups. These users are the members of the stakeholder's group wherein each group, stakeholders are defined with their roles and responsibilities. It has been observed that these stakeholders may have conflicting requirements during the elicitation process, and detecting the discordances among the stakeholders is one of the primary issues of the requirements elicitation. Conflictions among stakeholders during the requirements elicitation might cause software failure. Thus, it is indispensable to detect the disagreement among the stakeholders during the requirements elicitation process so that a successful software product can be developed [2, 3].

Software requirements may be classified into functional requirements (FRs) and non-functional requirements (NFRs). The FRs is the functionality of a system, on the other hand, NFRs describe the non-behavioral aspect of a system like performance, reliability, cost, availability, etc. These requirements have a significant role during the analysis process. For example, a customer wants a high-performance product that has several functions but low cost. On the other hand, developers want to develop an expensive product that requires fewer development efforts. Based on the cost, both customer and the developer have varying opinion on the same requirement [4].

Various methods have been developed in requirements engineering for the detection of the discordances among the stakeholders. For example, Kaiya et al. [5] developed a method for the analysis of the goals in which two attributes' values, i.e., *preference matrix* and *contribution value*, were attached to the nodes of the goals. In their work, the authors introduced two types of conflicts on goals, i.e., (a) *"confliction between goals"* and (b) *"conflict on a goal between two stakeholders."* The possibility of conflictions among the goals can be identified as (i) when the contribution value of two goals that are connected with an edge is negative and (ii) when there is a larger variance in the diagonal elements of the preference matrices.

In another study, Kaiya et al. [6] improved the method to detect the discordances among stakeholders. Motivated by the work of Kaiya et al. [5, 6], Mohammad et al. [7] developed a method in which *fuzzy preference matrix* and *fuzzy contribution value* were used for the analysis of the goals. In their work, the disagreements among the stakeholders were analyzed using *a fuzzy preference matrix* and very few stakeholders were used during the analysis of the goals in goal-oriented requirements analysis. As per our knowledge, there is no study that detects the requirements discordances among the stakeholders for large-scale software projects under fuzzy environment. Therefore, the objective of this paper is to develop a method for the requirements discordances among the stakeholders for large-scale software system under fuzzy environment.

The contributions of our work include the following:

1. A method has been developed for the detection of the requirements discordances among stakeholders in large scale software system
2. The stakeholders have been identified based on their roles and responsibilities for library information system (LIS).

The rest of this paper is structured as follows: The related work on requirements discordances among stakeholders is discussed in Sect. 31.2. The proposed method for the detection of requirements discordances among stakeholders under fuzzy environment is presented in Sect. 31.3. The steps of the proposed method are explained by considering the stakeholders of an LIS in Sect. 31.4. Finally, the conclusion and future work are discussed in Sect. 31.5.

31.2   Related Work

Various stakeholders are involved during the requirements elicitation process. To capture the complete set of requirements, there is a need to scale up the requirements elicitation process so that a large number of stakeholders can be invited for a critical analysis of the elicited requirements. Only a few studies have focused on this issue. For example, Huang and Mobasher [8] applied data mining and recommender systems to scale up the requirements process. In their work, the authors focused on different issues when scaling up the elicitation process like "information overload," "redundant requirements," "conflicting opinions," "unmanageable discussion," etc. Among these issues, detection of conflicting opinions in large-scale software projects is a key research issue because thousands of stakeholders are involved in the analysis of the software requirements.

Lim and Finkelstein [3] proposed a StakeRare method for eliciting the requirements for large-scale software projects. In their work, the authors have considered 60 stakeholders' groups and 30,000 students and staff. These stakeholders have some differing and conflicting requirements. One of the limitations of their work was that crisp data was used for detecting the opinions among the stakeholders. Keeping in view the significance of detecting the conflict among stakeholders, few methods have been developed in the literature on software engineering. For example, Kaiya et al. [5] developed an "*attributed goal-oriented requirements analysis method*" (AGORA) in which a preference matrix is used to capture the opinions of three stakeholders, i.e., customer, administrator, and developer. In their work, crisp values were considered for the requirements' analysis.

The characteristics of discordances include the following, i.e., missing, inconsistent, and discordant. In another study, Mohammad et al. [7] developed a *"fuzzy attributed goal-oriented software requirements analysis"* (FAGOSRA) method in which fuzzy preference matrices were used for the detection of the confliction among the stakeholders. One of the limitations of the FAGOSRA method was that only a few stakeholders were involved during the analysis of the goals. To deal with the large set of stakeholders, we present a method for the detection of requirements discordances in which $L^{-1}R^{-1}$ arithmetic principle has been used to address the vagueness and impreciseness during the analysis of the requirements of LIS. In our work, the stakeholders of LIS have been identified based on their role and responsibility.

31.3 Proposed Method

This section discusses the proposed method to detect requirements discordances among stakeholders for large-scale software project under fuzzy environment. The work flow of the proposed method is shown in Fig. 31.1. The proposed method includes the following:

- **Step 1:** Identification of stakeholders for large-scale software project
- **Step 2:** Elicitation of the different types of requirements
- **Step 3:** Capturing the opinions of stakeholders on the requirements using linguistic variables
- **Step 4:** Detection of requirements discordances among the stakeholders

The explanation of the steps of the proposed method is given below:

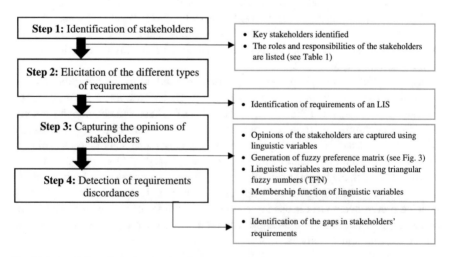

Fig. 31.1 Work flow diagram of the proposed method

Step 1: Identification of stakeholders for large scale software project

Stakeholders are the key sources of the requirements because requirements are elicited based on the stakeholders' needs. Variety of stakeholders participate in the software development process. For the success of the software project, the stakeholders should be recognized based on their roles and responsibility. For the large-scale software development, the stakeholders should be classified into groups and each group should include those stakeholders who have the same roles and responsibilities.

Step 2: Elicitation of the different types of requirements

This step elicits the requirements using a group of requirements elicitation techniques. Traditional method and goal-oriented method are used to elicit the various types of requirements like functional requirements (FRs) and non-functional requirements (NFRs) of the system.

Step 3: Capturing the opinions of stakeholders on the requirements using linguistic variables

The opinions of stakeholders for the requirements are captured in a preference matrix in which linguistic variables specify the stakeholders' preferences on a requirement. In the preference matrix, the stakeholders specify the preference on the requirements but the stakeholders also predict the preference of the other stakeholders. For example, in the AGORA method, -10, 0, and $+10$ values were used during the analysis of the requirements, and these values indicate dislike, unconcern, and preferable, respectively. In real-life applications, stakeholders apply linguistic variables for the requirements' evaluation. Therefore, their opinions are captured using linguistic variables in the preference matrix, which is referred to as fuzzy preference matrix.

Step 4: Detection of requirements discordances among the stakeholders

The "canonical representation of multiplication operation" (CRMO) associated with $L^{-1}R^{-1}$ inverse arithmetic principle and the "graded mean integration representation" [9–11] on TFNs has been used to model the linguistic variables of fuzzy preference matrix. In our work, the triangular fuzzy number models the linguistic variables. After that, the variance of each stakeholder in the fuzzy preference matrix is computed to detect whether there is any disagreement among the stakeholders on a particular requirement or not.

Let $Z_1 = (l_1, p_1, r_1)$ and $Z_2 = (l_2, p_2, r_2)$ be two triangular fuzzy number (TFNs) as shown in Fig. 31.2.

Here,

Fig. 31.2 Representation of two TFNs

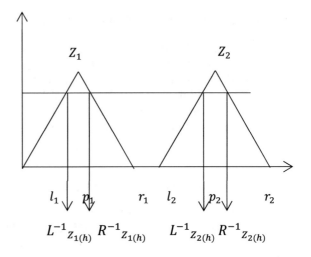

$$L_Z = \frac{x-1}{p-1}, l \le x \le p, \text{ and } R_Z(x) = \frac{x-r}{p-r}, p \le x \le r;$$

$$L_Z^{-1}(h) = l + (p-l)h \quad 0 \le h \le 1; \quad R_z^{-1}(h) = r + (p-r)h \quad 0 \le h \le 1$$

Here, $L_Z(x)$ and $R_Z(x)$ are the function L and R of the fuzzy number Z, respectively. $L_Z^{-1}(x)$ and $R_Z^{-1}(x)$ are the inverse functions of $L_Z(x)$ and $R_Z(x)$ at level h.

The multiplication of Z_1 and Z_2 at h-level can be computed as [9]:

$$Z_1(h) \times Z_2(h) = \left(L_{Z_{1(h)}}^{-1} L_{Z_{2(h)}}^{-1}, L_{Z_{1(h)}}^{-1} R_{Z_{2(h)}}^{-1}, R_{Z_{1(h)}}^{-1} L_{Z_{2(h)}}^{-1}, R_{Z_{1(h)}}^{-1} R_{Z_{2(h)}}^{-1} \right) \tag{31.1}$$

The graded mean integration representation of W_1 and W_2 at h-level is given as below:

$$P(Z_1 \times Z_2) = \int_0^1 \int_0^1 \int_0^1 \left[\left(h_{Z_1} L_{Z_{1(h)}}^{-1} \right) \left(h_{Z_2} L_{Z_{2(h)}}^{-1} \right) + \left(h_{Z_1} L_{Z_{1(h)}}^{-1} \right) \left(h_{Z_2} R_{Z_{2(h)}}^{-1} \right) \right.$$

$$\left. + \left(h_{Z_1} R_{Z_{1(h)}}^{-1} \right) \left(h_{Z_2} L_{Z_{2(h)}}^{-1} \right) + \left(h_{Z_1} R_{Z_{1(h)}}^{-1} \right) \left(h_{Z_2} R_{Z_{2(h)}}^{-1} \right) \right]$$

$$\times \frac{h_{Z_1 Z_2} dh_{Z_1} dh_{Z_2} dh_{Z_1 Z_2}}{\int_0^1 h_{Z_1} dh_{Z_1} \int_0^1 h_{Z_2} dh_{Z_2} \int_0^1 h_{Z_1 Z_2} dh_{Z_1 Z_2}} \tag{31.2}$$

After simplification, we have the following results:

$$P(Z_1 \times Z_2) = \left(\frac{l_1 + 4p_1 + r_1}{6} \right) \times \left(\frac{l_2 + 4p_2 + r_2}{6} \right) \tag{31.3}$$

Equation (31.3) will be applied to model the linguistic variables and also to analyze the fuzzy preference matrices of stakeholders.

31.4 Case Study

Various types of systems have been used in the literature of software engineering as a part of case studies like institute examination system, library management system, pricing system, online national election voting system, etc. [12]. In this work, we have considered the stakeholders of the LIS. This section explains the steps of the proposed method for the detection of requirements discordances among the stakeholders of LIS.

Step 1: In our work, we have identified more than one thousand stakeholders and among these stakeholders, we have considered 30 key stakeholders of LIS like head librarian, library administrator, deputy librarian, assistant librarian, etc. The roles and responsibilities of these stakeholders are exhibited in Table 31.1.

Step 2: In this step, the requirements of LIS have been identified using traditional requirements elicitation method which includes analysis of existing documents and goal-oriented method. As a result, following requirements for LIS have been identified:

- **FR1:** Login/signup facility for students and staff of the university.
- **FR2:** Searching for book/research article by the students and the staff.
- **FR3:** Reference number generation for the books.
- **FR4:** Verification of student by return email for library membership card.
- **FR5:** Sending of email/SMS to the user who has not yet returned the book on time and the corresponding late submission fine.
- **FR 6:** Based on the scale of requirements and availability, recommending the books to the head librarian for acquisition.
- **FR7:** Suggestion/recommendation of an alternate book title to the user in case the reference book is not available.
- **FR8:** Generating the list of books based on their requirements, books with very high requirements, and books which are not issued in an academic year.
- **FR9:** Provision for writing review for a book.
- **FR10:** Provision for downloading the e-book and the research articles.
- **Step 3**: In this step, the opinions of the stakeholders have been captured using linguistic variables for various requirements of the LIS. Here, we considered one of the requirements, i.e., *verification of student by return email for library membership card* (FR4). Five stakeholders, i.e., $S1$: head librarian, $S2$: library administrator, $S24$: developer, $S25$: tester, and $S28$: software architect, are selected characterized by their roles and responsibilities for the analysis of the requirement FR4. The opinions of these five stakeholders for FR4 are captured in a fuzzy preference matrix, which is shown in Fig. 31.3.

Table 31.1 Identified key stakeholders of an LIS

Stakeholders	Roles	Responsibilities
S1	Head librarian	Supervises the overall functioning of library
S2	Library administrator	Managing finance and resource of the library
S3	Deputy librarian	Responsible for library system, media, acquisition, cataloging, etc
S4	Assistant librarian	Facilitating the established policies of library
S5	Personal assistant	Assisting the task of head librarian
S6	Library assistant	Facilitating daily activities of librarian
S7	Cataloging librarian	Managing catalog enquiry facility
S8	Senior tutor librarian	Providing consultation, teaching, and liaison with schools
S9	Library attendant	Assists the users of the library
S10	Acquisition manager	Order new books for the library
S11	Archivist	Preserve important documents and records of library
S12	Conservationist	Assessment of library materials to stabilize the structure of library in order to sustain as long as possible
S13	System manager	Ensures proper working of the system
S14	Technician	Manages the media unit services of library
S15	Dean	Forwarding the requests of the books to head librarian with justification
S16	Head	Finalize the names of the books as per the syllabus and forward it to dean
S17	Class representative of student	Updating the information about the books like number of copies of a book, unavailability of the books, etc
S18	Data entry operator	Enters and updates data associated with the library
S19	Multi-tasking staff	Performs daily tasks associated with the library
S20	Inventory manager	Manages the inventories of the library
S21	Publishers-External	Maintaining the books of the external publishers
S22	Publishers-Internal	Maintaining the books of the internal publishers

(continued)

Table 31.1 (continued)

Stakeholders	Roles	Responsibilities
S23	Software vendors	Responsible to train the library staff about the latest software relates to library
S24	Developer	Responsible for developing the system
S25	Tester	Responsible to test the functionalities of the system
S26	Requirement analyst for functional requirements	Responsible for analyzing the functional requirements of the system
S27	Requirement analyst for non-functional requirements	To analyze the system's non-functional requirements
S28	Software architect	Responsible for the overall system architecture
S29	Cost analyst	Responsible for analyzing the cost associated with the system
S30	Software Requirements Modeler	Responsible to model the different system requirements

Fig. 31.3 Opinions of stakeholders on FR4

	S1	S2	S24	S25	S28
S1	VH	H	H	M	H
S2	VH	VH	L	M	VL
S24	H	M	VH	H	H
S25	H	M	H	H	H
S28	VH	L	H	H	VH

In our work, the following linguistic variables are used for capturing the opinions of stakeholders on FR4: Very Low (VL), Low (L), Medium (M), High (H), and Very High (VH). These linguistic variables are modeled using triangular fuzzy numbers because it is easy in understanding and computation. The TFNs for these linguistic variables are given below: VL $= (0, 0, 0.25)$, $L = (0, 0.25, 0.5)$, $M = (0.25, 0.5, 0.75)$, $H = (0.5, 0.75, 1.0)$, and VH $= (0.75, 1.0, 1.0)$.

The visual representation of these variables is exhibited in Fig. 31.4. In Fig. 31.4, $\mu_A(x)$ represents the membership value of fuzzy set A for element x.

Step 4: To identify the gap in understanding from the fuzzy preference matrix (FPM) of Fig. 31.3, we first apply the Eq. (31.3) and then the variance of each column of the stakeholders is calculated. We apply the following algorithm to check the disagreement among the stakeholders, see Fig. 31.5:

As a result, we identify that the variance of columns $S1$, $S2$, $S24$, $S25$, and $S28$ as 0.013, 0.073, 0.069, 0.018, and 0.12, respectively. From these values, it is clear that library administrator, developer, and software architect have different interpretations of FR4. Therefore, it is the responsibility of requirements analysts, i.e., $S26$ and $S27$ to analyze the requirement FR4 in more detail and decompose it into more concrete

Fig. 31.4 Membership
function of linguistic
variable VL, L, M, H, and
VH for analyzing the FR4

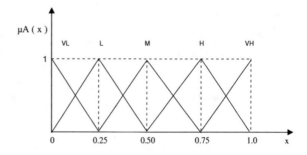

Fig. 31.5 Algorithm for
detection of disagreement
among stakeholders

```
if (variance of an FPM = = 0 or LOW)
{
print "Stakeholders mutually understand the goal"
}
else
{
print "There is disagreement among the stakeholders"
}
```

sub-requirements so that the gap in stakeholders' understanding can be filled. It is easy to recognize and improve the misleading requirements using preference matrices and the rationales.

31.5   Conclusion and Future Work

The paper explains a method to detect discordances among the stakeholders on software requirements during software analysis and development. The proposed method includes the following steps: identification of stakeholders for a large-scale software project, elicitation of the different types of requirements, capturing the opinions of stakeholders on the requirements using linguistic variables, and detection of requirements discordances among the stakeholders. In our work, in the initial phase, we have identified 30 key stakeholders for the LIS based on their roles and responsibilities and 10 FRs. To understand the process of detecting the discordances among the stakeholders, we have considered the requirement FR4 and five stakeholders $S1$, $S2$, $S24$, $S25$, and $S28$ during the analysis. The opinions of these stakeholders were captured in a fuzzy preference matrix. Based on the analysis, it is found that library administrator, developer, and software architect have different interpretations of FR4 because the values of the variances for these stakeholders are quite large as compared to the other two stakeholders. Such analysis helps in identifying misleading requirements from the set of the elicited requirements. Under this situation, requirements' analysts describe the requirements in more detail by decomposing it into sub-requirements so that the gap of misunderstanding among the stakeholders can be filled. Future research prospects include:

1. To develop a method for negotiating with the stakeholders during the conflictions
2. To develop a supporting tool for the detection of discordance among the stakeholders

References

1. Zhang, L., Tian, J.H., Jiang, J., et al.: Empirical research in software engineering—a literature survey. J. Comput. Sci. Technol. **33**, 876–899 (2018)
2. Sadiq, M.: A fuzzy set-based approach for the prioritization of stakeholders on the basis of the importance of software requirements. IETE J. Res. **63**(5), 616–629 (2017)
3. Lim, S.L., Finkelstein, A.: StakeRare: using social networks and collaborative filtering for large-scale requirements elicitation. IEEE Trans. Softw. Eng. 38(3), 707–735
4. Sadiq, M., Devi, V.S.: Fuzzy-soft set approach for ranking the functional requirements of software. Expert Syst. Appl. 193, 116452, ISSN 0957-4174
5. Kaiya, H., Horai, H., Saeki, M.: AGORA: attributed goal-oriented requirements analysis method.In: Proceedings IEEE Joint International Conference on Requirements Engineering. pp. 13–22
6. Kaiya, H., Shinbara, D., Kawano, J., et al.: Improving the detection of requirements discordances among stakeholders. Requirements Eng. **10**, 289–303 (2005)
7. Mohammad, C.W., Shahid, M., Hussain, S.Z.: Fuzzy attributed goal oriented software requirements analysis with multiple stakeholders. Int. J. Inf. Technol. **13**, 1–9 (2021)
8. Cleland-Huang, J., Mobasher, B.: Using data mining and recommender systems to scale up the requirements process. In Proceedings of the 2nd International Workshop on Ultra-Large-Scale Software-Intensive Systems (ULSSIS '08). Association for Computing Machinery, New York, NY, USA, 3–6 (2008)
9. Chen, S.H., Hsieh, C.H.: Graded mean integration representation of generalized fuzzy numbers. In: Proceedings of the 6th International Conference on Fuzzy Theory and Its Applications, Taiwan, pp. 1–6 (1998)
10. Chou, C.C.: The canonical representation of multiplication operation on triangular fuzzy numbers. Comput. Math. Appl. **45**, 1601–1610 (2003)
11. Chou, C.C.: The Representations of Multiplication Operation on Fuzzy Numbers and Application to Solving Fuzzy Multiple Criteria Decision Making Problems, pp. 161–169. PRICAI, LNAI, Springer (2006)
12. Sadiq, M., Parveen, A., Jain, S.K.: Software requirements selection with incomplete linguistic preference relations. Bus. Inf. Syst. Eng. **63**, 669–688 (2021)

Chapter 32
Blockchain-Enabled Data Sharing Framework for Intelligent Healthcare System

Ashish Kumar and Kakali Chatterjee

Abstract The rapid growth of health care 4.0 for sharing of health data between various stakeholders helps to save lives and improve treatment process. IoT-based intelligent healthcare system is a part of health care 4.0 which can be used in real-time health monitoring, sharing patient data for treatment, telecare medicine, etc. Normally this type of system suffers with different cyber-attacks like data interception, data stealing, unauthorized access, etc. To protect from such vulnerabilities, this paper proposes a blockchain-based framework for intelligent healthcare system where data sharing between multiple stakeholders is done in a secure manner. The proposed framework is implemented, and performance is evaluated using solidity.

32.1 Introduction

Intelligent healthcare system (IHS) is integrated with health care 4.0 which is introduced to industry 4.0. In healthcare domain, it builds the concept of personalized and virtualization of health data. This concept extents a patient-oriented system on data sharing among different stakeholders of healthcare system. In IoT-based healthcare system, many medical devices are used to capture patient data for real-time monitoring and data sharing in collaborative environment for treatment and diagnosis purpose. Any insecure technique adopted for health care may lead data breaches through which hackers can gain access of patient's private data. To ensure data privacy of electronic healthcare system, blockchain-based solutions are available [1]. Blockchain is used in two ways in IHS. Firstly, to store information within smart contracts and secondary as a decentralized storage platform [2]. For example, X-ray

A. Kumar (✉) · K. Chatterjee
Department of Computer Science and Engineering, National Institute of Technology Patna, Patna, Bihar 800005, India
e-mail: ashishk.ph21.cs@nitp.ac.in

K. Chatterjee
e-mail: kakali@nitp.ac.in

V. Bhateja et al. (eds.), *Evolution in Computational Intelligence*, Smart Innovation, Systems and Technologies 326, https://doi.org/10.1007/978-981-19-7513-4_32

files are stored as swarm, and hash of the file is stored in the smart contract. The blockchain is an immutable, durable ledger where the transactions are recorded after consensus among the peers and it cannot be altered or removed [3]. Ahmed et al. [4] presented a comprehensive survey of blockchain technique applied in healthcare sector. However, the work does not cover the difficulty to manage the records and fetch it at any time. Miler et al. [5] found that learning healthcare system function is already in blockchain technology. Tripathi et al. [6] reviewed the problems in accepting the digital technology-based healthcare system in aspect of technology and security. Hussein et al. [7] have surveyed a lot of article related to blockchain, health care, and electronic healthcare records in detail. They have more focused on the latest framework, solutions, models, approaches, algorithm, and contracts. From the literature work, some of the identified issues are mentioned below:

- Techniques required to gather and contribute the healthcare data.
- Difficulty in creating smart contract when healthcare data is not achievable due to some reason.
- Data acceptance in blockchain network of current healthcare data.
- Patient's less knowledge related to benefits of blockchain in health care.
- Government's less interest in promoting blockchain in healthcare system [9].
- Healthcare system focuses on conditions and emotional health of patient while automation is improved in blockchain-based data centric solution.

To solve the identified issues with blockchain, in this paper a framework for IHS has been developed which mainly works for data sharing in collaborative environment, thus the contribution of the work in this paper as follows:

1. This paper identifies several security and challenges in intelligent healthcare system.
2. A blockchain-based framework has been proposed which maintains the differential privacy of patient data.
3. The security analysis demonstrates to achieve the desired level of privacy.

The remaining parts of the paper are constructed as follows: In Sect. 32.2, we have discussed about the blockchain-based intelligent healthcare system. Then in Sect. 32.3, we have explained differential privacy in IHS. In Sect. 32.4, we have explained the proposed framework.1.

32.2 Blockchain-Enabled Intelligent Healthcare System

In general healthcare system, patients have to suffer a lot of problems for getting good doctors or hospitals and even if the patient found the good doctors, then also they have to face a lot of problems to get an appointment along with that a patient has to carry all the necessary documents that is so problematic things for any patient.

So, keeping all these problems into account, the intelligent health care (IHS) has been developed for low cost, easy, and efficient services in real-time monitoring as shown in Fig. 32.1. The working of the proposed IHS is described by Fig. 32.1. The important features of IHS are given below:

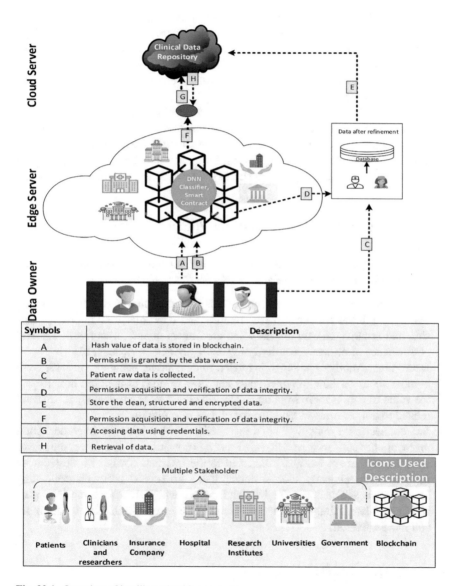

Fig. 32.1 Overview of intelligent health care system

32.2.1 Features of Intelligent Healthcare System

The intelligent healthcare system has many features. Among those features, a few features are discussed below:

- Smart Self Decision: IHS can take decision by its own, for example, it will show the doctor's or hospital's details according to the patient's disease. For example, if any patient has diabetes, then IHS will show only the doctors related to diabetes. Or if any patients have diabetes as well as tuberculosis, then it will show the doctors related to both the disease. And also, if any doctor is found malicious, then it can eliminate the doctor.
- Real-Time Monitoring: Whenever patient will upload the data related to their health in IHS, doctors can see those data in real time. For example, if any patient will upload their blood pressure report, then doctor can see that report in no delay.
- Intelligent Operation: While suggesting for the doctors, IHS will also select the good doctors based on the previous successful work done by doctors and hospitals.

32.2.2 Challenges of Intelligent Healthcare System

There are several issues related with data management and data sharing in the proposed IHS. The challenges related to IHS are discussed below:

- Network Security: In network security, it has unprecedented network attack potential as IIoT infrastructure involves Internet and data transmission. The ownership and security of data are the main challenges of IHS, along with that bad performance of network security brings more privacy disclosing [13].
- Data Value: Another challenge is related to data value, generally, in IHS, there is more focus on how to collect data and how to process data but how to find the value of data is more important. The effectiveness and efficiency of IIoT are related to how to mine potential value of user information.
- Interconnection Protocol: The third challenge is related to interconnection protocol, the foundation of IHS, is IoT and it is not sufficient to connect the sensor with the equipment but also connected to the Internet. For the wireless device network with relatively short distance, there are several protocols competing locally, such as Bluetooth, Zigbee, and thread, which must face the interoperability problem. Along with, the protection of user privacy has been paid more and more attention.

32.3 Differential Privacy in IHS

Differential privacy (DP) is a strongly accepted model which ensures that a single record does not affect the result of analysis on a certain data set [8]. Hence, a patient's data privacy will not be affected after participation in data collection process as it does not make much difference in finding output. Dwork and Roth first proposed differential privacy model which approximates a deterministic function $f()$ and DB_1, DB_2 (two different databases) different at most in one record given in equation below.

$$\ln\left(\frac{pr[Q(DB_1) \in S]}{pr[Q(DB_2) \in S]}\right) \leq \forall S \subseteq \text{Range}(Q)$$

Another model is found in [9] for exploring expanding differential privacy. The main goal is to perform data analysis of the whole data set within disclosing the information of a sample. It is used in multiparty data publishing methods. There are four major properties of DP such as.

- Sequential synthesis applied when DP algorithm is composed by several algorithms.
- Parallel synthesis for multiple disjoint data set.
- Transformation invariance given that $S(.) = S_2(S_1(.))$.
- Convexity for both algorithms A_1 and A_2 satisfying DP.

Implementing DP by adding noise is one of the major techniques to provide privacy. Laplace mechanism is suitable for this purpose.

1. Laplace Mechanism: Δf is defined as $f : DB \Rightarrow R^{DB}$, where DB is data set, then the random algorithm $S(DB) = f(DB) + Z$ will provide differential privacy. $Z \rightarrow L(\Delta f/\epsilon)$ is the random noise, which follows the Laplace distribution with scale parameter $\Delta f/\epsilon$ [10].

$$S(DB) = f(DB) + \left(L_1\left(\frac{\Delta f}{\epsilon}\right) \cdot L_2\left(\frac{\Delta f}{\epsilon}\right) \ldots \ldots L_d\left(\frac{\Delta f}{\epsilon}\right)\right)^T$$

2. Index Mechanism: D is the input data set for the algorithm M, entity object $r \in Range$ will be the output, availability function is $q(DB, r)$, and sensitivity of function $q(DB, r)$ is defined by Δq. All the probable values can be normalized to get the corresponding probability value, if the algorithm S is proportional to the probability of $exp\left(\frac{\epsilon q(DB,r)}{2\Delta q}\right)$.

There are several challenges implementing DP in health care 4.0. They are explained below.

- Proper migrate the differential privacy algorithm from data publishing privacy is an important challenge.
- In real world for large public health data sets where disease counts are many with number of variables name, age, sex, location, etc.
- It is possible to release differentially query responses with small errors. But when the counts are low, the amount of adding noise significantly affects the results.
- Present IHS is supported by edge training model for fast processing. Here, the protection of sensitive data using differential privacy is a big challenge.

- In IHS, collaborative environment where the node co-operation is inevitable while sharing data. In this case implementing DP is a big problem in healthcare industry.

32.4 Working of the Proposed Framework

In this section, we have explained our proposed IHS using blockchain technology. This framework has several parts like patients, D-app, blockchain network, doctors, and hospitals. IoT sensor will collect the data and share data to D-app, then D-app will share the data to blockchain and then according to the request, blockchain will deal with that request. The working of our proposed framework is described in Fig. 32.2.

32.4.1 Working of Proposed Intelligent Healthcare System

In our proposed system, we have divided the entire framework in three phases. The first phase contains the patient data and D-app where we collect the data from IoT sensors. In the next phase, we have blockchain network with cloud that stores the

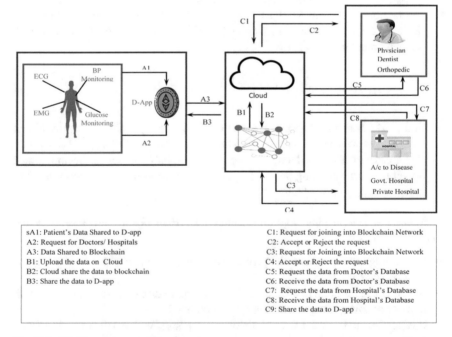

sA1: Patient's Data Shared to D-app
A2: Request for Doctors/ Hospitals
A3: Data Shared to Blockchain
B1: Upload the data on Cloud
B2: Cloud share the data to blockchain
B3: Share the data to D-app

C1: Request for joining into Blockchain Network
C2: Accept or Reject the request
C3: Request for Joining into Blockchain Network
C4: Accept or Reject the request
C5: Request the data from Doctor's Database
C6: Receive the data from Doctor's Database
C7: Request the data from Hospital's Database
C8: Receive the data from Hospital's Database
C9: Share the data to D-app

Fig. 32.2 Working of proposed framework

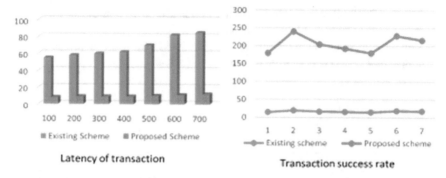

Fig. 32.3 Analysis and result of our proposed framework

data related to patient, doctors, and hospitals [11]. In the third phase, we have doctors and hospitals that is connected to blockchain network.

Phase 1:

Step1: The sensors collect the data from human body on the basis of sensors which is placed on human body.

Step2: After collecting the data from the patient, data is shared to D-app to know the requirements of patients.

Step 3: Now, D-app will share the data to blockchain network for getting the appropriate solution for the patient requirement.

Phase 2:

Step 1: Now, after getting the data from the D-app, blockchain will upload the data to cloud.

Step 2: Blockchain will share the data to D-app after completion of the request.

Phase 3:

Step 1: Doctors like physician and dentist will send the q request to blockchain network for joining network.

Step 2: After getting the request from the doctor, admin and hospital will verify the doctor's record and add or reject them into blockchain network.

Step 3: Hospitals, a/c to disease, govt. hospitals will send q request to blockchain network for joining the network.

Step 4: After getting the request from hospitals, admin will verify and add/reject in blockchain network.

Step 5: If patients want to connect with a special doctor, so based on the re-quest from patient, blockchain network requests for doctor's data.

Step 6: After getting the request, if that doctor will be in database, database will provide the doctors to blockchain network.

Step 7: If patients want to connect with a particular hospital, then based on patient's request, blockchain n/w will send request for hospitals data.

Step 8: After getting the request, if that hospital will be in database, database will provide the hospitals to blockchain network.

Step 9: Now, after executing all the request, blockchain will provide the data to D-app, and D-app will provide the service to patient.

32.4.2 Algorithm

Algorithm 1: Algorithm for Patients Health Record

1: **Procedure Input:** {(Create smart contract (C_I), Blockchain Network (B_I), Adding miner (M_I), Create Transaction (T_I), Patient data (P_I), Doctor's data (D_I), Hospital's data (H_I), Admin data (A_I))}.

2: **Output:** {Update patient (U_P), doctor's (U_D), hospital's data (U_H)}.

3: **Initialization:** Defining role for miners and non-miner for adding and verifying patient's, doctor's and hospital's data.

4: BEGIN: (M_I, C_I)

5: while true do.

6: Create transaction $(T_I) \leftarrow [P_I, D_I, H_I]$

7: Do this.

8: if $(A_I), (D_I)$ verifies P_I then.

9: Add (P_I) to (B_I).

10:     update (U_P).

11: else.

12: reject P_I

13: if $(A_I), (H_I)$ verifies (D_I) then.

14: Add (D_I) to (B_I).

15:      update U_D

16: else.

17: reject D_I

18: if (A_I) verifies (H_I) then.

19: Add (H_I) to (B_I).

20: update (U_H).

21: else.

22: reject (H_I).

23: if P_I request (B_I) then.

24: (B_I) sends $(D_I), H_I$ to P_I

25: Broadcast updated $(U_P), (U_D), (U_H)$.

32.5 Implementation and Results

We have implemented smart contract, blockchain technology using solidity and remix ide tool. We have taken help from GitHub smart contract using solidity for implementing of our model. The source code of our project comes under the Creative Common Attribution 4.0 International License (CC-BY-4.0), in which it can measure the different properties like throughput of transaction, latency of transaction, resource consumed by the source code. For our simulation work, we have used intel (R) core (TM), i5-8250U CPU @1.8GHZ, RAM is 8 GB (7.89 usable), 64-bit windows operating system, × 64-based processor, and 500 GB SSD. To implement our proposed model, we have added the 1000 records of the patients, doctors, and hospitals. The sharing rate of record is 100, 200, 300, ……, 1000 records per second. We have done the first experiment to determine the success rate of records, which defines the number of records added successfully into blockchain. The second experiment that we have done is how many requests are successfully serviced by the blockchain. In both of the implementation, we have considered the throughput, latency of the records adding, and request serviced in the blockchain network. So, by our experiment we have observed that the throughput is nearly 11.5 times and service time is 7 times less than the scheme [14].

32.6 Security Analysis of the Proposed Framework

This section describes how the framework protects the patient data from available threats. Detail description is given below:

- Unauthorized Access Prevention: The proposed framework is based on blockchain concepts where D-app is used to build the smart contract. The smart contract can be used to store patient medical record and provide restricted access. Every time the smart contract verifies the identities of the stakeholder and validates the information before storage.
- Protection from data tampering: In this framework, a smart contract is also used to prevent any hostile node from acting illegally. Any illogical activity by a stakeholder can be identified in smart contract and the immutability property helps to prevent data modification.
- Protection from data loss: In this framework, the distributed ledger distributes all blockchain information over numerous nodes. In case of data loss at any node, it may be simply recovered from other nodes.
- Providing data with trust: The data utilized in the blockchain network was confirmed by more than 51% of miners using the consensus method. As a result, reliable medical data have been preserved throughout the proposed system [12].
- Differential privacy protection: Patient records are very sensitive information and any leakage of record with identity disclosure may threat their lives. Hence in this framework, the original record table is altered with contingency table where simple noise has been added to prevent disclosure.

Table 32.1 Cost evaluation based on gas consumed and time taken

Contract	Gas consumed (gwei)	Time taken (sec)
addDoctor	74,138	0.00717142
removeDoctor	20,691	0.00547684
ClaimPaymentApproval	17,801	0.00547689
addPatient	24,228	0.00524243
subsequentPayment	20,108	0.00547684
dueAmountUser	23,801	0.0013333
addHospital	23,801	0.0020108
removeHospital	24,243	0.0054138
doctorCheckPatientDetailsForApproval	12,586	0.0000
patientApproval	133,332	0.0017801

32.7 Conclusion

This paper mainly highlights on many security risks and obstacles in the intelligent healthcare system. Because of its immutability, blockchain technology can basically provide a solution to the mentioned challenges. We presented a framework for data sharing platform where differential privacy is maintained. The suggested methodology secures data storage by utilizing blockchain technology. At the medical server, all data acquired from medical sensors are saved in the form of blockchain. As a result, the stored block of data is not editable by any attacker, and the data's privacy is preserved. It also improves the security and privacy of the present healthcare system and services.

References

1. Hathaliya, J.J., Tanwar, S.: An exhaustive survey on security and privacy issues in Healthcare 4.0. Comput. Commun. **153**, 311–335 (2020)
2. Shahnaz, A., Qamar, U., Khalid, A.: Using blockchain for electronic health records. IEEE Access **7**, 147782–147795 (2019)
3. Nakamoto, S.: Bitcoin: a peer-to-peer electronic cash system. Decentralized Bus. Rev. 21260 (2008)
4. Ahmad, S.S., Khan, S., Kamal, M.A.: What is blockchain technology and its significance in the current healthcare system? A brief insight. Curr. Pharm. Des. **25**(12), 1402–1408 (2019)
5. Gross, M.S., Miller Jr R.C. Ethical implementation of the learning healthcare system with blockchain technology. Blockchain Healthc. Today Forthcoming (2019)
6. Tripathi, G., Ahad M.A., Paiva, S.: S2HS-A blockchain based approach for smart healthcare system. Healthcare. vol. 8, no. 1. Elsevier, (2020)
7. Hussien, H.M., Yasin, S.M., Udzir, S.N., Zaidan, A.A., Zaidan, B.B.: A systematic review for enabling of develop a blockchain technology in healthcare application: taxonomy, substantially analysis, motivations, challenges, recommendations and future direction. J. Med. Syst. **43**(10), 1–35 (2019)

8. Jiang, B., Li, J., Yue, G., Song, H.: Differential privacy for industrial internet of things: opportunities, applications, and challenges. IEEE Int. Things J. **8**(13), 10430–10451 (2021)
9. Kumar, A., Krishnamurthi, R., Nayyar, A., Sharma, K., Grover, V.: A novel smart healthcare design, simulation, and implementation using healthcare 4.0 processes. IEEE Access **8**, 118433–118471 (2020)
10. Dwork, C., Roth, A.: The algorithmic foundations of differential privacy. Found. Trends Theoret. Comput. Sci. **9**(3–4), 211–407
11. Zhao, S., Li, S., Yao, Y.: Blockchain enabled industrial Internet of Things technology. IEEE Trans. Comput. Soc. Syst. **6**(6), 1442–1453 (2019)
12. Lu, Y.: Blockchain and federated learning for privacy-preserved data sharing in industrial IoT. IEEE Trans. Ind. Inf. **16**(6), 4177–4186 (2019)
13. Tariq, N., Qamar, A., Asim, M., Khan, FA.: Blockchain and smart healthcare security: a survey. Procedia Comput. Sci. **175**, 615–620 (2020)
14. Singh, A., Prabha, P., Chatterjee, K.: Security of IoT-based E-healthcare system: a blockchain solution. In: Artificial Intelligence and Sustainable Computing. Springer, Singapore, 227–237 (2022)

Chapter 33
Comparative Analysis of MQTT and CoAP Using Wireshark

Vanlalsiama Ralte and R. Chawngsangpuii

Abstract The computer age brings with it many devices which rely on the Internet for data transfer which increase the overall number of devices that needs to be connected to the World Wide Web. With the increase in the number of devices comes, the increase in the amount of data being passed and transmitted. For communication to take place between such IoT devices, different types of communication protocol exist. In this paper, we will compare and analyze the two most commonly used communication protocols, the Message Queue Telemetry Transport and the Constrained Application Protocol (CoAP). MQTT uses TCP/IP and CoAP uses UDP and both of which are application layer protocols. Both of which will be compared with the help of a special packet analyzer called Wireshark in which we will be able to determine which protocol performs better in terms of latency, throughput and bandwidth in low data traffic.

33.1 Introduction

This section explains the need and motivation in writing this paper.

33.1.1 Objective

As the Internet of Things (IoT) has grown in popularity through the years, the number of devices or things connected to the Internet has also grown to a great extent [1]. In many scenarios, it is beneficial and necessary to know which communication protocol is best suited for our implementation of IoT devices [2]. This experiment

V. Ralte (✉) · R. Chawngsangpuii
Department of Information Technology, Mizoram University, Aizawl, India
e-mail: siamralte@gmail.com

R. Chawngsangpuii
e-mail: sangpuii_77g@hotmail.com

V. Bhateja et al. (eds.), *Evolution in Computational Intelligence*, Smart Innovation, Systems and Technologies 326, https://doi.org/10.1007/978-981-19-7513-4_33

is done to show and compare the most commonly used protocol for communication between heterogeneous devices on the Internet.

33.1.2 Motivation

An IoT ecosystem consists of many heterogeneous devices which produce large amounts of data and information. These data or messages contain information or data which could be critical and requires the least amount of time to process. Many protocols such as MQTT, CoAP, AMQP, XMPP and DDS exist but we will be comparing only MQTT and CoAP as they are the most commonly used and most suitable protocol for lightweight/small devices and also constrained resources in terms of storage, power consumption and computing power. These two protocols have been used in many areas for different purposes. For example, MQTT has been used for devices like refrigerators, microwaves, televisions, vehicle, etc., for connecting to the Internet. CoAP has been used for applications ranging from smart cities, smart energy supply, intelligent lighting systems, product tracking, etc. These protocols helps us to take and store the monitored data from the connected devices in the network without any type of interaction between humans or interaction between human and computer interaction that can provide ways in improving our way of life, earnings or environments. Smart devices like fitness bands that can track heart rate, monitors pulse, track sleeping patterns and calculates as well as stores the amount of distance traveled or walked in a day. Some IoT devices use sensors to monitor water levels for smart homes, whereas some for monitoring crops by identifying the moisture of the soil, perform timely irrigation and identifying the types of diseases, etc. Effective and efficient IoT tools can enable us to use and control these devices from other remote places from our smart devices (phones, tablets). It is difficult to show any area in industries that does not gain from this IoT expansion. Shops and business owners can have easy access to data regarding inventory and sales without having to wait for periodic reports carried out monthly or yearly, they can also get accurate data and information regarding details of customers in real time. They can check the data to make more informative and reliable decisions, which can add value to their business.

33.1.3 Problem Statement

MQTT and CoAP have been analyzed and comparisons have been made between the communication protocols using different tools and methods. This paper will test and analyze the CoAP and MQTT protocol using Wireshark packet analyzer. Wireshark checks and analyzes packets and saves the network traffic on the device where the software is installed. The data can also be analyzed on the spot (real time) or may be analyzed from the saved data without the need for any Internet connectivity.

Implementation of CoAP protocol will be done in Cooja Contiki network emulator simulator with virtual machine workstation, and implementation of MQTT protocol using Paho Python MQTT client-subscribe.

33.2 Literature Study/Related Work

The authors in [3] compared the CoAP and MQTT protocols and using a network emulator which checks the throughput, inter-arrival time and latency using different scenarios which shows their strengths and weakness.

The authors in [4] presented and compared protocols such as CoAP, MQTT, IETF, IBM and HTML by checking security, reliability and energy consumption.

The authors in [5] presented a world-wide implementation of IoT using a central cloud. Cloud which is implemented using Aneka is also presented and the need for convergence of WSN is also discussed.

The authors in [6] proposed an adaptive transmission mechanism for social-oriented smart phone which could improve the throughput and network connectivity for use in IoT. The results they produced show the effectiveness of the proposed method.

The authors in [7] designed a small testbed that is used for collecting real-time environmental data for CoAP, MQTT and XMPP for testing their package creation and transmission time. In this test, CoAP and MQTT showed almost equal performance whereas XMPP had low performance as compared to the two.

The authors in [8] made comparisons on LoRaWAN, SigFox, RPMA and NB-IoT to test their features, effective range, bandwidth and applications.

The authors in [9] implemented two client applications based on CoAP and MQTT to test latency, bandwidth, energy consumption and resource usage. Here, CoAP had lower latency whereas MQTT had higher bandwidth.

The authors in [10] analyzed CoAP, MQTT and HTTP on response time, latency and throughput. They concluded that CoAP was the overall best and MQTT is better for small scale applications and HTTP is best for handling sensitive data.

The authors in [11] made performance evaluations of AMQP, CoAP and MQTT on throughput, packet loss and message size. The results show that CoAP gave the best result.

The authors in [12] focused on E-health systems using CoAP and MQTT where tests range from heartbeat, blood pressure, ECG and patient temperature. The result showed that MQT was better than CoAP.

33.3 Design and Analysis

The section explains how the experiment was done on two protocols namely MQTT and CoAP communication protocols.

33.3.1 MQTT

The Message Query Telemetry Transport protocol is used for IoT devices and is used for interaction and exchange of data between two or more devices [13]. The publish and subscribe mode is used for this protocol in which a broker provides or publishes data or message to a client if it is subscribed to a particular topic. Here, the broker acts as an intermediary between devices and splits the transmitted data into labels before sending the information [14].

The publish-subscribe model helps in the interaction of more than one clients with other clients using a broker and without the need for them to directly interconnect.

The broker receives messages regarding multiple topics that are published by multiple clients. The broker is a type of server that acts as a middleman which receives the data and information and distinguishes them according to their topics. The broker then transmits these received messages to the clients who have subscribed to the particular topics [15].

This experiment is implemented by using Paho Python. The Paho Python client has a class for client that supports two versions of the MQTT protocol namely version 5 and version 3. It includes functions called helper functions which allow publishing a one-time message to an MQTT server easy and less complicated. Here, we create a simple experiment to subscribe a topic called temperature so as to get random temperature values which would be given by a normal sensor as shown in Figs. 33.1 and 33.2. The test result is then captured with the use of Wireshark. It is a tool for checking and saving data packets. It stores the flow of traffic of the network on the device using the Wireshark tool and allows users to analyze the data real-time or by storing it for use when no Internet connectivity is available. Figure 33.3 shows the output of the experiment to subscribe a topic called temperature which would be equivalent to values given by a normal sensor. Wireshark helps in filtering the logs before saving the data or at the time of analyzing the packet so as to ease the network tracing capability [16].

33.3.2 CoAP

A client–server protocol is the basis of the Constrained Application Protocol. The CoAP protocol allows packet transferred using CoAP to be viewed or seen by multiple client nodes which are linked to a particular CoAP server [9]. Here, the task of the server is to transfer the information to the clients depending upon their logic which does not allow them to reply to the server. State transfer model supported applications can use this.

Fig. 33.1 Subscriber code and output using Paho Python

The CoAP was implemented with the use of Cooja Contiki network emulator simulator as shown in Fig. 33.4. Cooja is a network simulator used with Contiki. Cooja enables motes or notes of Contiki to be simulated regardless of the size of the network. It is specifically designed to simulate wireless sensor networks and is most widely used to simulate IoT network applications virtual machine workstation was also used for this experiment. These are software used by virtual machines that can run on 32-bit and 64-bit computers to enable more than one operating system to run on a specific non-virtual host computer. A virtual machine is able to run one instance of the desired type of operating system at a particular time which may be a Microsoft Windows, Ubuntu, Linux, etc. The testing is done with the help of a Chrome browser extension called Copper4Cr that helps us to interact with the IoT world as shown in Figs. 33.5 and 33.6. The results are then analyzed using Wireshark.

Fig. 33.2 Publisher code and output using Paho Python

33.4 Flowchart

See Fig. 33.7.

33.5 Implementation and Result

This section explains the results and analysis done and obtained from the experiments done on the two protocols to check latency, bandwidth and throughput.

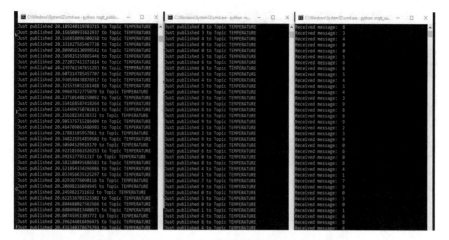

Fig. 33.3 Experiment generated to subscribe a topic called temperature

Fig. 33.4 Creation of server and client nodes using motes in cooja—the contiki network emulator and VMW are workstation

33.5.1 Implementation

Figure 33.8 shows the round trip time (RTT) of the packets. Here, the Y-axis determines the round trip time measured in milliseconds and the X-axis determines the time measured in seconds. Latency can be said to have similarities with round trip time where latency can be defined as the period between the transmissions of a packet from sender to receiver, whereas RTT can be defined as the time taken for a packet to go from the sender's end to the receiver's end and vice versa. [10]. Here, we should keep in mind that latency is not exactly half the time taken of the round trip time as delay could be asymmetric between the sender and the receiver.

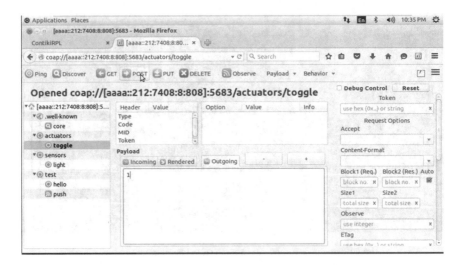

Fig. 33.5 Performing methods of CoAP protocols using Copper4Cr extension

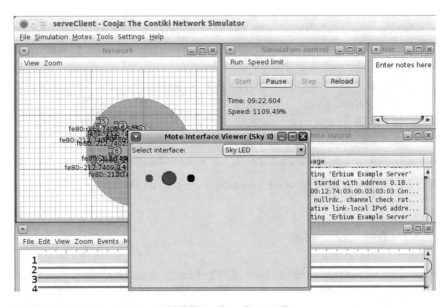

Fig. 33.6 Controlling actuators on LED lights from Copper4Cr

Figure 33.9 shows the rate at which packets are transmitted each second which would be used to determine the throughput. Here, the Y-axis determines the packet transferred per 1 s and the X-axis determines the time measured in seconds. Wireshark IO graphs display the overall traffic which is measured in bytes, i.e., rate per second or packets. By default the x-axis indicated the tick interval/sec, and y-axis indicated the packets for every tick (per second). In simple terms, throughput is the measure

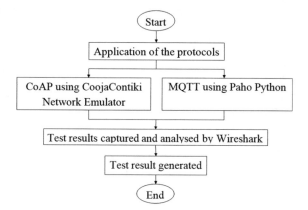

Fig. 33.7 Flowchart of the system

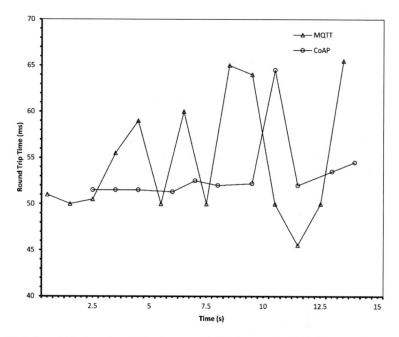

Fig. 33.8 Round trip time determining latency on MQTT protocol and CoAP protocol

of the speed in which the data on your network travels which determines how much information is actually delivered in a in a given time period. By comparison, if bandwidth is the total amount of data/information, then throughput would determine how much of that data makes it to the end point. [17].

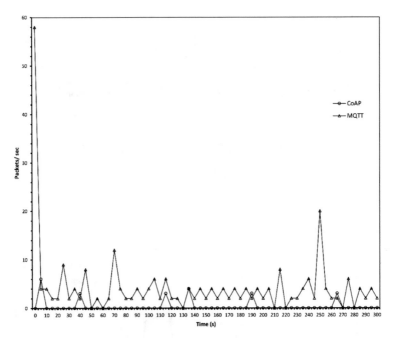

Fig. 33.9 I/O graph determining throughput on MQTT protocol and CoAP protocol

Figure 33.10 shows the statistics including the bandwidth which is measured in bits per second and expressed as bitrate. Bandwidth is measured as the transferrable amount of data in a given network from sender's to receiver's endpoint within a specified time [18].

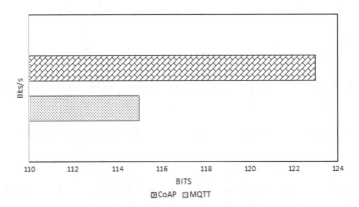

Fig. 33.10 Protocol hierarchy statistics for MQTT protocol and CoAP protocol showing bandwidth in Bits/s

33.5.2 Comparison Results

See Figs. 33.8, 33.9 and 33.10.

33.6 Implementation and Result

33.6.1 Conclusion

As from the experiment conducted on MQTT and CoAP on latency, throughput and bandwidth, we can come to conclude that MQTT has less delay as compared to CoAP when the message size is small. In terms of bandwidth, MQTT has a higher bandwidth but CoAP provides a much better bandwidth usage and utilization. When it comes to throughput, CoAP still has better throughput as compared to MQTT.

33.6.2 Future Work

Based on the test scenario and the experiments done, we can conclude that the experiments and comparisons can also be extended by including relevant communication protocols like AMQP, XMPP, HTTP and DDS to the experiment. Other works may include a more in-depth analysis of the protocols by using more advanced tools and real-life implementation of the IoT protocols to get a much more accurate view of the results. From reviewing other research papers, we understood that by using different test beds and tools for different scenarios for the different individual protocols, we may get different outcomes that may be slightly different from each other to get more desirable results.

References

1. Moraes, T., Nogueira, B., Lira, V., Tavares, E.: Performance comparisom of IoT communication protocols. In: IEEE International Conference on Systems, Man and Cybernetics (SMC) (2019)
2. Pohl, M., Kubela, J., Bosse, S., Turowski, K., Performance evaluation of application layer protocols for Internet-of-Things. In: 6th International Conference on Enterprise Systems. IEEE (2018). https://doi.org/10.1109/ES.2018.00035
3. Ouakasse, F., Rakrak, S.: A comparative Study of MQTT and CoAP applicaiotn layer protocols via. Performance evaluation. J. Eng. Appl. Sci. **13**(15), 6053–6061. ISSN: 1816-949X (2018)
4. Karagiannis, V., Chatzimisios, P., Vazquez-Gallego, F., Alonso-Zarate, J.: 'A survey on application layer protocols for the Internet of Things.' Trans. IoT Cloud Comput. **3**(1), 11–17 (2015)

5. Gubbi, J., Buyya, R., Marusic, S., Palaniswami, M.: Internet of Things (IoT): A vision, architectural elements, and future directions. Future Gener. Comput. Syst. **29**(7), 1645–1660. https://doi.org/10.1016/j.future.2013.01.010 (2013)
6. Ning, Z., Xia, F., Hu, X., Chen, Z., Obaidat, M.S., Social-oriented adaptive transmission in opportunistic internet of smartphones. In: IEEE Trans. Ind. Informat. **13**(2), 810–820. https://doi.org/10.1109/TII.2016.2635081 (2016)
7. Okay, F.Y., Ozdemir, S.: Comparative Analysis of IoT Communication Protocols. IEEE (2018). https://doi.org/10.1109/ISNCC.2018.8530963
8. Chaudhary, H., Tank, B., Patel, H.: Comparative analysis of Internet of Things (IoT) based low power wireless technologies. https://doi.org/10.17577/IJERTV7ISO10001 (2018)
9. Hamdani, S., Sbeyti, H.: A Comparative study of CoAP and MQTT communication protocols. IEEE (2019). 978-1-7281-2827-6/19/ 31.00
10. Sasirekha, S., Swamynathan, S., Chandini, S., Keerthana, K.: Analysis of application layer protocols in internet of things. 550–561. Springer Nature Singapore Pte Ltd., ICACDS 2016, CCIS 721 (2017). https://doi.org/10.1007/978-981-10-5427-3_56
11. Costa e Silva Nogueira B.: Performance comparison of IoT communication protocols. IEEE (2019). https://doi.org/10.1109/SMC.2019.8914552
12. Zorkany, M., Fahmy, K., Yahya, A.: Performance evaluation of IoT messaging protocol implementation for E-health systems. IJACSA **10**(11) (2019)
13. Okay, F.Y., Ozdemir, S.: Comparative analysis of IoT communication protocols. IEEE (2018). 978-15386-3779-1/18/31.00
14. Karagiannis, V., Chatzimisios, P., Vazquez Gallego, F., Alonso-Zarate, J.: A survey on application layer protocols for the internet of things. Centre Technologic de Telecomunicacions de Catalunya (CTTC), Spain
15. Nastase, L., Sandu, I.E., Popescu, N.: An experimental evaluation of application layer protocols for the internet of things. Stud. Inform. Control **26**(4), 403–312 (2017). ISSN: 1220-1766, eISSN: 1841-429X
16. Zorkany, M., Fahmy, K., Yahya, A.: Performance evaluation of IoT messaging protocol implementation for E-health systems. (IJACSA) Int. J. Adv. Comput. Sci. Appl. **10**(11) (2019)
17. Sultana, T., Wahid, K.A.: Choice of application layer protocols for next generation video surveillance using internet of video things. IEEE **7**, 2169–3536 (2019)
18. Yu, S.Y., Browniee, N., Mahanti, A.: Comparative performance analysis of high-speed transfer protocols for big data. In: 38th Conference on Local Computer Networks. LCN (2013). https://doi.org/10.1109/LCN/2013.6761252

Chapter 34
A Survey of Energy-Aware Server Consolidation in Cloud Computing

Sakshi Sagar, Anita Choudhary, Md. Sarfaraj Alam Ansari, and Mahesh Chandra Govil

Abstract In Cloud Data Centers (CDC), dynamic consolidation of Virtual Machines (VMs) is an efficient technique to increase resource usage and energy efficiency. Energy efficiency has become a new frontier in the virtualized cloud computing paradigm. Data centers must decide how, which, and when VMs must be merged onto a few physical servers, which is a difficult task. Server consolidation necessitates VMs migration, which has a direct influence on service response time. It means that overloading of server results in resource scarcity and application performance deterioration. Most of the existing solutions for VMs consolidation are based on the heuristic methods. The drawbacks of these methods are they produce sub-optimal outcomes and prevent the precise description of a Quality of Service (QoS) target. Further, the requirement for energy efficiency has grown as the number and scale of CDC has grown. Virtual machine migration is widely utilized cloud computing technology that is thus focused in this work to conserve energy. We conduct an in-depth study of the existing literature on that address server consolidation and energy consumption issues and provide optimal solutions. We conclude the survey by describing key challenges for future research on constructing effective and accurate server consolidation techniques with power modeling.

34.1 Introduction

After the evolution of cloud, the large number of applications is exported on the cloud environment. The continuous increasing demand for cloud resources has brought about the foundation of small to large-scale CDCs in which thousands of servers are installed to provide the guaranteed services over the Internet. But there is a issue

S. Sagar · Md. S. A. Ansari (✉) · M. C. Govil
National Institute of Technology Sikkim, Ravangla, India
e-mail: sarfaraj@nitsikkim.ac.in

S. Sagar
e-mail: b180057@nitsikkim.ac.in

A. Choudhary
Engineering College Ajmer, Kiranipura, Rajasthan, India

© The Author(s), under exclusive license to Springer Nature Singapore Pte Ltd. 2023
V. Bhateja et al. (eds.), *Evolution in Computational Intelligence*, Smart Innovation, Systems and Technologies 326, https://doi.org/10.1007/978-981-19-7513-4_34

with large-scale CDCs that these CDCs consume a huge amount of electricity to keep running of servers and cooling of equipment [1]. Due to continuous increasing cloud applications, it becomes difficult to optimize utilization of cloud resources while satisfying the Service Level Agreement (SLA) [2]. SLA is an agreement between service providers and consumers to ensure the delivery of QoS. SLA considers service pricing, obligations, and penalties in case of violations in the agreed services. In 2013, it was estimated that approximately 260 million Watts of electricity used up by Google data centers that much amount of electricity can give power to more than 200,000 homes continuously [3, 4]. In 2014, about 75% of cost had been contributed by infrastructure and energy whereas IT had contributed only 25% of cost to the overall operating cost of CDC [5]. Virtualization uses VM migration to scale a CDC by relocating VMs within and across CDCs to fulfill different resource management goals such as fault tolerance, power saving, maintenance of server provisioning, and load balancing.

Within CDC, inefficient resource management practices are underutilized system resources. To provide availability of high service and application QoS, CDCs are typically overprovisioned. To deliver guaranteed services, resources are overprovisioned, and in a typical data center, 30% of servers run without any load or in dead mode and their utilization is nearly 5–10% that leads to consumption of considerable energy without being utilized efficiently [6]. Dynamic consolidation of VMs is one way for improving data center resource consumption that has been proven to be effective [7, 8]. The VMs are periodically reassigned on the basis of present resource demand using live migration concerning to reduce the total of active physical servers, named as hosts, needed to service the workload. To decrease static power and reduce total energy consumption, idle hosts are converted to low-power modes with quick transition times. When resource demand increases, the hosts are restarted. This strategy has two main goals: reducing energy consumption and improving the system's QoS, which are combined to form an energy-performance trade-off.

Data center energy consumption can be divided into two categories: computing and cooling, with the latter accounting for 43% of total energy consumption [9]. On the other hand, server compute energy use of underutilized resources represent a large portion of total energy consumption, especially in cloud systems [10]. As a result, server consolidation is frequently utilized to improve computing energy efficiency [11] and functionalities by consolidating workloads onto fewer servers and turning off inactive servers (or set them into sleep form). Workload balancing, on the other hand, becomes a significant issue [12], to reduce cooling energy draw due to the skewed temperature distribution (i.e., cooling must target the hottest server). This is accomplished by uniformly dividing workload among servers concerning to decrease the greatest temperature and eliminate hot spots. From the standpoint of workload scheduling, the study yields two opposing goals: Although load consolidation seeks to reduce the total of active servers in order to conserve computing energy, consolidation onto fewer servers can evolve in hot spots, which necessitates more cooling energy. Although load balancing seeks to minimize hot spots, more active servers functioning at reduced utilization can waste processing power. Our major contributions in this paper can be summarized as follows:

1. We perform a comprehensive review of state-of-the-art server consolidation and resource allocation techniques and pointing out the strengths, weaknesses, and critical issues that require further research.
2. We briefly discuss the QoS parameters that considered for optimization.
3. Specific gaps are identified, and the research challenges are highlighted which improve the performance of server consolidation.

Our paper deals all these issues, and the remainder of this work is arranged in the following manner. Section 34.2 presents the state-of-the-art work. Section 34.3 presents the QoS parameters. Challenges in VM migration is briefly described in Sect. 34.4. Lastly, Sect. 34.5 brings the report to a close by outlining future research directions.

34.2 Background

As surveyed in [8, 13], a lot of studies have been focused in the field of cloud computing to minimize data center power usage. Many alternative strategies have been developed and created to reduce power waste, both with and without VM placement. Live VM migration has been studied extensively for a long time, and several strategies for migrating a live VM from one active node to another active node have been devised. It has been shown to be a useful strategy for controlling data center energy efficiently. The majority of current VM migration strategies for management of energy in CDC are complex, as they typically require VM consolidation or VM placement procedures at a greater level of implementation.

34.2.1 Virtual Machine Migration Approaches

Virtualization allows you to move VMs from one physical host (source) to another (destination). Virtual machine migration is a vital technique for data center and cluster administrators since it provides for a clean segregation of hardware and software. By moving a virtual machine, you can prevent process-level migration issues. Load balancing, energy savings, and effective resource consumption are all possible with virtual machine migration. There are two types of virtual machine migration methods: hot (live) migration and cold (non-live) migration, while moving, the virtual machine remains operational and does not change its state. In a hot (live) migration, the user does not experience any service disruption. During the VM memory transfer procedure, a hot VM migration sequence ensures on-going service availability to the hosting environment [14–16]. In cold migration, the VM's state is lost, and the user can observe the service disruption. The status of a virtual machine to deploy is transferred during the live migration procedure. The local file system and memory contents make up the state. It is not necessary to transfer the local file system. The virtual machine is first paused, then its state is transmitted, and finally, the virtual machine is restarted at the target host.

Sapuntzakis et al. [14] showed how to transfer the state of a functioning node on the network quickly, including memory, disks state, and CPU registers. A capsule is a physical state that contains the whole operating system and current programs. They have implemented methods for reducing the total amount of data delivered over the network. The copy on write disks keeps track of only the modifications to capsule drives, demand paging, and unused memory obtains only the blocks that are required, and hashing prevents transferring blocks that already present at the receiver side. Clark et al. [17] were the first to suggest the basic concept of a live migration method. After marking all pages as dirty, the Hypervisor uses an algorithm to iteratively transmit filthy pages over the network till the volume of material left to transfer is below a specific limit or a maximum number of iterations is reached. The transferred pages are then marked as clean by Hypervisor. Because the VM functions during live migration, previously transferred memory pages may get dirty during iteration, necessitating re-transfer. At a certain point on the source, the VM is halted to prevent additional memory writes and move any remaining pages. The VM restarts at the destination server after transmitting all of the memory contents. Luo et al. [18] proposed the whole live migration technique that transmits the virtual machine's entire run-time state, including memory data, CPU state, and local disk storage. They devised a three-phase migration technique and also an incremental migration strategy for migrating the virtual machine directly to the main machine in a very short period. Bradford et al. [19] described a solution for facilitating transparency, live broad area relocation of virtual machines with durable state stored locally. This method is visible to the migrated VM, and it does not disrupt open network connections to and from the VM during wide area connectivity. Migration, this, ensures that the VM's local durable state is consistent at both the source and destination after migration, and it can manage workloads that require a lot of writing.

Migration of VMs in real time is possible. Methods are categorized as either pre-copy or post-copy. Pre-copy [17, 20] replicates the original memory data to the target server first and then sends newly dirty pages in cycles repeatedly. The VM on the server side is closed down when the iterations reaches the specified levels. To restart the VM on the target server, the residual data is replicated. For example, the dismissal conditions for ending the recursive phase can be established as follows: (1) the transfer round attains a specified value; (2) the residual size of data is lesser than a specified value; (3) the proportion between the size of data transfer and the assigned size of storage space is greater than a specified value, and so on. Because various workloads have varied memory muddying characteristics, no one stopping criterion is appropriate for all VMs. As a result, multiple termination criteria are usually mixed to provide a complete strategy that can respond to as many VMs as possible. Because of its reliability, pre-copy is widespread used in prevailing VMMs/hypervisors including such VMware [15], Xen [21], and KVM [22]. At minimum, one site (the origin site) has all the migrated VM's data during migration. When a migration breaks halfway through (e.g., owing to a network outage) or the VM crashes to restart on the target server, the VM can keep running on the origin server without losing data. The data of the relocated VM is released by the origin server until it is properly restarted on the target server. It does, however, have two major issues. (1) The repetitive duplicating procedure not only gene-rates a lot of network traffic, but it also takes a lot of time to migrate. (2) The migration procedure cannot be completed when the memory tainting speed is greater than the total network bandwidth for VM migration. It therefore makes no logical sense to use repetitive copying to reduce the amount of data left on the sender side. This condition is known as the migration convergence issue, and it causes a significant amount of migration disruption or network congestion.

34.2.2   Server Consolidation Approaches

Server consolidation, a broad phenomenon in the data center sector, is a technology that data center owners are employing to save a significant amount of money and capital. This method implements and gene-rates newly configured VMs on existing server, allowing them to better utilize their processing capabilities while using less power. Server virtualization maximizes hardware efficiency by grouping more applications and services upon lesser hardware, allowing applications and services to securely coexisting on the same physical servers with many operating systems running at the same time in the cloud.

A huge scale CDC used a huge quantity of electricity, which resulted in a hefty operating expense. The author [23] presented a resource management system that employs virtualization to minimize consumption of energy and deliver QoS. Because application performance degrades as workload demands change, the authors [24] employed the virtualization approach to consolidate servers and handle the condition that influences application performance. The authors' concept of a set predefined threshold restricts resource use to a maximum. The important VM performance indicators are examined. In relation to these important indicators, the necessary consolidation is carried out in order to maximize performance. If a resource exceeds a predetermined threshold, the system may be forced to shift a VM to another host, resulting in a SLA breach. Authors have developed two types of load balancing algorithms, push and pull, for the identical load balancing challenge [25], which execute essential VM migrations. When the load of host is high, the VM migrates to the push technique, which means it works in moderate to high load conditions. Pull technique, on the contrary, is used when the load of host is low or underutilized, and it performs VM migration, which means it operates under medium to low load conditions. The CPU utilization was matched to a predetermined particular threshold value in the host over- and underutilization detections.

34.2.3   Resource Allocation

In cloud computing, resource allocation entails scheduling and resource provisioning while taking into account SLA, available infrastructure, cost, and energy variables. The resource allocation strategy (RAS) is concerned with incorporating cloud
provider tasks in order to use and allocate resources within the confines of the cloud
environment in order to cater the requirements of the cloud application. The order
and timing of resource allocation are also factors in determining an appropriate RAS.
The following criteria should be avoided by an ideal RAS:

1. When two applications try to access the same resource at the same time, resource
   contention occurs.
2. Scarcity of resources occurs when resources are scarce.
3. When resources are isolated, a resource fragmentation problem occurs. (There
   will be sufficient resources, but they will not be capable to be allocated to the
   required application.)
4. Overprovisioning of resources occurs when an application receives more resources
   than it requires.
5. When the application is given limited resources than the demand, this is known
   as under-provisioning of resources.

Authors [26] utilize the idea of "skewness" to examine the imbalance of server
usage in order to dynamically allocate resources on the basis of application demands
and to assist green computing by keeping an optimal number of servers in operation (CPU, memory, etc.). By reducing skewness, different forms of workloads
can be mixed more effectively, and the total use of host resources rises. Mitigation for hot spot selects the VM whose removal will reduce the host's skewness. In
other work, [27] tackled another challenge of VM placement to hosts in an energy-
efficient manner, proposing various ways of VM to acceptable host mapping. They
have offered three different VM selection policies for this: "migration minimization
policy," "highest potential growth policy," and "random choice policy." While coping
with migration author evaluated that CPU utilization threshold change continually
due to regularly changing workload therefore in their subsequent study authors [28]
have proposed two sort of adaptive techniques: Median Absolute Deviation (MAD)
and Inter Quartile Range (IQR) (IQR). These strategies are used to identify host
overload because if the current host utilization crosses the higher threshold the host
is deemed overloaded. The adjustable threshold notion outperforms the static threshold theory in a dynamic cloud environment. The authors have also developed various
approaches for predicting future load, such as Local Regression (LR) and Robust
Local Regression (RLR). When the only projected value is greater than 100%, the
host is regarded over-utilized using these approaches. The authors of [27] introduced
the Modified Best Fit Decreasing (MBFD) algorithm for VM placement. It is a type of
bin packing method that arranges VMs in decreasing sequence of CPU consumption
and then assigns them to the host with the smallest power spike.

34.2.4   Energy Consumption

In cloud setups, servers take a higher portion of the energy, and their energy consumption changes depending on usage. The inter-relationships among resource utilization, workloads performance, and energy consumption are studied by Srikantaiah et al. [29]. Authors treat the consolidation problem as a modified bin packing problem and outline the research challenges for finding an effective solution. To reduce energy consumption and provide QoS, Beloglazov and Buyya [23] proposed a resource management system that uses the concept of virtualization. Dynamic VM migrations are used to optimize the resource utilization and service performed by avoiding server overload situations. The consolidation problem can be treated as a modified bin packing problem and outlines the research challenges for finding an effective solution.

For reduction in energy consumption within suitable performance bounds, [30] authors used DVFS and dynamic network shutdown (DNS) approach and proposed an energy-aware resource-efficient workflow scheduling (EARES-D) algorithm. This approach minimizes the energy consumption, energy cost and CO_2 emission while increases the providers profit. They also consider some of the QoS parameters like response time, energy cost, cloud provider profit, and resource utilization.

For overload and underload detection, fuzzy logic is used by Salimian et al. [31] and adaptive fuzzy threshold-based algorithm is proposed. They provided the solution of VM consolidation sub-problem. In a comparison of above works, our work includes the fuzzy logic for VM selection with four linguistic variables of VM like RAM utilization, correlation coefficient, the standard deviation of CPU utilization, and SLA violation for taking the most feasible decision.

34.3   QoS Parameters

QoS relates to a network's capacity to achieve maximum bandwidth while also managing other network factors such as uptime, latency, and error rate. The control of various network resources by providing priorities to specific types of data (video, file, and audio) is included in QoS. Three essential components which are required for a basic implementation of QoS while dealing with VM migration are:

1. QoS within a single network node.
2. Controlling end-to-end traffic throughout the network with QoS policy and management functions.
3. Finding strategies for managing QoS between network parts from node to node.

There are numerous methods for providing high-quality service to cloud applications. Some of the approaches used to accomplish this goal include admission control, scheduling, dynamic resource supply, service provisioning which are:

1. Scheduling: Scheduling addresses issues that arise as a result of service provision between the service provider and the consumer.
2. Control of Admission: The primary goal of admission control is to ensure high performance. At admission control time, the Infrastructural Provider (IP) must take into account the additional requirements, as well as the basic computing and networking requirements, that may need to be introduced to run time to make it more adaptable. When compared to standard standards, these flexible requirements might be quite considerable in many circumstances.
3. Resource allocation: The process of assigning resources available to the cloud application is known as dynamic resource provisioning. If resource allocation is not handled correctly, services will suffer the consequences.
4. Average Packet Data Throughput: This is one of the most essential parameters from the perspective of the client, and it is described as the rate at which packets are delivered in a network.
5. Latency: Latency is a term that relates to network problems, such as congestion, that might affect the total amount of time it takes to travel.
6. Drop Rate: It is defined as the ratio of aberrant disconnects to all disconnects (both normal and pathological) and assesses the network's incapacity to sustain a connection.

34.4 Challenges

Despite the fact that cloud computing has been adopted by the industries, enterprises, and academias, the examination on cloud is still at its beginning age. Many existing issues have not been completely considered in on-going works, while new challenges continue rising up out of industry applications.

1. QoS-based resource selection and provisioning: Resource provisioning and scheduling are key challenging issues in cloud computing because cloud resources are provisioned and released on-demand and the goal of a service provider is to efficiently use cloud resources and ensure the SLA.
2. Server consolidation: It is considered as a better solution to deal with both host overload and underload problems. Any solutions developed for server consolidation have to provide answers to three research questions—(i) when to migrate VM(s)? (ii) which VM(s) to migrate? (iii) where to migrate VM(s)?
3. VM migration: The advantage of VM migration is to overcoming overload and underload problems. But overload detection and initiating of VM migration lacks the agility to handle sudden changes in workload. Additionally, the in-memory state ought to be exchanged reliably and effectively.
4. Energy-aware dynamic resource allocation: Energy efficiency is another challenging issue in cloud computing because inefficient use of cloud resources leads to needless energy consumption.

5. Traffic management: Examination of data transfer and controlling is important tasks at data centers.
6. Resources availability: It affects the performance of the migration and total migration time. It can also help to make a better decision, such as when to migrate VM and how to deal with server resource allocation servers.

34.5 Conclusion and Future Work

In recent years, wide adoption of ICT and exponential growth of Internet users have significant contribution in increases of world energy consumption [32], and impact of digital economy is expected to increase more over next year's [32, 33]. The large-scale CDC consumes a large amount of electrical energy for operations and cooling purposes. It also emits a large amount of CO_2. Electricity consumed by CDCs globally is approximate 1.1–1.5% of total electricity consumption [4]; thus, CDCs are known as energy hungry. Other than the economic issues, increase in energy consumption has huge negative impact on environment and has become a serious concern for both industries and governments. In this paper, we have discussed about server consolidation and related concepts to it. VM migration is key for server consolidation. Virtualization is among the most important approaches used in CDCs to utilize resources effectively. However, it has hot-spot and cold-spot issues, which might detract from its strengths. To alleviate these issues, hosted VMs should be moved from overloaded or underloaded servers to even more suitable server. As a result, in CDCs, the performance of the VM migration technique is critical. Prediction VM migrations attempt to carry out more efficient migrations depending on the present condition of cloud resources and the expected state of cloud resources. Different prediction VM migration strategies have been suggested in the literature to improve the migration technique in CDCs, such as limiting the number of VM migrations, reducing migration time by enhancing different kinds of migration processes, or restricting shifting pages that will be dirty in the coming days by predicting future workloads of VMs.

VM migration techniques confront a wide range of issues. Because of the diversity of cloud resources, the unexpected nature of workloads, the severity of SLA violations, and system workload, VM memory size and VM migration schemes must be robust, resource-aware, and algorithmically cheap. Huge VM memory lengthens migration times and impacts service availability. Application performance is improved by integrating optimization approaches such as memory content reduction, fine-grained redundancy, and variable write throttle measurement. Lastly, security is a critical concern during the VM migration process, which can be mitigated by (1) prohibiting affected entities from accessing VMM, (2) separating VM boundaries, and (3) safeguarding network connections.

Therefore, in order to efficiently utilize cloud resource, a novel algorithms and approaches needs to be designed and developed that efficiently handle energy issues at CDC.

References

1. Luo, J., Rao, L., Liu, X.: Eco-IDC: trade delay for energy cost with service delay guarantee for internet data centers. In: 2012 IEEE International Conference on Cluster Computing, pp. 45–53. IEEE (2012)
2. Kaushar, H., Ricchariya, P., Motwani, A.: Comparison of SLA based energy efficient dynamic virtual machine consolidation algorithms. Int. J. Comput. Appl. **102**(16), 31–36 (2014)
3. Server, S.: Storage servers (2013)
4. Koomey, J.: Growth in Data Center Electricity Use 2005 to 2010. A Report by Analytical Press, Completed at the Request of the New York Times, vol. 9, p. 161 (2011)
5. Belady, C.L.: In the data center, power and cooling costs more than the it equipment it supports. Electron. Cool. **13**(1), 24 (2007)
6. Uddin, M., Shah, A., Alsaqour, R., Memon, J.: Measuring efficiency of tier level data centers to implement green energy efficient data centers. Middle-East J. Sci. Res. **15**(2), 200–207 (2013)
7. Nathuji, R., Schwan, K.: Virtualpower: coordinated power management in virtualized enterprise systems. ACM SIGOPS Oper. Syst. Rev. **41**(6), 265–278 (2007)
8. Beloglazov, A., Buyya, R., Lee, Y.C., Zomaya, A.: A taxonomy and survey of energy-efficient data centers and cloud computing systems. In: Advances in Computers, vol. 82, pp. 47–111. Elsevier (2011)
9. Barroso, L.A., Clidaras, J., Hölzle, U.: The datacenter as a computer: an introduction to the design of warehouse-scale machines. Synth. Lect. Comput. Archit. **8**(3), 1–154 (2013)
10. Lee, Y.C., Zomaya, A.Y.: Energy efficient utilization of resources in cloud computing systems. J. Supercomput. **60**(2), 268–280 (2012)
11. Murtazaev, A., Oh, S.: Sercon: server consolidation algorithm using live migration of virtual machines for green computing. IETE Tech. Rev. **28**(3), 212–231 (2011)
12. Zhou, R., Wang, Z., Bash, C.E., McReynolds, A.: Data center cooling management and analysis—a model based approach. In: 2012 28th Annual IEEE Semiconductor Thermal Measurement and Management Symposium (SEMI-THERM), pp. 98–103. IEEE (2012)
13. Kaur, T., Chana, I.: Energy efficiency techniques in cloud computing: a survey and taxonomy. ACM Comput. Surv. (CSUR) **48**(2), 1–46 (2015)
14. Sapuntzakis, C.P., Chandra, R., Pfaff, B., Chow, J., Lam, M.S., Rosenblum, M.: Optimizing the migration of virtual computers. In: 5th Symposium on Operating Systems Design and Implementation (OSDI 02) (2002)
15. Choudhary, A., Govil, M.C., Singh, G., Awasthi, L.K., Pilli, E.S., Kapil, D.: A critical survey of live virtual machine migration techniques. J. Cloud Comput. **6**(1), 1–41 (2017)
16. Huang, D., Ye, D., He, Q., Chen, J., Ye, K. (2011) Virt-LM: a benchmark for live migration of virtual machine. In Proceedings of the 2nd ACM/SPEC International Conference on Performance Engineering, pp. 307–316
17. Clark, C., Fraser, K., Hand, S., Hansen, J.G., Jul, E., Limpach, C., Pratt, I., Warfield, A.: Live migration of virtual machines In: Proceedings of the 2nd Conference on Symposium on Networked Systems Design and Implementation, vol. 2, pp. 273–286. USENIX Association (2005)
18. Luo, Y., Zhang, B., Wang, X., Wang, Z., Sun, Y., Chen, H.: Live and incremental whole-system migration of virtual machines using block-bitmap. In: IEEE International Conference on Cluster Computing, pp. 99–106 (2008)
19. Bradford, R., Kotsovinos, E., Feldmann, A., Schiöberg, H.: Live wide-area migration of virtual machines including local persistent state. In: Proceedings of the 3rd International Conference on Virtual Execution Environments, pp. 169–179 (2007)
20. Nelson, M., Lim, B.H., Hutchins, G.:. Fast transparent migration for virtual machines. In: USENIX Annual Technical Conference, General Track, pp. 391–394 (2005)
21. Barham, P., Dragovic, B., Fraser, K., Hand, S., Harris, T., Ho, A., Neugebauer, R., Pratt, I., Warfield, A.: Xen and the art of virtualization. ACM SIGOPS Oper. Syst. Rev. **37**(5), 164–177 (2003)

22. Kivity, A., Kamay, Y., Laor, D., Lublin, U., Liguori, A.: KVM: the Linux virtual machine monitor. Proc. Linux Symp. **1**(8):225–230 (2007)
23. Beloglazov, A., Buyya, R.: Energy efficient resource management in virtualized cloud data centers. In: 10th IEEE/ACM International Conference on Cluster, Cloud and Grid Computing, pp. 826–831. IEEE (2010)
24. Khanna, G., Beaty, K., Kar, G., Kochut, A.: Application performance management in virtualized server environments. In: 2006 IEEE/IFIP Network Operations and Management Symposium NOMS 2006, pp. 373–381 (2006)
25. Forsman, M., Glad, A., Lundberg, L., Ilie, D.: Algorithms for automated live migration of virtual machines. J. Syst. Softw. **101**, 110–126 (2015)
26. Xiao, Z., Song, W., Chen, Q.: Dynamic resource allocation using virtual machines for cloud computing environment. IEEE Trans. Parallel Distrib. Syst. **24**(6), 1107–1117 (2012)
27. Beloglazov, A., Abawajy, J., Buyya, R.: Energy-aware resource allocation heuristics for efficient management of data centers for cloud computing. Future Gen. Comput. Syst. **28**(5), 755–768 (2012)
28. Beloglazov, A., Buyya, R.: Optimal online deterministic algorithms and adaptive heuristics for energy and performance efficient dynamic consolidation of virtual machines in cloud data centers. Concurr. Comput. Pract. Exp. **24**(13), 1397–1420 (2012)
29. Srikantaiah, S., Kansal, A., Zha, F.: Energy aware consolidation for cloud computing. In: Power Aware Computing and Systems (2008)
30. Cao, F., Zhu, M.M., Wu, C.Q.: Energy-efficient resource management for scientific workflows in clouds. In: IEEE World Congress on Services, pp. 402–409 (2014)
31. Salimian, L., Esfahani, F.S., Nadimi-Shahraki, M.-H.: An adaptive fuzzy threshold-based approach for energy and performance efficient consolidation of virtual machines. Computing **98**(6), 641–660 (2016)
32. Van Heddeghem, W., Lambert, S., Lannoo, B., Colle, D., Pickavet, M., Demeester, P.: Trends in worldwide ICT electricity consumption from 2007 to 2012. Comput. Commun. **50**, 64–76 (2014)
33. Gelenbe, E., Caseau, Y.: The impact of information technology on energy consumption and carbon emissions. Ubiquity, pp. 1–15

Chapter 35
Frequent Itemset Mining by Fuzzification of Purchase Quantity

Renji George Amballoor and Shankar B. Naik

Abstract Association rules and frequent itemsets are generated based upon merely the existence of item in a transaction. Information about the quantities purchased is ignored. Both the patterns do not contain any information about the quantity of items purchased. The study in this paper is an attempt to generate frequent sets of items of different quantities and mine associations between them using the concept of fuzzy sets. The patterns generated not only contain information about the items but also contain purchase quantity-wise item information.

35.1 Introduction

Market basket analysis (MBA) is increasingly used for the study of patterns in consumer behaviour especially in an ecosystem rich in real-time data on transactions. The business intelligence captured from the consumer behaviour in the online marketplace can significantly contributes towards the sales, marketing, pricing, inventory, profit scaling activities, etc., of the firm. It is widely accepted that the challenge encountering the business decision makers is no more a situation of data poverty but the identification and use of knowledge discovery methods [1]. The Knowledge Discovery in Database (KDD) process helps in converting low-level database in consumer behaviour into high-level strategic knowledge for creating, retaining, and expanding the market share [2]. The KDD process on transactional data can identify patterns in the form of frequent itemsets and association rules [3]. These itemsets and association rules do not reveal any information about the quantity and size of the items.

In this study, each item X is represented in three ways as X_{low}, X_{medium} and X_{high} based upon their quantity of purchase using the concept of fuzzy sets. Linguistic(If-Then rules) and verbal terms (low, medium of high) become a better descriptor in describing consumer behaviour with non-standard preferences than in terms Boolean logic [5]. The fuzzy logic can be a better alternative when the actual consumer

R. G. Amballoor · S. B. Naik (✉)
Directorate of Higher Education, Government of Goa, Alto-Porvorim, Goa, India
e-mail: xekhar@rediffmail.com

behaviour cannot be constrained to data point with a crisp membership of 0 (no) or 1 (yes) [3].

This study aims to propose an algorithm to generate frequent itemsets and association rules by considering the purchase quantities of items in transactions. Both patterns generated depict relation between the quantities of items purchased.

35.2 Related Work and Motivation

Traditionally, structural equation modelling (SEM) was used to estimate consumer behaviour, but it failed to provide results which are realistic to help managers in their decision making [4]. The wave particle duality in terms of rational and irrational decisions makes the estimation of consumer behaviour using the existing methodology becomes incommodious. The involvement of heuristics and biases in decision making makes the understanding of consumer behaviour difficult especially in uncertain scenarios.

Most of the algorithms generate frequent itemsets and association rules which contain information based upon the occurrence of an item in a transaction [6–8]. The algorithms presented in [9, 10] have used the concept of fuzzy sets to mine frequent itemsets and association rules.

The patterns generated by these algorithms do not have information about the quantities of items purchased. Patterns containing information about the quantity purchased are useful in deciding the size of production and packaging of items. This information is also useful on grouping items of different quantities.

The algorithm proposed in [11] generates association rules considering sales amount using fuzzy sets. However, only one kind of quantity label of an item is considered. The other quantities labels which are not considered may also be frequent.

35.3 Fuzzy Frequent Itemset Mining

35.3.1 Preliminaries

Let $I = \{x_j / 1 \le j \le m\}$ be the set of m literals representing items. Let D be the set of N transaction. A transaction T_i, contains the set of items purchased along with their quantity purchased in the transaction.

An itemset X is a set it items such that $X \subseteq I$.

35.3.2   Fuzzy Frequent Iteset Mining

35.3.2.1   Fuzzification of the Purchase Quantities

Different items are purchased in different quantities. The maximum purchase quantity is not same for all the items. Hence, the quantities purchased for each item are normalized to lie in the range [0, 1]. The normalized purchase quantity value of an item x is calculated as

$$x_{pn} = \frac{x_p}{x_{pmax}} \tag{35.1}$$

where, x_p is the purchase quantity in the current transaction and x_{max} is the maximum quantity of item x purchased in all the transaction of D.

Based upon the quantity of purchase, the items are categorized into low (l), medium (m), and high (h). The fuzzification is done using the database shown in Fig. 35.1. The membership function for *Low* fuzzy set for normalized purchase quantity x_{pn} is given as

$$\mu_l = \frac{0.5 - x_{pn}}{0.5} \ \forall x_{pn} \leq 0.5 \tag{35.2}$$

$$= 0 \ \forall x_{pn} > 0.5 \tag{35.3}$$

The membership function for *Medium* fuzzy set for normalized purchase quantity x_{pn} is given as

$$\mu_m = \frac{x}{0.5} \ \forall x_{pn} \leq 0.5 \tag{35.4}$$

$$= \frac{1 - x}{0.5} \ \forall x_{pn} > 0.5 \tag{35.5}$$

The membership function for *High* fuzzy set for normalized purchase quantity x_{pn} is given as

$$\mu_h = 0 \ \forall x_{pn} \leq 0.5 \tag{35.6}$$

$$= \frac{x - 0.5}{0.5} \ \forall x_{pn} > 0.5 \tag{35.7}$$

The membership function in Fig. 35.1 maps the normalized purchase quantity of the item to the membership.

A new database D_{fuzzy} is derived from database D, wherein the items and their purchase quantities in transactions are replaced by the labels representing the product and its category. For example, a product $P1$ with purchase quantity with normalized value 1 can be replaced as $P1_h$.

Fig. 35.1 Membership
function database

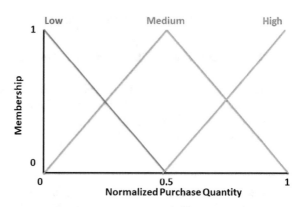

35.3.2.2 Frequent Itemset Generation

The frequent itemsets are generated using the Ariori algorithm. An itemset represented as X in database D will exist in database D_{fuzzy} in the form of three itemsets, namely X_l representing low quantity purchases of X, X_m representing medium quantity purchases of X, and X_h representing high quantity purchases of X. Support of itemset X_s in D_{fuzzy}, denoted as $supp(X_s)$ is the count of the number of occurrences of the itemset in D_{fuzzy}. An itemset is frequent if its support is greater than the minimum support threshold s_0.

35.3.2.3 Association Rule Mining

Association rules are generated based upon the itemsets generated in section 35.3.2.2. The association rules generated are of the form $X_{s1} \rightarrow Y_{s2}$, where $s1$ and $s2$ represent the fuzzified quantities represented as l, m and h. Support of am association rule $X_{s1} \rightarrow Y_{s2}$ is the support of the itemset $X_{s1} U Y_{s2}$. Confidence of the association rule $X_{s1} \rightarrow Y_{s2}$ is $\frac{supp(X_{s1} U Y_{s2})}{supp(X_{s1})}$.

An association rule is a heavy association rule if its support is greater than s_0 and confidence is not less than the minimum confidence threshold $minconf$.

35.4 Experiments

Experiments were conducted to compare the association rules generated using the proposed algorithm and Apriori algorithm. The algorithms were implemented using C++ and executed on a computer system having 8 GB RAM and Windows Operating System. The datasets considered are synthetic datasets generated using the IBM synthetic generator. The dataset has 250 K transactions consisting of 20 items. The results are listed in Table 35.1.

Table 35.1 Experimental results for $s_0 = 0.5$ and $minconf = 0.5$

Algorithm	Frequent itemset	Association rules	Execution time
Proposed	16	8	0.6 s
Apriori	10	8	0.14 s

Table 35.2 Comparison of the type of patterns

Algorithm	Frequent itemset	Association rules
Proposed	$\{P1_l, P5_h\}$	$\{P1_l, P5_h\} \rightarrow \{P3_m\}$
Apriori	$\{P1, P5\}$	$\{P1, P5\} \rightarrow \{P3\}$

It is observed that the number of frequent itemsets generated using the proposed approach is more as compared to the number of frequent itemsets generated using the Apriori approach. However, the number of association rules generated using both the approaches is the same.

The type of patterns generated by the proposed and the Apriori algorithms are presented in Table 35.2. The patterns generated by the proposed algorithm have items with quantity information added as suffix.

The proposed algorithm requires more time than the Apriori algorithm. This is because of the time requires for the fuzzification of purchase quantities.

35.5 Limitations of the Study

1. The experiments were done on synthetic datasets.
2. The proposed algorithm is not suitable for datastreams.
3. The proposed algorithm requires more time than the Apriori algorithm.

35.6 Conclusion and Future Work

In this study, an algorithm is proposed to generate frequent itemsets and association rules based upon the quantity of item purchased. The frequent itemsets and the association rules contain quantity-wise representation of items. Besides having information about the items, the frequent itemsets contain quantities of items as their elements. The association rules show the relationship between various quantities of items.

These patterns help understand trends not only about the items but also about the quantities of items purchased. This information has application in deciding the quantities and size of production and packaging of items.

This study is limited to fuzzification of the purchase quantity attribute of item. In future, the study will be extended to other attributes such as utility, etc., and address the limitations identified.

References

1. Casillas, J., Martinez-Lopez, F.J.: Mining uncertain data with multi objective genetic fuzzy systems to be applied in consumer behaviour modelling. Expert Syst. Appl. **36**, 1645–1659 (2009)
2. Mitra, S.: Data mining in soft computing framework: a survey. IEEE Trans. Neural Netw. **13**(1), 3–14 (2002)
3. Fuzzy set based frequent itemset mining: an alternative approach to study consumer behaviour
4. Laurent, G.: Improving the external validity of marketing models: a plea for more qualitative input. Int. J. Res. Mark. **17**, 177–182 (2000)
5. Djuris, J., Ibric, S., Djuris, Z.: Neural computing in pharmaceutical products and process development. In: Djuris, J. (eds.) Computer Aided Applications in Pharmaceutical Technology, Springer (2013)
6. Zaki, M., Parthasarathy, S., Ogihara, M., Li, W.: New algorithms for fast discovery of association rules. In: Proceedings 3rd International Conference on Knowledge Discovery and Data Mining (KDD'97), pp. 283–296. AAAI Press, Menlo Park, CA, USA (1997)
7. Han, J., Pei, J., Yin, Y.: Mining frequent patterns without candidate generation. ACM Sigmod Rec. **29**(2), 1–12 (2000)
8. Agrawal, R., Srikant, R.: Fast algorithms for mining association rules. In: Proceedings 20th International Conference Very Large Data Bases, vol. 1215, pp. 487–499. VLDB (1994, September)
9. Wu, T.Y., Lin, J.C.W., Yun, U., Chen, C.H., Srivastava, G., Lv, X.: An efficient algorithm for fuzzy frequent itemset mining. J. Intell. Fuzzy Syst. **38**(5), 5787–5797 (2020)
10. Cui, Y., Gan, W., Lin, H., Zheng, W.: FRI-miner: fuzzy rare itemset mining. arXiv preprint arXiv:2103.06866 (2021)
11. Dogan, O., Kem, F.C., Oztaysi, B.: Fuzzy association rule mining approach to identify e-commerce product association considering sales amount. Complex Intell. Syst. 1–10 (2022)

Chapter 36
Deep Learning-Based Identification of Vegetable Species: System for Specially Abled Persons

Siddhartha Sinha, Arnab Banerjee, Vishal Kumar Patel, Akshay Kisku, Khushi Singh, Debrik Chakraborty, and Nibaran Das

Abstract An automatic recognition system is developed to identify six different vegetable species, *Solanum melongena* (eggplant), *Abelmoschus esculentus* (okra), *Solanum tuberosum* (potato), *Raphanus sativus* (radish), *Solanum lycopersicum* (tomato), *and Daucus carota* (carrot). It helps the specially abled persons to buy these vegetables in the market independently in minimum time. The images of the vegetables were taken from various vegetable markets in Durgapur (West Bengal) at different times of the day. The proposed JUDVLP-BCRP: Vegdb.v1 dataset consists of 1800 images with 300 images in each species. After preprocessing the images, some popular deep learning architectures such as MobileNetV2, DenseNet121, and Xception with transfer learning techniques have been used to classify the species accordingly. Unfreezing the top 20% layers of the pre-trained networks and fine-tuning for 40 epochs, 100% accuracy is achieved using DenseNet121 and Xception network individually. The system produces a stable encouraging result in identifying vegetable species and leads to a viable system design.

36.1 Introduction

India became the second largest producer of vegetables and fruits worldwide, after China [1]. Major vegetable producing states in India are Uttar Pradesh, West Bengal, Madhya Pradesh, Bihar, Gujarat, Maharashtra, and Odisha. Around 40 different fresh vegetable crops are cultivated in India, and 546 vegetable varieties are found. Due to the lack of knowledge about these species, some people face problems identifying them correctly. Despite having a morphological understanding of the vegetable species, different specially abled persons may face challenges in buying them in the market as they have to interact with the seller. It is often impossible for normal people

S. Sinha · A. Banerjee (✉) · V. K. Patel · A. Kisku · K. Singh · D. Chakraborty
Dr. B. C. Roy Polytechnic, Durgapur, West Bengal 713206, India
e-mail: arnab.banerjee@bcrec.ac.in; researchwork.arnab@gmail.com

A. Banerjee · N. Das
Jadavpur University, Kolkata, West Bengal 700032, India
e-mail: nibaran.das@jadavpuruniversity.ac.in

to understand the signals used by specially abled persons, making them dependent on some other persons in the market, or they have to bring someone with them. Also, the entire process takes more time to complete. So, developing an application using computer vision and machine learning to detect the vegetable species and generate an auditory output through the smartphone is the utmost needed study. This study will also help a specific age group person (6–12 years) to train themselves to detect different vegetable species without getting help from someone.

This study proposes a recognition system for identifying six different vegetables using the three best performing deep learning networks: DenseNet121, Xception, and MobileNetV2 and generating an auditory output. The proposed system will work in the actual market situation and helps specially abled persons. A dataset named JUDVLP-BCRP: Vegdb.v1 with 1800 images is prepared in this study to conduct the experiments with those collected images.

The rest of the study is organized as follows: in Sect. 36.2, the literature review is presented, the dataset collection process and pre-processing steps are described in Sect. 36.3, Sects. 36.4 and 36.5 describes the proposed methodology and results, and the conclusion is presented in Sect. 36.6.

36.2 Literature Survey

In recent years, several works have been done on fruit recognition and classification [2–4]. Most of these works are based on traditional feature processing and machine learning. Very few studies have been reported on vegetable identification. Zhang et al. [5] proposed a recognition system of fruits and vegetables using the color and texture feature. The advancement of deep neural networks and convolutional neural networks (CNN) creates a strong wing in image recognition and classification. Some of the works on vegetable identification have been done using different CNN in recent years. In 2016, Singla et al. [6] proposed a food/non-food classification and food recognition system using GoogLeNet architecture. They have collected the data from an existing dataset, social media, smartphones, etc. In 2018, Femling et al. [7] proposed a system to identify the fruits and vegetables in the retail market from the images captured by the video camera. Different CNNs were applied to classify the images and predict the price according to the weight of the object. A total of ten species were used: apple, avocado, banana, bell pepper, clementine, kiwi, orange, pear, potato, and tomato. Most of the images were taken from the ImageNet dataset [8], and for each species, 30 images were taken using a video camera. Using MobileNet, the Top 3 accuracy of 97% was achieved, with difficulties in predicting clementines and kiwis. An improved VGG deep learning model was proposed in 2018 by Li et al. [9], combining the outputs of the first two fully connected layers in the original VGG model. This improved network was applied in classifying ten different vegetable species, such as broccoli, pumpkin, cauliflower, mushrooms, cucumber, Chinese cabbage, tomato, eggplant, garden pepper, and carrots. During the dataset collection, 50% of the total dataset was collected from the ImageNet dataset, 30% was collected using Web Crawling, and the authors took 20% data. Different augmentation process was applied to increase the number of images under each species. Using the improved VGG network, 96.5% accuracy was achieved on the test dataset. In 2021, Mia and Chakraborty [10] proposed a method to recognize local vegetables using gray level co-occurrence matrix and statistical features. Using the SVM classifier average accuracy of 97.97% was achieved. Most importantly, as the images in the datasets were primarily taken from the ImageNet dataset, many of those will not meet the input domain in the retail market, leading to confusing the system while predicting the images in the actual situation.

36.3 Dataset

Six vegetable species were taken into consideration in this study, *Solanum melongena* (eggplant), *Abelmoschus esculentus* (okra), *Solanum tuberosum* (potato), *Raphanus sativus* (radish), *Solanum lycopersicum* (tomato), *and Daucus carota* (carrot). Images were taken from different markets in Durgapur (West Bengal) at different times of the day. All the variations of these species found in the market were considered during the collection of the samples. A total of 1800 images were collected in different lighting, illumination, angle variations, and environments. Mainly the dataset collection was done under two different conditions: 1. The images of the species were directly collected from the market and 2. Vegetables were purchased from the market, and then the images were taken in the home environment. Standard smartphone cameras such as Realme X, Oppo F11 Pro, Redmi Note7/8 Pro, and Oppo A9 were utilized to capture the images. After collecting the images, the images are cropped so that a single object (vegetable species) will be present per image, and the size of the images is set to 224 × 224. There are notable varieties present in the background of the images. Global contrast normalization was applied to normalize the color distribution of the samples. After the preprocessing, the dataset is divided into the training, validation, and testing parts in a 60:20:20 ratio. The proposed dataset is named JUDVLP-BCRP: Vegdb.v1 as it was developed in the collaboration of DVLP Lab, Jadavpur University and Dr. B. C. Roy Polytechnic, West Bengal. Some of the images from the proposed dataset are presented in Fig. 36.1.

36.4 Proposed Methodology

This section presents brief details about the base learners with the specific customizations related to the problem under this study. Three popular deep learning architectures, MobileNetV2, DenseNet121, and Xception, are used to extract the features from the images in the JUDVLP-BCRP: Vegdb.v1 dataset. As the number of images in the dataset is not enough to run a deep learning algorithm from scratch, the transfer learning technique is employed in this study. Pre-trained networks on ImageNet [8]

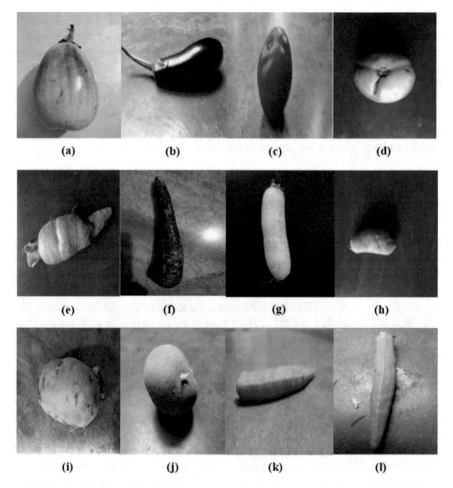

Fig. 36.1 Some images of JUDVLP-BCRP: Vegdb.v1 dataset: **a** and **b**—*Solanum melongena* (eggplant), **c** and **d**—*Solanum lycopersicum* (tomato), **e** and **f**—*Daucus carota* (carrot), **g** and **h**—*Raphanus sativus* (radish), **i** and **j**—*Solanum tuberosum* (potato), and **k** and **l**—*Abelmoschus esculentus* (okra)

dataset are used in this study to recognize the vegetable species. In the pre-trained models, the last fully connected layer consists of 1000 nodes to classify thousand different objects present in the ImageNet dataset. As per the problem, the last fully connected layer is deleted, and a customized fully connected layer with six nodes is introduced and trained for 100 epochs with a learning rate of 0.0001 on the JUDVLP-BCRP: Vegdb.v1 dataset. Adam optimizer and sparse categorical cross-entropy are used during the training. The base convolution part of the pre-trained network is frozen during the training, and only the customized, fully connected layer with six nodes is trained. After the training is complete, the last (top) 10% layers are unfrozen, and the network is fine-tuned for another 40 epochs with a learning rate of 0.00001.

Again, the last (top) 20% layers are unfrozen and fine-tuned for 40 epochs with a learning rate of 0.00001. During the fine-tuning process, Adam optimizer and sparse categorical cross-entropy are used. The networks are fine-tuned and tested on the test dataset using these experiment protocols. In Fig. 36.2, the proposed technique is pictorially presented.

MobileNetV2 [11]: MobileNetV2 is an updated version of MobileNetV1, a convolution neural network architecture that runs effectively on mobile devices. It has 32 filters in the initial block and 19 residual bottleneck layers. Mainly this architecture is based on the depth wise separable convolution that acts as the main building block. The linear bottlenecks were used between the network layers, and the shortcut connections were utilized between the bottlenecks. Depth wise separable convolution helps make the network deeper and, at the same time, reduces the parameter and computation.

DenseNet121 [11]: DenseNet121 is a convolutional neural network architecture consisting convolutional layer, pooling layer, dense blocks, and transition layer. Here, each layer is connected with the deeper layers in the architecture. Each layer gets the input from the previous layer and passes the generated feature map to all the layers next to it. This network needs fewer parameters than the traditional CNNs and can be trained effectively.

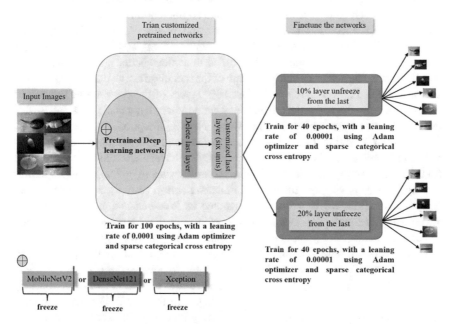

Fig. 36.2 Proposed approach for identifying the six vegetable species using MobileNetV2, DenseNet121, and Xception architecture with transfer learning technique

Xception [11]: Xception is another convolutional neural network that is the extreme version of the inception architecture. First, the different filters are applied on each depth map, and 1×1 convolution is used across the depth. This operation is very similar to the depth wise separable convolution operation [12]. Also, nonlinearity is not introduced after those two operations like the inception architecture. Basically, the inception modules are replaced by the depth wise separable convolution operation, which acts as a catalyst in the classification performance of the network.

36.5 Result and Discussions

The prepared dataset with 1800 images of six vegetable species is used in this study to conduct experiments. Under each species, 300 images are present with a 60:20:20 ratio for training, validation, and testing. The pre-trained MobileNetV2, DenseNet121, and Xception on the ImageNet challenge are taken, and the last layer with 1000 nodes is removed. Then, a layer with six units is introduced as the final layer because in this study total of six vegetable species need to be recognized. All the layers except the final layer is frizzed so that the learned weights on the ImageNet can be used, and only the last layer will be trained as per the JUDVLP-BCRP: Vegdb.v1 dataset. The pre-trained networks are run for 100 epochs with a learning rate of 0.0001 and batch size of 16. Online data augmentations are applied during the training, such as horizontal flip, zoom, contrast, and rotation. After this training, using MobileNetV2, Xception, and DenseNet121, 98.33, 98.89, and 99.17% accuracy is achieved. These networks are then fine-tuned to achieve better performance in this specific study. The accuracy and loss graph of the models after the fine-tuning process are depicted in Fig. 36.3. When the last 10% layers of these pre-trained networks are unfrozen and the networks are retrained for another 40 epochs, 98.61, 99.44, and 99.44% accuracy are achieved using MobileNetV2, Xception, and DenseNet121. In the case of Xception, one *Solanum tuberosum (potato)* is wrongly recognized by *Solanum lycopersicum* (tomato). In the case of MobileNetV2, three *Raphanus sativus* (radish) are wrongly identified as *Solanum melongena* (eggplant), one *Raphanus sativus* (radish) is wrongly identified as *Daucus carota* (carrot), and one *Solanum lycopersicum* (tomato) is wrongly recognized as *Solanum melongena* (eggplant). When using DenseNet121, two *Raphanus sativus* (radish) are wrongly recognized as *Solanum melongena* (eggplant) and *Daucus carota* (carrot). Again, the last (top) 20% of layers are unfrozen, and the networks are fine-tuned for 40 epochs. After the fine-tuning, DenseNet121 and Xception network achieved 100% accuracy, and MobileNetV2 achieved 99.44% accuracy on the JUDVLP-BCRP: Vegdb.v1 dataset. In the case of MobileNetV2, after fine-tuning with the top 20% layer unfrizzed, one *Solanum lycopersicum* (tomato) is wrongly identified as *Solanum melongena* (eggplant). Confusion matrices of the fine-tuned models are presented in Fig. 36.4. Some statistical features such as precision, recall, and F1-score are calculated and species-wise, shown in Table 36.1, when the models are fine-tuned using the top 20% layers unfrozen.

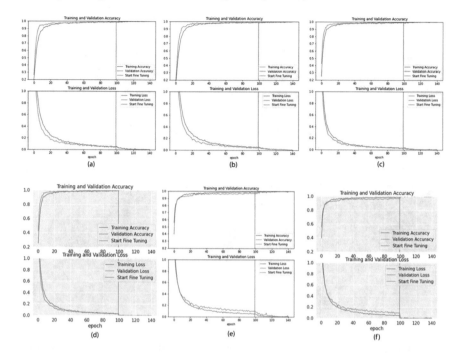

Fig. 36.3 Accuracy and loss graph of the fine-tuned deep neural networks, **a** DenseNet121 with 20% layers retrained, **b** DenseNet121 with 10% layers retrained, **c** MobileNetV2 with 20% layers retrained, **d** MobileNetV2 with 10% layers retrained, **e** Xception with 20% layers retrained, **f** Xception with 10% layers retrained

Very few studies in the literature directly involved classifying the vegetable species using deep learning models. The study by Femling et al. [7] applied different CNN models, and Li et al. [9] used an improved VGG model by fusing the last two fully connected layers and applied batch normalization to speed up the training process. As the dataset used in these studies is not available publicly, it is not possible to apply the proposed technique to their dataset and compare the results. The improved VGG model by Li et al. [9] is applied in the proposed JUDVLP-BCRP: Vegdb.v1 dataset, and 91.11% accuracy is achieved on the test data when the model is run for 400 epochs using Adam optimizer and sparse categorical cross-entropy loss function with a learning rate of 0.0001. In this study, the fine-tuned DenseNet121 and Xception pre-trained model with the last (top) 20% layers unfrozen achieved 100% accuracy on the test set and outperformed the MobileNetV2 (by 0.56%) and the improved VGG model (by 8.89%). All the misclassifications that occurred during the testing process are shown in Fig. 36.5. In the proposed study, the images collected from the vegetable market are used in the training process, leading to being a viable system in the retail market. It helps the specially abled persons by the auditory output.

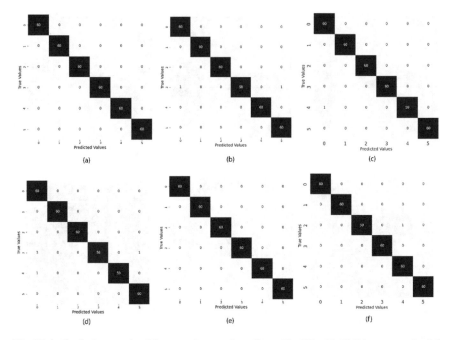

Fig. 36.4 Confusion matrix of fine-tuned networks, **a** DenseNet121 with 20% layers retrained, **b** DenseNet121 with 10% layers retrained, **c** MobileNetV2 with 20% layers retrained, **d** MobileNetV2 with 10% layers retrained, **e** Xception with 20% layers retrained, **f** Xception with 10% layers retrained

Table 36.1 Category wise precision, recall, and $F1$-score after fine-tuning the networks with top 20% layers unfrozen

Method	Statistics	Eggplant	Okra	Potato	Radish	Tomato	Carrot
DenseNet121	Precision	1.000	1.000	1.000	1.000	1.000	1.000
	Recall	1.000	1.000	1.000	1.000	1.000	1.000
	$F1$-score	1.000	1.000	1.000	1.000	1.000	1.000
MobileNetV2	Precision	0.984	1.000	1.000	1.000	1.000	1.000
	Recall	1.000	1.000	1.000	1.000	0.983	1.000
	$F1$-score	0.992	1.000	1.000	1.000	0.992	1.000
Xception	Precision	1.000	1.000	1.000	1.000	1.000	1.000
	Recall	1.000	1.000	1.000	1.000	1.000	1.000
	$F1$-score	1.000	1.000	1.000	1.000	1.000	1.000

36.6 Conclusion

Using the JUDVLP-BCRP: Vegdb.v1 dataset, the pre-trained architectures DenseNet121, and Xception were able to correctly identify all six types of vegetables

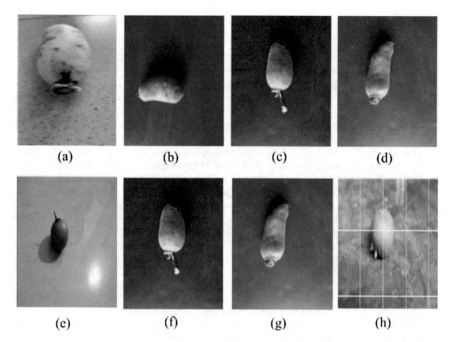

Fig. 36.5 Misclassification occurred during fine-tuning of pre-trained networks with the last 10% of layers frizzed, **a–d** correct category: *Raphanus sativus* (radish), using MobileNetV2 classified as **a-c** *Solanum melongena* (eggplant), **d** *Daucus carota* (carrot), **e** correct category: *Solanum lycopersicum* (tomato), using MobileNetV2 classified as *Solanum melongena* (eggplant), **f–g** *Raphanus sativus* (radish), using DenseNet121 classified as *Solanum melongena* (eggplant) and *Daucus carota* (carrot), **h** correct category: *Solanum tuberosum* (potato), using Xception classified as *Solanum lycopersicum* (tomato). Misclassification occurred during fine-tuning of pre-trained networks with the last 20% of layers frizzed, **e** correct category: *Solanum lycopersicum* (tomato), using MobileNetV2 classified as *Solanum melongena* (eggplant)

of the by unfreezing the final 20% of their pre-trained layers. The highest probability scores of these networks are used to generate the vegetable species name. This text output is then transformed into an auditory output that assists people with disabilities who want to recognize these species in the market. They can market vegetables in a minimum amount of time without any assistance from the others. In addition, the proposed study will help in the automatic sorting of these vegetables, which will benefit the packaging industry. The proposed approach gives 100% accuracy in recognizing the unseen images of these vegetables. We have to expand the system by introducing other varieties so that it can be a complete one. However, including all types of vegetables is difficult due to the fact that their availability is highly dependent on their geographical location. In this study, major varieties found in the various markets around Durgapur (West Bengal) are introduced in the proposed dataset. A larger number of images per category may produce in a more stable deep learning model. In addition, there is room to expand this dataset by including additional vegetable species commonly available in the Indian market.

References

1. FAO: India at a glance. Retrieved from https://www.fao.org/india/fao-in-india/india-at-a-gla nce/en/ (2022)
2. Ciptohadijoyo, S., Litananda, W.S., Rivai, M., et al.: Electronic nose based on partition column integrated with gas sensor for fruit identification and classification. Comput. Electr. Agric. **121**, 429–435 (2016)
3. Ninawe, P., Pandey, S.: A completion on fruit recognition system using K-nearest neighbors algorithm. Int. J. Adv. Res. Comput. Eng. Technol. **3**, 2352–2356 (2014)
4. Dubey, S.R., Jalal, S.: Fruit and vegetable recognition by fusing color and texture features of the image using matching learning. Int. J. Appl. Pattern Recognit. **2**, 160–181 (2015)
5. Zhang, Y., Wang, S., Ji, G., Phillips, P.: Fruit classification using computer vision and feedforward neural network. J. Food Eng. **143**, 167–177 (2014)
6. Singla, A., Yuan, L., Ebrahimi, T.: Food/non-food image classification and food categorization using pre-trained GoogLeNet model, ACM, Amsterdam, The Netherlands. In: International Workshop on Multimedia Assisted Dietary Management, pp. 3–11 (2016)
7. Femling, F., Olsson, A., Alonso-Fernandez, F.: Fruit and vegetable identification using machine learning for retail applications. In: 2018 14th International Conference on Signal-Image Technology and Internet-Based Systems (SITIS), pp. 9–15 (2018). https://doi.org/10.1109/SITIS. 2018.00013
8. Deng, J. et al.: Imagenet: a large-scale hierarchical image database. In: 2009 IEEE Conference on Computer Vision and Pattern Recognition, pp. 248–255. IEEE (2009)
9. Li, Z., Li, F., Zhu, L., Yue, J.: Vegetable recognition and classification based on improved VGG deep learning network model. Int. J. Comput. Intell. Syst. **13**, 559 (2020). https://doi.org/10. 2991/ijcis.d.200425.001
10. Mia, M.R., Chakraborty, N.R.: Computer vision based local vegetables recognition. In: 2021 IEEE 4th International Conference on Computing, Power and Communication Technologies (GUCON), pp. 1–6 (2021). https://doi.org/10.1109/GUCON50781.2021.9573704
11. Li Z. et al.: A survey of convolutional neural networks: analysis, applications, and prospects. IEEE Trans. Neural Netw. Learn. Syst. 1–21 (2021). https://doi.org/10.1109/TNNLS.2021.308 4827
12. Chollet, F.: Xception: deep learning with depthwise separable convolutions. In: Proceedings of the IEEE Conference on Computer Vision and Pattern Recognition, pp. 1251–1258 (2017)

Chapter 37
High-Utility Itemset Mining using Fuzzy Sets

Salman Khan, Tracy Almeida e Aguiar, and Shankar B. Naik

Abstract High-utility itemset mining aims to mine itemsets with high utilities. Utility of an itemset is the profit value associated with it. However, high-utility itemsets are generated considering only the occurrence of the items in the database. These patterns do not contain any information about the utilities of the items in the transaction. This paper aims to generate high-utility itemsets containing information about the utilities of the items.

37.1 Introduction

High-utility itemset mining (HUIM) aims to generate itemsets with high utilities. The utility of an item or itemset is the benefit it offers. For example, the utility can be the profit per item or the demand it has or can be the product of both.

Traditionally, high-utility itemsets (HUIs) are expressed in the form of items only. An item exists in an itemset just as an item name. They do not have any mention about the transactional utilities of the item. The items in HUIs are not maintained utility wise.

In order to do so, the utilities of an item in different transactions can be grouped into categories, such that each category reflects a range of utility values. Based upon the category to which the transactional utility belongs, the item in the transaction can be labelled with the category. For example, if the utility of the product *Cup* is low enough to belong to *Poor* category, then the item *Cup* in the transaction may be replaced as Cup_{Poor}. Thus, the HUIs will have items with labels which reflect their utilities. As the utilities are based on their quantities, these patterns help companies and retail stores to decide the quantities and size of production and packaging.

S. Khan (✉) · T. A. Aguiar
Rosary College of Commerce and Arts, Navelim, Salcete, Goa, India
e-mail: salmank.goa@gmail.com

S. B. Naik
Directorate of Higher Education, Government of Goa, Alto-Porvorim, Goa, India

In this paper, we propose an algorithm fuzzy set-based high-utility itemset mining (FSHUIM) to mine high-utility itemsets by employing the concept of fuzzy sets. Based on the membership function of each of the fuzzy sets, the utilities of items sold are classified as *Poor*, *Average* or *High*.

37.2 Related Work and Motivation

As proposed by Agrawal and Srikant, the goal of frequent itemset mining is to find frequent itemsets from a transactional database [1]. Various efficient algorithms such as Apriori, FP-Growth and Eclat are proposed to find frequent itemsets [1–3]. The patterns generated are in the form of association rule to assist a retailer in discovering the most items sold together. The association rules will help the retailer to realise the probability of a customer purchasing another item when he/she has already purchased an item. Here, the profit values are not considered in frequent itemsets mining, and all the items sold are given similar weights which does not provide the utility information for any retailer/business. These drawbacks are fulfilled by using high-utility itemsets mining where the volume of sales as a well as unit profit of each item are considered as weights for each item in the transactional databases.

Several algorithms have been proposed for mining high-utility itemsets [4–7]. Algorithms presented in [8, 9] employ the concept of fuzzy sets to generate HUIs.

These algorithms do not consider transactional utilities of items in HUIs generated. The concept of fuzzy sets can be used to convert the utilities into item label indicating the status of utility of the item in the transaction. In this paper, we propose a method to categorise each transactional utilities as *Poor*, *Average* or *High* depending upon the utilities of the items.

37.3 Fuzzy Set-Based High-Utility Itemset Mining

37.3.1 Preliminaries

Let D be the database with M columns where ith column pertains to the item x_i. The jth row in D represents transaction j such that the value in column i, represented as $D[j][i]$, denotes the quantity if item i is purchased in transaction j.

Let EU $=< e_1, e_2, ..., e_M >$ be the set of profit values called as external utilities such that e_i is the external utility of the item x_i. The utility of item x_i is $d[j][i] * e_i$.

Fig. 37.1 Membership function for item utility

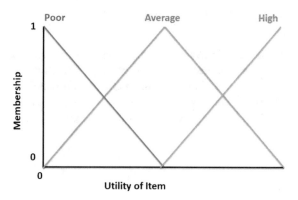

37.3.2 The Proposed Algorithm

37.3.2.1 Fuzzification of the Utilities

The items are categorised into poor (p), average (a) and high (h) based upon the utility value. The fuzzification is done using the database shown in Fig. 37.1. The membership function for *Poor* fuzzy set for utility value $u(x_p)$ is given as

$$\mu(x_p) = \frac{0.5 * u_{\max}(x) - u_x}{0.5 * u_{\max}(x)} \; \forall u(x) \leq 0.5 * u_{\max}(x) \tag{37.1}$$

$$= 0 \; \forall u(x) > 0.5 * u_{\max}(x) \tag{37.2}$$

The membership function for *Average* fuzzy set for utility value $u(x_a)$ is given as

$$u(x_a) = \frac{u(x)}{0.5 * u_{\max}} \; \forall u(x) \leq 0.5 * u_{\max} \tag{37.3}$$

$$= \frac{u_{\max} - x}{0.5 * *u_{\max}} \; \forall u(x) > 0.5 * u_{\max} \tag{37.4}$$

The membership function for *High* fuzzy set for utility $u(x_h)$ is given as

$$u(x_h) = 0 \; \forall u(x) \leq 0.5 * u_{\max} \tag{37.5}$$

$$= \frac{u(x) - 0.5 * u_{\max}}{0.5 * u_{\max}} \; \forall u(x) > 0.5 * u_{\max} \tag{37.6}$$

37.3.3 HUI Generation

The utility of every item in each transaction is calculated using the formula $D[j][i] * e_i$. Let x_{ji} be item i in transaction j. The utility of x_{ji} is calculated as $u(x_{ji}) = D[j][i] * e_i$. Using the membership functions as described in Eqs. (37.1), (37.3) and (37.5), the values of $\mu(x_{jip})$, $\mu(x_{jia})$ and $\mu(x_{jih})$ are calculated.

The total category-wise utilities are calculated as

$$\mu(x_{ip}) = \sum_{j=1}^{N} \mu(x_{jip}) \tag{37.7}$$

$$\mu(x_{ia}) = \sum_{j=1}^{N} \mu(x_{jia}) \tag{37.8}$$

$$\mu(x_{ih}) = \sum_{j=1}^{N} \mu(x_{jih}) \tag{37.9}$$

The items x_{ip}, x_{ia}) and x_{ih} are derived items of the original item x_i. The resultant high-utility itemsets will be in terms of the derived items x_{ip}, x_{ia}) and x_{ih} and not x_i. A derived item is a high-utility item if its utility is greater than the minimum threshold support value s_0. Only, the derived high-utility items are considered to be part of the high-utility itemsets.

The transactional utility of an itemset X in a transaction j is calculated as

$$\mu(X_j) = \sum \mu(x_{jiy}) \; \forall x_{jiy} \in X_j \text{ and } y = p, a \text{ or } h \tag{37.10}$$

The total utility of an itemset X is calculated as

$$\mu(X) = \sum_{j=1}^{N} \mu(X_j) \tag{37.11}$$

An itemset X of derived items is considered as high-utility itemset if $\mu(X) \geq s_0$.

37.4 Experiments

Experiments were conducted to compare the HUIs generated using the proposed algorithm and HUIM [4]. The datasets considered are synthetic datasets. The dataset has 200K transactions consisting of 10 items. The results are listed in Table 37.1.

The proposed algorithm generates more HUIs as compared to that of HUIM. The execution time of the proposed algorithm is more because of the extra step

Table 37.1 Experimental results for $s_0 = 0.5$

Algorithm	HUI count	Execution time
Proposed	21	0.16 s
Apriori	15	0.22 s

of fuzzification. However, the HUIs generated by the proposed algorithm contain derived items. These derived items have information about the status of the utility in them.

37.5   Conclusion and Future Work

The focus of the study in this paper is on generation of high-utility itemsets from transactional databases. The transactional utilities of items are fuzzified using the concept of fuzzy sets to derive utility level-wise items. The HUIs consist of these derive items. Thus, the HUIs contain information about the various levels of utilities of items.

The experiments were conducted on synthetic datasets. The proposed algorithm has requires more time than the traditional ones and is suitable only for static databases. Our future work will address these issues.

References

1. Agrawal, R., Srikant, R.: Fast algorithms for mining association rules. In: Proceedings 20th International Conference Very Large Data Bases, vol. 1215, pp. 487–499. VLDB (1994, September)
2. Zaki, M., Parthasarathy, S., Ogihara, M., Li, W.: New algorithms for fast discovery of association rules. In: Proceedings 3rd International Conference on Knowledge Discovery and Data Mining (KDD'97), pp. 283–296. AAAI Press, Menlo Park, CA, USA (1997)
3. Han, J., Pei, J., Yin, Y.: Mining frequent patterns without candidate generation. ACM Sigmod Rec. **29**(2), 1–12 (2000)
4. Chan, R., Yang, Q., Shen, Y.D.: Mining high utility itemsets. In: Proceedings of the IEEE International Conference on Data Mining, pp. 19–26. Melbourne, FL (2003)
5. Liu, Y., Liao, W.K., Choudhary, A.: A fast high utility itemsets mining algorithm. In: Proceedings of the 1st International Workshop on Utility-Based Data Mining, pp. 90–99 (2005, August)
6. Fournier-Viger, P., Wu, C.W., Zida, S., Tseng, V.S.: FHM: faster high-utility itemset mining using estimated utility co-occurrence pruning. In: International Symposium on Methodologies for Intelligent Systems, pp. 83–92. Springer, Cham (2014, June)
7. Tseng, V.S., Wu, C.W., Shie, B.E., Yu, P.S.: UP-growth: an efficient algorithm for high utility itemset mining. In: Proceedings of the 16th ACM SIGKDD International Conference on Knowledge Discovery and Data Mining, pp. 253–262 (2010, July)

8. Wu, J.M.T., Srivastava, G., Wei, M., Yun, U., Lin, J.C.W.: Fuzzy high-utility pattern mining in parallel and distributed Hadoop framework. Inf. Sci. **553**, 31–48 (2021)
9. Yang, F., Mu, N., Liao, X., Lei, X.: EA-HUFIM: optimization for fuzzy-based high-utility itemsets mining. Int. J. Fuzzy Syst. **23**(6), 1652–1668 (2021)

Chapter 38
Experimental Analysis of Oversampling Techniques in Class Imbalance Problem

Shweta Sharma, Jaspreeti Singh, and Anjana Gosain

Abstract The abstract should summarize the contents of the paper and should Class Imbalance is consistently being faced by real-world datasets, where one class have a high number of instances (i.e., majority class) while the other one is having a smaller number of instance (minority class). Current data mining and machine learning algorithm struggle to handle imbalanced data by reducing the error rate while ignoring the minority class. To tackle this issue, various strategies that range from modifying the training data to modifying the algorithm itself. The most widely used approach is data resampling, either oversampling the minority class instances or undersampling the majority class instances. This paper aims at the performance analysis of various oversampling techniques. The resampling technique includes oversampling based on synthetic minority oversampling technique (SMOTE) and clustering. The experimental results are obtained using 15 publically available UCI datasets with varying imbalance ratios. It also addresses the issues and challenges of SMOTE technique when applied alone. The performance of these techniques is assessed through performance matrices like ROC-AUC, *F*-measure and Recall. Through experimental analysis, it is found that the cluster-based oversampling technique outperforms the SMOTE-based techniques.

38.1   Introduction

The imbalance of class distribution has been a great concern for the machine learning research community for the past two decades. Imbalance occurs when the Number of instances in the minority class is much smaller than that in the majority class. This

S. Sharma (✉) · J. Singh · A. Gosain
USICT, Guru Gobind Singh Indraprastha University, New Delhi, India
e-mail: shwetabhardwajj15@gmail.com

J. Singh
e-mail: jaspreeti_singh@ipu.ac.in

A. Gosain
e-mail: anjana_gosain@ipu.ac.in

© The Author(s), under exclusive license to Springer Nature Singapore Pte Ltd. 2023   415
V. Bhateja et al. (eds.), *Evolution in Computational Intelligence*, Smart Innovation,
Systems and Technologies 326, https://doi.org/10.1007/978-981-19-7513-4_38

imbalance issue is prevalent in many real-world applications such as fraud diagnosis [1], sentiment analysis [1, 2], medical diagnosis, face recognition, text classification [5] and many more.

Different researchers have proposed different approaches to solve the class imbalance problem which can be categorized as data-level approach, algorithm-level approach and cost-sensitive approach. The data-level approach introduces a preprocessing step in the data distribution through oversampling the minority class and/or undersampling the majority class. Oversampling generates the novel samples in the minority class while we discard some instances from the majority class in undersampling. Oversampling is regarded to be more advantageous than undersampling as it emphasizes on cloning or adding up more instances, which might lose some important findings from the given samples [4].

The simplified oversampling approach is the random oversampling, which randomly selects the candidate minority instance and replicates them until it reaches the desired level of the minority class, but this often results in overfitting. To overcome this problem, Chawla et al. proposed SMOTE, which generates synthetic instances along a line segment.

However, SMOTE was negatively impacted by the overgeneralization issue as it blindly generates the instances without considering the majority instances. To overcome this problem, different Researchers proposed different versions of SMOTE like Adaptive Synthetic Sampling (ADASYN) [6], borderline-SMOTE [8], Safe-level SMOTE [10], etc. In general, it is essential to consider instances which are hard-to-learn for oversampling as they encase crucial information for learning the model. Recently, clustering-based techniques like cluster-SMOTE, Agglomerative Hierarchical Clustering (AHC), Proximity Weighed Synthetic Oversampling (ProWsyn) [11], Adaptive semi-unsupervised weighted oversampling (A-SUWO) [12], Density-based Synthetic Minority Oversampling Technique (DBSMOTE) [13], Cluster-based Synthetic Oversampling (CBSO) [17] have gained significant heed as they focus on areas that truly need the instance generation.

In this paper, we have analyzed ten widely used oversampling techniques, namely, SMOTE, Borderline-SMOTE, ADASYN, Safe-level SMOTE, Random SMOTE, AHC, CBSO, Cluster-SMOTE, DBSMOTE and ProWSyn. We have compared the performance of these techniques on 15 publicly available UCI datasets with varying imbalance ratio, and it was found that ProWysn, which assigns instances based on the proximity distance had outperformed the other algorithms.

The remaining part of this paper is organized as follows. Section 38.2 presented an overview of the methods to handle traditional data with imbalance problems. In Sect. 38.3, we provide our findings into the existing problem with the SMOTE and its variants and try to provide an effective solution with cluster-based techniques. Section 38.4 presents the analysis and the experimental design, while the conclusions are discussed in Sect. 38.5.

38.2   The Resampling Techniques to Address Class Imbalance Problem

Several machine learning approaches have been developed in the past two decades to cope with imbalanced data, namely, data-level approach, algorithm-level approach and hybrid approach.

38.2.1   Data-Driven Approach

The data-level approach works in modifying the collected imbalanced training dataset to make it suitable for standard learning classifiers. This approach uses sampling techniques like oversampling, undersampling, or feature selection [5].

38.2.1.1   Oversampling

It is one of the most widely used sampling approaches to deal with class imbalance problems. There are various oversampling techniques used rigorously, including random and synthetic oversampling. In random sampling, the instance of the minority class is replicated randomly until they achieve equal distribution of classes. While synthetic sampling works, by generating new instances through interpolation among the minority class instances. SMOTE was the first technique that came up with this idea and ensured no overfitting.

38.2.1.2   Undersampling

Undersampling eliminates instances of the majority class to equate the distribution. The balanced dataset may discard some potentially significant samples by removing instances from the majority class and thus receive low classifier performance. Neighborhood-based undersampling [3] were proposed by Goyal, which attains high accuracy while identifying the defective instances.

38.2.1.3   Feature Selection

Feature selection is another preprocessing step that draws attention to class imbalance. The primary goal is to choose a subset of k (adaptively selected parameter) features from the entire feature space to boost the classifier's performance. It can be categorized into filter, wrapper and embedded methods [9].

38.2.2 Algorithm-Level Approach

In Algorithm-level methodology, we want to equalize the dissemination of classes. This method is usually enumerated into two batches: cost-sensitive learning, a popular learning paradigm branch [5] based on the reduction of misclassification costs and ensemble learning.

38.2.3 Hybrid Approach

Hybrid approach aggregates both, the features of data-level and algorithm-level methods [15]. The major objective of hybrid approach is to eliminate the limitations of aforementioned methods. In this paper [16] a hybrid method is proposed based on k-means clustering and genetic algorithm.

38.3 Review of Oversampling Techniques

In 2002 [4], Chawla proposed a new technique, the synthetic minority oversampling technique (SMOTE), which has proven to set a standard benchmark in the field of class imbalance. It generates synthetic samples by interpolating between a candidate minority instance and its nearest neighbors (Fig. 38.1).

38.3.1 SMOTE and Its Variants

In the past two decades, various extensions and upgradations have been performed with SMOTE, but still, they face numerous issues:

- **Small Disjunct, Noise and Lack of Data**

  SMOTE operates throughout the dataset, and hence generates synthetic instances in every region whether it is noisy and might adversely affects the performance

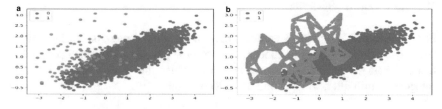

Fig. 38.1 a Original dataset, **b** oversampled dataset using SMOTE

of classifier. To deal with these complexities, we should investigate the nature of outliers before applying oversampling.

- **The Use of Feature Space Rather than data Space**

  SMOTE selects data points that are closely related in the feature space, and it is being used only with the features that are continuous to synthesize the data. Hence there is a need to design an algorithm to minimize the feature space to produce artificial samples by increasing the mutual dissimilarity within the class.

- **Ignorance of Minority Class**

  The synthetic instances are generated without considering the majority class, which further leads to ambiguous data points if overlapping exists among the classes.

- **The Correlation Variation**

  At the time of instance creation, different values of the k-nearest neighbors are tested in the SMOTE procedure. If we take a minimal k (say $k = 1$), it generates the sample close to original data points. This variation of value influences the type of data point created, which may impact the model's performance. The selection of k should be a sensible choice.

- **High Dimensional Data**

  Most of the SMOTE-generated samples are assigned only on the line segments connecting the K-neighbors, results in an unrealistic graph shape, where edges are filled with data points, and internal portions are void of them. This kind of problem is accentuated even more in higher dimensions [9].

38.3.2 Clustering-Based Approach

Recently clustering techniques have gained significant heed as they tend to address the within-class imbalance. This approach works by decomposing the dataset into various small clusters and using these sampling methods to undersample or oversample the data points. To overcome the limitations of the aforementioned problems in the previous section, various cluster-based techniques have been proposed. Generally, the clustering-based algorithms are designed to reduce the generation of overlapping synthetic new instances.

Cluster-SMOTE was the first attempt to address this issue. It first clusters the minority class into n sub-clusters and then applies SMOTE to each of them. In 2012, DBSMOTE [13] was proposed by chumphol, inveigled by B-SMOTE, works in the overlapping regions and oversample this region. Synthetic instances are created along the minimized path from each minority instance to a pseudo-centroid of a cluster of minority without being operated within a noise region.

Admin et al. proposed a technique called A-SUWO [12] in 2015 for handling imbalanced distribution of datasets, which clusters the minority instances using a semi-supervised hierarchical clustering approach.

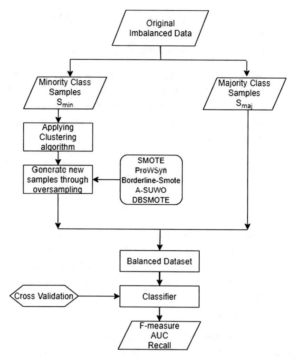

Fig. 38.2 Framework of cluster-based oversampling

Clustering algorithms [14] are generally designed to reduce overlapping and to lower the misclassification rate. Moreover, they avoid the problems found in SMOTE technique by increasing the density of the datapoints in the minority class as shown in the Fig. 38.2. This shows the procedure followed by clustering-based oversampling in accordance with the SMOTE technique. The given dataset is being decomposed into minority and majority class. An efficient clustering algorithm is being employed to overcome the complexities like outliers. Within the created clusters, new samples are generated through an oversampling technique. Newly added instances are combined with the majority class subset to provide a balanced distribution. Finally, the classifier is trained by using tenfold cross-validation and classification results are found.

38.4 Experimental Analysis

In this section, we have presented the experimental results from various oversampling approaches. The comparison is evaluated among five SMOTE Techniques and five Cluster-based techniques which are implemented on scikit-learn python library, a freely available machine learning library for python. Its features like classification, regression and clustering enables it to perform better. The datasets [7] are collected

Table 38.1 Description of dataset

Dataset	Examples	Features	Imbalance ratio
Isolate	7797	617	12:1
Satimage	6435	36	9.3:1
Libras_Move	360	90	14:1
Abalone	4177	8	9.7:1
Wine_quality	4898	11	26:1
Breast_cancer	102,294	117	163.19
Yeast1	1484	8	2.46
Yeast1vs7	947	8	13.87
Vehicle0	846	18	3.23
Vowel0	988	13	10.1
Segment0	2308	19	6.01
Pima	768	8	1.9
Page_block	5472	10	8.77
Glass1	214	9	10.39
Yeast6	1484	8	39.15

from UCI machine learning repository to show the analysis on classifiers like Decision tree and Logistic regression. Table 38.1 shows the characteristics of the dataset, such as the Number of total samples, Number of features and minority class sample.

The parameter of each classifier were optimized over a small set of values using fivefold stratified cross-validation that is, training data is divided into 5 sets, and cross-validation performs classification on subsets of training data to identify the best parameter. In this study, the performance measures used to compare the results are F-measure, Recall and Area under receiver operating characteristic curve.

Oversampling methods is implemented with imblearn python library executes on various datasets ranging from 2 to 163. The balanced data is oversampled until the class imbalance ratio becomes 1:1. The simulation results of AUC, F-measure and Recall are shown in Tables 38.2 and 38.3, in which each result is found using fivefold stratified k-fold cross-validation. It is found that cluster-based techniques like cluster-SMOTE, DBSMOTE, ProWysn outperforms SMOTE-based techniques in terms of AUC, F-measure and Recall. Clustering algorithm is generally designed to reduce the generation of overlapping synthetic new instances unlike SMOTE. It will take care of the fact that no two minority clusters are merged if a majority sample exists between them and thus avoids generating overlapping synthetic instances.

Cluster-based approaches gives significance to hard to learn instances as performing clustering on minority class helps in identifying difficulties engrafted in data. It can also be used for fitting more specialized classifiers so as to decompose the classification problem and analyze independently each part of the problem and provide better results. Our results also indicates that compared to SMOTE-based

Table 38.2 Performance of oversampling techniques on real world datasets using decision tree classifier

Dataset	Metric	Oversampling techniques									
		SMOTE	B-SMOTE	ADASYN	Safe-level	Random SMOTE	AHC	CBSO	Cluster-SMOTE	DBSMOTE	ProWsyn
Isolate	AUC	0.86	0.831	0.829	0.860	0.804	0.80	0.868	**0.878**	0.89	0.856
	F	0.49	0.50	0.503	**0.508**	0.472	0.49	0.474	0.484	0.440	0.471
	Recall	0.88	0.78	0.77	0.85	0.930	0.69	0.914	**0.932**	0.874	0.88
Satimage	AUC	0.819	0.828	0.814	0.78	0.819	0.828	0.864	**0.878**	0.56	0.79
	F	0.469	0.417	0.46	0.426	0.469	**0.505**	0.447	0.484	0.23	0.514
	Recall	0.810	0.915	0.89	**0.745**	0.810	0.738	0.93	0.932	0.13	0.686
Libras_Move	AUC	0.798	0.766	0.780	0.54	0.831	0.623	0.844	**0.857**	0.545	0.838
	F	0.695	**0.696**	0.761	0.166	0.667	0.394	0.727	0.692	0.166	0.695
	Recall	0.727	0.545	0.727	0.090	0.727	0.272	**0.727**	0.727	0.521	0.722
Abalone	AUC	0.805	0.767	0.776	0.78	0.502	**0.805**	0.764	**0.805**	0.784	0.786
	F	0.365	0.343	0.343	0.39	0.324	0.365	0.325	**0.435**	0.341	0.397
	Recall	0.97	0.92	0.971	0.833	0.863	0.970	0.95	0.970	**0.972**	0.823
Wine_quality	AUC	0.72	0.661	0.71	0.63	0.547	0.727	0.727	0.66	0.614	**0.736**
	F	0.137	0.194	0.31	0.216	0.156	0.165	0.165	0.126	**0.582**	0.576
	Recall	0.743	0.435	**0.717**	0.33	0.102	0.66	0.66	0.56	0.256	0.666
Breast_cancer	AUC	0.834	0.712	0.785	0.685	0.692	0.825	0.787	0.816	0.814	**0.846**
	F	0.438	0.418	0.454	0.368	0.456	0.245	0.264	**0.594**	0.356	0.576
	Recall	0.846	0.465	0.766	0.462	0.402	0.762	**0.786**	0.568	0.466	0.766
Yeast1	AUC	0.745	0.681	0.825	0.728	0.540	0.790	0.782	**0.805**	0.764	0.812
	F	0.365	0.343	**0.454**	0.439	0.326	0.365	0.382	0.365	0.342	0.397

(continued)

Table 38.2 (continued)

Dataset	Metric	Oversampling techniques									
		SMOTE	B-SMOTE	ADASYN	Safe-level	Random SMOTE	AHC	CBSO	Cluster-SMOTE	DBSMOTE	ProWsyn
Yeast1vs7	Recall	0.827	0.92	0.842	0.823	0.863	0.820	**0.952**	0.942	0.942	0.823
	AUC	0.744	0.767	0.776	0.78	0.5	0.805	0.764	0.805	**0.865**	0.786
	F	0.454	0.343	0.343	0.39	0.32	0.365	0.325	0.365	0.341	0.397
Vehicle0	Recall	0.621	0.92	0.971	0.833	0.863	0.970	0.95	0.970	0.972	0.823
	AUC	0.821	0.766	0.844	0.494	0.831	0.623	**0.882**	0.857	0.545	0.838
	F	0.661	0.600	0.750	0.186	0.667	0.394	0.727	**0.785**	0.166	**0.7012**
Vowel0	Recall	0.712	0.592	0.727	**0.790**	0.727	0.272	0.727	0.727	0.09	0.750
	AUC	0.821	0.742	0.782	0.721	0.522	**0.824**	0.744	0.805	0.784	0.798
	F	0.368	0.452	0.423	0.439	0.414	0.368	0.415	**0.454**	0.410	0.461
Segment0	Recall	0.916	0.902	0.961	0.812	0.863	0.970	0.95	**0.970**	0.942	0.802
	AUC	0.902	0.852	0.863	0.63	0.547	0.727	**0.910**	0.66	0.614	0.736
	F	**0.627**	0.594	0.584	0.216	0.156	0.165	0.165	0.126	0.246	0.176
Pima	Recall	0.781	0.635	0.717	0.33	0.102	0.660	**0.66**	0.560	0.256	0.666
	AUC	0.838	0.766	0.850	0.54	0.831	0.623	**0.882**	0.857	0.545	0.838
	F	0.695	0.666	0.761	0.166	0.667	0.394	0.727	**0.785**	0.166	0.695
Page_block	Recall	0.727	0.545	0.727	0.090	0.727	0.272	0.727	0.727	0.09	**0.750**
	AUC	0.72	0.66	0.680	0.589	0.489	0.682	0.727	0.680	**0.782**	0.762
	F	0.137	0.194	0.312	0.341	0.286	0.258	0.415	**0.624**	0.456	0.272
	Recall	0.743	0.435	0.655	0.425	0.508	0.621	**0.712**	0.562	0.324	0.701

(continued)

Table 38.2 (continued)

| Dataset | Metric | Oversampling techniques | | | | | | | | | |
		SMOTE	B-SMOTE	ADASYN	Safe-level	Random SMOTE	AHC	CBSO	Cluster-SMOTE	DBSMOTE	ProWsyn
Glass1	AUC	0.828	0.786	0.820	0.562	0.811	0.613	0.844	0.857	0.540	**0.838**
	F	0.725	0.586	0.691	0.528	0.577	0.494	0.702	**0.781**	0.412	0.685
	Recall	0.722	0.615	0.720	0.620	0.687	0.314	0.480	0.727	0.209	0.714
Yeast6	AUC	0.86	0.831	0.829	0.86	0.80	0.80	0.868	**0.882**	0.89	0.856
	F	0.49	0.50	0.503	0.508	0.47	0.49	**0.498**	0.484	0.440	0.471
	Recall	0.88	0.78	0.77	0.85	0.93	0.69	**0.914**	0.932	0.874	0.890

The bold numbers represent the best score among the comparative algorithms

Table 38.3 Performance of oversampling techniques on real world datasets using logistic regression classifier

Dataset	Metric	Oversampling techniques									
		SMOTE	B-SMOTE	ADASYN	Safe-level	Random SMOTE	AHC	CBSO	Cluster-SMOTE	DBSMOTE	ProWsyn
Isolate	AUC	0.943	0.929	0.936	0.921	0.936	0.921	0.936	0.934	0.912	**0.948**
	F	0.845	0.803	0.839	0.790	0.836	0.848	0.834	0.843	**0.858**	0.844
	Recall	0.907	0.889	0.895	0.870	0.895	0.858	0.895	0.889	0.834	**0.901**
Satimage	AUC	0.671	0.696	0.683	0.527	0.677	0.541	0.701	**0.740**	0.548	0.669
	F	0.267	**0.290**	0.277	0.141	0.273	0.162	0.286	0.285	0.802	0.270
	Recall	0.745	0.758	0.751	0.130	0.745	0.117	0.803	**0.758**	0.620	0.712
Libras_Move	AUC	0.902	0.909	0.798	0.642	0.902	0.727	0.902	0.902	0.755	**0.942**
	F	0.857	0.900	0.902	0.166	0.857	0.625	**0.920**	0.857	0.166	0.857
	Recall	0.818	0.818	0.818	0.252	0.818	0.454	0.818	0.818	**0.902**	0.818
Abalone	AUC	**0.971**	0.798	0.798	0.720	0.798	0.514	0.797	0.777	0.798	0.784
	F	0.363	0.364	0.364	0.368	0.364	0.019	0.362	0.342	0.367	**0.369**
	Recall	0.931	0.911	0.950	0.607	0.950	0.502	**0.950**	0.921	0.941	0.901
Wine_quality	AUC	0.734	0.742	0.728	0.67	0.749	0.55	0.761	0.707	0.730	**0.763**
	F	0.163	0.163	0.56	0.297	0.171	0.173	0.176	0.159	0.168	**0.179**
	Recall	0.692	0.717	0.692	0.384	0.717	0.102	0.743	0.615	0.666	**0.743**
Breast_cancer	AUC	0.842	0.821	0.892	0.883	0.880	0.859	**0.892**	0.880	0.885	0.890
	F	0.733	0.696	0.682	0.738	0.724	0.722	0.739	0.759	0.725	**0.764**
	Recall	0.689	0.650	0.751	0.130	**0.745**	0.688	0.803	0.117	0.6	0.712
Yeast1	AUC	**0.943**	0.929	0.936	0.92	0.936	0.921	0.936	0.934	0.921	0.912
	F	0.845	0.803	0.839	0.79	0.836	0.848	0.834	0.843	**0.858**	0.844

(continued)

Table 38.3 (continued)

Dataset	Metric	Oversampling techniques									
		SMOTE	B-SMOTE	ADASYN	Safe-level	Random SMOTE	AHC	CBSO	Cluster-SMOTE	DBSMOTE	ProWsyn
Yeast1vs7	Recall	0.907	0.889	0.895	0.87	0.895	0.858	0.895	0.889	0.834	**0.901**
	AUC	0.890	0.880	0.889	0.524	0.890	0.726	0.898	0.900	0.580	**0.926**
	F	0.602	0.456	0.852	0.189	0.822	0.781	**0.924**	0.812	0.169	0.844
	Recall	0.710	0.818	0.792	0.355	0.800	0.542	0.812	0.850	**0.890**	0.802
Vehicle0	AUC	0.790	0.793	0.683	0.781	0.783	0.604	0.822	**0.824**	0.825	0.817
	F	0.412	0.362	0.277	0.420	0.370	0.262	0.670	0.640	0.645	**0.663**
	Recall	0.645	0.592	0.751	0.634	0.597	0.417	0.780	**0.688**	0.653	0.702
Vowel0	AUC	0.943	0.929	0.936	0.92	**0.936**	0.921	0.936	**0.936**	0.912	0.921
	F	0.845	0.803	0.839	0.79	0.836	0.848	0.834	0.843	**0.858**	0.844
	Recall	0.907	0.889	0.895	0.87	0.895	0.858	0.895	0.889	0.834	**0.901**
Segment0	AUC	0.734	0.742	0.728	0.67	0.749	0.55	**0.752**	0.707	0.730	**0.702**
	F	0.163	0.163	0.56	0.297	0.171	0.173	0.176	**0.260**	0.168	0.179
	Recall	0.692	0.717	0.692	0.384	0.717	0.102	0.743	0.615	0.666	**0.788**
Pima	AUC	0.769	0.757	0.683	0.767	0.769	0.541	0.811	**0.822**	0.813	0.825
	F	0.696	**0.589**	0.277	0.607	0.596	0.162	0.649	0.660	0.669	0.658
	Recall	0.745	0.758	0.751	0.130	0.745	0.117	0.803	**0.758**	0.610	0.712
Page_block	AUC	0.734	0.742	0.728	0.67	0.749	0.55	0.761	0.707	0.730	**0.763**
	F	0.163	0.163	0.56	0.297	0.171	0.173	0.176	0.159	0.168	**0.179**
	Recall	0.692	0.717	0.692	0.384	0.717	0.102	0.743	**0.782**	0.666	0.743

(continued)

Table 38.3 (continued)

Dataset	Metric	Oversampling techniques									
		SMOTE	B-SMOTE	ADASYN	Safe-level	Random SMOTE	AHC	CBSO	Cluster-SMOTE	DBSMOTE	ProWsyn
Glass1	AUC	0.842	0.825	0.822	0.829	0.822	0.827	**0.841**	0.822	0.641	0.818
	F	0.645	0.640	0.624	0.630	0.645	0.629	0.640	0.670	**0.825**	0.663
	Recall	0.907	0.889	0.814	0.807	0.895	0.858	**0.895**	0.889	0.834	0.880
Yeast6	AUC	0.823	0.829	0.886	0.802	0.912	0.704	0.882	0.823	0.912	**0.924**
	F	0.445	0.781	0.819	0.761	0.536	0.814	0.810	0.810	0.688	**0.864**
	Recall	0.607	0.849	0.604	0.572	0.592	0.680	0.665	**0.901**	0.812	0.702

The bold numbers represent the best score among the comparative algorithms

Fig. 38.3 AUC results for decision tree

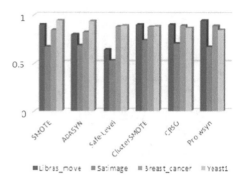

techniques, cluster-based techniques works well as shown in Fig. 38.3, with higher imbalance ratio like isolate, libras_move and wine_quality datasets.

38.5 Conclusion

Most related studies attempt on solving class problem by emphasizing a specific region, for instance, the safe region [10] or by creating the decision boundaries, but they ignore the other complexities that are inherited with the dataset. In this paper, 5 oversampling strategies based on the clustering technique are demonstrated. The research objective is to find the applicability of using cluster center and their nearest neighbors to represent all the data samples of the minority class. For this we have done an experimental study on 15 UCI datasets with varying class imbalance ratio ranging from 2 to 163.

The clustering-based oversampling techniques outperforms the SMOTE techniques as they generally tend to reduce the generation of overlapping between synthetic new instances. Also it will take care of the fact that no two minority clusters are merged if a majority data point exists.

References

1. Ali, A., Shamsuddin, S.M.: Classification with class imbalance problem. Int. J. Adv. Soft Compu. Appl. **7**(3) (2015)
2. Krawczyk, B.: Learning from imbalanced data: open challenges and future directions. Prog. Artif. Intell. **5**, 221–232 (2016)
3. Goyal, S.: Handling class-imbalance with KNN (neighbourhood) under-sampling for software defect prediction. Artif. Intell. Rev. **55**(3), 2023–2064 (2022)
4. Chawla, N.V., Bowyer, K.W., Hall, L.O., Kegelmeyer, W.P.: SMOTE: synthetic minority over-sampling technique. J. Artif. Intell. Res. **16**, 321–357 (2002)
5. Haixiang, G., Li, Y., Shang, J., Mingyun, G., Yuanyue, H., Gong, B.: Learning from class-imbalanced data: review of methods and applications. Expert Syst. Appl. **73** (2016). https://doi.org/10.1016/j.eswa.2016.12.035

6. He, H., Bai, Y., Garcia, E.A., Li, S.: ADASYN: adaptive synthetic sampling approach for imbalanced learning. In: Proceedings of the International Joint Conference on Neural Networks (IJCNN), pp. 1322–1328 (2008)

7. Gosain, A., Sardana, S.: Handling class imbalance problem using oversampling techniques: a review. In: 2017 International Conference on Advances in Computing, Communications and Informatics (ICACCI), pp. 79–85 (2017)

8. Han, H., Wang, W.Y., Mao, B.H.: Borderline-SMOTE: a new over-sampling method in imbalanced data sets learning. In: Advances in Intelligent Computing, pp. 878–887. Springer, Berlin (2005)

9. Zheng, Z., Wu, X., Srihari, R.: Feature selection for text categorization on imbalanced data. ACM SIGKDD Explor. Newsl. **6**(1), 80–89 (2004)

10. Bunkhumpornpat, C., Sinapiromsaran, K., Lursinsap, C.: Safe-Level-SMOTE: safe-level-synthetic minority over-sampling technique for handling the class imbalanced problem. In: Proceedings of the 13th Pacific-Asia Conference on Advances in Knowledge Discovery and Data Mining, pp. 475–482 (2009)

11. Barua, S., Islam, M.M., Murase, K.: ProWSyn: proximity weighted synthetic oversampling technique for imbalanced data set learning. In: Advances in Knowledge Discovery and Data Mining, pp. 317–328. Springer Berlin (2013)

12. Nekooeimehr, I., Lai-Yuen, S.K.: Adaptive semi-unsupervised weighted oversampling (A-SUWO) for imbalanced datasets. Expert Syst. Appl. **46**, 405–416 (2016)

13. Bunkhumpornpat, C., Sinapiromsaran, K., Lursinsap, C.: DBSMOTE: density based synthetic minority over-sampling technique. Appl. Intell. **36**(3), 664–684 (2012)

14. Chawla, N.V.: Data mining for imbalanced datasets: an overview. In: Data Mining and Knowledge Discovery Handbook, pp. 853–867

15. Leevy J.L. et al.: A survey on addressing high-class imbalance in big data. J. Big Data 5–42 (2018)

16. Sharma, S., Bellinger, C.: Synthetic oversampling with the majority class: a new perspective on handling extreme imbalance. In: IEEE International Conference on Data Mining, pp. 447–456 (2018)

17. Barua, S., Islam, M.M., Murase, K.: A novel synthetic minority oversampling technique for imbalanced data set learning. In: International Conference on Neural Information Processing (ICONIP), pp. 735–744 (2011)

Chapter 39
Head Pose Estimation and Validation of Medical Imaging Data for Image Guided Stereotactic Brain Surgeries

Abhilash Bhardwaj⊙**, P. P. K. Venkata, Soumitra Kar, Dinesh M. Sarode, and Pranav K. Gaur**

Abstract Computer assisted brain surgeries (also called stereotactic brain surgeries) are gaining prominence due to their effectiveness in handling the brain related ailments. As the surgeon cannot see the operative zone directly, due to the minimalistic invasive approach used in these surgeries, these procedures are heavily dependent on the medical imaging data obtained either through Magnetic Resonance Imaging (MRI) or Computed Tomography (CT). In these surgeries, where the process of high accuracy localization of the tumors is crucial, the pose of the patient during the medical imaging procedure becomes utmost important. Though the head pose related information is embedded in the DICOM medical images as metadata, assertion of the correctness is highly desired owing to the clinical impact. As not much work was found in this direction, we propose a novel way to estimate head pose of the patient from medical imaging data. Our proposed method uses a modified deep residual network to estimate the head pose, i.e., yaw, pitch and roll angles of medical imaging data with 4° Mean Absolute Error (MAE), nearly 50% improvement over our experiments using a modified AlexNet architecture. In light of unavailability of a labeled dataset for medical imaging head pose, the work reports a data generation algorithm with a dataset of 150 k samples. Reported experiments show 22% and 50% increase in accuracy by using data-driven and combined data-model-driven strategy

A. Bhardwaj (✉) · P. P. K. Venkata · D. M. Sarode · P. K. Gaur
Computer Division, BARC, Mumbai, India
e-mail: abhilashb@barc.gov.in

P. P. K. Venkata
e-mail: panikv@barc.gov.in

D. M. Sarode
e-mail: dinesh@barc.gov.in

P. K. Gaur
e-mail: pranav@barc.gov.in

S. Kar
SESSD, BARC, Mumbai, India
e-mail: skar@barc.gov.in

A. Bhardwaj · S. Kar
HBNI, Mumbai, India

respectively. The estimated head pose can be used for validating MRI metadata, computing deviation and assistance in patient to MRI registration.

39.1 Introduction

Medical imaging data are acquired for various purposes like diagnosis, therapy planning, surgeries, visualization and research [1, 2]. MRI and CT are two common types of medical imaging techniques used for these purposes [3, 4]. Unlike other organs of the human body, brain is highly complex and sensitive. Even a tiny mistake during surgery can lead to disastrous effect [5]. Due to the critical nature of brain, surgeons tend to use minimalistic invasive approaches [6], where even the laparoscopic cameras are also avoided. Hence, the biggest problem faced by the neurosurgeons is the visibility inside the patient skull while performing the brain related procedures. To have visibility of the brain interiors, surgeons are heavily dependent on the medical imaging data both for diagnosis as well as for surgery planning.

For surgery planning and execution, surgeons need to know the head pose of medical imaging data. This information is generally entered by radiologist manually and exported as metadata. In some scenarios, we may encounter situation where pose information in medical imaging data is missing, corrupted or deviated from original value. The *contribution* of our work is to address this situation by estimating head pose from medical imaging data. This work independently estimates the head pose associated with medical imaging data, which in turn can be used to correct, augment or compute deviation from metadata of MRI. As no such dataset or strategy is publicly available, this work addresses this limitation by presenting a way to generate dataset for training and validation of the model. The outcomes of this work can be used to, estimate, correct and validate the head pose information, missing value imputation, establish correspondence with the actual patient head pose, i.e., registration. Rest of the paper is organized as follows. We first describe the previous works related to head pose estimation followed by the details of proposed approach, i.e., dataset, architecture and challenges. The experimentation part describes implementation and training. The results, conclusion and future work at the end complete the paper.

39.2 Related Works

Human face pose estimation using regular RGB images is now a well-established process but the same from medical imaging is still an unexplored area. To the best of our knowledge, there is no publicly available annotated dataset or reported work related to head pose estimation of medical imaging data. To start with, we explored various works on human head pose estimation to understand the nature of problem. Girshick et al. [7] presents an in-depth study of relatively shallow networks trained using a regression loss on the AFLW dataset. In KEPLER [8] the authors presented a modified GoogleNet architecture which predicts facial keypoints and pose jointly. Apart from these, nonlinear regression methods for head pose estimation were also used on labeled training set to create nonlinear mappings from images to poses by Xia et al. [9]. Liu et al. [10] generated a realistic head pose dataset using rendering techniques and evaluated their CNN-based method on synthetic as well as real data. Ahn et al. [11] proposed a multitask convolutional network for face detection, bounding box refinement and head pose estimation. Chang et al. [12] used landmark-free head pose estimation to regress 3D head pose using a simple CNN. Work from Gu et al. [13] uses a VGG network to regress the head pose Euler angles. Ruiz et al. [14] presented a multi-loss CNN-based approach to predict head pose using regular RGB images. The approach and experiments reported in this work are inspired from CNN techniques used for human head pose estimation.

39.3   Materials and Methods

39.3.1   Dataset

We have acquired 50 samples of MRI for training and testing purpose. We have generated 3D head volume by extracting iso-surfaces from MRI data. Further 2D snapshots (like normal RGB image) were generated by orienting head volume at different angles, i.e., different combinations of yaw, pitch, roll using our in-house software [15] and the algorithm given in Table 39.1. Different step sizes have been chosen while generating the snapshots to maintain diversity in the dataset.

We have generated 150 k 2D snapshots for training and 1 k handpicked samples including extreme cases (e.g., noisy snapshot containing distorted faces, faces with extreme rotations, etc.) to evaluate model generalization capability.

39.3.2   Data Analysis and Challenges

Before model development and training iterations, we concentrated on data analysis to identify difficulty and differences between normal human face and 2D snapshot of MRI head volume. We identified that unlike human head, 3D head volume generated from MRI contains only half of human head, hence its 2D snapshot has lesser features compared to human face. By deeply analyzing the 2D snapshots we identified that surface smoothness, texture and characteristics of generated snapshots have high variance [16]. The observed samples containing noisy and distorted faces are depicted in Fig. 39.1. This exploratory analysis helped us to identify challenges as well as to make dataset more robust, generalized and exhaustive by specifically including noisy samples in the dataset.

Table 39.1 Algorithm for dataset generation

Step 1	Load Brain MRI DICOM file
Step 2	Generate 3D head volume by reading DICOM file
Step 3	Initialize Stepsize
Step 4	Initialize Euler Angles yaw ← − 90 pitch ← − 90 roll ← − 90

Step 5	While yaw < = 90
	    pitch ← − 90
	    while pitch < = 90
	      roll ← − 90
	      while roll < = 90
	        orient head volume at yaw, pitch, roll angle
	        save 2D image of head volume along with angles
	        roll ← roll + Stepsize
	      end of while
	      pitch ← pitch + Stepsize
	    end of while
	    yaw ← yaw + Stepsize
	end of while

Fig. 39.1 Normal and noisy snapshot containing distorted face

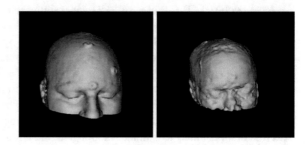

39.3.3   Evaluation Metric

This work estimates yaw, pitch and roll angles associated with head pose of medical imaging data. It attempts to minimize mean absolute error across all the three angles as given in Eq. (39.1).

$$\text{MAE} = \frac{\sum_{i=1}^{n} |y_i - x_i|}{n} \tag{39.1}$$

Equation (39.1): Formula to calculate Mean Absolute Error, in our case $n = 3$ (yaw, pitch, roll), y (predicted angle) and x (true angle).

39.3.4 Solution Architecture

As shown in Fig. 39.2, raw 2D snapshot of 3D head volume is generated. We identified that cropping out the region of interest containing only head is not feasible with available face detectors. Face detectors are trained over complete human face, while generated 3D head from MRI data has only half of it. Here we applied a loose cropping mechanism to crop out area of interest. The loose cropping mechanism analyzes and produces a rectangular bounding box containing region of interest. The cropped 2D snapshot is resized to 64 × 64 pixels and fed as input to the backbone CNN network. The proposed CNN network contains a modified deep residual network, ResNet50 [17]. End layer of ResNet50 have been modified and connected as fully connected layer with linear activation function to three output nodes. These three output nodes correspond to yaw, pitch and roll angles. When a cropped and resized 2D snapshot passes through the backbone network, it produces three values, which are further used to calculate the MAE. We have used ADAM optimizer to optimize loss function, i.e., MAE, calculated using predicted values and ground truth.

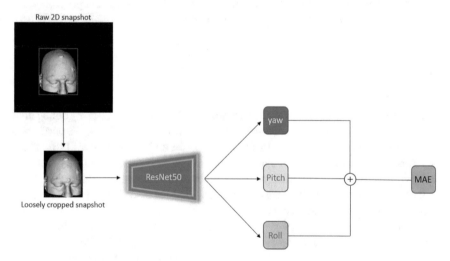

Fig. 39.2 Medical imaging head pose estimation pipeline

Fig. 39.3 Comparison of validation loss for different configurations of modified AlexNet

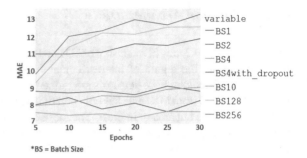

39.4 Experimentation

39.4.1 Implementation

Proposed pipeline is implemented in Python3. We have used Tensorflow2 to develop the deep learning model. OpenCV has been used for data loading and preprocessing operations. We have performed model training over NVIDIA Quadro M4000 GPU.

39.4.2 Training

In the very first experiment, we trained a modified AlexNet architecture by connecting the tail of network to three output nodes to predict yaw, pitch and roll. We performed multiple experiments by varying batch-sizes, epochs and dropout layers [18]. These experiments resulted in a significant difference between validation and training loss along with a problem of saturated model accuracy, which can be observed in Fig. 39.3. Data-driven experiments by infusing a larger dataset also did not eliminate the drawback of overfitting and accuracy saturation.

A thorough analysis of these experiments motivated us to use deep residual network ResNet50. The experiments were performed using modified ResNet50 architecture with the same setup like earlier one. This model-driven strategy performed significantly better than earlier approach, depicted in Fig. 39.4. This proposed approach also eliminated the drawback of overfitting and accuracy saturation to a great extent and achieved state of the art results over 2D MRI snapshots. Table 39.2 presents summary of experiments.

39.5 Result

The proposed modified ResNet50 network achieved 50% more accuracy than earlier AlexNet version. This work performed remarkably well over distorted and noisy

Fig. 39.4 Comparison of validation loss of modified AlexNet and modified ResNet50

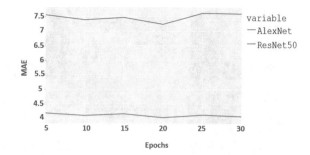

Table 39.2 Summary of training experiments

Architecture	Dataset size (k)	Train loss (MAE)	Validation loss (MAE)
Modified AlexNet	50	1.98	9.8
Modified AlexNet	150	3.48	7.6
Modified ResNet50	150	1.8	4.0

Table 39.3 Comparison with human head pose estimation approaches

Approach	MAE (in degree)
3DMM (uses depth data) [19]	2.0
Papzow et al. [20]	3.13
FSA-Caps-Fusion [21]	3.9
Multi-loss ResNet50 [14]	4.8
Proposed approach for medical imaging data	**≈ 4.0**

samples by achieving an accuracy of 4° MAE over test samples. Achieved accuracy is in line with human head pose estimation approaches using regular RGB images. Table 39.3 shows a comparison between human head pose estimation approaches and our proposed approach for medical imaging data. Figure 39.5 shows some of the test samples with predictions.

39.6 Conclusion

In this work, we have shown that the proposed method can achieve an acceptable MAE of 4° in terms of estimating the head pose from the medical imaging data. This method gave appreciable results even in case of partial and noisy snapshots. This work also presents a simple yet effective algorithm to generate dataset for medical

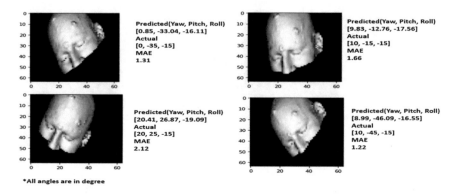

Fig. 39.5 Predictions over 2D medical imaging snapshots with ground truth

imaging head pose estimation. We have also reported the experiments on the model and data-driven strategy and its effect on performance. The estimated head pose can be used for validating head pose in MRI metadata, computing deviation and assistance in patient to MRI registration.

39.7   Future Works

Experiments reported in this work suggest room for improvement. By diversifying dataset and augmenting more and more training samples over the same architecture can be one direction of enhancement. Obtaining large number of medical imaging samples have its own limitations, hence generative networks can be used further to generate huge synthetic dataset. Model hyper-parameter tuning and customizing the loss function to target vulnerabilities of the model are other prominent areas that can be targeted to boost the performance.

References

1. Hu, S., Kang, H., Baek, Y., El Fakhri, G., Kuang, A., Choi, H.: Real-time imaging of brain tumor for image-guided surgery. Adv. Healthc. Mater. (2018)
2. Udupa, J., Herman, G.: 3D imaging in medicine. CRC Press, Boca Raton (2000)
3. Filler, A.: The history, development and impact of computed imaging in neurological diagnosis and neurosurgery: CT, MRI, and DTI. Nat. Precedings (2009)
4. Alexander, E., III., Moriarty, T., Kikinis, R., Black, P., Jolesz, F.: The present and future role of intraoperative mri in neurosurgical procedures. Stereotact. Funct. Neurosurg. **68**(1–4), 10–17 (1997)
5. Kondziolka, D., Firlik, A.D., Lunsford, L.D.: Complications of stereotactic brain surgeries. Neurol. Clin. **16**(1), 35–54 (1998)

6. Shamir, R., Freiman, M., Joskowicz, L., Shoham, M., Zehavi, E., Shoshan, Y.: Robot-assisted image-guided targeting for minimally invasive neurosurgery: planning, registration, and in-vitro experiment. In: Lecture Notes in Computer Science, pp. 131–138 (2005)
7. Girshick, R., Donahue, J., Darrell, T., Malik, J.: Rich feature hierarchies for accurate object detection and semantic segmentation. In: Computer Vision and Pattern Recognition (2014)
8. Kumar, A., Alavi, A., Chellappa, R.: Kepler: keypoint and pose estimation of unconstrained faces by learning efficient H-CNN regressors. In: 12th IEEE International Conference on Automatic Face and Gesture Recognition (FG 2017), pp 258–265 (2017)
9. Xia, J., Cao, L., Zhang, G., Liao, J.: Head pose estimation in the wild assisted by facial landmarks based on convolutional neural networks. IEEE Access **7**, 48470–48483 (2019)
10. Liu, X., Liang, W., Wang, Y., Li, S., Pei, M.: 3D head pose estimation with convolutional neural network trained on synthetic images. In: Proceedings of the IEEE International Conference on Image Processing (ICIP), pp. 1289–1293 (2016)
11. Ahn, B., Choi, D.G., Park, J., Kweon, I.S.: Real-time head pose estimation using multi-task deep neural network. Robot. Auto. Syst. **103**, 1–12 (2018)
12. Chang, F.J., Tran, A.T., Hassner, T., Masi, I., Nevatia, R., Medioni, G.: Faceposenet: making a case for landmark-free face alignment. In: IEEE International Conference on Computer Vision Workshop (ICCVW), pp. 1599–1608 (2017)
13. Gu, J., Yang, X., De Mello, S., Kautz, J.: Dynamic facial analysis: from bayesian filtering to recurrent neural network (2017)
14. Ruiz, N., Chong, E., Rehg, J.: Fine-grained head pose estimation without keypoints. In: IEEE/CVF Conference on Computer Vision and Pattern Recognition Workshops (CVPRW) (2018)
15. Bhutani, G., Dwarakanath, T.A., Lagoo, K.D., Moiyadi, A., Venkata P.P.K.: Neuro-registration and navigation unit for surgical manipulation. In: 1st International and 16th National Conference on Machines and Mechanisms (iNaCoMM2013). IIT Roorkee (2013)
16. https://www.dicomstandard.org/current
17. He, K., Zhang, X., Ren, S., Sun, J.: Deep residual learning for image recognition. In: IEEE Conference on Computer Vision and Pattern Recognition (CVPR) (2016)
18. Krizhevsky, A., Sutskever, I., Hinton, G.: ImageNet classification with deep convolutional neural networks. Commun. ACM **60**(6), 84–90 (2017)
19. Yu, Y., Mora, K.A.F., Odobez, J.M.: Robust and accurate 3D head pose estimation through 3DMM and online head model reconstruction. In: 2017 12th IEEE International Conference on Automatic Face and Gesture Recognition (FG 2017), pp. 711–718. IEEE (2017)
20. Papazov, C., Marks, T.K., Jones, M.: Real-time 3D head pose and facial landmark estimation from depth images using triangular surface patch features. In: Proceedings of the IEEE Conference on Computer Vision and Pattern Recognition, pp. 4722–4730 (2015)
21. Yang, T.Y., Chen, Y.T., Lin, Y.Y., Chuang, Y.Y.: FSA-net: learning fine-grained structure aggregation for head pose estimation from a single image. In: Proceedings of the IEEE Conference on Computer Vision and Pattern Recognition, pp. 1087–1096 (2019)

Chapter 40
Application of Intelligent Programmed Genetic Algorithm for Price Forecasting in Green Day-Ahead Market of Indian Energy Exchange

Lalhungliana, Devnath Shah, and Saibal Chatterjee

Abstract The recent introduction of new power exchange product called Green Day-Ahead Market in Indian Energy Exchange with sole motive to promote short-term power trade generated from renewable energy sources is crucial for promoting green energy. Precise forecasting of market clearing price in this market is important for generating companies to strategically submit their sell offers in Day-Ahead Market which can maximize their net profit. Therefore, this work presents an advanced forecasting model for precise forecasting of market clearing price by training an ANN with its synaptic weight updated using the recently proposed Intelligent Programmed Genetic Algorithm. The performance of the proposed model is validated by comparing its result with that obtained using a simple Genetic Algorithm, Simulated Annealing, Particle Swarm Optimization, and Biogeography-Based Optimization in terms of Mean Absolute Error. Results clearly demonstrate the superiority of the proposed model in precise forecasting of market clearing in the Green Day-Ahead Market.

40.1 Introduction

Electricity price forecasting based on a time series model refers to the process of predicting future electricity prices using past data. With climate change and growing political pressure, many countries around the world devoted themselves to increasing the share of renewable energy in their energy sector. Indian Energy Exchange also launch a new product called Green Day-Ahead Market (GDAM) to promote short-term trading of power generated from renewable sources like hydro, wind, and solar.

Lalhungliana (✉) · S. Chatterjee
National Institute of Technology Mizoram, Aizawl, Mizoram, India
e-mail: hremuan78@gmail.com

S. Chatterjee
e-mail: saibalda@ieee.org

D. Shah
Madanapalle Institute of Technology and Science, Madanapalle, Andrapradesh, India

© The Author(s), under exclusive license to Springer Nature Singapore Pte Ltd. 2023   441
V. Bhateja et al. (eds.), *Evolution in Computational Intelligence*, Smart Innovation,
Systems and Technologies 326, https://doi.org/10.1007/978-981-19-7513-4_40

This product is in addition to the existing Day-Ahead Market (DAM) which provides a platform for trading quarter-hourly power products a day ahead of their delivery. Market participants can submit their offers and bids in a similar manner as that of DAM. First GDAM is cleared and then DAM. The uncleared offer and bid in GDAM get carried away DAM, i.e., they are further considered for clearing in DAM. This is very advantageous for market players with significant renewable generators. For generating companies participating in DAM, it is very important to precisely forecast the market clearing price in GDAM as it helps them to formulate their bidding strategy in such a way to maximize their individual surplus. Over the years, many generating companies use this data to optimally formulate their bidding strategy so as to maximize the individual surplus. Author in [1] presents a comprehensive review on data-driven models for energy price forecasting in the last decade. In literature, a number of work exist that focuses on price and load forecasting in electricity market [2–9].

The performance of Artificial Neural Network can be enhanced by its training algorithm which updates its synaptic weight in each iteration. Researchers around the world prefer to use evolutionary optimization algorithms like Genetic Algorithm (GA), Simulated Annealing (SA), and Particle Swarm Optimization (PSO) to update the synaptic weight iteration-wise to obtain a better forecast of considered quantity. Recently, author in [10] proposed an improved version of Genetic Algorithm called Intelligent Programmed Genetic Algorithm which uses an additional diversity creating operator apart from pre-existing crossover and mutation. That operator injects additional diversity using some pre-defined set of rules which is determined deterministically based on type of optimization problem in hand. Advantage is that the optimization algorithm reaches global optima with minimum number of objective function evaluations or iterations. IPGA is claimed to be almost five time faster as compared to existing GA in terms of number of number of function evaluation or iterations. Therefore, this paper presents a forecasting technique using ANN with its weight optimized using IPGA to forecast four quarter hourly price in Green Day-Ahead Market of Indian Energy Exchange. The obtained simulation results are compared with that obtained using ANN optimized with GA [11], SA, PSO [12], and Biogeography-Based Optimization [13] in terms of Mean Absolute Error (MAE).

The rest of the work is organized as follows: A brief description of simple ANN is covered in Sect. 40.2. Section 40.3 briefly explains the flow of IPGA. The working of Green Day-Ahead Market in Indian Energy Exchange is introduced in Sect. 40.4. Data collection and MATLAB Simulation is covered in Sect. 40.5. Simulation Results and Analysis is covered in Sect. 40.6. The final conclusion along with some future work is pointed in Sect. 40.7.

40.2 Artificial Neural Network (ANN)

For complex function approximation using available data set, ANN is highly suitable. It has the ability to learn any function using past data. Over the years application of

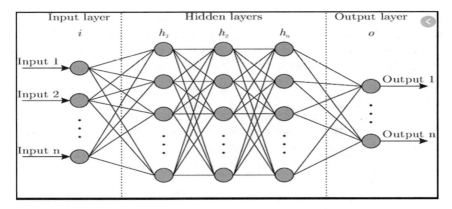

Fig. 40.1 Layers of ANN. Schematic diagram of typical artificial neural network

ANN for approximating time series function from time series data is on rise [14]. General ANN consists of three layers, namely input layer, hidden layer, and output layer as depicted in Fig. 40.1. Depending on the complexity of problem in hand, the number of hidden neuron could be increased or decreased.

Any two neurons in a neural network are connected by its synaptic weight between them. The input–output relation of ANN is defined by Eq. (40.1).

$$Y = \theta \left(\sum_{i=1}^{n} W_{ij} X_j - B \right) \tag{40.1}$$

Here, Y is output vector, W_{ij} is the weight matrix, θ is activation function, and B is an associated bias. The weight of ANN is adjusted in an iterative process such that output error get reduces over time. Back propagation learning algorithm is earliest and most popular algorithm for adjusting synaptic weight of an ANN. But to get very good results using Back propagation method initial guess must be very close to optimal solution otherwise there is possibility that the solution may converges to sub-optimal point. Evolutionary optimization algorithm can be used to get the optimal weight of ANN such that output error should have minimum value.

40.3 A Brief Introduction of Intelligent Programmed Genetic Algorithm

In [10], author proposed Intelligent Programmed Genetic Algorithm, which is a modified form of Visualized Interactive Genetic Algorithm [15]. This algorithm apart from the existing probabilistic genetic operator, i.e., crossover and mutation, for creating diversity in the current population, a new deterministic operator is also introduced for creating diversity in the current population [10]. This operator generates a new solution in the region with a higher value of the objective function and helps GA to converge to the optimal solution with a lesser number of objective function evaluations (iterations). Intelligent Programmed Genetic Algorithm is given.

Algorithm 1: IPGA [10]	
Input:	**Population** Size *NP,* Crossover Probability C_p, Mutation Probability M_p, Decision variable dimension *D,* Lower and upper bounds of problem being solved [X_{min}, X_{max}]
Output:	**Optimal** solution of an optimization problem
Step 1:	**Randomly** initialize first generation population P_1 between its [X_{min}, X_{max}]
Step 2:	**Evaluate** fitness value
Step 3:	**While** termination criterion is not met
Step 4:	**Set** iteration count, it $=$ it $+$ 1
Step 5:	**Copy** current population into each of three new variable $$P_{it}^{Cross} \leftarrow P_{it}$$ $$P_{it}^{Mut} \leftarrow P_{it}$$ $$P_{it}^{IPGA} \leftarrow P_{it}$$
Step 6:	**Apply** crossover operator to P_{it}^{Cross} and generate new solutions i.e $$P_{it}^{Cross} \leftarrow CrossoverOperator(P_{it}^{Cross})$$
Step 7:	**Apply** mutation operator to P_{it}^{Mut} and generate new solutions i.e $$P_{it}^{Mut} \leftarrow MutationOperator(P_{it}^{Mut})$$
Step 8:	**Apply** IPGA operator as described in Algorithm-1 to P_{it}^{IPGA} and generate new solutions i.e $$P_{it}^{IPGA} \leftarrow IPGAOperator(P_{it}^{IPGA})$$
Step 9:	**Evaluate** objective function value of each new solution
Step 10:	**Merge** current and all newly generated solutions i.e $$P_{it}^{New} = [P_{it}, P_{it}^{Cross}, P_{it}^{Mut}, P_{it}^{IPGA}]$$
Step 11:	**Sort** all solutions on the basis of objective function value and keep best *NP* solution
Step 12:	**Check** termination criterion
Step 13:	EndWhile
Step 14:	**Terminate** and return Solution-1 as best solution

40.4   Brief Introduction of Green Day-Ahead Electricity Market [16]

In 2016 under the Paris agreement, Government of India committed itself to increase the contribution of non-fossil electric power by 40%, i.e., a total generation of 450 GW by 2030. To accelerate the trading of power generated from renewable energy sources, the Union Minister of Power and New and Renewable Energy has launched the new market segment called Green Day-Ahead Market (GDAM) at Indian Energy Exchange which currently operates in conjunction with existing Day-Ahead Market. This market will allow renewable energy sellers to further trade their power generation apart from conventional Day-Ahead Market.

Figure 40.2 gives the flow diagram for power trade in Green Day-Ahead Market at IEX. The timing of various activities performed at GDAM is listed in Table 40.1. The operation starts with invitation of integrated offer/bids for GDAM and DAM from market participants between the time periods 10.00 AM and 12.00 PM. The market operator provides options to all traders to transfer uncleared offer/bids in GDAM to DAM. After that, the power exchange clears the market and determine common market clearing price and market clearing volume by 12.20 PM. By 01.00 PM, exchange communicates with National Load Dispatch Centre for the required transmission corridors. By 02.00 PM, NLDC confirms the availability along with actual scheduling transmission corridors. Thereafter, the exchange adjusts the clearing price and volume based on transmission congestion and re-determine the clearing price. Generally, market splitting (a market-based congestion alleviating technique) is used to alleviate transmission congestion. By 02.45 PM, the exchange issues the final cleared offer/bids to its market participants. Finally, by 05.30 PM, the exchange conveys acceptance of all scheduling to GDAM and DA.

Nature of market clearing price in GDAM for three days, i.e., from April 10, 2022 to April 12, 2022 is plotted in Fig. 40.3. This is helpful as before forecasting any quantity its long-term and short-term behavior would be helpful in designing the forecasting model. From Fig. 40.3, it is clear that the MCP is clamped at Rs. 12,000 per MWh. This is because there is shortage of supply from renewable sources between 06.00 PM and 06.00 AM. The market price falls rapidly and fluctuate around Rs. 6000 per MWh. In this work, the main focus is to forecast the market clearing price in this region only as there is no point in forecasting for rest as that is fixed at maximum price.

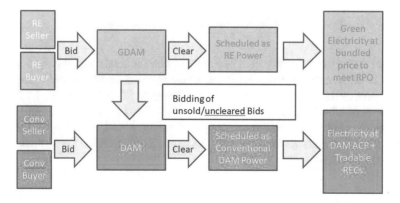

Fig. 40.2 Trading process of electric power in green day-ahead electricity market of Indian energy exchange

Table 40.1 Time line of various activities performed at GDAM [16]

Sl. No.	Time interval	Activities performed by power exchange
1	10.00 AM–12.00 PM	Exchange invites Integrated bids for G-DAM and DAM with option for transfer of uncleared renewable bids from GDAM to DAM
2	By 12.20 PM	Exchange determines clearing price and volume of G-DAM and DAM sequentially and provisional obligations of the Members
3	By 01.00 PM	Exchange communicates combined transmission corridor required for G-DAM and DAM to NLDC
4	By 02.00 PM	NLDC confirms available transmission corridor for scheduling
5	By 02.30 PM	Exchange determines final clearing price and volume first for G-DAM and then for DAM based on transmission corridor availability given by NLDC
6	By 02.45 PM	Exchange issues final obligation to Members
7	By 03.00 PM	Exchange files application for scheduling of G-DAM and DAM to NLDC
8	By 05.30 PM	NLDC conveys acceptance of scheduling of G-DAM and DAM

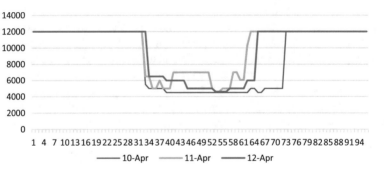

Fig. 40.3 MCP in GDAM from April 10, 2022 to April 12, 2022 [16]

Apart from quarter hourly market clearing price for single day, this work is also interested in forecasting daily MCP for a given period in price variable range. A typical plot of daily MCP for a time period between 12.00 PM and 12.15 PM for the month of March 2022 is plotted in Fig. 40.4. It is evident that the daily MCP has significant chaotic behavior as compared to Fig. 40.3 Forecasting daily MCP would be challenging task and for that, this paper proposes ANN based model with its synaptic weight optimized IPGA.

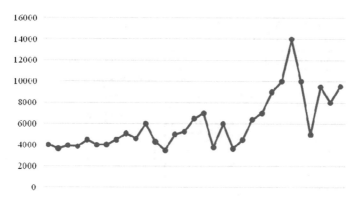

Fig. 40.4 Daily MCP for the month of March 2022 (time period: 12.00–12.15 PM) [16]

40.5 Numerical Simulation

The data for training Artificial Neural Network is obtained from websites of Indian Energy Exchange. For forecasting quarter hourly price of a particular trading day, data from March 1, 2022 to March 20, 2022 is extracted from GDAM section of IEX. For forecasting daily MCP for the month of March 2022, data for the month of January and February 2022 is extracted from the same platform. The program for all algorithms were written in MATLAB 2016(a) [17] and run on a PC with a core-i5 processor operating at 2.60 GHz and 4 GB of RAM. Number of input, hidden and output neuron is taken as 4, 10 and 1, respectively. The schematic diagram is shown in Fig. 40.5. Mean Absolute Error is considered as main objective function for forecasting the considered quantity. Maximum number of function evaluation is the termination criterion. Forecasted value of quarter hourly MCP are obtained for March 21, 2022 [16]. Forecasting is done for all 96 time slots and plotted to compare the results with actual value. Parameter setting for all algorithms are listed in Table 40.2.

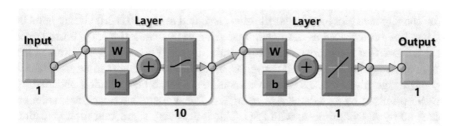

Fig. 40.5 Artificial neural network model in MATLAB [17]

Table 40.2 Parameter setting of GA and PSO based optimized ANN

GA parameter	Value	PSO parameter	Value
Total population size	50	Total population size	50
No. of iteration	50	No. of iteration	50
Type of crossover	Uniform crossover	Inertia weight	0.5
Crossover percentage	0.7	Inertia weight damping rate	0.99
Type of mutation	Normal distribution	Personal learning coefficient	1
Mutation percentage	0.4	Global learning coefficient	2
Mutation rate	0.02		
Mutation step size	0.1		
SA parameter	Value	BBO parameter	Value
Total population size	50	Total population size	50
Maximum iteration	50	No. of iteration	50
Initial temperature	0.1	Migration rate	0.5
Temp. reduction rate	0.99	Immigration rate	0.5
Neighbors per individual	5	Mutation probability	0.02
Mutation rate	0.5		
IPGA parameter	Value		
No. of iteration	100		
Total population size	50		
Max. function evaluation	5000		
Type of decision variable	Real		
Type of crossover	SBX ($\eta_c = 15$)		
Type of mutation	Normally distributed		
Dimensional reduction technique	Principal component analysis		

40.6   Simulation Results and Analysis

The convergence plot for all algorithms is presented in Fig. 40.6. Maximum number of iterations considered as termination criterion in this work (i.e., 50). From figure it is evident that the proposed ANN optimized with IPGA model performs much better as compared to model. It converges much faster to the best solution obtained by other algorithm. It takes only five iterations to reach best solution obtained by ANN optimized by Simulated Annealing and four iterations to reach best solution obtained by ANN optimized with PSO. This is also very significant for application which require forecasted value with very little time frame.

The final value of objective functions, i.e., Mean Absolute Error obtained by all forecasting models are listed in Table 40.3. From table, it is clear that the proposed model performs best with MAE value of 992.31 which is around 23.33% lower than that obtained using ANN optimized with BBO. Comparing with ANN optimized

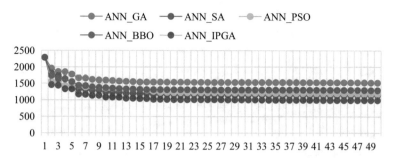

Fig. 40.6 Convergence plot of quarter hourly MCP for all optimization algorithms

Table 40.3 Quarter hourly MCP forecasting comparison in terms of MAE for all five models

Algorithm	MAE
ANN trained by GA	1519.29
ANN trained by SA	1142.31
ANN trained by PSO	1185.08
ANN trained by BBO	1294.29
ANN trained by IPGA	992.31

with GA which settles with MEA value of 1519.29, the IPGA-based model performs better by 34.69%.

The actual and forecasted value of quarter hourly MCP in GDAM using proposed model along with other existing model for March 21, 2022, is obtained and plotted in Fig. 40.7. Here, also it is evident that the proposed model gives much accurate results as compared to other techniques.

The convergence plot for forecasting daily MCP for a period between 12.00 PM and 12.15 PM using proposed algorithm along with the existing model is plotted in

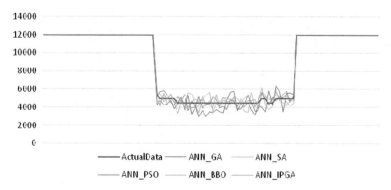

Fig. 40.7 Forecasted and actual value of quarter hourly MCP for March 21, 2022, using all five models

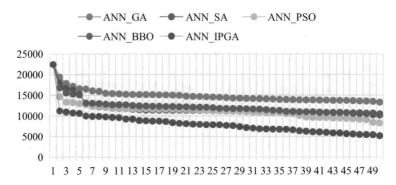

Fig. 40.8 Convergence plot of quarter hourly MCP for all optimization algorithms

Fig. 40.8. From Figure, it is evident that daily MCP obtained using the proposed fore-casting model performs much better and quickly converges to near optimal solution. On the other hand, the other algorithms gets trapped in local optima and results in suboptimal solution. The proposed algorithm takes only 25 iterations count to reach the best solution obtained by ANN optimized with PSO (Second best performing algorithm). This further reiterates that the proposed algorithms is much faster as compared with rest of the existing system.

Final value of objective functions is listed in Table 40.4. The final value of MAE obtained by the proposed algorithm is 5166.66. As compared to ANN optimized with BBO, the proposed model has better performance of 48.33% while comparing with ANN optimized with SA its performance is better by 50.58%. The worst performing model is ANN optimized with GA with MAE of 13,306.98.

Actual and forecasted value of daily MCP for the month of March 2022 is obtained and plotted in Fig. 40.9. It is evident from figure that the proposed model forecast the daily MCP with much accuracy as compared with the existing model.

Table 40.4 Daily MCP forecasting comparison in terms of MAE for all five models	Model used	Mean absolute error
	ANN trained by GA	13,306.98
	ANN trained by SA	10,458.71
	ANN trained by PSO	8270.36
	ANN trained by BBO	10,165.78
	ANN trained by IPGA	5168.66

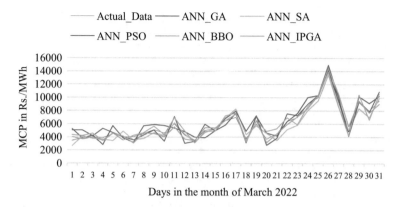

Fig. 40.9 Forecasted and actual value of daily MCP using all five models for the month of March 2022 (Time Period: 12.00–12.15 PM)

40.7 Conclusions and Future Scope

India Energy Exchange recently launch it new product called its market participants with an online market platform for short-term power trade generated from renewable source including hydro, wind and solar. This increase the strategic importance of market clearing price in GDAM for all generating companies as they can use this data to strategically submit their offers/bids in DAM which could maximize their individual surplus. Therefore, this paper proposes a new forecasting technique based on Artificial Neural Network with its synaptic weight updated using recently proposed IPGA equipped with additional diversity creating operator for precisely forecasting MCP in Green Day-Ahead Market of Indian Energy Exchange. The proposed model is tested for obtaining the forecasted value of quarter hourly and daily MCP in GDAM. The performance of the proposed model is compared with that obtained using four other optimization algorithm including GA, SA, PSO, and BBO. Results from Figs. 40.6, 40.7, 40.8 and 40.9 and Tables 40.3 and 40.4 clearly validates the superiority of the proposed algorithm. In future, the proposed model could be further modified by using concept of wavelets transform and could be used to forecast MCP for whole month.

References

1. Lu, H., Ma, X., Ma, M., Zhu, S.: Energy price prediction using data-driven models: a decade review. Comput. Sci. Rev. **39**, 100356 (2021)
2. Bhatia, K., Mittal, R., Varanasi, J., Tripathi, M.M.: An ensemble approach for electricity price forecasting in markets with renewable energy resources. Util. Policy **70**, 101185 (2021)
3. Gundu, V., Simon, S.P.: PSO–LSTM for short term forecast of heterogeneous time series electricity price signals. J. Ambient Intell. Humaniz. Comput. **12**(2), 2375–2385 (2021)

4. Fraunholz, C., Kraft, E., Keles, D., Fichtner, W.: Advanced price forecasting in agent-based electricity market simulation. Appl. Energy **290**, 116688 (2021)
5. Memarzadeh, G., Keynia, F.: Short-term electricity load and price forecasting by a new optimal LSTM-NN based prediction algorithm. Electr. Power Syst. Res. **192**, 106995 (2021)
6. Bas, E., Egrioglu, E., Kolemen, E.: Training simple recurrent deep artificial neural network for forecasting using particle swarm optimization. Granul. Comput. **7**(2), 411–420 (2022)
7. Zhang, F., Fleyeh, H., Bales, C.: A hybrid model based on bidirectional long short-term memory neural network and Catboost for short-term electricity spot price forecasting. J. Oper. Res. Soc. **73**(2), 301–325 (2022)
8. Tonyali, O., Bayram, D.: Forecast for market clearing price with artificial neural networks in day ahead market. Balk. J. Electr. Comput. Eng. **9**(4), 398–403 (2021)
9. Singh, S.N., Mohapatra, A., et al.: Data driven day-ahead electrical load forecasting through repeated wavelet transform assisted SVM model. Appl. Soft Comput. **111**, 107730 (2021)
10. Devnath, S., Chatterjee, S.: An intelligently programmed genetic algorithm with advanced deterministic diversity creating operator using objective surface visualization (Accepted and in Press, Manuscript Id. EVIN-D-19–00139). Evol. Intell. (2020)
11. Elsayed, S.M., Sarker, R.A., Essam, D.L.: A new genetic algorithm for solving optimization problems. Eng. Appl. Artif. Intell. **27**, 57–69 (2014)
12. Wang, Y., Yao, M.: A new hybrid genetic algorithm based on chaos and PSO. In: 2009 IEEE International Conference on Intelligent Computing and Intelligent Systems, vol. 1, pp. 699–703 (2009)
13. Shah, D., Chatterjee, S.: Optimal VAR dispatch using elitist multi-objective biogeography based optimization. In: Proceedings of 2019 3rd IEEE International Conference on Electrical, Computer and Communication Technologies (ICECCT2019) (2019)
14. Shah, D., Chatterjee, S.: Forecasting in Indian day-ahead electricity market using ANN optimized with biogeography based optimization algorithm. In: Proceedings of 2019 3rd IEEE International Conference on Electrical, Computer and Communication Technologies (ICECCT2019) (2019)
15. Hayashida, N., Takagi, H.: Visualized IEC: interactive evolutionary computation with multidimensional data visualization. In: 26th Annual Confjerence of the IEEE Industrial Electronics Society, vol. 4, pp. 2738–2743. IECON (2000)
16. IEX: Official websites available: https://www.iexindia.com/, in Indian Energy Exchange.
17. M. Inc., Matlab documentation (2019)

Chapter 41
Predicted New Adaptive Linear and Nonlinear Signal in Different QAM Using SRRC Filter

Preesat Biswas, Suman Sahu, Ishan Sen Chowdhury, Khemlal Dhruw, Hemant Kumar Binjhekar, and M. R. Khan

Abstract This paper describes optimizing the bit error rate using minimum power, less echo system, and getting high data rate compressing to various types of equalizer in the signal system, like a signal does not work properly due to bit error rate (BER), signal-to-noise ratio (SNR), burst error, and signal power spectrum. When we increase the data rate QAM also increases with SRRC filter connected with an adaptive filter, all these things rectify *"Predicted New Linear and Nonlinear Signal in Different QAM using SRRC Filter"* simulated in MATLAB-2019 a process for the future generation.

41.1 Introduction

The ultimate goal of digital communication is the secure transmission of data at the highest data rates achievable, however, when high-speed data is communicated via a communication channel, inter symbol interference (ISI) occurs. A delayed spread is formed throughout all multi-path objects when a signal is delivered from source to destination (IOS), resulting in ISI and a corrupted signal received is described in this paper "Predicted New Adaptive Linear and Nonlinear Signal in Different QAM using SRRC Filter". As a result, we must make advancements on the receiver side to lessen the influence of ISI in order to ensure reliable communication.

To combat ISI, we must apply equalization techniques on the receiver side for trustworthy communication. Equalization is a technique for compensating ISI caused

P. Biswas (✉)
Electronics and Communication Engineering, Dr. C. V. Raman University, Bilaspur, India
e-mail: preesat.eipl@gmail.com

S. Sahu · I. S. Chowdhury · K. Dhruw · H. K. Binjhekar · M. R. Khan
Electronics and Telecommunication Engineering, Government Engineering College, Jagdalpur, India
e-mail: mrkhan@gecjdp.ac.in

© The Author(s), under exclusive license to Springer Nature Singapore Pte Ltd. 2023   453
V. Bhateja et al. (eds.), *Evolution in Computational Intelligence*, Smart Innovation, Systems and Technologies 326, https://doi.org/10.1007/978-981-19-7513-4_41

by multi-path time-varying or scattered channels. As a result, an equalizer's primary objective is to counter the effects of the channel on the sent signal in order to duplicate the input signals at the receiver side [1].

41.2 Method Uses

This model shown in Fig. 41.1 represents the system model in which an input is provided. These signal goes through series to parallel converter then to block finite impulse response (FIR) filter weight vector are filter. Then, it goes to QAM system, SRRC filter system, then it goes to parallel to series converter, mechanism for sectoring, the output of which is contaminated by addictive Gaussian noise which is the error signal. The equalizer's role is to reduce the channel effect from transmitted information and also to rearrange. It leads to the original signal from the receiver's input. At the receiver side, we are arranging Least_Mean_Square (LMS) and recursive mean square (RMS) in cascade system.

A. Quadrature_Amplitude_Modulation (QAM)

In the digital or analogue modulation concept, QAM is defined as the modulation techniques and demodulation from receiver which can be used depending upon input signal.

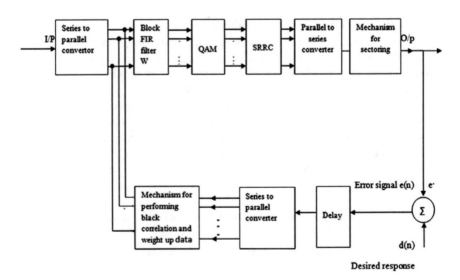

Fig. 41.1 Block diagram of transmission system of adaptive linear and nonlinear system

B. *Square-Rooted Raised Cosine Signals (SRRC)*

SRRC—The incoming signal which is filtering and they are not affected by AWGN which easily pass through the channel model noise is in receiver and this is Gaussian noise and if you want more information on that so the input signal is convoluted with the transmitted filter is involved with the received filter and the noise that's in the receiver amplifier in the receiver goes through the required filter and its components here why do we need these filters, you can send 1 a 0, you have to send a uniform.

C. *Equalization methods.*

There are two types of equalizer methods:

- Simple linear equalization-suboptimal
- Noisy nonlinear equalizer for serve and channel.

i. *Simple Linear Equalizer*

The most common linear form of channel equalization used in applications to reduce ISI is a linear equalizer. A transversal filter is also known as a FIR filter, can be used to build a linear equalizer. In this equalizer, the present and previous values of the message data are linearly scaled by the adjustable filter coefficient and summed to form the output as shown in the picture.

ii. *Noisy Channel with Nonlinear Equalizer for Serve*

Nonlinear equalizers are employed when the channel distortions are too high for a linear equalizer to reduce the effect of channel impairments. The performance of linear equalizers in channels with nulls is ineffective, which is why nonlinear equalizers are preferred over linear equalizers. Long impulse response and noise enhancement in these areas is an issue. The fundamental cause of this issue is that noise and linear filtering are treated in linear filtering, resulting in noise augmentation. Based upon the type of nonlinear equalizer, they are classified in two types which are as follows:

(a) Decision_Feedback_Equalizer-DFE
(b) Maximum_Likelihood_Sequence_Estimation-MLSE.

(a) Decision Feedback Equalizer (DFE)

DFE is a straight forward nonlinear equalizer, when it uses high amplitude distortion. The feed-forward portion is just a linear equalizer, it is full protected device in the feedback system and the output comes from the decision device. The output decision drives the feed–back system.

D. Signal-to-Noise Ratio (SNR)

The signal-to-noise ratio (SNR) is a measure of how much signal is received compared to how much noise is received. The SNR is calculated as follows:

$$SNR. = 10. \log_{10}\left(E_b / N_0.\right) dB \qquad (41.1)$$

The normalized SNR is defined as (E_b/N_0). SNR per bit, or normalized SNR, is the ratio of energy per bit to noise power spectral density. A high signal-to-noise ratio is good because it means that we are getting more signal and less noise, When the SNR is high, the signal strength is also high, but when it is low or very low, the signal is completely distorted, and we cannot retrieve the original signal.

E.   Bit Error Rate (BER)

When the source sends the bits stream through the communication channel, the number of bits mistakes is the number of bits that have been changed as a result of noise, interference, distortion, or bit synchronization problems.

As a result, BER refers to the number of bits that are corrupted or destroyed while being transmitted from source to destination, or the number of bits error per unit time. Error probability can also be used to express BER (POE). Formula of bit error ratio

$$BER = \text{No. of Errors} / \text{Total no. of bits sent}$$

BER is influenced by a number of parameters, including bandwidth, ratio of signal-to-noise (SNR), transmission medium, and transmission speed.

F.   Burst Error

In telecommunication, an uninterrupted succession of symbols is transferred across the transmitting data channel, the first and last symbols are in mistake whenever they are viewed at the receiving end and there is not a continuous sequence. The burst error is the difference between two "m" symbols. The error burst's guard band is represented by the numeric value "m."

G.   Least Mean Square (LMS)

Widrow and Hoff implemented the least mean square algorithm for the first time in 1950. This is one of the most basic and extensively used implementation methods. Because of the fairly simple solving mechanism used, it is a part of the stochastic algorithm (Fig. 41.2).

H.   Recursive Least Square (RLS)

The RLS algorithm solves the least squares problem recursively at each iteration when a new data sample is available the filter tap weights are updated and this leads to savings in computations more rapidly and convergence is also achieved (Fig. 41.3).

41.3   Mathematical Equations

The component measures the exponentially weighted error between the planned responsed (4) and the actual response of the filter. The tap-input vector u is related to $y(i)$ using the Eq. (41.2).

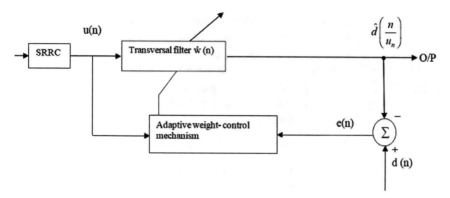

Fig. 41.2 Least mean square (LMS) with SRRC Filter

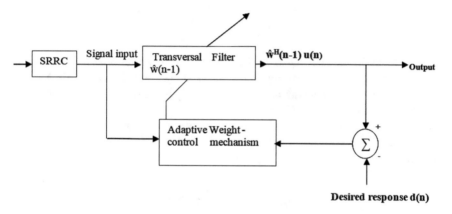

Fig. 41.3 Recursive least square (RLS) with SRRC filter

$$y(i) = W^H(n) \cdot u(i) \tag{41.2}$$

The regularizing term

$$\delta \cdot \lambda^{n} \cdot \| W(n) \|^2 = \delta \lambda^{n} \cdot W^H(n) \cdot W(n) \tag{41.3}$$

The regularization parameter is defined as positive real numbers. For the factor, the regularizing term is solely determined by the tap-weight vector [3]. The regularizing term is added in the cost function to smooth the solution and stabilize the recursive least squares problem. In a strict sense, the term $\delta \cdot \lambda^{n} \cdot \| W(n) \|^2$ to the cost function's favourable influence is forgotten with time. Second, and more importantly, the regularizing term should take the form $\delta \| DF \cdot (\hat{w}) \|^2$, where $F \cdot (\hat{w})$ is the RLS algorithm's input–output map and D [3] is a "rough" from of regulation, for two reasons.

For instance, the exponential weighting factor λ is in the interval; $0 < \lambda < 1$; So, for λ less than unity, λ^* approaches to zero for big n., which is beneficial for effective adding of $\delta \cdot \lambda^{n\cdot} \|W(n)\|^2$ to the cost function's favourable influence is forgotten with time. Second, and more importantly, the regularizing term should take the form $\delta \|\mathrm{DF} \cdot (\hat{w})\|^2$, where $F \cdot (\hat{w})$ is the RLS algorithm's input–output map and D is the differential operator. Nonetheless, the regularizing term in Eq. (41.2) is frequently used in the RLS algorithm formulation.

A.   The Normal Equation Reformulation

Expanding Eq. (41.3) and collecting terms, we discover that inserting the regularizing term $\delta \cdot \lambda^{n\cdot} \|W(n)\|^2$ in the cost function—has the same impact as reformulating the M-by-M time-average correction, tap—matrix—in out vector—.

$$\phi \cdot (n) = \sum_{i=1}^{n.} \lambda^{n.-1} \cdot u(i) \cdot u^H(i) + \delta \cdot \lambda^{n\cdot} I \tag{41.4}$$

I is the M-by-M identity matrix in this equation. It' i worth noting that adding the regularizing term makes the correlation matrix $\phi \cdot (n)$ non-singular at all stages of the computation, starting with $n. = 0$. A diagonally loaded correlation matrix is one that has been adjusted as in Eq. (41.5).

By the rule of equalization, M-by-1 time-average vector $z(n)$ are not effected when arrange between in the tap inputs of FIR filter and desired response with it. The equation are

$$z \cdot (n) = \sum_{i=1}^{N.} \lambda^{N.-1} \cdot u(i) \cdot d^*(i) \tag{41.5}$$

The usage of the "*" indicates complex conjugation and pre-windowing is assumed once more.

According to the least squares method, the M-by-1 tap-optimum weight's value vector $\hat{w}(n)$, when the cost function is defined by the fact that it reaches its minimal value by normal equation. For the recursive least-squares problem, the normal equation is written in matrix from as [4].

$$\phi \cdot (n) \cdot \hat{w}(n) = z(n) \tag{41.6}$$

B.   Recursive Computation of $\mathbf{z}(n)$ and $\phi(n)$ [4]

We may write Eq. (41.3) by separating the term corresponding to $i = n$. from the rest of the summation on the right-hand side.

$$\phi \cdot (n) = \lambda \cdot \left[\sum_{i=1}^{N.} \lambda^{N.-1} \cdot u(i) \cdot u^H(i) + \delta \cdot \lambda^{N.} I \right] u(n) \cdot u^H(n) \tag{41.7}$$

The analytic method for dealing with the correct regularizing term $\delta \left\| DF \cdot \left(\hat{w} \right) \right\|^2$ is based on the concept of a function space, which is a normed space of functions. A continuous function is represented by a "vector" in such a space of many (strictly speaking, infinitely many) lesions. This geometric figure establishes an interesting link between the liner differential operator and matrices (Lanczos 1964).

By definite in, the correlation matrix $\phi \cdot (n - 1)$ is identical to the expression inside the brackets on the right-hand side of Eq. (41.9). As a result, we get the following recursion for updating the value of the inputs' correlation matrix

$$\phi \cdot (n) = \lambda \cdot \phi \cdot (n - 1) \cdot u(i) \cdot u^H(i) \tag{41.8}$$

Here, $\phi \cdot (n - 1)$ is the "old" value of the correlation matrix, and the matrix product $u(n) \cdot u^{H \cdot}(n)$ plan the role of a "correlation" term in the updating operation. Note that the recursion of Eq. (41.10) holds, irrespective of the correlation matrix of the tap inputs [8]:

Similarly, we may use Eq. (41.8) to derive the correlation vector between the tap inputs and the desired response.

$$z \cdot (n) = \lambda \cdot z \cdot (n - 1) + u(n) \cdot d^*(n) \tag{41.9}$$

To compute the least-square estimate for $\hat{w}(n)$ the tap-weight vector in accordance with Eq. (41.7), we have to determine $\phi \cdot (n)$ the inverse of the correlation matrix. In practice, however, we usually try to avoid performing such an operation, as it can be quite time consuming, particularly if the number of tap weight, M, is high. Also, we would like to be able to compute the least square estimate $\hat{w}(n)$ for the tap-weight vector recursively for $n = 123....\infty$.

C.   The Adaptive Noise Canceller with Single Weight

The adaptive noise canceller with single-weight and dual-input is a primary signal $d(n)$ used to represent the two inputs, and it consists of a data containing signal element and an additive noise component, and $u.(n)$, which is associated with the disturbance but has no distinguishable contribution to the content signal. The goal is to use the reference signal's qualities in respect to the primary signal to reduce interference at the adaptive noise canceller's output.

$$(n) = \left[\frac{1}{\lambda \cdot \hat{\sigma}_{n.}^1 (n - 1) + \left[u(n)^{2\cdot} \right]} \right] \cdot u(n) \tag{41.10}$$

D.   The Learning Curve

There are different kinds of errors in the RLS algorithm: a priori estimation error $j(n)$ and a posteriori estimation error $e_{0.}(n)$. We discover that the mean-square value of these two errors varies differently with time n, given the initial condition. When $N. = 1$, the mean-square value of $J.(n)$ becomes large equivalent to the mean-square value of the intended response $d(n)$ and then decays as n increases. The mean-square value of $\xi(n)$, on the other hand [2], is tiny at $n = 1$ and climbs with rising n until it reaches a point where $e_{0.}(n)$ equals $\xi(n)$ for high n. As a result, using $\xi \cdot (n)$ as the error of interest produces a learning curve for the RLS method that is similar to that of the LMS algorithm.

$$J' \cdot (n) = E \cdot \left[|\xi \cdot (n)|^2 \right] \tag{41.11}$$

The prime in the symbol $J'(n)$ is distinguish the mean-square value of $\xi(n)$ from that of $e(n)$. Eliminating the desired response $d(n)$, we may express the a priori estimation error

$$\xi \cdot (n) = e_{0.}(n) + \left[W_{0.} \hat{W} \cdot (n - 1) \right]^H \cdot u(n) = e_{0.}(n) + \varepsilon^H \cdot (n - 1)u(n) \tag{41.12}$$

The outer product $\varepsilon(n - 1)\varepsilon^H(n - 1)$ of the weight-error vector at time step n fluctuates at a slower than the outer product $u(n)u^H(n)$. Applying Kushner's direct a = averaging method.

$$E \cdot \left[u^H(n) \cdot \varepsilon^H(n - 1) \cdot u(n) \right] \approx \mathrm{tr}\left[E \cdot \left\{ \left[u(n) \cdot u^H(n) \cdot \varepsilon(n - 1) \cdot \varepsilon^H(n - 1) \right] \right\} \right]$$
$$= \mathrm{tr} \cdot [R \cdot K(n - 1)] \tag{41.13}$$

After substituting above equations into Eq. (41.14) yields [5]

$$E \cdot \left[u^H(n) \cdot \varepsilon(n - 1) \cdot \varepsilon^H(n - 1) \cdot u(n) \right] \approx \frac{1}{n} \cdot \sigma_o^2 \mathrm{tr} \cdot \left[RR^{-1} \right] = \frac{1}{n} \cdot \sigma_o^2 \mathrm{tr} \cdot [I]$$
$$= \frac{M}{n} \cdot \sigma_o^2, n. > M \tag{41.14}$$

where M is the filter length, the third expectation is zero, for two reasons. First, viewing time step m as the present, weight-error vector depends on the past values of the input vector $u(n)$ and measurement noise $e_{0.}(n) \cdot u(n)$ and $e_{0.}(n)$ are statistically independent, and $e_{0.}(n)$ has zero mean. Hence, we may write

$$E \cdot \left[\varepsilon^H(n - 1) \cdot e_o^*(n) \right] = E \cdot \left[\varepsilon^H(n - 1) \cdot e_o^*(n) \right] \cdot E\left[e_o^*(n) \right] = 0 \tag{41.15}$$

As a result, the fourteenth expectation has the same mathematical form as seventeenth expectation, except for a trivial complex conjugation. Hence, the fourteenth expectation is also zero.

Using these results in Equation, we obtain the following result:

$$J' \cdot (n) \approx \sigma_o^{2.} + \frac{M}{n} \cdot \sigma_o^{2.}, n. > M \tag{41.16}$$

E. Adaptive Filtering

There are four types of filtering which are as follows:

1. Configuration of an adaptive system for identifying it
2. Noise cancellation that is adaptive
3. Linear regression with adaptive parameters.

Input, output, desired result, adaptive transfer function, and an error signal are all present in all of these above systems.

The difference between the original and the required signal is called as error signal [7]. The system identification and inverse system design both involve an unknown linear system transfer function that can receive an input and output a linear output in addition to these components.

So now we will talk about how to configure adaptable system identification.

In general, adaptive system identification is only responsible for the unknown determination of the transfer function, and the same input is applied to the adaptive filter configuration, unknown system from which the outputs are compared, and adaptive system identification configuration, and the results will produce an error, and this error will manipulate the adaptive weight of this configuration, after the number of iterations is performed and if the system is zero, Although conversion to zero is an ideal condition, it does not occur in practice; there will be a difference between the adaptive and unknown transfer functions, but the error will be tiny; also, in the adaptive system's order will influence the minimum error that the system can achieve.

If there are insufficient coefficient in the adaptive system to model the unknown system, it is set to under specified, which may cause the error to convert to a nonzero error instead of zero. If the adaptive filter is over specified, the error will convert to zero but conversion time will increase.

Adaptive system identification configuration now starts to implement the filer algorithm. So here, we are going to take signal which is sinusoidal using some noise and then we will perform filtering where we will filter and then we will get desired output/result (Fig. 41.4).

F. FIR Low Pass Filter in Time Domain and Frequency Domain

In this, we are going to show you how we are going to apply low pass and band pass FIR filter in MATLAB by using two different technique the first technique is by filtering the noisy signal in time domain by using convolution theorem technique and second technique is by filtering the noise signal in frequency domain by using multiplication techniques both techniques will produce the same results. Our noisy signal is in time domain, so we are going to filter this signal by using impulse response of the filter and then we are going to convolve both the signals. Now in the frequency domain, the first thing to do in frequency domain signal from time domain signal

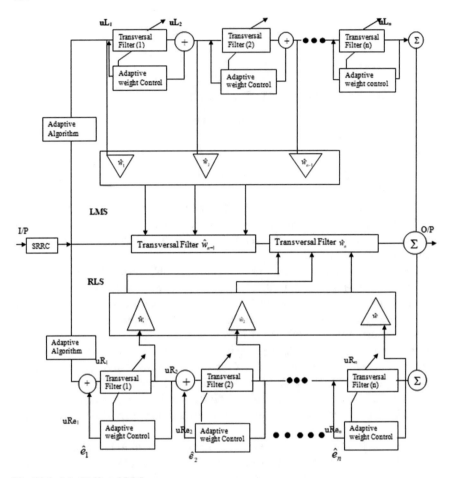

Fig. 41.4 Join LMS and RLS

so that you can see in this particular time domain signal. There are three different frequencies so the aim is to filter this signal and retain the frequency signal.

G. AWGN

AWGN is a crucial factor to decide the performance of a communication system. Meaning:

A—Additive

While modelling the impact of noise to the transmitter signal. Here, X is the output, Y is the input, and N is the noise [6].So, what do we mean by it is additive, like we are not modelling it like this, we are not multiplying noise to the transmitted signal, we are not doing this. Actually, we are adding the noise to the transmitted signal. So, if I have to model it like this, $X = Y + N$. Here, X is the output, Y is the input [4], and N is the noise.

W—White

It has uniform power across whole frequency band that means if I have to draw the power spectral density, is the power per unit of frequency that will be in flat line. So, in theory or literature, you might have noticed that this power spectral density is $N_0/2$. We will talk about this more but this is a flat line.

G—Gaussian

As these are random variables, therefore, we have to define their statistical parameters like probability density function (PDF). The probability density function (PDF) of noise is Gaussian in nature with zero mean. We also say this as bell shaped curve.

N—Noise—To assess the effectiveness of a digital communication system which can be indicated by bit error rate (BER)—thermal noise plays a crucial role here.

41.4   Results and Discussion

In transmission system, data rate 10^{-7} bits/sample using the 128-QAM and 512-QAM with SRRC filter in adaptive feedback the simulated signal as shown in Figs. 41.5, 41.6, 41.7, and 41.8.

In transmission system, data rate 10^{-7} bits/sample using the 1024-QAM with SRRRC filter in adaptive feedback the simulated signal as shown in Figs. 41.9 and 41.10.

Fig. 41.5 LSM equalizer 128-QAM

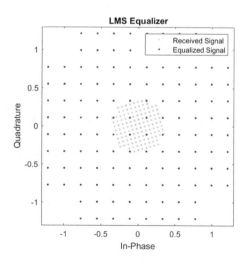

Fig. 41.6 RLS equalizer
128-QAM

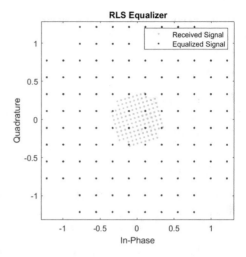

Fig. 41.7 LMS equalizer
512-QAM

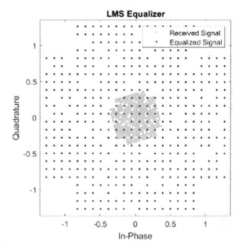

These signal which are unwanted of estimating error those are goes to feedback system where the signal both LMS error estimation and RMS error estimation with 256-QAM, 1024-QAM having 20,000 sample rate and path delay is 500 and radiant is $= 10^{-7}$ bits/sample as shown in Figs. 41.11, 41.12, 41.13, and 41.14. Error estimate of RLS equalizer in 1024-QAM having 20,000 sample rate, path delay is 500 and radiant is equal to 10^{-7}.

Figure 41.15 shown LMS equalizer in 1024-QAM and Fig. 41.16 RLS equalizer in 1024-QAM having 12,000 sample rate, path delay is 150 and radiant is equal to 10^{-7}.

Fig. 41.8 RLS equalizer
512-QAM

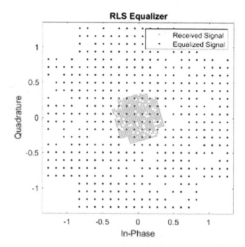

Fig. 41.9 LMS equalizer
1024-QAM

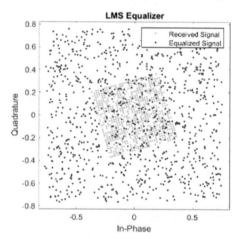

Fig. 41.10 RLS equalizer
1024-QAM

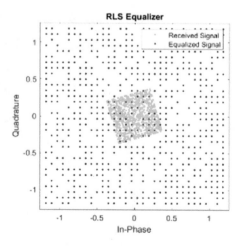

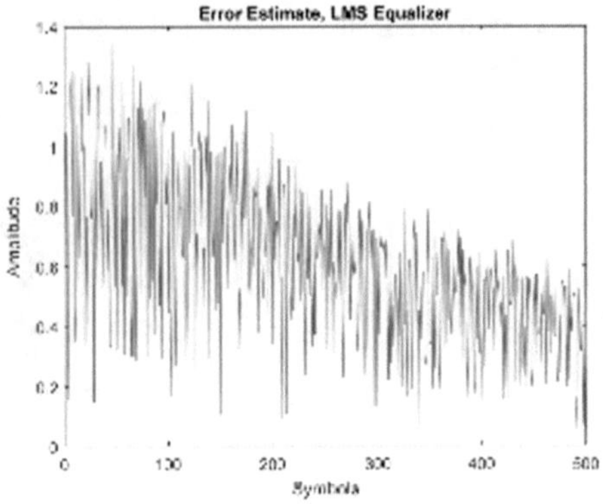

Fig. 41.11 Error estimate
LMS equalizer 128-QAM

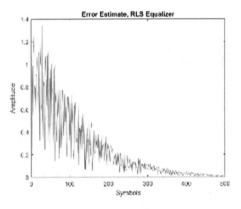

Fig. 41.12 Error estimate
RMS equalizer 128-QAM

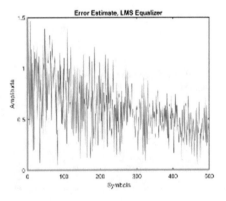

Fig. 41.13 Error estimate
LMS equalizer 1024-QAM

Fig. 41.14 Error estimate, RLS equalizer 1024-QAM

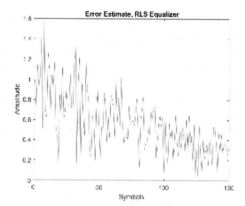

Fig. 41.15 LMS equalizer 1024-QAM

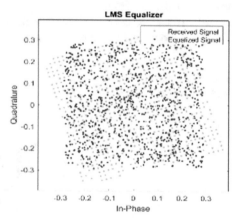

Fig. 41.16 RLS equalizer 1024-QAM

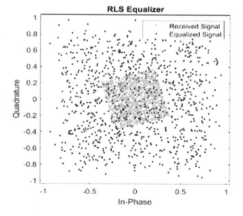

Figure 41.17 is shown error estimate of RLS equalizer in 1024-QAM, Fig. 41.18 is shown RLS equalizer in 1024-QAM. Figure 41.19 is shown error estimate of LMS equalizer in 1024-QAM. Figure 41.20 is shown RLS equalizer in 1024-QAM having 15,000 sample rate, path delay is 200 and radiant is equal to 10^{-7}.

Table 41.1 depicts the system error with delay and system with delay removed, which is best output by using sample 256-QAM, the result system error with delay is 7969 bits, having error vector magnitude (EVM) of 57.0306, with system error delay of 8320 bits, and, system error with delay removed is bits, having error vector magnitude (EVM) of 45.5840, with system error delay of 9357 bits.

H. MSE

The MSE tells us about we have choose a reprogram to a set of points, as it does by testing the distance from these points to points to the regression line these distances are called the error and those error are squared to remove any negative signs the error also gives more weight to large distances and it is called mean square error because

Fig. 41.17 Error estimate RLS equalizer 1024-QAM

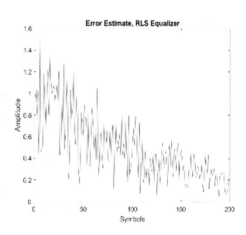

Fig. 41.18 RLS equalizer in 1024-QAM

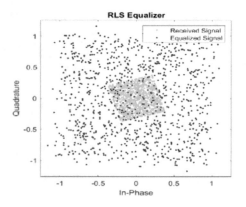

Fig. 41.19 Error estimate
RLS equalizer 1024-QAM

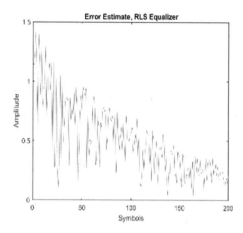

Fig. 41.20 RLS equalizer
1024-QAM

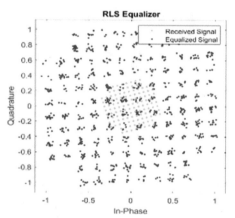

Table 41.1 System with delay and system with delay removed

S No.	QAM	SymErr with delay	Evm with delay	SymErr delay removed	Evm delay removed
1.	M = 16	7462	56.5685	5814	45.4060
2.	M = 32	7694	53.8874	5464	44.7394
3.	M = 64	7850	57.1914	7711	45.2427
4.	M = 128	7926	51.6756	8320	45.4118
5.	M = 256	7941	52.8042	8995	45.1867
6.	M = 512	7963	52.7391	9216	44.7358
7.	M = 1024	7969	57.0306	9357	45.5840

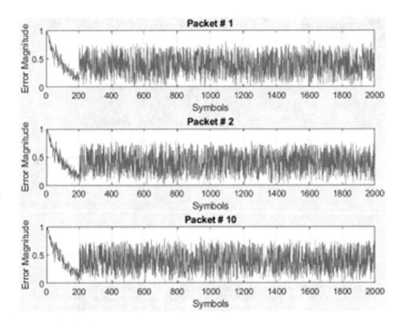

Fig. 41.21 Error magnitude versus symbols in 1024-QAM

we are finding the average of a set of errors positive errors the embassy the better the forecast.

Figure 41.21 shown error magnitude versus symbols in 1024-QAM, Fig. 41.22 shown error magnitude versus symbols in 1024-QAM. Figure 41.23 shown angular error over time and time-varying channel without retaining in 1024-QAM. Figure 41.24: scatter plot of 1024-QAM in quadrature versus in-phase (Fig. 41.25).

In the figure that is Fig. 41.26 shows the error magnitude versus symbols in 1024-QAM. Error magnitude versus symbols in 1024-QAM.

In the figure that is Fig. 41.26 shows the error magnitude versus symbols in 1024-QAM. Error magnitude versus symbols in 1024-QAM. Figure 41.25 shown angular error over time and time-varying channel without retaining in 1024-QAM. Figure 41.26 shown scatter plot of 1024-QAM in quadrature versus in-phase (Fig. 41.28).

The error output of the equalized signals, delay, and delay removed, is plotted in Fig. 41.27. The equalization converges during the 10,000-symbol training period for the signal that has had the delay removed. To account for uncovered output and system latency between equalization output and transmitted symbols, skip the first 500 symbols while demodulating symbols and determining symbol faults. Reconfiguring the equalizer to account for system delay improves signal equalization and lowers symbol errors and EVM.

Figure 41.29 shown average coefficient trajectories for simulation and theory—coefficient value versus time index. Figure 41.30 shown MSE (SIM.), final MSE, and

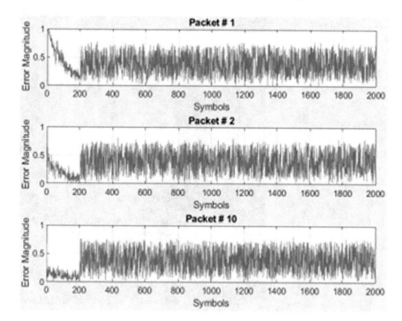

Fig. 41.22 Error magnitude versus symbols in 1024-QAM

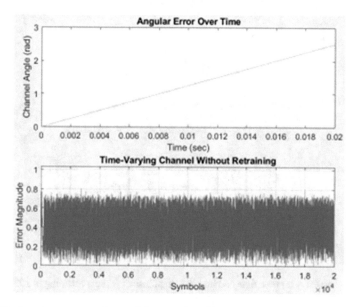

Fig. 41.23 Angular error over time and time-varying channel without retaining in 1024-QAM

Fig. 41.24 Scatter plot of
1024-QAM versus in-phase

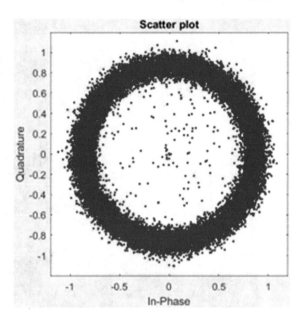

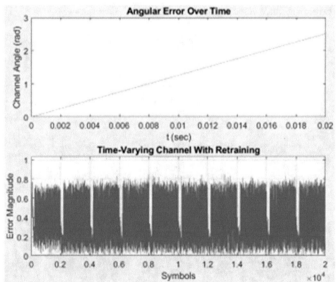

Fig. 41.25 Angular error over time and time-varying channel without retaining in 1024-QAM

Fig. 41.26 Scatter plot of 10,244 QAM in quadrature versus in-phase

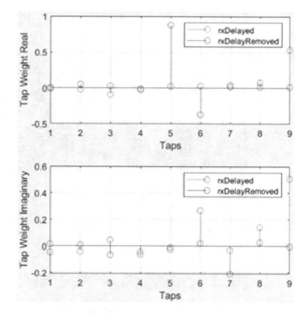

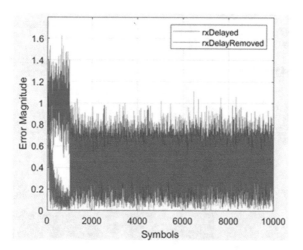

Fig. 41.27 Tap-weight real and imaginary with delay and delay removed

Min. MSE are all measures of mean square error performance. Figure 41.31: Sum of squared coefficient errors. Figure 41.32 shown the FIR filter's system identification.

The graphs above show the values for the minimal MSE, simulated MSE, predicted MSE, and final MSE. Figure 41.30 shows the final MSE value as the sum of the minimum and excess MSE values.

The anticipated and simulated MSEs have the same trajectory. The steady-state (final) MSE is reached by both of these trajectories (Fig. 41.31).

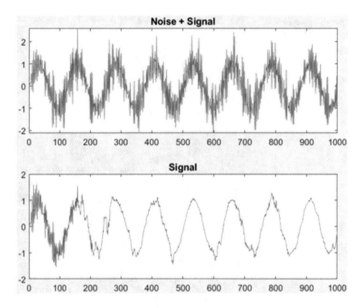

Fig. 41.28 Noise plus signal and equalized signal

Fig. 41.29 Average
coefficient trajectories for
simulation and
theory—coefficient value
versus time index

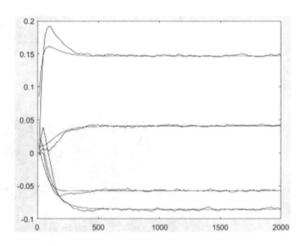

We have used the "msepred" and "msesim" functions, provided by "msepred" and "msesim" in Fig. 41.32, we compared the total of the squared coefficient errors, i.e. at each time instant, the anticipated MSE value and the simulated MSE value. The coefficient covariance matrix's trace provides these values.

The values of the LMS filter that is taken to resemble the unknown system (FIR filter) are represented by the weights vector w in the above figures, which are Figs. 41.31 and 41.32. To confirm convergence, the numerator of the FIR filter is compared, and the estimated weights of the adaptive filter is calculated. The predicted

Fig. 41.30 MSE (Sim.),
final MSE, and min. MSE
are all components of mean
square error performance

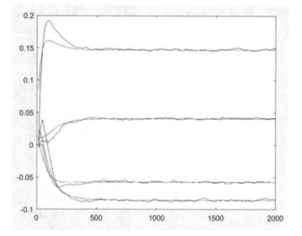

Fig. 41.31 Sum of squared
coefficient errors

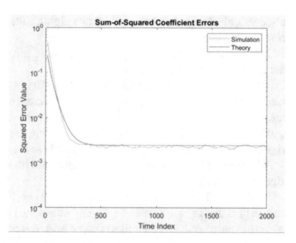

Fig. 41.32 System
identification of an FIR filter

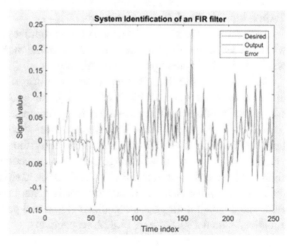

Fig. 41.33 Result of noise cancellation and actual signal

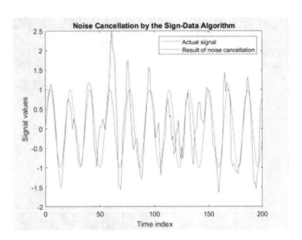

filter weights do not match the real filter weights, indicating that the prior signal plot results were correct.

The intended input is the sign-data procedure, and the noise is correlated noise in the above picture, i.e. Figure 41.34.

We have established the LMS filter as our processing aim step size. The coefficients and mu values we choose influence whether the adaptive filter can eliminate noise from the signal route, as we said earlier in the sections.

We built a default filter in Fig. 41.35 by setting the filter coefficients to zero. The sign-data method does not respond well to this strategy in most circumstances.

We started with the noise filter coefficients and tweaked them slightly to force the algorithm to adapt (Fig. 41.36).

Figure 41.37 shows the received source grid and Fig. 41.38 shows the equalized source grid having ndlrb value of 20, and similarly, Figs. 41.39 and 41.40 show the received source grid and equalized source grid, respectively, having ndlrb of 110.

For the given above figure, i.e. Figure 41.33 noise is the correlated noise and the desired input is the sign-data algorithm.

As a result, in simulation the adaptive filter removes the noise from the signal path when we set for coefficients and "mu" determination. Set the LMS filter processing with weighted and step size signa at initially.

Arranging by the default filter, it set the filter to be zero. The expected value those are closer it and most likely case is that the algorithm remains maintained and converges to a filter solution which removes the noise.

In Figs. 41.33, 41.34, and 41.35, we have started with the values used in the noise filter and modified them slightly so that the algorithm can be adapted.

Table 41.2 depicts the percentage of RMS error vector magnitude of pre-equalized signal and percentage of RMS-EVM post-equalized signal, which has the best output by using the NDLRB value of 110, the result system shows the percentage of RMS error vector magnitude of pre-equalized signal and percentage of RMS-EVM post-equalized signal 139.144 and 17.822, respectively.

Fig. 41.34 Result of noise
cancellation and actual
signal by sign error

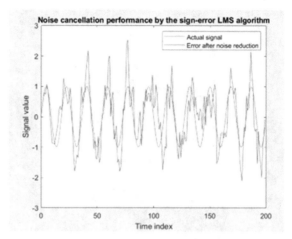

Fig. 41.35 Result of actual
signal after error reduction

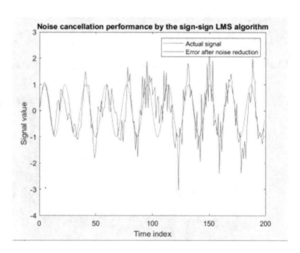

41.5   MMSE Equalization

The equalizer MMSE function is used to balance the effects of the channel on the
received resource grid. This function uses the estimated channel and noise to equalize
the received resource grid. The function returns the equalized grid. The equalized grid
has the same dimensions as the original transmitted grid before OFDM modulation.

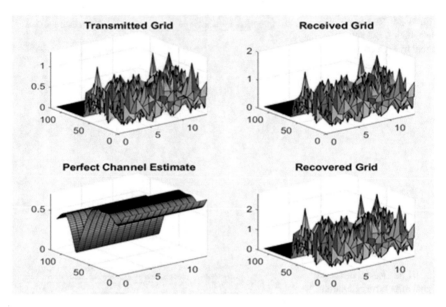

Fig. 41.36 Shows the transmitted grid, received grid, perfect channel estimate, and recovered grid

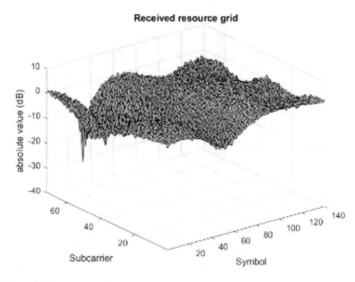

Fig. 41.37 Received source grid

41.6 Analysis of MMSE Equalization

The equalized resource grid is compared to the resource grid that was received. It is estimated the difference between the transmitted and equalized grids, as well as

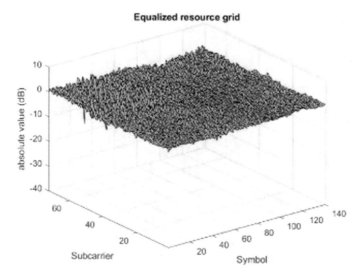

Fig. 41.38 Equalized source grid

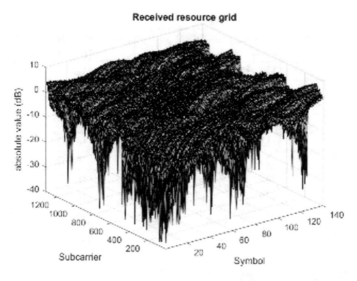

Fig. 41.39 Received source grid

the transmitted and received grids. This results in two matrices (of the same size as the resource arrays) containing the error for each symbol. Within the received and equalized grids are plotted on a logarithmic scale to make scrutiny easier. These diagrams show how channel equalization reduces the inaccuracy in the received resource grid dramatically.

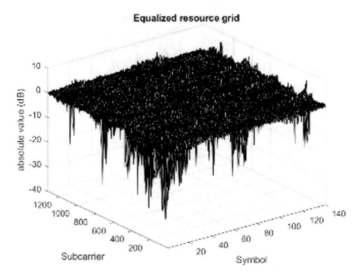

Fig. 41.40 Equalized source grid

Table 41.2 Post- and pre-equalized signal

S. No.	NDLRB	Percentage RMS EVM pre-equalized signal	Percentage RMS EVM post-equalized signal
1.	6	119.939	12.903
2.	10	120.112	14.461
3.	15	124.133	15.598
4.	20	125.813	16.940
5.	30	122.026	18.222
6.	60	125.131	18.463
7.	90	133.087	18.494
8.	100	136.351	18.034
9.	105	137.430	18.105
10.	110	139.144	17.822

41.7 Results and Discussion

The result from Table 41.1, the system error with delay and system with delay removed, which is best output by using sample 256-QAM, the result system error with delay is 7969 bits, having error vector management (WVN) of 57.0306, with system error delay of 8320 bits, and system error with delay removed is bits, having error vector management (EVM) of 45.5840, with system error delay of 9357 bits.

Table 41.2 depicts the percentage of RMS error vector magnitude of pre-equalized signal and percentage of RMS-EVM post-equalized signal, which has the best output by using the NDLRB value of 110, the result system shows the percentage of RMS error vector magnitude of pre-equalized signal and percentage of RMS-EVM post-equalized signal 139.144, and 17.822, respectively.

41.8   Conclusion

In this paper, "Predicted New Linear and Nonlinear Signal in Different QAM using SRRC Filter", we have compared and simulated the performance of various types of equalizers in the signal system of mobile communication systems. Where a signal is not properly working due to bit error rate (BER), burst error, and signal power spectrum, the system mode in which an input is provided and data goes through series to parallel converter then to block FIR filter W then QAM, SRRC, parallel to series converter, mechanism for sectoring, the output of which is contaminated by addictive Gaussian noise which is the error signal. The equalizer's role is to reduce the channel effect from transmitted information and also to regain it, leading in the original signal at the receiver's output, i.e. the desired output.

As a result, the simulation and result show how the equalization outperforms the other using MATLAB-2019a.

References

1. Paliwal, S., Grover, D.K., Krayla, J.: Comparison of linear and non-linear equalizer using the maltlab. Commun. Appl. Electron. (CAE) **4**(1) (2016). ISSN: 2394-4714. Foundation of Computer Science FCS, New York
2. Borisagar, K.R., Thanki, R.M., Sedani, B.S.: Speech Enhancement Techniques for Digital Hearing Aids. Springer Science and Business Media LLC (2019)
3. Borisagar, K.R., Thanki, R.M., Sedani, B.S.: Chapter 3 Introduction of adaptive filters and noises for speech. Springer Science and Business Media LLC (2019)
4. Rangayyan, R.M.: Biomedical Signal Analysis. Wiley (2015)
5. Song, S.: Reduced complexity self-tuning adaptive algorithms in application to channel estimation. IEEE Trans. Commun. (2007)
6. Bouguerra, F., Saidi, L.: An efficient ANN interference cancelation for high order modulation over Rayleigh fading channel. J. Telecommun. Inf. Technol. (2019)
7. Sahoo, S., Barapatre, Y.K., Sahoo, H.K., Nanda, S.: FPGA implementation of fuzzy sparse adaptive equalizer for indoor wireless communication systems. Appl. Soft Comput. (2021)
8. Haykin, S.: Adaptive systems for signal process. In: Electrical Engineering and Applied Signal Processing Series (2000)

Chapter 42
Comparative Analysis of Wind Power Forecasting Using LSTM, BiLSTM, and GRU

Manisha Galphade⬛, **V. B. Nikam**⬛, **Biplab Banerjee**⬛, **and Arvind W. Kiwelekar**⬛

Abstract Artificial intelligence (AI) technology is rapidly evolving, and its applications in big data analytics and practices are becoming increasingly diverse. Deep learning-based time series prediction, such as LSTM, GRU, and BiLSTM, is a promising area for future big data analytics and techniques. Forecasting wind power is an important but difficult element of time series data analysis. The type of time series data used, as well as the underlying context, are the most important aspects that influence the performance and accuracy of time series data analysis and forecasting methodologies. This research examined the state-of-the-art techniques LSTM, BiLSTM, and GRU for wind power forecasting. The results reveal that BiLSTM-based modeling outperforms ordinary LSTM-based models in terms of prediction. Furthermore, based on time horizons, input characteristics, calculation time, error measurements, and other factors, this study produced a guideline for wind power forecasting process screening, allowing wind turbine/farm operators to pick the most relevant prediction approaches.

M. Galphade (✉) · V. B. Nikam
Department of Computer Engineering and IT, Veermata Jijabai Technological Institute, Mumbai, India
e-mail: galphademanisha@gmail.com

V. B. Nikam
e-mail: vbnikam@it.vjti.ac.in

B. Banerjee
Center of Studies in Resources Engineering, IIT Bombay, Mumbai, India
e-mail: bbanerjee@iitb.ac.in

A. W. Kiwelekar
Department of Information Technology, Dr. Babasaheb Ambedkar Technological University, Lonere, Raigad, India
e-mail: awkiwelekar@dbatu.ac.in

42.1 Introduction

Wind energy is a clean and sustainable type of energy that is becoming more popular across the world. Wind turbine farms, on the other hand, can generate grid instability due to the uncontrollability of the wind, and the cost of cycling traditional power units to compensate for this fluctuation is roughly 157 million dollars per year in places with just 35% integration [1]. In the power sector, or the electricity market, forecasting allows power supply businesses to quickly detect demand trends that may arise in the future. As a result, predicting essentially assists people in being prepared for the impending circumstance. In addition, corporations might become optimistic or bearish on the market's electrical demands, causing their prices to fluctuate correspondingly. For time series issues, such as wind power forecast, there are a variety of techniques available, including mathematical and statistical modeling. Machine learning techniques, on the other hand, may be more robust and accurate in terms of robustness and accuracy since they are very adaptive and do not need modeling a turbine environment [2].

Deep learning [3] approaches primarily consist of multiple neural networks inspired by brain neurons. RNN, which is a broad name for a succession of neural networks capable of processing sequence data [4], is the ideal technique to deal with time series data [5] such as meteorological data. With gradients, RNN has a lot of trouble dealing with long-term dependencies [6]. The long short-term memory network was designed to overcome the problem of long-term reliance, and it performed well in dealing with time series data concerns. It has the ability to forecast new data points by effectively using temporal information. It is been used in natural language processing [7], to predict stock market movements [8], and even in medicine [9].

This research shows three strategies for predicting wind turbine power production using wind speed, wind direction, and theoretical power based on the turbine's power curve as an input parameters: Long short-term memory (LSTM), bidirectional long short-term memory (BiLSTM), and gradient recurrent unit (GRU). Then, using particular performance metrics, all three models are compared to one another, and analysis is performed based on the results of these parameters. Furthermore, depending on the results, the optimal model is proposed.

42.2 Description of Algorithm

42.2.1 LSTM

LSTM [10] is a modified version of RNN, which is based on time series data. It is a special type of circulatory neural system that decides how to depend on data for a long period of time. Memory cells are used instead of RNN's the hidden layers, as seen in Fig. 42.1

Fig. 42.1 Internal structure
of LSTM cell

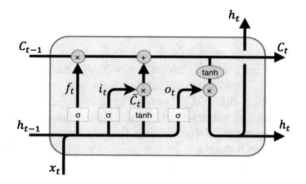

LSTMs learn behavior of temporal correlations using various gate units such as input gate (i_t), forget gate (f_t), and output gate (o_t), along with the tanh and sigmoid activation functions. The output of each gate is 1 or 0, which describes the amount of data each gate allows, calculated by the sigmoid function. Input gate decides which information is worth to remember and finally output gate decides output of each cell. If forget gate output is one, then the complete information will be preserved, otherwise, it will be discarded. The computation process is described as follows:

Equation (42.1) calculated the output vector h_t

$$\text{Output} : h_t = o_t * \tanh(C_t) \tag{42.1}$$

where C_t is cell state, o_t is output gate vector, expressed using Eqs. (42.2) and (42.3)

$$\text{Output Gate} : o_t = \sigma\left(W_o \cdot [h_{t-1}, x_t] + b_o\right) \tag{42.2}$$

$$\text{Cell Update} : C_t = f_t * C_{t-1} + i_t * \widehat{C_t}) \tag{42.3}$$

$$\text{Process Input} : \widehat{C_t} = \tanh\left(W_c \cdot [h_{t-1}, x_t] + b_c\right) \tag{42.4}$$

At each step C_t is calculated to update the memory cell using tanh function so as to give better training performance. f_t and i_t are forget gate and input gate, respectively, calculated using Eqs. (42.5) and (42.6). W_o, W_c, W_i and W_f are weight matrix corresponding to each gate.

$$\text{Forget Gate} : f_t = \sigma\left(W_f \cdot [h_{t-1}, x_t] + b_f\right) \tag{42.5}$$

$$\text{Input Gate} : i_t = \sigma\left(W_i \cdot [h_{t-1}, x_t] + b_i\right) \tag{42.6}$$

42.2.2 BiLSTM

Because the only inputs it has seen are from the past, LSTM only saves information from the past. This flaw may be overcome by adopting bidirectional LSTMs, which process inputs in two directions: past to future and future to past as shown in Fig. 42.2. BiLSTM [11] saves information from the past as well as the future. The hidden and cell states are computed in the forward component, identical to a normal unidirectional LSTM, whereas the backward component computes them by taking the input sequence in reverse-chronological order, beginning from time step Tx to 1. The computation process of BiLSTM [12] is described using Eqs. (42.7–42.13)

$$\text{Forget Gate} : f_t = \sigma\left(W_f \cdot \left[h_{t-1}, x_t\right] + b_f\right) \tag{42.7}$$

$$\text{Input Gate} : i_t = \sigma\left(W_i \cdot \left[h_{t-1}, x_t\right] + b_i\right) \tag{42.8}$$

$$\text{Process Input} : \widehat{C_t} = \tanh\left(W_c \cdot \left[h_{t-1}, x_t\right] + b_c\right) \tag{42.9}$$

$$\text{Cell Update} : C_t = f_t * C_{t-1} + i_t * \widehat{C_t}) \tag{42.10}$$

$$\text{Output Gate} : o_t = \sigma\left(W_o \cdot \left[h_{t-1}, x_t\right] + b_o\right) \tag{42.11}$$

$$\text{Output} : h_t = o_t * \tanh(C_t) \tag{42.12}$$

$$y_t = f\left(\overrightarrow{h_t}, \overleftarrow{h_t}\right) \tag{42.13}$$

where $\rightarrow$ and $\leftarrow$ represent the result of forward and backward layer output, can be sum, multiplication, concatenation, or average function.

42.2.3 GRU

The GRU [13] model and the LSTM model are quite similar. To manage the flow of information, they both use the notion of gates. The value of the hidden cell unit h_t^i is always the combination of new input data and previous information, shown through the Eq. (42.14).

$$h_t^i = \left(1 - z_t^i\right)h_{t-1}^i + z_t^i \hat{h}_t^j \tag{42.14}$$

At the same time, the update gate u_t^i decides whether the unit should update the activation function or keep the proportion and quantity of existing activation

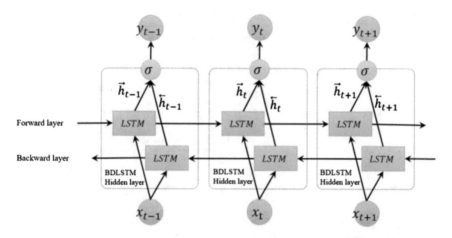

Fig. 42.2 Architecture of a BLSTM network

functions the same. The following is the update gate u_t^i:

$$u_t^i = \sigma(W_z x_t + u_z h_{t-1})^j \tag{42.15}$$

The GRU [14] model cannot regulate the range of state updates, therefore each computation updates all of the states once. The candidate activation function $\hat{h}_t^i$ is calculated in the same way as a simple RNN calculation, and its computational function is

$$\hat{h}_t^j = \tanh(W x_t + u(r_t \otimes h_{t-1}))^j \tag{42.16}$$

Among which r_t is the reset gate, $\otimes$ is the vector product. The contents of the input sequence can be read when the reset door (r_t^i) is closed, while the previous state is forgotten. The following formula is used to compute the reset gate:

$$r_t^i = \sigma(W_r x_t + u_r h_{t-1})^j \tag{42.17}$$

The tanh function has been well-developed and frequently utilized in several studies. Assuming that the input of the model is expressed as, $X = x_1, x_2, \ldots, x_T$, the process is shown in Fig. 42.3

Fig. 42.3 GRU architecture

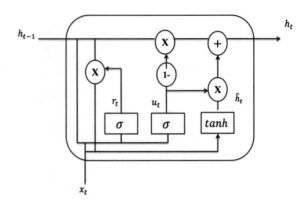

42.3 Performance Evaluation Parameter

42.3.1 Mean Square Error (MSE)

The MSE is defined as the error's second moment (around the origin), and so includes the predictor's variation as well as the bias's variance. MSE may be expressed mathematically as Eq. (42.18).

$$\text{MSE} = \frac{1}{n} \sum_{t=1}^{n} (\hat{Y}_t - y_t)^2 \tag{42.18}$$

42.3.2 Root Mean Square Error (RMSE)

It is the discrepancy between a model's predicted values and the observed values. The RMSE is a measure of ACCURACY that is widely used to evaluate predicting mistakes across various predictive models for the same data collection. RMSE is mathematically represented as Eq. (42.19). The square root of mean square error is usually positive, therefore a number near 0 indicates improved performance.

$$\text{RMSE} = \sqrt{\frac{1}{n} \sum_{t=1}^{n} (\hat{Y}_t - y_t)^2} \tag{42.19}$$

where $\hat{Y}_t$ is predicted value, y_t actual value.

42.3.3 Mean Absolute Error (MAE)

The total of all absolute errors is the mean absolute error [15]. It makes no distinction between positive and negative directions.

$$\text{MAE} = \frac{1}{n} \sum_{t=1}^{n} \left| \hat{Y}_t - y_t \right| \tag{42.20}$$

42.3.4 Mean Absolute Percentage Error (MAPE)

It is a measure of a forecasting method's accuracy [16]. It simply means accuracy expressed as a percentage.

$$\text{MAPE} = \frac{1}{N} \sum_{w=1}^{N} \frac{\left| \hat{Y}(w) - Y(w) \right|}{\hat{Y}(w)} * 100 \tag{42.21}$$

42.4 Results and Discussion

The simulation results are given and analyzed in this section using a tabular comparison analysis for all three models: LSTM, BiLSTM, and GRU. The models are all ran for the same amount of epochs and have the same input vector size, hidden layers, and output vector size. The implementation code was written in Python 3.4 while for all deep learning models Keras library was utilized. The loss function of LSTM was the mean-squared error and optimizer was 'Adam'. Dataset is splitted into two parts: training and testing. The wind power forecasting model is trained with 70% of the samples, and the remaining samples are utilized to test the model. The forecasting models' performance is measured using the MAE, MAPE, MSE, and RMSE.

Yalova Wind Farm dataset is used for the experimentation purpose which is located at west Turkey with a total power of 54,000 kW, it includes 36 wind turbines. Wind turbine data such as wind speed, wind direction, and theoretical power based on the turbine's power curve and generated power were measured and saved using a SCADA system. The data is stored at 10 min resolution. The dataset is available at https://www.kaggle.com/berkerisen/wind-turbine-scada-dataset in CSV format for one year period from January 1, 2018, to December 31, 2018. Before the input features hit the deep learning layer, they are scaled to a range of zero to one using the min-max scalar. Equation (42.22) shows the formula for min-max scalar.

Table 42.1 Prediction performance comparison

Model	MSE	RMSE	MAE	MAPE
LSTM	122,621.94	350.17	181.05	30.06
BiLSTM	119,318.32	345.42	179.11	30.35
GRU	130,636.76	361.44	195.54	33.46

$$X_{\text{scaled}} = \frac{X_i - \min(X)}{\max(X) - \min(X)} \tag{42.22}$$

The smaller the values of RMSE, MAE, and MAPE are the more accurate the forecasting result is. Table 42.1 lists the comparison of MAE, RMSE, MAPE of the training results and the calculation time of each model.

As a result, it is clear that BiLSTM outperforms LSTM and GRU in terms of ultimate results and performance.

Execution time for CPU and GPU are represented in Fig. 42.4. When comparing the results from the CPU and GPU, it is clear that the GPU's processing capability boosts performance only when a complicated neural network is being trained. When the complex model BiLSTM is trained on GPU execution time is decreased by almost 28.83%.

Among all the experimental models, GRU has the largest error, and its MAE, MAPE, and RMSE are 195.54, 33.46, and 361.44, respectively. MAPE of LSTM is slightly less as compared to BiLSTM model. Comparison of all model using different evaluation parameters is shown in Fig. 42.5.

Figure 42.6 shows the actual versus predicted values by LSTM, BiLSTM, and GRU. BiLSTM models outperform standard unidirectional LSTMs, as evidenced by the findings. By traversing input data twice from left to right and then from right to left, BiLSTMs appear to be able to grasp the underlying context better.

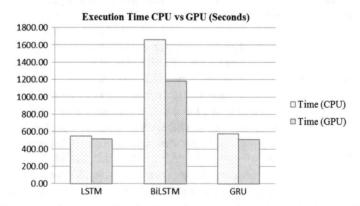

Fig. 42.4 Model execution time on CPU versus GPU

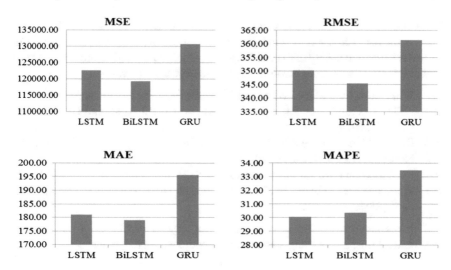

Fig. 42.5 Comparison of all model using different evaluation parameters

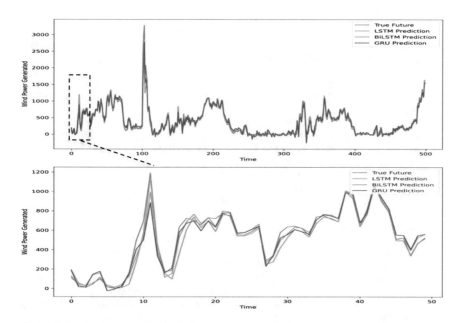

Fig. 42.6 Actual versus predicted wind power

42.5 Conclusion

This research presented the findings of an experiment in which the performance and accuracy of LSTM, BiLSTM, and GRU models, as well as execution time on CPU versus GPU, were studied and compared. The result shoes that BiLSTM may be applied for the prediction of wind power generation. We may conclude from the above research that the BiLSTM can fully use sequence past and future information. The following observations are drawn from the experimental results:

(i) When compared to the LSTM and GRU, the BiLSTM has lower MAE values.
(ii) LSTM has the lowest MAPE value compared to BiLSTM, and GRU has the greatest MAPE value.
(iii) BiLSTM has the lowest MSE, RMSE, and MAE values, whereas GRU has the highest, and LSTM has a middle value.
(iv) Execution time for simple network like LSTM and GRU is less than BiLSTM.
(v) When complex network BiLSTM is trained on GPU execution time is reduced by 28.83%.

This research can be further expanded to forecasting problems for multivariate and seasonal time series using some signal decomposition techniques.

References

1. Western wind and solar integration study | Grid modernization | NREL, https://www.nrel.gov/grid/wwsis.html. Accessed 06 Apr 2022
2. Qu, X., Kang, X., Chao, Z., Shuai, J., Ma, X.: Short-term prediction of wind power based on deep long short-term memory. In: 2016 IEEE PES Asia-Pacific Power and Energy Engineering Conference (APPEEC), vol. 2016, pp. 1148–1152 (2016). https://doi.org/10.1109/APPEEC.2016.7779672
3. Tang, Y., Huang, Y., Wu, Z., Meng, H., Xu, M., Cai, L.: Question detection from acoustic features using recurrent neural network with gated recurrent unit. In: 2016 IEEE International Conference on Acoustics, Speech and Signal Processing (ICASSP), vol. 2016, pp. 6125–6129 (2016). https://doi.org/10.1109/ICASSP.2016.7472854
4. Wang, H.Z., Wang, G.B., Li, G.Q., Peng, J.C., Liu, Y.T.: Deep belief network based deterministic and probabilistic wind speed forecasting approach. Appl. Energy **182**, 80–93 (Nov.2016). https://doi.org/10.1016/J.APENERGY.2016.08.108
5. Ghofrani, A., Suherli, M.: Time series and renewable energy forecasting. In: Time Series Analysis and Applications, pp. 77–92 (2017)
6. Zhang, Z., Qin, H., Yao, L., Lu, J., Cheng, L.: Interval prediction method based on long-short term memory networks for system integrated of hydro, wind and solar power. Energy Procedia **158**, 6176–6182 (Feb.2019). https://doi.org/10.1016/J.EGYPRO.2019.01.491
7. Wang, S., Jiang, J.: Learning natural language inference with LSTM. In: NAACL HLT 2016: Proceedings of the 2016 Conference of the North American Chapter of the Association for Computational Linguistics: Human Language Technologies, pp. 1442–1451 (2015). https://doi.org/10.48550/arxiv.1512.08849
8. Chen, K., Zhou, Y., Dai, F.: A LSTM-based method for stock returns prediction: a case study of China stock market. In: 2015 IEEE International Conference on Big Data (IEEE Big Data 2015), pp. 2823–2824 (2015). https://doi.org/10.1109/BIGDATA.2015.7364089
9. Lipton, Z.C., Kale, D.C., Elkan, C., Wetzel, R.: Learning to diagnose with LSTM recurrent neural networks. In: 4th International Conference on Learning Representations (ICLR 2016), Conference Track Proceedings (2015). https://doi.org/10.48550/arxiv.1511.03677
10. Siami-Namini, S., Tavakoli, N., Siami Namin, A.: A comparison of ARIMA and LSTM in forecasting time series. In: 2018 17th IEEE International Conference on Machine Learning and Applications (ICMLA), pp. 1394–1401 (2019). https://doi.org/10.1109/ICMLA.2018.00227

11. Schuster, M., Paliwal, K.K.: Bidirectional recurrent neural networks. IEEE Trans. Signal Process. **45**(11), 2673–2681 (1997). https://doi.org/10.1109/78.650093
12. Biswas, S., Sinha, M.: Performances of deep learning models for Indian Ocean wind speed prediction. Model. Earth Syst. Environ. **7**(2), 809–831 (2020). https://doi.org/10.1007/S40808-020-00974-9
13. Cho, K. et al.: Learning phrase representations using rnn encoder-decoder for statistical machine translation. In: Proceedings *of* the *2014* Conference on Empirical Methods in Natural Language Processing (EMNLP 2014), pp. 1724–1734 (2014). https://doi.org/10.48550/arxiv.1406.1078
14. Kisvari, A., Lin, Z., Liu, X.: Wind power forecasting—a data-driven method along with gated recurrent neural network. Renew. Energy **163**, 1895–1909 (Jan.2021). https://doi.org/10.1016/J.RENENE.2020.10.119
15. Willmott, C.J., Matsuura, K.: Advantages of the mean absolute error (MAE) over the root mean square error (RMSE) in assessing average model performance. Clim. Res. (2005). https://www.jstor.org/stable/24869236. Accessed 06 Apr 2022
16. Chai, T., Draxler, R.R.: Root mean square error (RMSE) or mean absolute error (MAE)?—Arguments against avoiding RMSE in the literature. Geosci. Model Dev. **7**(3), 1247–1250 (Jun.2014). https://doi.org/10.5194/GMD-7-1247-2014

Chapter 43
A Lightweight and Precise Information Retrieval System for Organisational Wiki

Kundan Kanti Saha, Sangram Ray, and Dipanwita Sadhukhan

Abstract Large organisations have vast amount of proprietary information in the form of internal wiki, which requires a fast and efficient information retrieval system to address employees' informational needs. This paper lays down the design and development of an information retrieval system (IRS) for answering employees' queries based on knowledge from the organisation's internal wiki website. Organisational Wikis are not for public access and hence protected with stronger security mechanisms. An IRS that can understand the access level of the user and accordingly access the internal WiKi pages to fetch the precise needed information related to the query has been developed. This IRS is web based and specialist in nature as it serves the organisation's employees with answers using the secure crawler developed here and the concise reference index built thereof. The query processing module of IRS uses natural language processing (NLP) tools. To enhance the speed of the process, a fast and secure database is maintained for indexing the most relevant information of the wiki by the crawler. The IRS serves updated answers from the wiki on the fly.

43.1 Introduction

Information retrieval system (IRS) is defined by Kumar and Sharma [16] as a software which finds relevant information from a vast space of knowledge to satisfy user needs. They further identify four main modules of the system as indexing, query processing, searching and ranking. Web crawlers read and download web pages in bulk. Web crawlers are an integral part of IRS. They automatically and periodically crawl website(s) and index them. In the process, they detect updates in the existing pages as well as web pages that have been added or deleted since the last crawl and updates them in the index. The IRS then display results pointed by these indexes. Crawlers are used for electronic documents like web pages, but no such bots are required for other forms of data storage like database. One of the motivations behind designing an IRS is the scale of the knowledge base. Organisational wiki is enormous, consisting

K. K. Saha (✉) · S. Ray · D. Sadhukhan
National Institute of Technology Sikkim, Barfung Block, Ravangla, Sikkim 737139, India
e-mail: kundansaha@gmail.com

V. Bhateja et al. (eds.), *Evolution in Computational Intelligence*, Smart Innovation, Systems and Technologies 326, https://doi.org/10.1007/978-981-19-7513-4_43

495

of millions of documents scattered over various data centres. The real-time scanning of the entire site upon every query is not possible as it will take hours to complete a single full search. Therefore, the wiki is crawled on regular intervals for updates and indexed in a database from which a pointer to the answer is obtained. Important terms, hereby referred as keywords, are identified from a document, processed using NLP and stored in the database. The wiki itself is distributed over several servers and geographical locations. Users from all over the world access the IRS. So, the database must be distributed in nature and reliable without compromising on the speed and efficiency.

43.2 Literature Survey

Tonon and Fusco [22] showed that data mining is viable for effective information retrieval for large size institutional repositories. Ceci et al. [7] developed a method to extract semantically machine readable data sets to create computational ontologies for a better information retrieval system. Estevão and do Rocio Strauhs [12] studied the ontology-based historical organisational memory as a possible solution for information management and retrieval in 11 institutions from diverse fields. Barbosa et al. [6] studied the corporate memory retrieval system in three large engineering firms in Brazil and concluded that information retrieval from corporate repositories are largely dependent on individual memory and some corporate content management system in a project-based organisation. Mahdi et al. [18] had suggested an agent-based information retrieval model for institutional setting where they found that the system performance is inversely proportional to the amount of concurrent cloud users. Kanoje et al. [14] explored the usage of user profiling-based recommender systems for creating IRS. Pouamoun and Kocabaş [21] proposed a further enhanced novel cloud-based distributed IRS utilising the broke-based P2P network. Cruz et al. [9] proposed a conversational agent-based IRS for academic institutions, wherein they divide the IRS into modules like user interface, NLP module, data collection and processing. Cruz and Guelpeli [10] explored the viability of using summarisation technique in Cassiopeia model for developing an institutional IRS.

43.3 Preliminaries

The current wiki website has a generic search engine producing myriads of result pages based on typical search engine rules. Although exhaustive, it does not consider the conciseness and consumption of retrieved information requires time. Every search operation is like finding a needle in haystack. The objective is to develop an efficient, secure and robust chatbot like IRS that answers user queries precisely based on the information extracted and processed from organisation's wiki, reducing the noise from irrelevant information. If the answer lies in the wiki and is accessible to the user based on their rights, the IRS reduces the time and effort of the searcher to find answer. If the answer is there, but the user doesn't have access, the chatbot would notify the same. Building an intelligent IRS is a multidisciplinary activity with the following processes.

- Constructing a targeted crawler that is robust, efficient and scalable.
- Building the IRS index from crawled information.
- Scheduling revisits of the wiki to check for updates, additions, deletions and modifications.
- Avoiding content that might create errors or unnecessary prolongation of the crawling process.
- Query processing using natural language processing (NLP) techniques.
- Producing unambiguous responses in accordance to the policies and security mechanisms of the search space.

The components of this system are shown in Fig. 43.1 and further discussed below.

43.3.1 Wiki Crawler

Web crawlers or web spiders are computer programmes that read web pages from websites specified as the allowed domains and extract information for further analysis and processing. These crawlers keep working round the clock in the background to detect and "read" any kind of changes, like additions, deletions and modifications in the website(s) and sends the data thus obtained to the information management system containing the index or database. For this project, a custom web crawler has been developed using the Scrapy tool for Python.

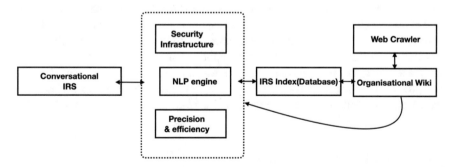

Fig. 43.1 Organisational information retrieval system for internal Wiki repository

43.3.2 Indexing Database

The data generated by the IRS crawler from the wiki are vast, fast-moving and distributed over many nodes. Using the traditional RDBMS would not satisfy the needs of searching and analysing it in an efficient and quick manner. While SQL and RDBMS are good for handling certain types of data, it requires spending a lot of time and effort in preprocessing the data. This preprocessing, although a one-time activity results in breaking down the data into several indices, tables, columns, etc. Searching and updating this data later is resource intensive and complicated. Looking at the volumes of data generated by organisations of modern times, big data solutions like Hadoop, NoSQL, Solr and Elasticsearch have come into existence. While NoSQL had its origins in 1960s, the term was coined in 2000s, and its extensive usage started then as Web 2.0 came into existence with social media, blogs and other internet resources that were not structured and required something more versatile than the tabular forms of data storage. Solr and ElasticSearch are full-text search databases based on Apache Lucene library, written by Doug Cutting in 1999 in Java, but later ported to other programming languages. Solr has all the capabilities and advantages of NoSQL and provides rich functionalities. But, it can exist only on a standalone server. ElasticSearch is a distributed full-text search-based database and search engine. For construction of the IRS knowledge base and searching through it, ElasticSearch is the most efficient option. ElasticSearch is a total open source, search engine solution. ElasticSearch was originally programmed in Java. As it is based on REST API, official ElasticSearch Clients are available for Java, .NET, C#, PHP, Python, Apache Groovy, Ruby, etc. ElasticSearch, being a big data type of database, has all its advantages like scalability, elasticity, real-time search and updating. In ElasticSearch, data are stored as a JSON document. But, at the core, the ElasticSearch shares a lot of similarities with the traditional SQL base database. Kononenko et al. [15] laid out the analogies between the two. At the top level in SQL lies the database, which corresponds to index in ElasticSearch. A database contains various schema. An ElasticSearch index contains mappings. Schema contains tables while mappings contain document type. Tables contains rows. Document types have documents. In addition to this, ElasticSearch has few other concepts like shards and replicas not present in SQL or RDBMS. Shards are pieces of the database divided horizontally between the machines on the network which gives ElasticSearch the distributed nature. Replicas are copies of the data used to prevent data loss and reduce the load on the primary host machine [5].

43.3.3 Conversational Agent

A chatbot is an information retrieval application that mimics the human being with text and/or voice responses to queries and commands. This chatbot is text based and serves responses to user queries using the domain knowledge from the IRS. The chatbot user interface (UI) was developed in Python and present in the organisation before this project started. However, the chatbot present before was a document retrieval system (documents like word, pdf, ppt and xls.) which served results from a document repository hosted in another sub-domain of the organisation. The IRS developed will be integrated into the existing UI so that users can search for documents as well as get textual answers to their queries from the wiki through the same platform.

43.3.4   NLP Engine

NLP is the field of study that is concerned with the interaction of computers and human languages, in particular processing large amount of text from languages spoken by humans. Computer, by virtue, is a digital device which understands the language of binary numbers, i.e. 0 and 1. In order to make it understand, process and respond in languages spoken by humans, several processes are involved. It includes but not restricted to, tokenisation, lemmatisation, pattern matching, intent extraction, reading comprehension, etc. For this IRS, the natural language toolkit (NLTK) [17] package for Python has been used. The IRS users will input their queries into the UI in plain English. Using straightforward mapping of user input to the chatbot index may not yield appropriate answers or any answer at all or too many answers. So, NLP is used to process the user query as well as processing data from the crawler before populating the index.

43.3.5   Integrating Security Policies

Security is one of the key factors for any organisation that has intellectual property. Organisations must keep its intellectual property safe and avoid them from falling into the hands of hackers or other external agents who are not stakeholders. This is an important reason for not opening the organisational wiki to external crawlers like Googlebot. Several security mechanisms are available in the current world to protect data. This IRS integrates some of those security mechanisms in order to restrict access to the IRS to only those individuals who are qualified to get the required information, depending on their business needs. The organisation uses multiple layers of security to protect the wiki content, and all those layers of security are integrated into the IRS.

43.3.6   Relevancy of Query Output

The current organisational wiki website contains millions of pages with enormous amount of data that are crucial for the organisation's success. It takes several minutes if not hours, for a user to search a query and find answers when using the built-in search engine. The users must go through dozens of pages to find what they are looking for. Also, the web pages in the knowledge base have robust and elaborate security and privacy mechanism which can restrict the users from accessing the data based on their access rights. The initial search results show all the pages, but often when the users click on one of the results, they find that the page is inaccessible to them, wasting time and effort in the process.

43.4 High-Level System Architecture

The IRS consists of two distinct activities. While both the activities can operate in parallel, first activity has to be completed at least once before the second can be initiated. This is the building of the IRS indexing database, summarised in Fig. 43.2. Once built, it can be scheduled for periodic updates as per the requirement. The second activity, which is the running of the conversational agent, can commence only after the IRS indexing database has been built. Initially, the ElasticSearch server is started. Once started, it can run without halting except for periodic maintenance and updates. ElasticSearch server can be started using the initialisation script that comes with the installation or from within the Python programme. Then, the Scrapy crawler is started from the initial page. The web pages in the domain are scraped, and important data are extracted according to the HTML cues where they are likely to be found. The data obtained from the crawling process are sent to the NLP engine for analysis, and the processed data are sent to the ElasticSearch for indexing. Then, duplicates are removed from the database, and the process is iterated for other eligible links which are obtained while crawling the pages. This process must be done with elevated permission, i.e. administrative privileges, so that all the resources on the wiki domain that come under the purview of the project can be indexed. The conversational agent module of the system consists of human–computer interaction as well as information retrieval. While designing the UI is not under the purview of the project (the project will be integrated with an existing chatbot of the organisation), all the background work for the Q&A system for the wiki user is going on in this part. As illustrated by Fig. 43.3, user queries which are in plain English are obtained from the IRS UI and transferred to the back end for further actions. These queries are processed using the same NLP tool that was used to analyse the scraped data in the first process. The processed data are searched in the index to extract the location of the complete answer in the wiki. The actual answers to the queries, obtained from the wiki, are sent to the UI as output.

43.5 Experimental Model

In this section, the developed system have been discussed in detail. At the heart of the system lies the creation and implementation of the index to the wiki. Therefore, the crawler and the indexing database developed have been detailed. Other components like NLP engine and security infrastructure have been discussed.

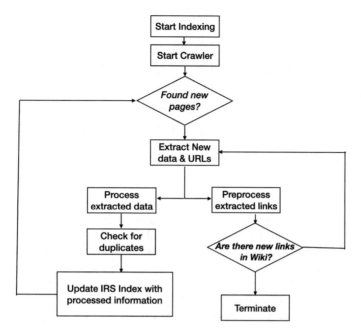

Fig. 43.2 Flowchart for crawling and indexing wiki

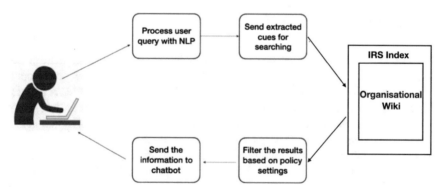

Fig. 43.3 Conversational agent or chatbot module of the IRS

43.5.1 Step I: Crawling the Wiki

The crawlers are programmes that start from a few starting points or uniform resource locators (URLs), read and index the content and extract hyperlinks and their properties from the pages crawled. They keep on crawling the web pages corresponding to the hyperlinks extracted. The crawler that is developed for indexing process uses depth-first Search [8] to repeatedly parse the different pages for hyperlinks. A crawler function is created which receives the link to be crawled as a parameter. Reg-

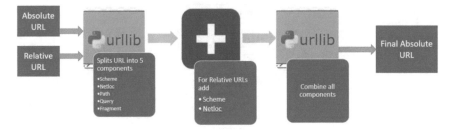

Fig. 43.4 Processing of extracted URLs

ular expression is then used to extract the links in the web page. Regular expression module of Python is a powerful tool to process, analyse and extract information from strings in Python, and findall is one of the key functions in it. This function returns all possible matches of the expression in the function. Here, findall function is used to extract URLs.

URLs are present in the wiki as absolute as well as relative form. Crawler returns URLs as it is in the page. But, to navigate to an extracted URL, absolute URL is required. This is obtained through URL correction as shown in Fig. 43.4.

43.5.2 Step II: *Filtering Content*

The wiki site has links to all sorts of resources including electronic documents in the domain and outside the domain having extensions such as .html, aspx, .doc, .pptx, .xls, .pdf and .zip. All these content types are not supported by the crawler. A crawler designed for internal web pages may not work on external web pages or documents in other formats in the site. Only, hypertext markup language (HTML) and active server page extended (ASPX) pages of the internal wiki domain are the targets of the crawler. Any other format of resources has been conveniently ignored for now. The crawler picks up cues like title, H1, H2, …, a href, strong and ul. Excluded content doesn't contain these cues.

43.5.3 Step III: *Information Extraction*

There are 3 main types of Python libraries to build a crawler. They are

- **Scrapy** has been used in this design primarily because it is more scalable, robust and a full-fledged spider [3].
- **BeautifulSoup** [23] is primarily used to extract specific elements from a web page.
- **Selenium** [13] was created to automate web browser interactions.

Scrapy is extensively documented and provides more freedom. Since a full-fledged domain-specific search engine crawler that is scalable and customisable was needed, Scrapy was a convenient option. CURL [1] provides methods to exchange data with web pages from the command line as well as scripts. It is used in various devices connected to the internet including IoT devices, PCs and laptops, mobiles and others. The copyright is owned by Daniel Stenberg, but anyone can use it freely, without restriction and modify the source code as per their requirement. CURL has been used to read data from wiki using the GET method of HTTP. Curl has also been used to send the username and password of the user for authentication after appropriate encryption. Request objects of Scrapy are used to crawl websites. Request objects are generated in spiders and passed across the system until they are captured in the Downloader and a Response object is sent back to the system. XPath [2] has been applied on response object to extract textual data from selective portions of the page, like titles, headers and paragraphs. XPath is an XML processing language which models the document as a tree of nodes. XPath axes on the other hand have been used to identify and extract elements based on their position in the document relative to another element. The target element could be a child, descendant, parent, ancestor or sibling of that particular element. XPath axes easily locate the relevant answer based on their distance from the keywords.

43.5.4   Step IV: *NLP Engine*

The organisational wiki already contains a custom search engine that processes the user query using pattern matching. The search engine matches the text patterns and yields result pages containing the patterns similar to other web search engines. The user must navigate to the result pages and read through the page to locate the answer. The IRS must quickly and concisely answer the user queries with as much accuracy as possible. In order to do so, the IRS needs to understand the context with the help of NLP. While English language packages in most natural language toolkits are enough to detect day-to-day keywords, a large organisation has several terms and acronyms which are internal to the organisation and meaningful only to the employees and stakeholders. The NLP corpus do not have these terms listed. Neither can these terms be ignored in the search query, nor can they be added to the corpus for protection of intellectual property. So, the IRS must be intelligent enough to detect them based on the context. NLP has been implemented by using the natural language toolKit developed by Loper and Bird [17]. The NLP system works as follows: First stage is the raw input which can be the output of the crawler or the user query from the chatbot. The input, which is in plain English, is then sent to the splitter where it is split into the individual words or tokens. This process, called tokenisation, is done in two stages. In the first stage, the entire text is divided into individual sentences or clauses, determined by the punctuation. Then, each of the obtained sentences is further divided into individual words. The tokens or words are then analysed to find out which part of speech (POS) they belong to. This process is called lemmatisation.

The main POS is adjective, adverb, verb and noun with noun being the default. Complete list of POS tags can be found at the University of Pennsylvania website [19]. Based on the POS, the base word or lemmatised word is obtained. The output, consisting of original keyword, lemmatised word(s) and POS is either sent to the index for indexing or query processing.

43.5.5 Step V: *Building the Index*

The IRS index is open to all user and serves as a pointer to the actual answer of the queries, wherein the answer are extracted from the wiki. Amongst all the candidate noSQL solutions, ElasticSearch was perfect for indexing. It provides simple and powerful REST API-based (Application programming interface) interface with most modern programming language for database creation and search. Various Python packages were used for development of crawler, UI and NLP tools, and ElasticSearch came with required functionalities in Python. ElasticSearch [11] is implemented into the IRS using ElasticSearch-py client. ElasticSearch is installed on the default port of 9200 using Java 7 or higher. ElasticSearch has its own query language called domain-specific language (DSL). Each query is a JSON object that can be processed by the ElasticSearch REST API to search in the specified index. The search is done using the lemmatised words of the NLTK.

43.5.6 Step VI: *Integrating Security Infrastructure*

Internal privacy policies and security mechanisms are essential to the IRS. All wiki pages are not accessible to all employees. Different employees have different access rights and can only view selected pages based on their rights. This is determined by the rank, department, role, and other factors. The IRS had to be designed such that results from only those pages that the searcher has access to would be displayed. The IRS had been designed with due consideration of the detailed security mechanism of the wiki. Basic authorisation scheme along with virtual private network is used with the help of SSO [20]. The full source code of the experimental system can be found at Github [4].

43.6 Experimental Results

50 different search terms were run on the system repeated on two different type of roles, i.e. engineering manager and software developer. The results adhered to the access level and security policies of the wiki. The search terms included a mix of 1–3 keywords, phrases containing 1–3 keywords along with some stop words and a few

```
Enter your query/ enter '-1' to exit :Fortville DDP

Fortville DDP
['DDP Support - ▓▓▓▓▓▓▓▓▓ - Intel Enterprise Wiki ']
['Fortville DDP Release process BKM', 'Adding new DDP package to FVL', 'FVL DDP technology Enabling
page', '▓▓▓▓▓▓ functions and their plan on ▓▓▓', '▓▓▓▓▓▓ Hash and Flow director Filters']
```

Fig. 43.5   Example output of the query "Fortville DDP"

```
Enter your query/ enter '-1' to exit :NVM with DPDK

2.2.1 Updating the NVM with a DPDK Driver
['DPDK performance on ▓▓▓▓▓▓ - ▓▓▓▓▓▓▓▓▓▓▓i - Intel Enterprise Wiki ']
['If all of the following are true:']
```

Fig. 43.6   Example output of the query "NVM with DPDK"

search queries that didn't have any relevance to the repository. The repository for simulation consists of pages related to Ethernet products of the organisation only. The algorithm successfully identified the keywords in 90% of the searches that included relevant keywords. All of the irrelevant search terms didn't produce any output. The output was restricted to upto 6 sentences for conciseness. 66% of the successful searches produced more than one results, while the remaining successful searches had single result. The time to generate output varied between 1 and 3 s. Figure 43.5 shows the output of the query "Fortville DDP." This query is simple and based on proprietary keywords. Both the words in the query are keywords. The output shows 1 result amongst 6 results, which is also the highest number of results that any of the tested queries produced. All the 6 results are relevant to both the keywords. The same query yields more than 100 suggested pages in the search engine box built into the wiki. Fig. 43.6 shows the output of the query "NVMe with DPDK." This query consists of two technologies, both of which were originally invented by Intel but are now widely used in the industry. The query includes a stop word which the algorithm successfully discarded, while identifying the 2 relevant keywords. The output is precise and closer to the intent of the searcher.

43.7   Conclusion

The IRS produces direct answers in response to the search query keyed in by a user, thereby reducing the time and effort of the searcher. This IRS takes care of the login and security issues. The searches were done with two different roles having different set of permissions, and the results were consistent with the access levels and security policies of the searcher profiles. The IRS performs faster by taking cues from the search terms and searching in the related domain and context. The crawler took around 4 h to crawl the allowed space in the wiki and create the database. Still it takes more time to search (1–3 s in the test cases, as compared to less than 0.1 s in case of existing search box of the wiki) because it extracts answers on the fly rather than from a cache. But, caching can be implemented for popular queries and the corresponding results of the index to reduce latency and remains a future opportunity. The IRS takes user queries as input, analyses them using NLP and outputs answers with help of the crawler and index, within 3 s. The information resulting from the queries is clear, concise and accurate. The number of answers in the output varied between 1 and 6, with nearly 33% of the tested search queries producing single answer. With better NLP techniques and more efficient intent extraction, this number may be further increased and remains an area for improvement.

Acknowledgements The authors wish to express special thanks of gratitude to Intel India which provided the necessary resources for experimentation.

References

1. Everything curl. https://curl.se/ (2020). Last accessed 20 Dec 2021
2. W3c xpath. https://www.w3.org/TR/xpath (2020). Last accessed 20 Dec 2021
3. Scrapy 2.3 documentation. https://docs.scrapy.org (2021). Last accessed 07 Apr 2021
4. Wikichatbot. https://github.com/kundansaha82/WikiChatbot (2021). Last accessed 27 Nov 2021
5. Akca, M.A., Aydoğan. T., Ilkuçar, M.: An analysis on the comparison of the performance and configuration features of big data tools solr and elasticsearch. Int. J. Intell. Syst. Appl. Eng. 8–12 (2016)
6. Barbosa, L.M., de Carvalho, R.B., Choo, C.W., França, Â.: Corporate memory processes in project-based organizations: a framework for engineering design firms. In: 17th International Conference on Intellectual Capital, Knowledge Management and Organisational Learning. ICICKM 2020, p. 45 (2020)
7. Ceci, F., Pietrobon, R., Gonçalves, A.L.: Turning text into research networks: information retrieval and computational ontologies in the creation of scientific databases. PLoS ONE **7**(1), e27499 (2012)
8. Cormen, T., Leiserson, C., Rivest, R., Stein, C.: Elementary graph algorithms. Introd. Alg. **1**, 540–549 (2009)
9. da Cruz, J.A., Nasser, M., Tuler, E., Carvalho, D., Rocha, L., Viana, M.: Creating an academic conversational agent for dynamic information retrieval. In: XVI Brazilian Symposium on Information Systems, pp. 1–8 (2020)
10. Cruz, L.A., Guelpeli, M.V.C.: Information retrieval in institutional repositories using the summarization technique derived from the selection of cassiopeia attributes. Braz. J. Dev. **6**(11), 88022–88041 (2020)
11. Divya, M.S., Goyal, S.K.: Elasticsearch: an advanced and quick search technique to handle voluminous data. Compusoft **2**(6), 171 (2013)
12. Estevão, J.S.B., do Rocio Strauhs, F.: Information retrieval in institutional repositories: proposal of an ontology for historical organizational memory. Qualit. Quantit. Methods Libr. **4**(4), 983–992 (2017)
13. Gheorghe, M., Mihai, F.C., Dârdală, M.: Modern techniques of web scraping for data scientists. Int. J. User-Syst. Interact. **11**(1), 63–75 (2018)
14. Kanoje, S., Mukhopadhyay, D., Girase, S.: User profiling for university recommender system using automatic information retrieval. Proc. Comput. Sci. **78**, 5–12 (2016)
15. Kononenko, O., Baysal, O., Holmes, R., Godfrey, M.W.: Mining modern repositories with elasticsearch. In: Proceedings of the 11th Working Conference on Mining Software Repositories, pp. 328–331 (2014)

16. Kumar, R., Sharma, S.: Information retrieval system: an overview, issues, and challenges. Int. J. Technol. Diffus. (IJTD) **9**(1), 1–10 (2018)
17. Loper, E., Bird, S.: Nltk: the natural language toolkit. arXiv preprint cs/0205028 (2002)
18. Mahdi, M., Ahmad, A., Ismail, R.: Agent based information retrieval in a cloud: a theoretical framework for an institutional repository. Int. J. Emerg. Technol. Adv. Eng. **4**, 447–450 (2014)
19. Marcinkiewicz, M.A.: Building a large annotated corpus of english: the Penn Treebank. Using Large Corpora, p. 273 (1994)
20. Parker, T.: Single sign-on systems-the technologies and the products (1995)
21. Pouamoun, A.N., Kocabaş, I.: Multi-agent-based hybrid peer-to-peer system for distributed information retrieval. J. Inf. Sci. 01655515211010392 (2021)
22. Tonon, L., Fusco, E.: Data mining as a tool for information retrieval in digital institutional repositories. Proc. CSSS **2014**, 180–183 (2014)
23. Zheng, C., He, G., Peng, Z.: A study of web information extraction technology based on beautiful soup. J. Comput. **10**(6), 381–387 (2015)

Chapter 44
Simulation of Monthly Runoff in Mahanadi Basin with W-ANN Approach

Gopal Krishna Sahoo, Aiswarya Mishra, Debi Prasad Panda, Abinash Sahoo, Sandeep Samantaray, and Deba Prakash Satapathy

Abstract The Mahanadi River basin is one of the biggest basins of India and serves as a lifeline for the region it passes through. However, the region often experiences an erratic rainfall and climate condition which affects the livelihoods of the people living nearby. Aiming to solve this problem, a novel approach is illustrated in this paper. Application of artificial neural network is a popular technique for prediction of various hydrological parameters as it provides fairly correct results. Here, an attempt has been made to incorporate wavelet transform in ANN, also known as wavelet artificial neural network (W-ANN), to further increase its scope and efficacy. The results were then evaluated using well-known statistical indices. It was concluded that W-ANN has better forecasting capacity than simple ANN model and it can be implemented for prediction of monthly runoff in similar basins and reservoirs.

44.1 Introduction

Rapid population development, urbanization, and industrial growth have raised the demand for water in many regions of the world. These same forces have changed watersheds and river systems, contributing to a higher damage to property and loss of life because of floods. It is becoming gradually important to carefully plan, manage, and develop water resource systems. Hydrologists have spent many years attempting to comprehend the correct prediction of runoff to estimate runoff for objectives such as flood control, water supply, fish and wildlife propagation, power generation, water quality, drainage, and irrigation. Runoff simulation methodologies range from conceptual-based, physical-based models to data-driven models. Conceptual

G. K. Sahoo · A. Mishra · D. P. Panda · S. Samantaray (✉) · D. P. Satapathy
Department of Civil Engineering, OUTR Bhubaneswar, Bhubaneswar, Odisha, India
e-mail: sandeep1139_rs@civil.nits.ac.in

D. P. Satapathy
e-mail: dpsatapathy@cet.edu.in

A. Sahoo
Department of Civil Engineering, NIT Silchar, Silchar, Assam, India

509

and physically based models, often known as process-driven models, are typically made up of analytical and empirical equations based on physical occurrences. Data-driven models, on the other hand, consider the links between runoff and related meteorological data without requiring an explicit understanding of the hydrological system's physical behaviour [1–5]. Because of recent advancements in computational intelligence, AI-based methodologies have dramatically increased possibilities of data-driven models.

Tokar and Johnson [6] indicated that for the forecast of the daily runoff as a function of snowmelt, temperature, and daily rainfall for River Little Patuxent in United States, ANN model outperformed regression and simple conceptual models in terms of training and testing accurateness. Tokar and Markus [7] used the ANN approach to estimate runoff in three basins in the United States as a function of temperature, rainfall, and snow equivalent and compared it to the conceptual water balance (Watbal) model. Outcomes indicated that ANN model was able of properly predicting runoff; it also offered a systematic method and reduced the time needed on model training when compared to the conceptual models. Ahmadi et al. [8] proved advantage of ANN models over SWAT and IHAC-RES model for daily, monthly, and annual runoff simulation using time series rainfall, temperature, and runoff data of Kan watershed, Iran. Poonia and Tiwari [9] evaluated the performance of two different types of ANN models (i.e. RBF network and the FFBP network) using antecedent precipitation and runoff data of the Hoshangabad catchment, India, and validated the better performance of the RBF network over the FFBP network. Mao et al. [10] compared the ANN model with the LSTM model for rainfall-runoff simulation at the upper Heihe River Basin, China. For monthly scale simulation, the ANN performed superior than LSTM model while on a daily scale LSTM shows better performance. Mitra and Nigam [11] implemented a feed-forward back-propagation (FFBP) method and Levenberg–Marquardt (LM) algorithm to develop monthly and annual rainfall-runoff correlations in Hoshangabad catchment of River Narmada. Their findings demonstrate that ANN accurately captured linearity of rainfall-runoff modelling.

Rajaee [12] presented and compared W-ANN with ANN, MLR, and traditional sediment rating curve models for predicting daily sediment load in River, Yadkin USA. When forecast accuracy of the models was compared, it was discovered that W-ANN was the best performing model in predicting sediment load. Abghari et al. [13] investigated several forms of mother wavelets as activation functions rather than the frequently utilized sigmoid for predicting pan evaporation on daily basis at Lar synoptic site in Iran. Mexican Hat W-NNs were found to be more accurate than polyWOG1 W-NNs in simulating pan evaporation. Venkata Ramana et al. [14] predicted monthly rainfall for Darjeeling rain gauging site, India using maximum and minimum temperature, and monthly rainfall data as inputs and concluded that WNN model shows better performance than ANN model. Ravansalar and Rajaee [15] used monthly electrical conductivity and discharge measurements to compare the W-ANN and ANN models for predicting electrical conductivity of River Asi in Antakya, Turkey. When compared to simple ANN, W-ANN is more proficient of simulating nonlinear correlation between inputs and output. Shafaei and Kisi [16] estimated short-term daily streamflow using W-ANN, ANN, and SVM for Vanyar site, on River Ajichai, Iran and authenticated ascendancy of the W-ANN against ANN and SVM. Sharghi et al. [17] used W-ANN, FFNN, and EANN for monthly rainfall-runoff modelling of West Nishnabotna and Trinity watershed, and deduced that W-ANN outperformed FFNN and EANN. The objective of the research is to predict runoff using hybrid W-ANN approach.

44.2 Study Area

Mahanadi River emerges in the Indian state of Chhattisgarh and runs about 851 km from west to east before discharging into Bay of Bengal with a drainage area of 141,589 km^2 in total. It lies between19° 8' and 23° 32' N latitudes, and 80° 28' and 86° 43' E longitudes. It is bordered by Central India hills on north, on east and south by Eastern Ghats, and on west by Maikala ranges. The mean yearly precipitation across the basin varies from 1200 to 1400 mm. Mahanadi is one of India's major rivers, and among peninsular rivers it ranks second to Godavari in terms of water potential and flood-producing capability (Fig. 44.1).

44.3 Methodology

44.3.1 ANN

ANN consists of several interlinked nodes that are categorized into three fundamental layers: input, hidden, and output layer. The input nodes do not undertake any computation but are utilized to distribute inputs into the network. In this type of network, since the information travels only in one direction (i.e. from input to hidden to output layer) it is known as a feed-forward network. Feedback between layers is possible with recurrent networks. The set of input and output node requirements in an ANN is problem-dependent while the set of hidden nodes needs to be evaluated on the basis of experiments only.

$$P_y = \sum_{i=1}^{n} w_{xy} v_y + w_{oy} \qquad (44.1)$$

v_y represents the number of inputs to the neuron x. w_{xy} is the weight (weight is a factor with which the values passing to the neuron are multiplied), and w_{oy} is the bias. A learning parameter (a single multiplier used to multiply all adjustments) determines how much each neuron's weights and bias are altered in the back-propagation method

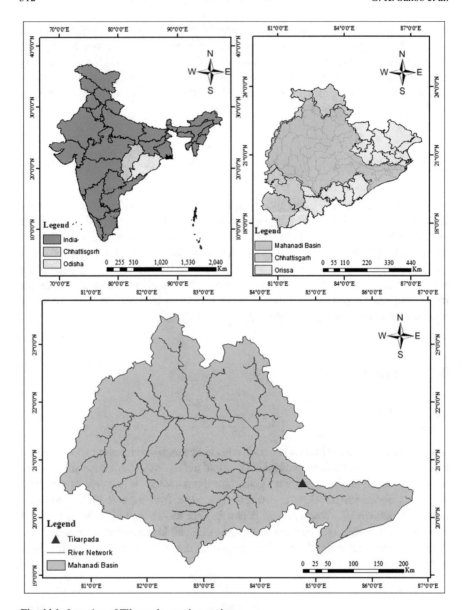

Fig. 44.1 Location of Tikarpada gauging station

44.3.2 W-ANN

The WT is used to evaluate data because of its ability to extract relevant information on time-frequency from transient and non-stationary signals. Frequency elements of signals are decomposed via wavelet. Data is decomposed into several frequency components by wavelet functions, and then every constituent is studied with a resolution equivalent to its scale. Compressed wavelet's lower scale points can extract signal's high-frequency constituents or rapidly changing details. The stretched version of a wavelet is represented by the higher scale, and corresponding coefficients indicate gradually changing feature of a low-frequency component. Wavelet transforms are classified into two types: discrete WT and continuous WT. In this work, DWT is employed since discrete time series data were used for hydrological applications.

For the DWT of the time series, Mallat's method was used in this work. The DWT of signal $q(t)$ is represented as

$$\omega_{i,j} = 2^{-i/2} \sum_{t=0}^{N-1} \gamma(2^{-i}t - j)q_t \tag{44.2}$$

where $\omega_{i,j}$ is the scale parameter's wavelet coefficient. The integers i and j govern the scale and time, and t is integer time stages.

44.4   Results and Discussion

Rainfall (P_t), maximum and minimum temperature $(T_{\max}$ and $T_{\min})$, and Humidity (H_t) data of 20 years are used to develop the proposed models for predicting target variable R_t. Total 240 available monthly data series is divided into training (168), i.e. 70% of dataset and rest for testing (72), i.e. 30%. Proposed models, i.e. ANN, and W-ANN are trained and tested to predict runoff, and outcomes are assessed using different statistical performance indices.
[18, 19].

$$\text{RMSE} = \sqrt{\frac{1}{N} \sum_{i=1}^{N} (p_i - o_o)^2} \tag{44.3}$$

[20, 21].

$$R^2 = 1 - \frac{\sum_{i=1}^{N} (p_i - o_o)^2}{\sum_{i=1}^{N} (\overline{p_i} - p_i)^2} \tag{44.4}$$

[21, 22].

$$\text{NSE} = \left[\frac{\sum_{i=1}^{N} (o_o - p_i)^2}{\sqrt{\sum_{i=1}^{N} (o_o - o_o)^2}} \right] \tag{44.5}$$

Table 44.1 Performance assessment results

Technique	Models	RMSE		R^2		NSE	
		Training	Testing	Training	Testing	Training	Testing
ANN	ANN * 1	10.991	14.3894	0.939	0.9202	0.9354	0.9165
	ANN * 2	10.4283	13.7796	0.9407	0.9216	0.9361	0.9173
	ANN * 3	9.0037	13.163	0.9413	0.9238	0.937	0.918
	ANN * 4	8.194	12.479	0.9421	0.924	0.9388	0.9199
W-ANN	W-ANN * 1	2.09	7.882	0.9824	0.9638	0.978	0.9745
	W-ANN * 2	1.5549	6.4094	0.983	0.9665	0.9794	0.962
	W-ANN * 3	1.0076	6.001	0.9847	0.967	0.9801	0.9638
	W-ANN * 4	0.895	5.2897	0.9869	0.9683	0.9846	0.9642

where p_i and o_o—predicted and observed runoff values, $\overline{p_i}$ and $\overline{o_o}$—mean of predicted and observed runoff values, and N—number of data points.

Evaluated outcomes of two runoff prediction models utilizing statistical indices and are tabularized in Table 44.1. The models are assessed based on R^2, NSE, and RMSE during training and testing phase. From the table and graphical representations, it is clearly evident that W-ANN is the best performing model than the ANN model. Based on Table 44.1, W-ANN * 4 with RMSE—5.2897, R^2—0.9683, and NSE—0.9642 predicts runoff better as compared to ANN * 4 with RMSE—12.479, R^2—0.924, and NSE—0.9199. Assessment of predicted runoff outcomes of W-ANN and observed runoff has least RMSE and higher R^2, NSE in testing and training phases in comparison with simple ANN model.

Trend in variation of prediction results of W-ANN model tried to follow the variation trend of actual runoff dataset, relatively more than simple ANN model (Fig. 44.2). Predicted runoff values of W-ANN model mimicked the actual runoff better than ANN (Fig. 44.2) which highlights its superiority and indicated better results over the individual model. For hydrological runoff prediction, pre-eminence of W-ANN model has demonstrated its feasibility, and for similar applications can also provide viability. In general, proposed hybrid W-ANN model gave better prediction performance than ANN models. NNs have no clear regulations in recognizing appropriate network structure. Hence, effective network structure can be attained by changing the number of neurons and layers (Figs. 44.3).

44.5 Conclusion

In this study, the application of neural network models for prediction of runoff in the Mahanadi River basin was evaluated. Here, the performance of classical ANN model and hybrid W-ANN model were demonstrated and compared. The proposed hybrid model proved to be more consistent than the standalone ANN model and

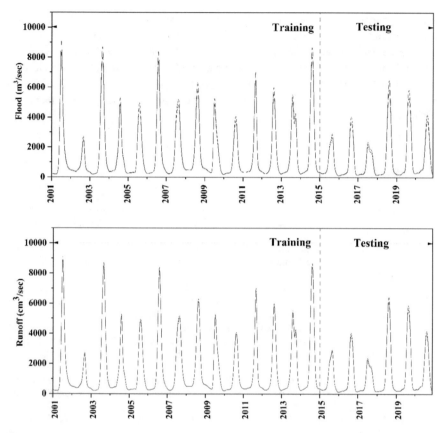

Fig. 44.2   Actual versus predicted values of runoff for Tikarpada station

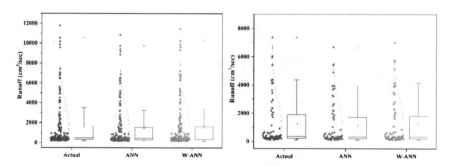

Fig. 44.3   Box plots showing actual and predicted values during training and testing

also, the wavelet model satisfactorily stimulates nonlinear relationships between input and output. The evaluation criteria for this paper were RMSE, R^2, and NSE and the respective values obtained were 5.2897, 0.9683, and 0.9642 for W-ANN model during testing phase. Wavelet usage as a transfer function at the structure of ANN caused an improvement in modelling of monthly runoff as a result of which W-ANN proved to be an effective technique for predicting monthly runoff due to its high accuracy.

References

1. Sahoo, A., Samantaray, S., Ghose, D.K.: Multilayer perceptron and support vector machine trained with grey wolf optimiser for predicting floods in Barak river, India. J. Earth Syst. Sci. **131**(2), 1–23 (2022)
2. Agnihotri, A., Sahoo, A., Diwakar, M.K.: Flood prediction using hybrid ANFIS-ACO model: a case study. In: Inventive Computation and Information Technologies, pp. 169–180. Springer, Singapore (2022)
3. Samantaray, S., Sumaan, P., Surin, P., Mohanta, N.R., Sahoo, A.: Prophecy of groundwater level using hybrid ANFIS-BBO approach. In: Proceedings of International Conference on Data Science and Applications, pp. 273–283. Springer, Singapore (2022)
4. Mohanta, N.R., Panda, S.K., Singh, U.K., Sahoo, A., Samantaray, S.: MLP-WOA is a successful algorithm for estimating sediment load in Kalahandi Gauge Station, India. In: Proceedings of International Conference on Data Science and Applications, pp. 319–329. Springer, Singapore (2022)
5. Sridharam, S., Sahoo, A., Samantaray, S., Ghose, D.K.: Assessment of flow discharge in a river basin through CFBPNN, LRNN and CANFIS. In: Communication Software and Networks, pp. 765–773. Springer, Singapore (2021)
6. Tokar, A.S., Johnson, P.A.: Rainfall-runoff modelling using artificial neural networks. J. Hydrol. Eng. 232–239 (1999)
7. Tokar, A.S., Markus, M.: Precipitation-runoff modelling using artificial neural networks and conceptual models. J. Hydrol. Eng. 156–161 (2000)
8. Ahmadi, M., et al.: Comparison of the performance of SWAT, IHACRES and artificial neural networks models in rainfall-runoff simulation (case study: Kan watershed, Iran). J. Phys. Chem. Earth 65–77 (2019)
9. Poonia, V., Tiwari, H.L.: Rainfall-runoff modeling for the Hoshangabad basin of Narmada river using artificial neural network. Arab. J. Geosci. **13** (2020)
10. Mao, G., et al.: Comprehensive comparison of artificial neural networks and long short-term memory networks for rainfall-runoff simulation. J. Phys. Chem. Earth 103026 (2021)
11. Mitra, S., Nigam, R.: An approach to utilize artificial neural network for runoff prediction: river perspective. J. Mater. Today: Proc. (2021)
12. Rajaee, T.: Wavelet and ANN combination model for prediction of daily suspended sediment load in rivers. J. Sci. Total Environ. **409**, 2917–2928 (2011)
13. Abghari, H., et al.: Prediction of daily pan evaporation using wavelet neural networks. J. Water Resour. Manag. **26**, 3639–3652 (2012)
14. Venkata Ramana R., et al.: Monthly rainfall prediction using wavelet neural network analysis. J. Water Resour. Manag. **27**, 3697–3711 (2013)
15. Ravansalar, M., Rajaee, T.: Evaluation of wavelet performance via an ANN-based electrical conductivity prediction model. J. Environ. Monit. Assess. **187** (2015)
16. Shafaei, M., Kisi, O.: Predicting river daily flow using wavelet-artificial neural networks based on regression analyses in comparison with artificial neural networks and support vector machine models. J. Neural Comput. Appl. **28**, 15–28 (2017)

17. Sharghi, E., et al.: Emotional ANN (EANN) and wavelet-ANN (WANN) approaches for Markovian and seasonal based modeling of rainfall-runoff process. J. Water Resour. Manag. **32**, 3441–3456 (2018)
18. Jimmy, S.R., Sahoo, A., Samantaray, S., Ghose, D.K.: Prophecy of runoff in a river basin using various neural networks. In: Communication Software and Networks, pp. 709–718. Springer, Singapore (2021)
19. Sahoo, A., Samantaray, S., Paul, S.: Efficacy of ANFIS-GOA technique in flood prediction: a case study of Mahanadi river basin in India. H2Open J. **4**(1), 137–156 (2021)
20. Mohanta, N.R., Patel, N., Beck, K., Samantaray, S., Sahoo, A.: Efficiency of river flow prediction in river using wavelet-CANFIS: a case study. In: Intelligent Data Engineering and Analytics, pp. 435–443. Springer, Singapore (2021)
21. Samantaray, S., Sahoo, A., Agnihotri, A.: Assessment of flood frequency using statistical and hybrid neural network method: Mahanadi river basin, India. J. Geol. Soc. India **97**(8), 867–880 (2021)
22. Samantaray, S., Biswakalyani, C., Singh, D.K., Sahoo, A., Prakash Satapathy, D.: Prediction of groundwater fluctuation based on hybrid ANFIS-GWO approach in arid Watershed, India. Soft Comput 1–23 (2022)

Chapter 45
Universal Dependency Treebank for Ho Language

D. Bankira, S. Panda, S. R. Dash, S. Parida, A.K.Ojha, and S. K. Dash

Abstract Language is the best way to share information. Each and every language has a grammatical structure to make complete sense. A sentence can be extracted with a dependency word sequence. The development of a large-scale dependency treebank for the Indian poor resource language is a great effort. The purpose of this paper is to build the first treebank of Ho language. The treebank contains 51 sentences of 315 tokens in the Ho language. The universal dependency guidelines have been followed for all the selected annotated sentences. The machine learning technique is used for morphological analysis of Ho treebank. The Ho language resources can be enhanced to create annotated treebanks. It will help in creation of language information systems for morphological analysis and cross-linguistic training. In the preliminary stages, we are using a machine learning approach to build Ho parsers. The linguistic study of the Ho universal dependency treebank is briefly mentioned in the conclusion section of this paper.

45.1 Introduction

A universal dependency treebank is a cross-lingual treebank annotation, whose main objective is to provide guidelines that apply to typologically different languages. Treebank annotation guidelines are suitable for computer parsing [1]. Parsing is an

D. Bankira · S. Panda
KISS Deemed to be University, Bhubaneswar, Odisha, India

S. R. Dash (✉)
KIIT Deemed to be University, Bhubaneswar, Odisha, India
e-mail: sdashfca@kiit.ac.in

S. Parida
Silo AI, Helsinki, Finland

A.K.Ojha
National University of Ireland, Galway, Ireland

S. K. Dash
National Institute of Technology, Aizawl, Mizoram, India

effort to establish an agreement to resolve a sentence into the part and describe the character of natural language using grammar. It determines the sentence structure, grammar-based approaches, which can identify attributes of phrases, such as tense, case, number, etc. [2]. The collection of annotated parse trees is called treebanks, which can be implemented in machine learning techniques to build most accurate parsers for natural languages. A treebank is a linguistically annotated corpus, which has followed a set of rules and grammar. The term 'treebank' was formulated by Geoffrey Leech, it mentioned the tree structure as the most common way to represent grammatical analysis. This paper's main objective is to develop the initial annotated corpus of Ho language texts. The text structure requires different works in different annotation measures. The corpus that is composed of various sub-corpora differs in the annotation level. The predicted annotation levels are (1) It is the normal form of lemma and identifies part of speech for each word in lemmatized texts, (2) Every word has morphologically tagged texts, which are given with the POS and lemma for a complete set of morphological properties, and (3) Syntactically tagged texts have a syntactic structure assigned to each sentence in addition to the morphological annotations at the word level. The purpose of this research is to create a linguistic parser. To perform language morphology parsing analysis, the machine system utilizes cross-lingual learning. The main objective is to create a large-scale treebank for language with limited resources. The unavailability of such resources is a major challenge to the creation of high-quality natural language techniques and applications for low-resource languages (LRLs). The UD treebank contains 51 sentences of 315 tokens in the Ho language. The universal dependency guidelines have been followed for all the selected annotated sentences. The Ho annotated dependency treebank will help to enhance the Ho language resources and develop language technological tools for cross-lingual training and morphological analysis. Using machine learning, we can create the first Ho parser. In the conclusion part, the paper briefly mentions the linguistic analysis of the Ho universal dependency (UD) treebank. In this technological age, the natural language is used as input during inference in business, government, and academia for communication with users and systems. Currently, natural language processing is repeatedly used with the help of machine learning approaches. Machine learning requires training data, but it is insufficient for low resource languages. Due to insufficient data and unsatisfactory performance to get the output of NLP, it can be possible to solve with cross-lingual learning. Cross-lingual learning is a conceptual framework for the transfer of knowledge from one human language to another human language. The transformation of information can help us to overcome the insufficient data in the target language and make machine learning models for low resource languages [3].

45.1.1 About Ho Language

India is a tribal-dominated country, where 8.6% (as per census 2011) tribal population are residing in various parts of the country. It is believed that tribals are the original inhabitants of the earth [4]. The 'Ho' is one of the sizeable tribal groups in India. The meaning of 'Ho' in Ho language (*Ho Hayam)* is 'human beings'. The Ho tribe belongs to the Austroasiatic family of the Munda group. It is believed that the original place of Ho is the *Chota Nagpur* Plateau of Jharkhand state. The 'Ho' speaking tribes are generally found in the states of Odisha, Jharkhand, Bihar, West Bengal, Chhattisgarh, and Assam and are also found in the country of Bangladesh and Nepal. As per the census report of 2011, Ho speaking people are more than 1,421,418 [5]. This 'Ho' language is spoken by the Ho, Kol, Munda, and Kolha tribe of India. The written system of the Ho language is called "Warang Chiti". This Warang Chiti script was discovered by *Ott Guru* Kol Lako Bodra in the twentieth century. The Warang Chiti script has 32 alphabets. It has both capital letters and small letters. The first letter is a sacred letter of Ho script, i.e. ⴲ(ong), 10 are vowels and 21 are consonants. In Ho grammar, alphabets are classified into four categories, i.e. (1) *Muulu Chiti* (first letter) ⴲ (2) *Ipan Borong* (vowels) ⴸ, ⴹ, ⴺ, ⴼ, ⴳ, Z(ang, ah, uih, iyuh, aye, o) (3) *Dobdi Borong* (vowels) Y, ⴴ, ⴱ, ⵁ(yoh, yah, e, u) (4) *Bengan Borong* (consonant) ⵝ, ⴲ, ⵀ, ⴶ, E, ⴱ, ⴵ, ⵃ, L, ⵌ, ⴱ, Z, ⵂ, ◊, U, ⴵ, ⴱ, 5, T, ⴱ, ⴽ(pronounce as Ng, Ga, Ka, Nj, Ja, Cha, Na, Da, Ta, Na, Da, Th, M, B, P, H, L, La, R, Sh, S, respectively). The digits of the Ho language are 0, ⴵ, ⴱ, ⵉ, 4, ⵤ, W, ⴵ, ⴷ, ⴹ.

45.1.2 Grammar of Ho Language

The Ho language has its grammar, which is structurally and functionally different from other languages. In Ho grammar, if the gender of the subject changes, the verb remains unchanged. However, in Hindi, there is a change in the verb with the subject, for example—*Laxmi ati hai* (female), *Raju ata hai* (male). In Ho—*Laxmi senoh tana* (female), *Raju senoh tana* (male). Similarly, if we will look at the structure of English sentences of simple present tense, e.g. I go to the Jungle, it's word sequence is "S+V+O", but in Ho sentence—*Anj Buru Senanj,* it's word sequence is "S+O+V+ S suffix". If we write in English sentence of simple present tense, i.e. I go (In this sentence no object) in Ho—"*Anj senanj*". So the word sequence in Ho is "S+V+S suffix". Generally not changeable of primary root and monosyllabic, which can be used as noun, verbs, or adjective. There are three numbers, i.e. singular, dual, and plural, e.g. *anj, aling, abu,* etc. The singular is always the original root, to which the suffixes *king* and *ko* are added to form the dual and plural, respectively. This only takes place in the case of nouns denting animate objects. In Ho, the grammatical different between genders is replaced by the distinction between animate and inanimate objects. Particularly, noun denotes a male or a female being does not affect the construction of a sentence. Noun denoting family relations insert *te* before the dual and plural suffixes, e.g. *misi*—a younger sister, *misiteking*—two younger sisters, *misteko*—younger sisters. Simple postposition of *a*ʹ *(ah)* is a relation either of ownership or complement parts, e.g. *Baguna*ʹ *ti*—Bagun's hand. *Re* (re) is the locative of rest indicating the place or time at which something occurs, e.g. *hature*—in the village. The compound postposition *Ete* (ete) is used in the ablative case sign and indicates motion away from a definite place, e.g. *oah-ete*—from home. *Ete* is' also used with time and *ete* may be translated as 'since', 'from the time of', 'for', etc. *Pa' re* is the locative of rest like *re,* but it is indefinite as compared with the latter. *Paete* is similarly indicates indefinitely motion away from. *Ren* is the locative genitive case sign used only with nouns denting animate objects, e.g. *haturen*—villager's. *Rea* is forms the locative genitive of inanimate nouns only, e.g. *haturea'*—of the village. There is no article in Ho. The context must show in each case whether the definite or indefinite article is to be used in translating a Ho sentence into English. Ho adjectives are invariable; it is not affected by the gender or number of the noun they qualify. The comparative degree is formed by adding the ablative case sign, *i.e. ete* to the noun, which the comparison is made and the superlative degree is formed by the addition of *ete* suffix in English. There are four classes of adverbs, i.e. adverbs of time, place, quantity, and manner, e.g. *tising*—today, *nenre*—here, *isu purah*—very much, *lika* —almost, etc. There are some conjunctions in Ho, i.e. *ondo* (and), *mendo* (but), *redo* (if), *mente* (because), *chi* (or), etc. Interjections are *mar* (all right), *haina* (alas), etc. [6].

45.1.3 POS Tag

A part of speech (POS) tagger assigns the appropriate tag to word sequence such as nouns, pronouns, verbs, adverbs, and adjectives. The part of speech refers to the role a word plays in a sentence. In Ho, there are ten types of "part of speech, i.e. *Nue* (noun), *Horoh* (pronoun), *olong* (adjective), *beda* (verb), *beda-olong* (adverb), *sagaiyan-koboy* (preposition), *Kononja-koboy* (conjunction), *Haiyan-koboy* (interjection), *Hopolbeda* (RE-active verb), and *Apan-beda* (self-verb). Again the noun has divided into 7 (seven), i.e. *Basia-nue* (proper noun), *jatiya-nue* (common noun), *jumur-itad-nue* (collective noun), *ginisiya nue* (material noun), *dor-itad-nue* (abstract noun), *sagaiyan-nue* (relative noun), and *bedan-nue* (verbal noun). There are ten types of pronoun, i.e. personal pronoun, definite pronoun, indefinite pronoun, relative pronoun, self-pronoun, interrogative pronoun, distributive pronoun, possessive pronoun, emphatic pronoun, and reciprocal pronoun. Kind of adjective of ho are *gui-itad olong* (adjective of quality), *Jong-itad olong* (adjective of quantity), *leneka olong* (adjective of number), and *horohyan-olong* (adjective of pronoun). *Kumiya beda* (intransitive verb) and *gumiya-beda* (transitive verb) and also auxiliary verbs (*denga beda*) in Ho. Adverbs are *tayed-itad- beda-olong* (adverb of place), *uli-itad-beda-olong* (adverb of time), *chonol-itad-beda-olong* (adverb of manner), *umu-itad-beda-olong* (adverb of quantity), and *konohyan-beda-olong* (adverb of place).

45.2   Related Work

Various methods have been adopted for rich resource languages, but in the low resource languages, there is insufficient and unavailability of such a type of universal dependency treebank. Linguistic knowledge is required for any natural language processing (NLP) application. The knowledge can be prepared in the form of grammar, dictionaries, rules of word order in a sentence, etc. The Santal linguistic resources programme seeks to annotate language information in digital texts, with the annotated texts being used in machine learning to develop the resource [3]. Rolando used a UD tool to improve documentation of the language and develop language learning materials and NLP tools for the indigenous language of Bribri of Costa Rica [7]. Chun introduced the manual assessment and revision process for the phrase-structure to UD conversion of Penn Korean and KAIST treebanks with the statistics and the current issues relating the treebanks [8]. Raj et al. [9] discussed the development of treebanks for Braj and Magahi of Eastern Indo-Aryan language—the treebank is annotated with UPOS, lemma, morphological attributes, and universal dependency relations for languages spoken in India with extremely low resource levels. [9]. Ojha reports the development of the first dependency treebank for Bhojpuri, which is a low resource language using the annotation scheme of UD [10]. Kondratyuk proposed and evaluated universal dependency, a multilingual capability of providing annotations for any universal dependency treebank, and achieving advances in universal dependency parsing in a large number of languages while maintaining comparable tagging and lemmatization accuracy [11]. It reported that universal dependency treebank, making a treebank for a target low resources language, has become easier and tried to create the most out of the target language treebank by integrating a source-language parser as an advance in learning a neural network parser [12]. According to Rasooli, a cross-lingual dependency transformation model, which takes into account the difficulty of word sequence differences between the source and target languages and projection-driven reordering, improves the accuracy of non-European languages while maintaining high accuracies in European [13].

45.3   Methodology

All selected annotated sentences are converted in universal dependencies standard file structure known as CoNLL-U format, which has ten fields, i.e. ID, word, lemma, UPOS, XPOS, FEATS, HEAD, DEPREL, DEPS, and MISC. The universal part of speech (UPOS) tags are following the guideline of universal dependency. We are following part of speech (PoS) guidelines of the Bureau of Indian Standards (BIS) and Department of Information Technology Ministry of Communication and Information Technology, Government of India. The Ho language follows the part of speech tagset guideline. The universal dependencies are denoted by UD tags (see

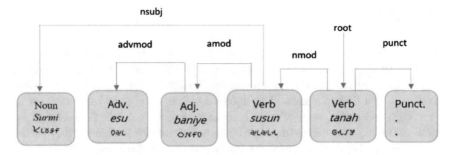

Fig. 45.1 Annotated treebank for Ho language

Table 45.1 Statistics of UPOS tags in the Ho PUD treebank

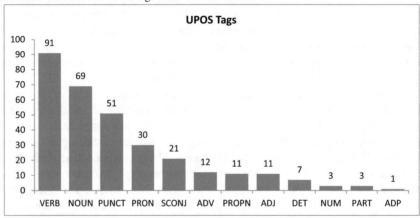

Fig. 45.1), which are a modified version of Stanford dependencies [3]. This dataset does not include the lemma or feats.

The Ho treebanks statistics are given in Table 45.1, and the dependency relations statistics are given in Table 45.2.

45.4 UD Data Format and Tools

All the data are encoded in a universal standard file format called CoNLL-U (Fig. 45.2).

Table 45.2 Statistics of used UD relation in Ho PUD treebank

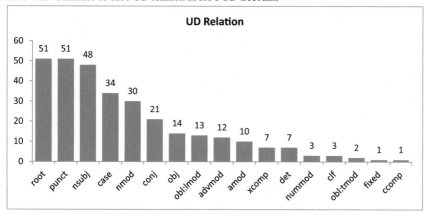

#text=5   ꞉ꞇꞇꞈ ꞽꞐ ꞈꞇꞎꞐꞇ ꞇꞈꞇ Ꞑ꞉Ꞑꞈ ꞇꞈꞏ.

#ID	Word	Lemma	UPOS	XPOS	FEATS	HEAD	DEPREL	DEPS	MISC
1	꞉ꞇꞇꞈ	–	NOUN	N_NN	–	5	obl:lmod	–	–
2	ꞽꞐ	–	SCONJ	CC_CCS	–	1	conj	–	–
3	ꞈꞇꞎꞐꞇ	–	NOUN	N NN	–	4	nmod	–	–
4	ꞇꞈꞇ	–	NOUN	N_NN	–	5	nsubj	–	–
5	Ꞑ꞉Ꞑꞈ	–	VERB	V_VM_VF	–	0	root	–	–
6	ꞇꞈꞏ	–	VERB	V_VM_VF	–	5	case	–	–
7	.	–	PUNCT	RD_PUNC	–	5	punct	–	–

#text=6   ꞷꞇꞽꞏ Ꞑꞇꞈ ꞐꞐꞏꞐ ꞇꞈꞐꞈꞇ ꞇꞈꞏꞽ.

#ID	Word	Lemma	UPOS	XPOS	FEATS	HEAD	DEPREL	DEPS	MISC
1	ꞷꞇꞽꞏ	–	NOUN	N_NN	–	4	nsubj	–	–
2	Ꞑꞇꞈ	–	ADV	RB	–	3	advmod	–	–
3	ꞐꞐꞏꞐ	–	ADJ	JJ	–	4	amod	–	–
4	ꞇꞈꞐꞈꞇ	–	VERB	V_VM_VF	–	5	nmod	–	–
5	ꞇꞈꞏꞽ	–	VERB	V_VM_VF	–	0	root	–	–
6	.	–	PUNCT	RD_PUNC	–	5	punct	–	–

Fig. 45.2 CONLL-U format example

45.5   Experiments and Result

The universal dependency (UD) annotation structure is used to manually annotate the Ho treebank. Using the universal dependency pipe open source tool, we built a Ho parser out of 315 tokens (51 sentences) [14]. For data separation, we used a cross-validation 90:10 average, where the batch size, hidden layer, dropout, and

Table 45.3 Accuracy of Ho parser

Tokenization	UPOS	XPOS	UAS	LAS
99.52%	99.19%	99.19%	73.99%	65.91%

learning rate were 10, 200, 0.10, 0.0200, and 0.005, respectively, and the other hyperparameters were randomized. The results are shown in Table 45.3.

45.6 Conclusion and Work in Future

The initial universal dependency treebank of Ho was presented in this paper for linguistic analysis and application in natural language processing (NLP), first for part of speech tagging, semantic analyser, parser, and machine translation and then using the UDPipe tool, we developed our first Ho parser.

Future work on universal dependency treebank research involves (1) collective more annotated data for training, validation, and development to improve the Ho treebank, (2) including a lemma for the Ho tokens, (3) perform a morphological analysis of the Ho language, and (4) test neural network models to complete the research.

References

1. Thomas, G.: Universal dependencies for mbyá guaraní. In: Proceedings of the Third Workshop on Universal Dependencies (udw, syntax fest 2019), pp. 70–77 (2019)
2. Magerman, D.M., Marcus, M.P.: Parsing a natural language using mutual information statistics. In: AAAI, vol. 90, pp. 984–989 (1990)
3. Dash, S., Sahoo, S., Mishra, B.K., Parida, S., Besra, J.N., Ojha, A.K.: Universal dependency treebank for Santali language. SPAST Abs. 1(01) (2021)
4. Ankathi, R., Gade, J.: Tourism and Indigenous People. Zenon Academic Publishing
5. Patnaik, J.K.: Adivasi, vol. 52, no. 1 and 2 (2012)
6. Burrows, L.: Ho Grammar (with vocabulary) Book. Shastri Indo-Canadian Institute, 156 Golf Links, New Delhi, India (1915)
7. Coto-Solano, R., Loáiciga, S., Flores-Solórzano, S.: Towards Universal Dependencies for Bribri. In: Proceedings of the Fifth Workshop on Universal Dependencies (UDW, SyntaxFest 2021), pp. 16–29 (2021)
8. Chun, J., Han, N.R., Hwang, J.D., Choi, J.D.: Building universal dependency treebanks in Korean. In: Proceedings of the Eleventh International Conference on Language Resources and Evaluation (LREC 2018) (2018)
9. Raj, M., Ratan, S., Alok, D., Kumar, R., Ojha, A.K.: Developing universal dependency treebanks for Magahi and Braj. arXiv Preprint (2022). arXiv:2204.12633
10. Ojha, A.K., Zeman, D.: Universal dependency treebanks for low-resource Indian languages: the case of Bhojpuri. In: Proceedings of the WILDRE5–5th Workshop on Indian Language Data: Resources and Evaluation, pp. 33–38 (2020)
11. Kondratyuk, D., Straka, M.: 75 Languages, 1 model: parsing universal dependencies universally. arXiv Preprint (2019). arXiv:1904.02099

12. Duong, L., Cohn, T., Bird, S., Cook, P.: Low resource dependency parsing: cross-lingual parameter sharing in a neural network parser. In: Proceedings of the 53rd Annual Meeting of the Association for Computational Linguistics and the 7th International Joint Conference on Natural Language Processing (vol 2: short papers), pp. 845–850 (2015)
13. Rasooli, M.S., Collins, M.: Low-resource syntactic transfer with unsupervised source reordering. arXiv Preprint (2019). arXiv:1903.05683
14. Straka, M., Straková, J.: Tokenizing, POS tagging, lemmatizing and parsing UD 2.0 with UDPipe. In: Proceedings of the CoNLL 2017 Shared Task: Multilingual Parsing from Raw Text to Universal Dependencies, pp. 88–99 (2017)

Chapter 46
Mental Health Prediction Among Students Using Machine Learning Techniques

Savita Sahu and Tribid Debbarma

Abstract Mental health problems in students are increasing worldwide. In this research, we consider some mental illnesses like stress, anxiety, Post-traumatic stress disorder (PSTD), Attention Deficit Hyperactivity Disorder (ADHD), and depression in students. If students mental health problem can be diagnosed early, they can treated in earlier stage. Presently, Machine Learning techniques are well suited for the analysis of medical data and the diagnosis of mental health problems. In this research work we apply Machine learning techniques are Logistic Regression, Decision Trees, Random Forests, KNN (K-Nearest Neighbors) Classifiers, and Neural networks, and compared their accuracy on different measures. Data sets are collected for training and testing the performance of the techniques. Twelve factors have been identified as being important for predicting mental health from the data set, including biological, psychological, and physical factors. By applying feature selection algorithms we have been able to improve the accuracy of the proposed model. We prove that the Neural Network model is the most accurate for this type of prediction.

46.1 Introduction

The World Health Organization (WHO) [1], estimates that one out of four people in the world will experience mental health problems at some point in their lives. The student community is also affected by mental health problems. Most of the time students are unable to realize they are dealing with mental illness. Students, teachers, and parents have to realize how common mental health problems in kids and teenagers are having and assist them when they need it. Mental health

---

Tribid Debbarma contributed equally to this work.

S. Sahu (✉) · T. Debbarma
Computer Science and Engineering, National Institute of Technology Agartala,
Agartala, Tripura 799046, India
e-mail: savitasahu63@gmail.com

T. Debbarma
e-mail: tribid@ieee.org

V. Bhateja et al. (eds.), *Evolution in Computational Intelligence*, Smart Innovation,
Systems and Technologies 326, https://doi.org/10.1007/978-981-19-7513-4_46

problem [2], is second in the world behind heart disease. Other than heart disease problems mental health is ahead of all other diseases. There is a very slow growth of mental health professionals in comparison to the growth of mental health patients.

There is a high prevalence of anxiety, depression, stress, and addictive disorders among students. Research conducted by the National Alliance on Mental Illness (NAMI) [3], indicates that one in four students is diagnosed with mental illness. The NAMI carried out a countrywide survey of university students living with mental health situations to learn about their experiences in college. NAMI designed the survey to listen at once from students about whether colleges are meeting with their students and considering their needs, and what enhancements are needed to support their educational enjoyment.

Here are a few reasons why students are affected by mental health diseases:

- Past experiences of sexual abuse by a family member or an outsider, or losing anyone close to you.
- Factors such as whether someone in the family has a mental health issue, whether they have friends.
- Economical factor such as financial problems.
- Education factors such as education, academic pressure, whether parents are happy with him/her or not.

Today mental health problems are very common among students so if any student experiences any of the following symptoms, the risk of mental illness increases:

- The problem facing at the time of sleeping.
- Sleeping time up and down.
- Mood swings.
- Psychological and physical problems such as depression, stress, and nervousness can negatively affect academic performance.
- Monetary strain can cause tension, depression, stress, and anxiety.
- Think about harming yourself.

In this study, we consider some mental health problem symptoms in students such as stress, depression, anxiety, Post-traumatic stress disorder (PSTD), and Attention Deficit Hyperactivity Disorder (ADHD).

- **Stress** can be particularly trouble college students. Instead of getting up early and having a routine as they did in high school, college students often stay up late to study or socialize. It's hard to get the vitamin D needed to fight off sad symptoms if they don't have morning classes.
- In among students, attention problems include failing to pay close attention to details, making careless mistakes, not listening when speaking directly to them, or having trouble reading and retaining basic information. Students who have academic difficulties often have problems with studying and reading as well as anxiety about exams rather than technical or practical skills. **Anxiety** [4], problems are

characterized by excessive worry, agitation, fatigue, sleep problems, and concentration problems. **ADHD (Attention deficit hyperactivity disorder)** [3], is a very common childhood condition, which can persist into adulthood.

- Many factors of university lifestyles contribute to risk elements of **depression** [5]. Many students are unprepared for university lifestyles. Today's students face excessive debt. In addition, they have fewer job prospects after commencement than preceding generations. These delivered issues can cause depression in college students.
- A person may develop **PTSD (Post-traumatic stress disorder)** [6], if they witness the unexpected or violent death of a family member or close friend, or if they suffer serious injury or harm. It is also possible to suffer from PTSD due to survivor guilt (feeling guilty after surviving a death). PTSD can affect anyone at any age. Symptoms of PTSD can appear immediately after a trauma or may not appear for a month or more afterwards. In some cases, PTSD develops long after a traumatic event.

46.2   Literature Review

To begin with one of the most recent research paper [7] by, Sofianita et al. Aiming at predicting depression, stress, and anxiety in higher education. The data set is collected from students of higher education institutions in Kuala Terengganu. To predict stress, depression, and anxiety, authors use a Naive Bayes Decision Tree, Logistic Regression, Neural Network and Support Vector Machine Algorithms and classify the most accurate model as a Decision Tree, Support Vector Machine, and Neural Network, respectively.

In [8], author Mengjun Lao, describes the mental health intelligent evaluation system based on the joint optimization model. The joint optimization process consists of the improvement of the Decision Network and improving the ANN (artificial neural network). In the Joint Optimization model, the lowest error rate and highest accuracy index are achieved. In this paper authors are used Student-Life and Reach Out Online forum Dartmouth College post, their methodology is applied to the student life data set. Currently, mental health is suffering from a high death rate and low productivity. The authors propose an intelligence evaluation process based on a joint optimization algorithm for mental health intelligence evaluation.

Further, in a research [9], by Sandhya P. et al. Proposed the model for the prediction of mental disorders for employees in the IT industry. The authors use the data set collected in the survey among IT professionals from different countries. Authors compare Logistic Regression, K Nearest Neighbors Classifier, Decision Tree, Random Forest, and Neural Network models. So the highest accuracy is achieved in a Neural Network that is used for prediction.

In [10], by Vidit Laijawala et al. Proceeds in the data mining for systematic review for prediction of mental health. Researchers collected data set from the online open-sourcing mental illness survey consisting of data of working individuals, then pre-

Table 46.1 Comparison of existing work

References	Techniques	Data set	Accuracy (%)
[7]	DT, NN, SVM, Logistic regression	NH survey	71
[8]	DT, NN, SVM, Logistic regression	Online communities	81
[13]	Naive Bayes, Random rorest, KNN, SVM	PSS questionnaire	85.71
[5]	Decision tree, SVM, ANN, BN	Mental status (GDS)	96.7
[11]	KNN, SVM, BN	Depression level	96.2

processed the data and encode it for better prediction. And apply Machine Learning algorithms Random forest and Decision Tree and the tool is applied in weka which is a Machine Learning algorithm directly in the data set from the java code. The authors predict mental health in people above the age of 18. Create a website that can predict the outcome according to the information provided by the user once the model is built.

Apart from this in the research paper [4], by Anu Priya et al. Commenced with the problem of predicting anxiety, depression, and stress in modern life. Collect the data set from different cultures and communities using depression, anxiety, and stress scale questionnaires 21. Machine Learning techniques are Random Forest, Decision Tree, KNN, Naive Bayes, and Support Vector Machine. After applying these models authors identified the imbalance in the confusion matrix, so an f1 score measure was added and identify the best accurate model which is the random forest classifier, and the accuracy achieved by the authors is 71%.

The science of Machine Learning concerns how computers learn or acquire knowledge through data. As a field of study, Machine Learning is about giving computers the capability of learning without explicitly being programmed [5–8, 11–13]. Machine Learning can be divided into four types: Supervised Learning, Unsupervised Learning, Semi-supervised Learning, and Reinforcement Learning. As shown in Table 46.1, supervised learning is the most popular Data Mining technique for solving problems relating to mental health.

46.3 Proposed Model

Proposed a system with the primary goal of predicting the risk of mental illness in students according to the value obtained in the data set. Firstly, we collect real-time data set using some questionnaires. Then analyzed and pre-processed the data. Data capture in the real world may contain nonviable, noisy, irrelevant, and illogical content values. The procedure for obtaining useful results we clean the data, data cleaning is a crucial task when preparing data for its processing in machine learning algorithms after preprocessing we have 5840 data for this model then applied Machine Learning models in that data set. The steps of our proposed model are shown in Fig. 46.1.

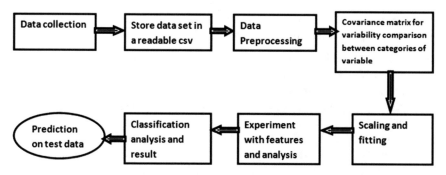

Fig. 46.1 Proposed model

We use covariance matrix for variability comparison between categories of variables than scaling and fitting wherein the feature scaling we're going to scale age because it is extremely different from the other ones. In the scaling part splitting the data set, the procedure by taking a data set and dividing it into two 70 and 30% subsets. Fitting the model is done using the first subset and is referred to as the training data set. And the second subset is not used to train the model instead, the input element of the model is fed into the data set, then predictions are generated and compared to the expected values. The second data set is referred to as the test data set. The train-test split is a technique used for evaluating the performance of a Machine Learning algorithm. We have planned to apply Logistic Regression, K-Nearest Neighbors Classifier, Decision Tree, Random Forest, Neural Network Classification algorithm for the classification of the data, experiment with these classification algorithms to predict the risk of mental health.

46.4 Data Set

The dataset consists of a survey conducted among students from different schools, colleges, and universities using Google form questionnaires. Questions contain information like age of student, gender, education, family past mental health history, sexual abuse by anyone of any age, whether he or she lost anyone close or not, Mental health consequence, physical health consequence and many more. The data set was taken from 6030 students. After data preprocessing we feed 5840 data in the model.

46.4.1 Description of the Data Set

The data set contain 14 features including the predicted attribute. In the target field we have 2 classes yes and nowhere 'yes' class means the student is having a risk of mental health, and 'No' means the student doesn't have the risk of mental health. Now some important features are described below:

 (i) Gender-male and female.
 (ii) Age-age of student in years.
 (iii) Education-primary, secondary, graduation, post-graduation.
 (iv) Family history-whether is the student's family or anyone having a mental health issue or not.
 (v) Sexually Abused-Whether a student is abused by anyone of any age.
 (vi) Lost anybody close-whether students lost one which is closed to them (yes, no).
 (vii) Do you have friends-have friends in their life or not (yes, no).
 (viii) Trouble sleeping history-having trouble sleeping (1 = yes, 0 = no).
 (ix) Are parents proud-r-Are his/her parents are proud or happy with you or not (yes, no).
 (x) Academic pressure-Whether the student having the pressure, burdened or not (yes, no).
 (xi) Financial problem-they have a financial issue in their family (yes, no).
 (x) Sleep hours-how many hours students take sleep.
 (xi) Mental health consequence-Whether the student feels any mental consequences like energy level, concentration, dependability (yes, no, maybe).
 (xii) Physical health consequence- Physical health covers healthy diet, healthy weight, dental health, sleep, flu, addictive (yes, no, maybe).

Target Attribute: Risk of mental illness: whether the student has a risk of mental health issues.

46.5   Experimental Setup

46.5.1   Data Preprocessing

Then we do data preprocessing for a Machine Learning model. In the data preprocessing first step is data profiling, in data profiling, we analyze and review data and its quality. It starts with comparing the existing data. The next step is data cleaning, in data cleaning eliminating the noisy data filling the missing data and unnecessary columns like timestamp, country, etc., and removing the duplicate data.

46.5.2   Covariance Matrix, Variability Comparison Between Categories of Variables

Here we can see in Fig. 46.2 correlation matrix of mental health in the student's data set. Correlation matrix: statisticians and data analysts measure the correlation

Fig. 46.2 Corelation heat map

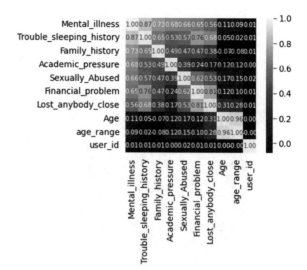

of two variables to find an insight into their relationships. The correlation value between two of its attributes forms a matrix which is called a correlation matrix. Correlation values can be calculated in several ways. The most popular one is the Pearson Correlation Coefficient. Nevertheless, it should be noticed that it measures the only linear relationship between two variables. In other words, it may not be able to reveal nonlinear relationships. The value of Pearson correlation ranges from −1 to +1, where ± describes a perfect positive/negative correlation and 0 means no correlation.

46.5.3   Scaling and Fitting

Ensuring and validating the data is suitable for feature engineering. In Machine Learning, feature scaling signifies the end of data preprocessing. It's a technique for keeping a dataset independent variables inside a certain range here we scale the age in the range 10–14, 15–19, 20–24, 25–28. To put it another way, feature scaling narrows the range of variables so that we may compare them on equal footing. In the fitting of the model we split the data into two sets. The first set is the train set which is used to train the model. The second set is testing data that is used for testing the prediction model. In the next step, we encode the data converted categorical attributes to numeric values by using the label encoder for python and applying a scalar normalization method, in the mental health record. Then find important feature which feature is relevant for prediction with random forest model selection [14], shown in Fig. 46.3. It can help to better understand a solving a problem and also it can improve the model efficiency.

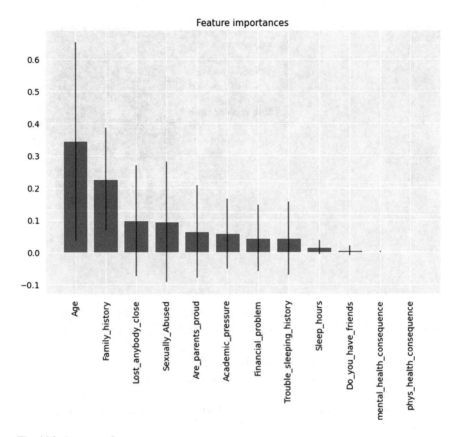

Fig. 46.3 Important features

46.6 Experiment and Analysis

This section represents the result of the modeling algorithm and visualization of the analysis in the modeling evaluation phase we identified which algorithm is best suitable for our data set based on their accuracy and also doing experiments with features and finding the probability of mental health according to all feature values. Figures 46.4 and 46.5 represent the probability risk of mental health in the students according to the gender blue color shows females and the red color shows a male in Fig. 46.4 graph we can see the probability of the students mental health according to the age group based on the gender were according to this data set 20–24 age group of the student probability of mental health risk is very high. Here we can assume that at this age students are going to their college or university they are varied about their future and also they are unfamiliar with the new life at university. And Fig. 46.5 we can see the probability of risk of mental health according to the education. Where we can see a risk of mental illness in the college or university students is very high.

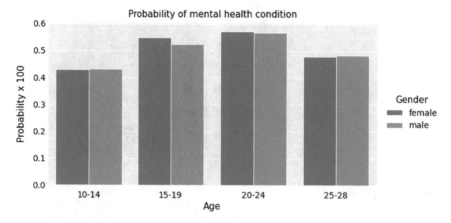

Fig. 46.4   Probability of mental health based on age group

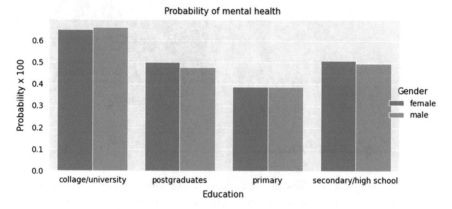

Fig. 46.5   Probability of mental health based on education

46.7   Result and Discussion

46.7.1   Analysis

Figure 46.6 shows distribution of the mental health disease where we have 7 classes ['ADHD' class-0], ['No' class- 1], ['PSTD' class-2], ['anxiety class-3], ['depression' class-4], ['stress' class-5]. According to the research, nearly half of the students are healthy and do not have any mental health issues. The number of students suffering from depression and stress is exceptionally high. Anxiety, ADHD, and PSTD, on the other hand, are less common than stress and sadness. According to Fig. 46.6, is shows depression is more in the students and also we can see in Fig. 46.5 that graduate students are having more mental health problems. They are apprehensive about their future because they are still adapting to their new university life.

Fig. 46.6 Distribution of
mental health disease

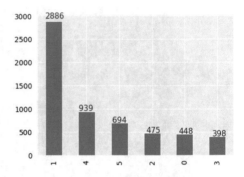

Fig. 46.7 Confusion matrix

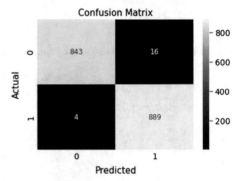

46.7.2   Evaluation Process

For the evaluation process classification accuracy, classification error, precision, false
positive, cross validated accuracy, ROC and confusion matrix are evaluated. Confusion matrix: Table that describes the performance of a classification model. With
Fig. 46.7 we can see most accurate model performance based on confusion matrix.
  Confusion matrix table of testing data set described below:

  True Positives (TP): we correctly predicted that 843 have mental health problems.
  True Negatives (TN): we correctly predicted that 889 don't have mental health
  problems.
  False Positives (FP): we incorrectly predicted that is 16 they do have mental issues
  (a "Type I error")
  Falsely predict positive.
  False Negatives (FN): we incorrectly predicted that is 4 they don't have a mental
  issue (a "Type II error") Falsely predict negative.

Table 46.2 Comparison on machine learning algorithms

Algorithm	Accuracy (%)	Error	False positive	Precision (%)	CV AUC (%)
Logistic regression	98.970	0.00970	0.0186	98.23	98.86
Random forest	99.03	0.0114	0.0186	98.23	98.82
Decision tree	92.58	0.0376	0.0721	93.48	92.87
*K*NN classifier	98.57	0.0142	0.0209	98.00	98.98
Neural network	99.03	0.0625	0.0588	99.00	99.03

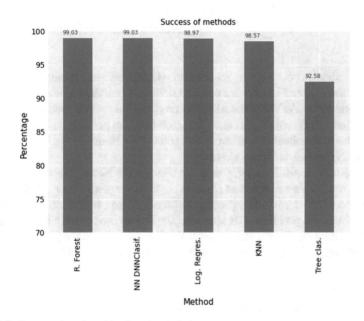

Fig. 46.8 Success plot of machine learning techniques

46.7.3   Classification Comparison

From Table 46.2, we can identify the best accurate algorithm which is a Neural Network, having bit high precision, cross-validation accuracy (CV AUC), and classification accuracy so we applied this algorithm in test data for prediction it will give 99% accuracy in the testing data set (Fig. 46.8).

The Receiver Operator Characteristic (ROC) curve is a binary classification issue evaluation metric. It's a probability curve that displays the True Positive Rate (TPR) against the False Positive Rate (FPR) at different threshold levels, thereby separating the 'signal' from the 'noise'. The Area Under the Curve (AUC) is a summary of the ROC curve that measures a classifier's ability to distinguish between classes.

Fig. 46.9 ROC curve

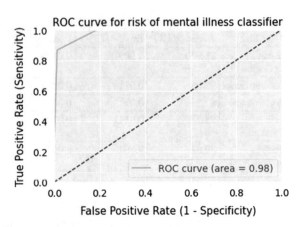

The AUC indicates how well the model distinguishes between positive and negative classes. According to Fig. 46.9 we obtain a 0.98 ROC curve area which is achieved using Neural Network. When the AUC reaches 0.5–1, there is a good likelihood that the classifier will be able to tell the difference between positive and negative class values because the classifier can recognize more True Positives and True Negatives than False Negatives and False Positives.

Here with Table 46.2, and Fig. 46.8 we can see the Neural Network is the best prediction model. Some important features required for predicting mental health are the Age of the student, Whether parents are proud, Losing anybody close, Academic pressure, Financial problems, Trouble sleeping history, and Family history. A CSV file is prepared with the user Id number as the first column and the predicted column is a risk of mental illness where we can identify the risk of mental illness in students.

46.8 Conclusion and Future Work

Nowadays, many expert systems are used in medicine to predict illnesses in the early stage to provide effective and efficient treatment. This research could be useful for mental health professionals, parents, and teachers so, they can identify the situation of mental health in the students. Mental health impacts their behavior, thoughts, and academic growths. The most common factors discovered in this research are: lack of parents support, financial problem, education environment, trouble in sleeping, family past history, mental health consequence, physical health consequence, lost some one closed to them, etc. The results were obtained using many Machine Learning algorithms that predict the risk of mental health accurately, and we were able to obtain highly accurate model Neural Network with accuracy 99.03%. We hope that the overall accuracy in this field of interest will be improved through further studies and also they can try to add more classes, features, and data that can be included to compute and increase the efficiency and accuracy of the proposed methodology.

References

1. Sumathi, M.R., Poorna, B.: Mental health prediction in children using machine learning techniques. Int. J. Adv. Comput. Sci. Appl. **7**(1) (2016)
2. http://blogs.fortishealthcare.com/mental-health-india-wake-up-call
3. A-survey-report-on-mental-health.   https://www.nami.org/Support-Education/Publications-Reports/Survey-Reports/College-Students-Speak
4. Priyaa, A., Garga, S., Tiggaa, N.P.: Predicting anxiety, depression and stress in modern life using machine learning algorithms. In: International Conference on Computational Intelligence and Data Science (ICCIDS) (2019)
5. Hou, Y., Xu, J., Huang, Y., Ma, X.: An application to predict depression in the university based on reading habits. In: ICSAI, pp. 1085–1089 (2017)
6. Ge, F., Li, Y., Yuan, M., Zhang, J., Zhang, W.: Children and adolescents exposed to earthquakes and their risk factors for posttraumatic stress disorder: a longitudinal study using a machine learning approach. J. Affect. Disord. **264**, 483–493 (2020)
7. Mutalib, S., Shafiee, N.S.M., Abdul-Rahman, S.: Mental health prediction in higher study student using machine learning. Turk. J. Comput. Math. Educ. (TURCOMAT) (2021)
8. Luo, M.: Research on students' mental health based on data mining algorithms. Hindawi J. Healthc. Eng. (2021)
9. Sandhya, P., Kantesaria, M.: Prediction of mental disorder for employees in IT industry. Int. J. Innov. Technol. Explor. Eng. (IJITEE) 8(6S) (2019)
10. Laijawala, V., Aachaliya, A., Jatta, H., Pinjarkar, V.: Data mining for systematic review for prediction of mental health. In: Proceedings of the 3rd International Conference on Advances in Science and Technology (ICAST) (2020)
11. Spyrou, I.M., Frantzidis, C., Bratsas, C.: Methodologies of Classification Compared Control and Signal Processing in Biomedicine, pp. 118–129 (2016)
12. Suhaimi, N.M., Abdul-Rahman, S., Mutalib, S., Hamid, N.H.A., Ab Malik, A.M.: Learning Machine Learning Algorithms to Predict Graduate-on-Time. vol. 1100, pp. 130–141 (2019)
13. Sabourin, A.A., Prater, J.C., Mason, N.A.: Mental health in pharmacy student. Dept. Pharm. **11**(3), 243–250 (2013)
14. Dimitriadis, S.I., Liparas, D., Tsolaki, M.N., Alzheimer's Disease Neuroimaging Initiative: Random forest feature selection for disease prediction. J. Neuron Sci. Method **302**, 14–23 (2018)
15. Bhakta, I., Sau, A.: Prediction of Depression Among Senior Citizens Using Machine Learning Classifiers
16. Fayez, M.A.: Diagnose mental health using new machine learning optimization technique. Department of ECE, Institute of Science, Altınbaş University, Istanbul Turkey **12**(13), 809–815 (2021)
17. Alonso, S.G., De La Torre-Díez, I., Hamrioui, S., López-Coronado, M., Barreno, D.C., Nozaleda, L.M., et al.: Machine learning techniques in mental health. J. Med. Sci. **42**(161) (2018)
18. Ramírez-Gallego, S., Krawczyk, B., García, S., Woźniak, M., Herrera, F.: Survey of data prepossessing neurocomputing current and future setuation. Neurocomputing **239**, 39–57 (2017)
19. https://towardsdatascience.com/machine-learning-basics-with-the-k-nearest-neighbors-algorithm
20. Navyasri, M., RajeswarRao, R., DaveeduRaju, A., Ramakrishnamurthy, M. et al.: Robust features for emotion recognition from speech by using Gaussian mixture model classification. In: International Conference and Published Proceeding in SIST Series, vol. 2, pp. 437–444. Springer (2017)

Chapter 47
Application of Support Vector Machine Integrated with Grasshopper Optimization for Runoff Prediction: A Case Study

Aiswarya Mishra, Gopal Krishna Sahoo, Debi Prasad Panda, Abinash Sahoo, Shaswati S. Mishra, Sandeep Samantaray, and Deba Prakash Satapathy

Abstract An adaptive data analysis approach for decomposing yearly rainfall series in a rainfall-runoff model based on a support vector machine (SVM) for the Mahanadi River basin is presented in this study. The SVR hyper parameters were determined using the grasshopper optimization algorithm (GOA). When compared to traditional data mining approaches, SVR produced considerable improvements in training, calibration, and validation. SVM is based on the structural risk minimization principle, which reduces a bound on a generalized risk (error) rather than the empirical risk reduction principle used by standard regression approaches. It effectively employs a convex quadratic optimization problem, resulting in a solution that is always unique and globally optimum. Evaluation parameters R^2, MSE, and WI with values 0.9465, 5.992, and 0.9528, respectively, indicate that SVM-GOA model had better prediction capability than the standard SVR model. Hence, the hybrid SVM-GOA model outperforms the classical SVM model in terms of prediction accuracy, making it a useful tool for the monthly runoff prediction.

A. Mishra · G. K. Sahoo · D. P. Panda · D. P. Satapathy
Department of Civil Engineering, OUTR Bhubaneswar, Bhubaneswar, Odisha, India
e-mail: dpsatapathy@cet.edu.in

A. Sahoo
Department of Civil Engineering, NIT Silchar, Silchar, Assam, India

S. S. Mishra
Department of Philosophy, Utkal University, Bhubaneswar, Odisha, India

S. Samantaray (✉)
Department of Civil Engineering, NIT Srinagar, Jammu and Kashmir, India
e-mail: samantaraysandeep963@gmail.com

© The Author(s), under exclusive license to Springer Nature Singapore Pte Ltd. 2023
V. Bhateja et al. (eds.), *Evolution in Computational Intelligence*, Smart Innovation, Systems and Technologies 326, https://doi.org/10.1007/978-981-19-7513-4_47

47.1   Introduction

The modeling and forecast of runoff in watersheds is required for many real-world applications including environmental disposal, conservation, and water resource management. However, because to the interdependence of runoff on the underlying meteorological and physiographic factors, the runoff modeling process is a dynamic, nonlinear, and complex hydrological event to simulate. Runoff models can be categorized as physically based (knowledge-driven) or data-driven depending on the governing processes. Physically based models entail a comprehensive interconnection of numerous physical processes that govern a system's hydrologic behavior. Data-driven models, on the other hand, are mostly based on observations (measured data) and attempt to define the system reaction from those data using transfer functions [1–3]. Data-driven modeling techniques have been popular as a substitute for physically based models in recent years, as they overcome some of the constraints associated with physically based approaches [4–9].

Botsis et al. [10] compared the performance of SVR and multilayer feed-forward neural network (MFNN) to simulate rainfall-runoff relationship for a mountainous watershed located in Northern California. SVR exhibited better results in developing the relationship between runoff and rainfall. Bell et al. [11] incorporated SVM model with RBF kernel for prediction of runoff in the Folsom dam situated in American river. The results of the experiments indicate that SVM is a good model for forecasting runoff. Chu et al. [12] examined the application of SVM for mid-term and long-term forecasting of runoff in the Yellow River basin, China. SVR produced better efficiency than auto-regression model and RBFNN and is a reliable data mining technique for mid-term and long-term runoff forecast. Sarzaeim et al. [13] applied ANN, GP, and SVM for runoff estimation in the Aidoghmoush Basin, Iran. They concluded that SVM outperformed the other methods and proved to be the most viable model for runoff prediction. Sharifi implemented ANNs, local linear regression, an ANFIS, and SVM models to construct a model for the prediction of daily runoff of Amameh watershed, Iran. SVM showcased its superiority over other models in terms of efficiency and accuracy.

Barman et al. [14] used day type, load, temperature, and humidity of the same hour of the similar day's data for short-term load forecasting in Assam, India. The results demonstrated the ascendancy of GOA-SVM model over GA-SVM and PSO-SVM for short-term load forecasting. Alizadeh et al. [15] simulated monthly streamflow in Iran's Kraj river basin using SVM-GOA model. Their findings showed that SVR-GOA algorithm outperformed SVM model with non-optimal parameters. Alrashidi et al. [16] hybridized SVR with GOA and utilized the Boruta algorithm for feature selection (SVR-GOA-BA), and compared its performance to ANN, DT, KNN, RF, SVM, SVM-PSO-BA, SVM-COA-BA, and SVM-NNA-BA for global solar radiation prediction in the Saudi cities of Dhahran, Riyadh, and Jeddah. The findings revealed that the SVR-GOA-BA performed best comparably to existing hybrid algorithms for global solar radiation prediction. Goodarzizad et al. [17] used ANN-GOA algorithm to determine most critical factors influencing construction labor productivity in Iran among 19 factors and discovered that the most influencing factors are labor experience, skill, and motivation, pay, site accidents, proper supervision, and weather conditions. Fattahi and Hasanipanah [18] calculated fly rock distance in mine blasting for three Malaysian quarry sites and discovered that the ANFIS-GOA model outperformed the ANFIS-CA model in terms of modeling efficiency. Panahi et al. [19] demonstrated that SVM-GOA outperformed SVM-PSO and standalone SVM model for flood modeling in Qazvin Plain, Iran, by considering Geospatial data of that area. This study aims at application of robust SVM-GOA model for monthly runoff prediction considering data from Salabheta gauging station of Mahanadi River Basin.

47.2 Study Area

The Mahanadi River Basin consists of two provinces: Chhattisgarh ($75,136$ km^2) and Orissa ($65,580$ km^2) (Fig. 47.1). It begins in Chhattisgarh and flows approximately 851 km before emptying into the Bay of Bengal. The Mahanadi majorly flows through the state of Odisha. The weather is tropical, with a hot and humid monsoon environment. During the Indian summer monsoon season, the basin is regularly flooded (June to September). Mahanadi is primarily rainfed, and the water supply varies greatly throughout the year. The average annual rainfall in the basin is 1572 mm, with a mean annual discharge of 67 billion cubic meters (BCM). While floods occur in the lower Mahanadi sub-basin during moist years, water shortage prevails across the basin during dry years. Furthermore, with swiftly expanding economic activities, the region will experience tremendous stress in the future.

47.3 Methodology

47.3.1 Support Vector Regression (SVR)

Vapnik [20] pioneered the use of SVR. When data points and their attributes are constrained, SVR can be utilized to tackle linear and nonlinear problems. Given a set of n data points for training, then

$$P = \left\{ (a_i - b_i) | a_i \in S^d, \, b_i \in (-1, \, 1) \right\}_{i=1}^{n}$$

where input and related output values are a_i and b_i, respectively, and d is defined as an input space dimension. The SVR method's main purpose is to describe a LR function

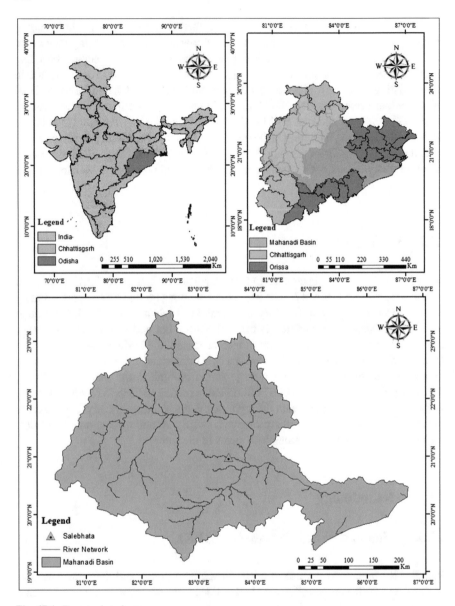

Fig. 47.1 Proposed study area

$$g(a) = \omega^T \psi(a) + v$$

where $g(a)$ represents the predicted values, $\psi(a)$ specifies a nonlinear function with regard to input variable matrix a; v and ω^T—reversed values of bias parameter and weight factor, respectively. To achieve maximum model accuracy, it is critical to precisely determine C, ε. Hence, we employed the sophisticated metaheuristic algorithm GOA to find optimum values for these parameters.

Grasshopper Optimization Algorithm (GOA)

The GOA proposed by Saremi et al. [21] addresses optimization problems, it mathematically replicates the behaviour of grasshopper swarms in nature. Search process in GOA is categorized into two parts: exploration and exploitation. Adults and nymph grasshoppers implement the search process in this algorithm. Adult grasshoppers travel quickly and across vast distances. As a result, they are used to explore the whole search space for improved food supply locations which means they carry out exploration. Nymph grasshoppers, on the other hand, are used to migrate and target a specific region or area locally. In optimization terms, this is referred to as exploitation GOA ensures a smooth balance amid exploitation and exploration, resulting in a mathematically simpler algorithm. Mathematical model, which depicts the grasshoppers' swarming behavior, is represented as

$$Y_i = I_i + G_i + W_i$$

where X_i— location of ith grasshopper, S_i—notion of social interaction, G_i—force of gravity exerted on ith grasshopper, and A_i—wind advection.

The mathematical model provided above cannot directly address the optimization challenges because the grasshoppers soon reach their comfort regions and swarms do not come together to definite spots. To avoid this difficulty, an altered version of this equation is presented below

$$Y_i^d = c \left(\sum_{\substack{j=1 \\ j \neq i}}^{N} c . \frac{(ub_d - lb_d)}{2} . s\left(\left|a_j^d - a_i^d\right|\right) . \frac{(a_j - a_i)}{x_{ij}} \right) + S_d$$

Lower and upper limits in dth dimension are denoted by lb_d and ub_d, respectively. S_d is the location of the best solution it has discovered so far.

To reduce repulsion zone, attraction zone, and comfort zone, decreasing coefficient c is applied. Because of the comfort zone parameter c, grasshoppers impose a progressive and seamless balance between discovery and exploitation. This feature enables GOA (Fig. 47.2) to accurately estimate the global optimum. In each cycle, the dynamic coefficient c may be calculated as follows:

$$c = c_{max} - l . \frac{c_{max} - c_{min}}{L}$$

where c_{min} and c_{max} represent minimum value and maximum value of coefficient c, l—current iteration, and L—maximum number of iterations.

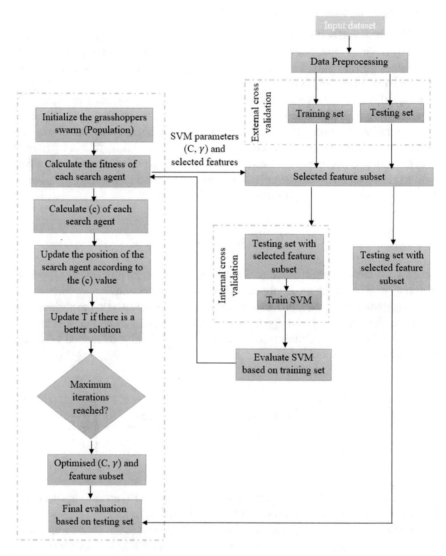

Fig. 47.2 Flowchart of SVM-GOA

47.4 Results and Discussions

To evaluate efficiency and consistency of proposed model, two statistical parameters are utilized. Runoff data of past 25 years were collected from Central Water Commission, Bhubaneswar. 80% of total dataset (240) are considered for training the models and rest 20% (60) are considered for testing [22–24].

$$R^2 = \left(\frac{\sum_{i=1}^{N} \left(P_i^{obs} - \overline{P_i^{obs}} \right) \left(Q_i^{pre} - \overline{Q_i^{pre}} \right)}{\sqrt{\sum_{i=1}^{N} \left(P_i^{obs} - \overline{P_i^{obs}} \right)^2 \sum_{i=1}^{N} \left(Q_i^{obs} - \overline{Q_i^{obs}} \right)^2}} \right)^2$$

$$WI = 1 - \left[\frac{\sum_{i=1}^{N} \left(P_i^{obs} - Q_i^{pre} \right)^2}{\sum_{i=1}^{N} \left(\left| Q_i^{pre} - \overline{P_i^{obs}} \right| + \left| P_i^{obs} - \overline{P_i^{obs}} \right| \right)^2} \right]$$

$$MSE = \frac{1}{n} \sum_{k=1}^{n} \left(P_i^{obs} - Q_i^{pre} \right)^2$$

P_i^{obs} and Q_i^{pre} are measured and predicted runoff, and $\overline{P_i^{obs}}$ and $\overline{Q_i^{pre}}$ are mean of measured and predicted runoff.

Table 47.1 Results of SVM and SVM-GOA

Model	Input scenario	Training			Testing		
		R^2	MSE	WI	R^2	MSE	WI
SVM:1	$Q_{t\#1}$	0.9349	13.04	0.9395	0.9077	15.917	0.9129
SVM:2	$Q_{t\#1}$, $Q_{t\#2}$	0.9355	12.8963	0.9401	0.9096	15.3892	0.915
SVM:3	$Q_{t\#1}$, $Q_{t\#2}$, $Q_{t\#3}$	0.937	11.3984	0.9426	0.9103	14.88	0.9166
SVM:4	$Q_{t\#1}$, $Q_{t\#2}$, $Q_{t\#3}$, $Q_{t\#4}$	0.9381	10.897	0.944	0.912	14.1896	0.9185
SVM:5	$Q_{t\#1}$, $Q_{t\#2}$, $Q_{t\#3}$, $Q_{t\#4}$, $Q_{t\#5}$	0.9393	10.032	0.9458	**0.9144**	**13.775**	**0.9217**
SVM-GOA:1	$Q_{t\#1}$	0.9644	3.4496	0.9697	0.9348	8.1143	0.9394
SVM-GOA:2	$Q_{t\#1}$, $Q_{t\#2}$	0.9689	2.9084	0.9734	0.9361	7.567	0.942
SVM-GOA:3	$Q_{t\#1}$, $Q_{t\#2}$, $Q_{t\#3}$	0.9705	2.38	0.9751	0.9379	7.0031	0.9435
SVM-GOA:4	$Q_{t\#1}$, $Q_{t\#2}$, $Q_{t\#3}$, $Q_{t\#4}$	0.972	1.9835	0.9773	0.9402	6.3887	0.9471
SVM-GOA:5	$Q_{t\#1}$, $Q_{t\#2}$, $Q_{t\#3}$, $Q_{t\#4}$, $Q_{t\#5}$	0.9748	1.007	0.979	**0.9465**	**5.992**	**0.9528**

The bold value defines best or prominent performane of model

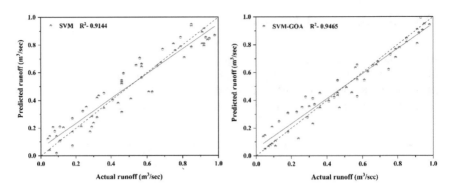

Fig. 47.3 Scatter plot between actual and predicted runoff

The perormance of all the model are given in Table 47.1. The scatter plot of observed versus predicted runoff by robust SVM-GOA and standalone SVM model in testing phase is shown in Fig. 47.3. A regression line, $y = ax + b$ has been incorporated for every model for assisting with understanding of these results where constants a and y-intercept (b) are utilized for assessing overall accuracy of the models. Clearly, hybrid SVM-GOA technique performed very well in predicting runoff for 3-month lag forecasting that is better than 1- and 2-month lag time. Noticeably, it is apparent that runoff value estimations by hybrid SVM-GOA method are closer to related observed values and in all plots follow same trend (Fig. 47.4). From fitted equations, it is also obvious that hybrid SVM-GOA model attains a and b values that lie close to 1 and 0, having higher R^2 (0.9465) for the 3-month compared to 1- and 2-month lag runoff forecasting.

47.5   Conclusion

The study demonstrates the application of standalone SVM and hybrid SVM-GOA model for monthly runoff forecast. By using GOA to optimize the hyper parameters of SVR, it resulted in estimation of global optima, rather than being stuck in the local optimum value. The target's fitness improved as the number of iterations increased, implying that estimation of global optimum improved proportionately to number of iterations. In addition, results of this work indicate that SVM-GOA achieved R^2—0.9465, MSE—5.992, and WI—0.9528 values, respectively, while the SVM model produced R^2—0.9144, MSE—13.775, and WI—0.9217 values, showcasing its accurateness in runoff prediction under changes in climate conditions that outperformed those of other machine learning algorithms.

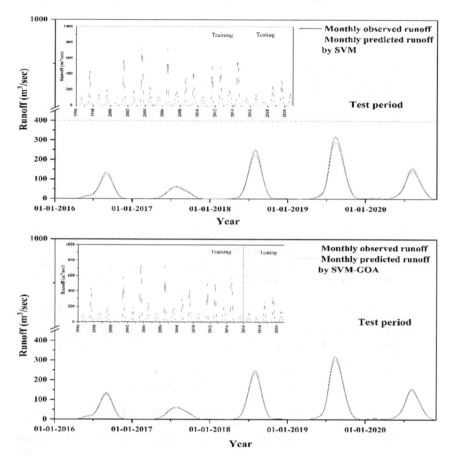

Fig. 47.4 Time series plot of actual versus predicted runoff by SVM and SVM-GOA

References

1. Mohanta, N.R., Patel, N., Beck, K., Samantaray, S., Sahoo, A.: Efficiency of river flow prediction in river using wavelet-CANFIS: a case study. In: Intelligent Data Engineering and Analytics, pp. 435–443. Springer, Singapore (2021)
2. Jimmy, S.R., Sahoo, A., Samantaray, S., Ghose, D.K.: Prophecy of runoff in a river basin using various neural networks. In: Communication Software and Networks, pp. 709–718. Springer, Singapore (2021)
3. Samantaray, S., Das, S.S., Sahoo, A., Satapathy, D.P.: Monthly runoff prediction at Baitarani river basin by support vector machine based on Salp swarm algorithm. Ain Shams Eng. J. **13**(5), 101732 (2022)
4. Samantaray, S., Biswakalyani, C., Singh, D.K., Sahoo, A., Prakash Satapathy, D.: Prediction of groundwater fluctuation based on hybrid ANFIS-GWO approach in arid watershed, India. Soft Comput. 1–23 (2022)
5. Samantaray, S., Sahoo, A., Satapathy, D.P.: Prediction of groundwater-level using novel SVM-ALO, SVM-FOA, and SVM-FFA algorithms at Purba-Medinipur, India. Arab. J. Geosci. **15**(8), 1–22 (2022)

6. Samantaray, S., Ghose, D.K.: Prediction of S12-MKII rainfall simulator experimental runoff data sets using hybrid PSR-SVM-FFA approaches. J. Water Clim. Change **13**(2), 707–734 (2022)

7. Samantaray, S., Das, S.S., Sahoo, A., Satapathy, D.P.: Evaluating the application of meta-heuristic approaches for flood simulation using GIS: a case study of Baitarani river Basin, India. Mater. Today: Proc. (2021)

8. Samantaray, S., Sahoo, A., Agnihotri, A.: Assessment of flood frequency using statistical and hybrid neural network method: Mahanadi river basin, India. J. Geol. Soc. India **97**(8), 867–880 (2021)

9. Samantaray, S., Sahoo, A.: Modelling response of infiltration loss toward water table depth using RBFN, RNN, ANFIS techniques. Int. J. Knowl.-based Intell. Eng. Syst. **25**(2), 227–234 (2021)

10. Botsis, D., Diamantaras, K., Latinopoulos, P.: Rainfall–runoff modeling using support vector regression and artificial neural networks. In: Proceedings of the 12th International Conference on Environmental Science and Technology (2011)

11. Bell, B., Wallace, B., Zhang, D.: Forecasting river runoff through support vector machines. In: Proceedings of the 11th IEEE International Conference on Cognitive Informatics and Cognitive Computing ICCI*CC 2012, pp. 58–64 (2012)

12. Chu, H., Wei, J., Li, T., Jia, K.: Application of support vector regression for mid- and long-term runoff forecasting in 'Yellow River Headwater' region. J. Procedia Eng. **154**, 1251–1257 (2016)

13. Sarzaeim, P., Bozorg-Haddad, O., Bozorgi, A., Loáiciga, H.A.: Runoff projection under climate change conditions with data-mining methods. J. Irrig. Drainage Eng. **143**, 04017026 (2017)

14. Barman, M., Choudhury, N.B.D., Sutradhar, S.: A regional hybrid GOA-SVM model based on similar day approach for short-term load forecasting in Assam, India. J. Energy **145**, 710–720 (2018)

15. Alizadeh, Z., Shourian, M., Yaseen, Z.M.: Simulating monthly streamflow using a hybrid feature selection approach integrated with an intelligence model. Hydrol. Sci. J. **65**(80), 1374–1384 (2020)

16. Alrashidi, M., Alrashidi, M., Rahman, S.: Global solar radiation prediction: application of novel hybrid data-driven model. J. Appl. Soft Comput. **112**, 107768 (2021)

17. Goodarzizad, P., Golafshani, E.M., Arashpour, M.: Predicting the construction labour productivity using artificial neural network and grasshopper optimisation algorithm. Int. J. Constr. Manage. (2021).

18. Fattahi, H., Hasanipanah, M.: An integrated approach of ANFIS-grasshopper optimization algorithm to approximate flyrock distance in mine blasting. J. Eng. Comput. (2021)

19. Panahi, M., Dodangeh, E., Rezaie, F., Khosravi, K., Le, H.V., Lee, M.J., Lee, S., Pham, B.T.: Flood spatial prediction modeling using a hybrid of meta-optimization and support vector regression modeling. J. Catena **199**, 105114 (2021)

20. Vapnik, V.N.: The Nature of Statistical Learning Theory. Springer, New York (1995)

21. Saremi, S., Mirjalili, S., Lewis, A.: Grasshopper optimisation algorithm: theory and application. Adv. Eng. Softw. **105**, 30–47 (2017)

22. Samantaray, S., Ghose, D.K.: Modelling runoff in an arid watershed through integrated support vector machine. H2Open J. **3**(1), 256–275 (2020)
23. Samantaray, S., Ghose, D.K.: Dynamic modelling of runoff in a watershed using artificial neural network. In: Smart Intelligent Computing and Applications, pp. 561–568. Springer, Singapore (2019)
24. Samantaray, S., Sahoo, A., Ghose, D.K.: Assessment of runoff via precipitation using neural networks: watershed modelling for developing environment in arid region. Pertanika J Sci Technol **27**(4): 2245–2263 (2020)

Chapter 48
Crop Prediction System with a Convolution Neural Network Model

Bhaskar Marapelli, Bereket Simon Balcha, Asebe Teka Nega, and Anil Carie

Abstract Most of the countries in this world are dependent on agriculture, and a lot of the population depends on their predator's knowledge in choosing the crop selection. However, as the years pass, change in agricultural land makes it difficult for the farmers in getting the expected yield. To improve the yield from the agricultural land, a change in the crop is recommended. Instead of using predator's knowledge for crop selection, we can make it easy and productive by using precision agriculture a new farming technique that recommends crops based on nutrition values in the land and weather conditions. In our work, we build a deep learning model using convolutional neural networks (CNNs) to predict a crop for selected land based on nutrition values in the land and weather conditions. The CNN model is trained and tested using research data with nutrition values and weather conditions that predict suitable crops for that land. Our model achieved an accuracy score of 93.69% on training data and 94.20% on testing data. The CNN model is compared with other baseline models.

48.1 Introduction

Agriculture plays a significant role in any nation's economy, and improving the yield from the agricultural land by selecting a suitable crop is an interesting research area. Every year, farmers choose the same crop for yield in their lands from the

B. Marapelli (✉) · A. T. Nega
Software Engineering Department, College of Computing and Informatics, Wolkite University, Wolkite, Ethiopia
e-mail: bhaskarmarapelli@wku.edu.et

A. T. Nega
e-mail: asebe@wku.edu.et

B. S. Balcha
Information Technology Department, College of Computing and Informatics, Wolkite University, Wolkite, Ethiopia
e-mail: bereket.simon@wku.edu.et

A. Carie
School of Computer Science and Engineering, VIT-AP University, Amaravathi, India

practice of their predators. As years pass with the change in the land (because of the use of the same crop, fertilizers, and climate change), the same crop yield quality and quantity may reduce [1]. Instead of depending on own knowledge, selecting a crop based on minerals in the land and weather conditions improves the quality and quantity of the crop yield [2]. Choosing a crop for specific land is a technique called precision agriculture that helps farmers in improving the yield from farmlands [3, 4]. With the advancements in modern technology as well as sensor devices, identifying nutrition in the land and the weather situation around the farmland could be used to select the right crop. The right crop for specific land could be suggested based on data that describe nutrition values in the land like nitrogen (N), phosphorus (P), potassium values (K), and weather situations like rainfall and temperature around the farm land [2]. Machine learning is a tool widely used for data analysis that helps in preparing prediction models with the available data [5, 6]. With historical data, machine learning models can be prepared to classify crops based on minerals in the land and weather situation.

The objective of our work is to, 1. Build a deep machine learning model that predicts an output as a suitable crop for input values minerals in the land and weather situation. 2. Build a Web-based interface that uses the build model and suggests suitable crops for farmers.

48.2   Related Work

There is a lot of research based on precision agriculture and machine learning models to suggest suitable crops for specific land. Here, we present some works in the following paragraphs.

Precision agriculture is a new development in farming that suggests site-specific farming. In [2], author proposed a crop selection method over a season based on weather, soil type, water density, and crop type and improves the net yield rate of the crop. In this method, the author considers the factors like production rate, market price, and government policies in selecting crops. In [4], author describes precision agriculture (PA) and the need for a software model to deliver direct advisory services to farmers using the most accessible technologies such as SMS and email. The building model is for the scenario in Kerala State, to deploy the model in other places where it needs little modifications. In [3], author studied precision agriculture for crop recommendation based on site-specific parameters. The author prepared a recommendation system based on an ensemble model with random tree, Naïve Bayes, CHAID, and K-nearest neighbor learners. The author concludes that the proposed model solves the problem of the wrong choice of a crop and increases productivity.

Machine learning models help in knowledge mining; the complex nonlinear behavior of output data can be predicted from input data by using machine learning models. In [7], author proposed a crop yield prediction system using machine learning algorithms using previous data. The author divides his work into three parts 1.

soil classification using soil nutrients data, 2. crop yield prediction using crop yield data, nutrients, and location data, and 3. fertilizer recommendation using fertilizer data, and crop and location data. The author worked on his model preparation with random forest and support vector machine learning models. In [8], author discusses different classification algorithms and their performance in yield prediction in precision agriculture with datasets collected over the years for the yield prediction of the soybean crop. The author concludes bagging algorithm predicts yield with minimum mean absolute error. In [9], author proposed an ensemble machine learning model with support vector machines and artificial neural networks for a crop recommendation system. The dataset is prepared from the soil testing lab, Playtest Laboratories soil testing lab, Pune, Maharashtra, India. The model proposed helps farmers around Maharashtra state India since only soil properties for that state are considered.

In general, farmers choose their crop based on their regular practice over years, which may cause a reduction in the quality and quantity of the yield. With the introduction of precision agriculture, change in crop is suggested based on soil conditions, and machine learning models help in suggesting crops based on soil conditions with available data. The previous works carried out have considered either soil conditions [7], seasonal weather conditions [2], and location-specific data [4, 9] but not both weather conditions and minerals in the land. In our work, we used datasets containing nutrition values in the land and weather conditions suitable for crops to suggest which crop is right for a land. We build a deep machine learning model that predicts suitable crops based on nutrition's values in the land and weather conditions and a Web-based interface that suggests crops for given minerals and weather conditions (Figs. 48.1 and 48.2).

Fig. 48.1 Model proposed

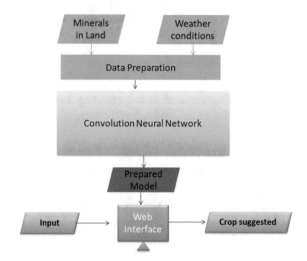

Fig. 48.2 Visualization of convolution neural network

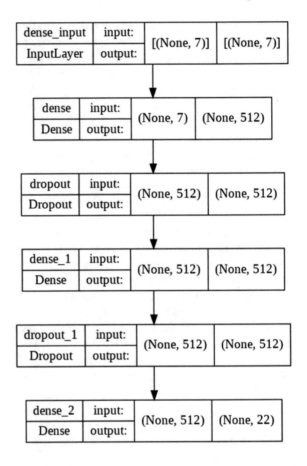

48.3 Methodology

The proposed research aims at suggesting a suitable crop for a specific land based on nutrition values in the land and environmental conditions by using a machine learning model.

Our methodology involves two steps

1. Build a deep machine learning model that predicts an output as a suitable crop for input values minerals in the land and weather situations.
2. Build a Web-based interface that uses the build model and suggests suitable crops for farmers.

48.3.1 Datasets

We have used a crop recommendation dataset that contains information about 22 unique crops, with land nutrition values and environmental conditions needed for the crops of 100 values for each crop. The dataset has been collected from Kaggle repository [10].

To uncover different factors that decide the crop growth, the dataset has been analyzed; Fig. 48.3 shows the ratio of nitrogen, phosphorous, and potassium in soil for different crops, and Fig. 48.4 shows temperature, humidity, and the amount of rainfall required for different crops. From the data analysis, we can understand that for a rice crop we need more nitrogen (N) than phosphorous (P) and potassium (K) in the land, we need high rainfall, high temperature, and humidity rages compared with other crops. Crops like apples and grapes need more potassium (K) than all other crops, and the coconut crop needs very low values of N, P, and K. The crop chickpea needs very low range values for rainfall, temperature, and humidity.

48.3.2 Model Preparation and Crop Recommendation

The proposed crop recommendation system has been built based on convolutional neural networks (CNNs) machine learning model. The dataset collected has been divided into training and testing data; we have used 70% of the data for training and 30% data for testing. Python's Skit-learn and Keras libraries have been used for preparing the model; Skit-learn libraries are helpful in loading the dataset, dividing the datasets into training and testing data, and Keras libraries are helpful in preparing deep learning machine learning models. The CNN model has been trained with

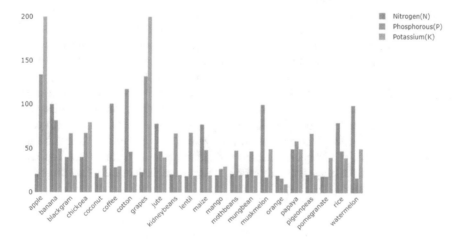

Fig. 48.3 Land mineral values for different crops

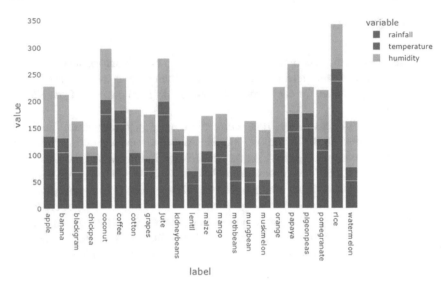

Fig. 48.4 Rainfall, temperature, and humidity for different crops

training data and tested with testing data; accuracy score has been used as a measurement for model validity. Our model performance has been compared with other works.

48.3.3 Experimental Setup

After trying different models, we have prepared our CNN model (Fig. 48.2 shows the visualization of the model) with 2 hidden layers, an input layer with 7 input values (nutrition values in the land like nitrogen (N) phosphorus (P), potassium values (K), PH value of land, weather data like rainfall and temperature and humidity), the first hidden layer with 512 neurons that uses 'relu' activation function followed by a dropout layer. The second hidden layer also contains 512 neurons and uses 'relu' activation function. The final layer contains 22 neurons that represent a number of classes (crop types) that uses the softmax activation function when compiling the model we have used 'categorical cross-entropy' as loss function, 'Adam' as activation function, and 'accuracy' as our metric for evaluating the model. We trained the model with 8 epochs 128 batch size; the validation has been done on test data. The proposed CNN model gave accuracy scores of 93% on train data and 94% on test data (Fig. 48.5 and Table 48.1).

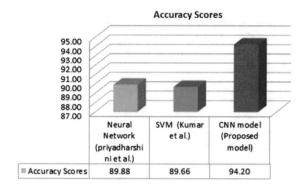

Fig. 48.5 Baseline models comparison accuracy scores

Table 48.1 Baseline models accuracy comparison with proposed models

Baseline model by Priyadharshini et al. [11]		
1	SVM	78%
2	Decision tree	81%
3	K-nearest neighbor with cross-validation	88%
4	Linear regression model	88.26%
5	Naive Bayes	82%
6	**Neural network**	**89.88%**
Baseline model by Kumar et al. [12]		
1	Decision tree	86.80%
2	Naive Bayes	88.00%
3	Random forest	88.00%
4	K-nearest neighbor	88.00%
5	Linear regression model	86.04%
6	**SVM**	**89.66%**
Proposed model		
1	CNN model 1 (256 neurons in hidden layer)	92.72%
2	**CNN model 2 (512 neurons in hidden layer)**	**94.20%**

48.4   Results and Discussion

We implemented our deep learning CNN model to predict suitable crops for a specific land with historical datasets. The dataset has been split into 70% training and 30% testing; the testing data have been used for validating the model. Figure 48.6 shows the accuracy and loss graph for training and testing data. Table 48.2 shows a comparison of model measures with training and testing data.

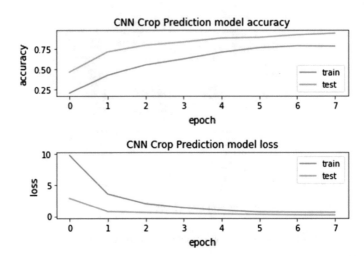

Fig. 48.6 Train and test accuracy and loss graph

Table 48.2 Train and test measures

	Training time		Testing time	
1	Accuracy	0.93	Accuraccy	0.94
2	Loss	0.18	Loss	0.18

48.4.1 Accuracy

To evaluate our model performance, we have used accuracy as the metric. Accuracy gives a ratio of correctly predicted values over total predictions. Our CNN model gives a 93.69% accuracy score with a well-balanced dataset.

$$\text{Accuracy} = \frac{\text{NTP} + \text{NTN}}{n} \tag{48.1}$$

where

NTP is number of samples in which crop names are correctly predicted,
NTN is number of samples in which crop names are wrongly predicted,
  n is the total number of samples in crop dataset.

48.4.2 Comparison with Other Models

We compared the performance of our proposed CNN crop recommendation model with baseline models proposed by Refs. [11] and [12], see Table 48.1. In [11], author presented crop recommendation system neural network model with 89.88% accuracy, author-verified other models like SVM, KNN, decision tree, Naive Bayes, and linear regression models, but the neural network model has the best accuracy score. In [12], author presented recommendation system for crop identification and pest control technique author concluded that the SVM algorithm suits the best. In [12], author-verified other models like KNN, decision tree, random forest, Naive Bayes, and linear regression models, but the SVM model accuracy score of 89.66% is high.

We have verified CNN model with 256 neurons in the hidden layer (CNN model1) and 512 neurons in the hidden layer (CNN model2); the CNN model gave an accuracy score of 92.72%, and the CNN model2 with 512 neurons in the hidden layer gave accuracy score 93.69% on training data and 94.20% on testing data. Figure 48.5 shows a comparison between baseline models and the proposed CNN model.

48.5 Conclusion

Precision agriculture helps in increasing the yield from the land by incorporating technology (machine learning) into agriculture. With the use of precision agriculture, farmers can choose the right crop for land that increases the chance of getting more yield. In this paper, a CNN model was proposed for crop recommendation based on nutrients in the land and weather conditions for a specific land. Our proposed CNN model is compared with baseline models to verify the performance, the neural network model by Ref. [11] accuracy 89.88% the SVM model by Ref. [12] accuracy 89.66% our proposed CNN model with 94.20% accuracy outperforms the baseline models.

References

1. Pathak, P., Phalke, R., Patil, P., Prakash, P., Bhosale, S.: Survey on crop suggestion using weather analysis (2020)
2. Chlingaryan, A., Sukkarieh, S., Whelan, B.: Machine learning approaches for crop yield prediction and nitrogen status estimation in precision agriculture: a review. Comput. Electron. Agric. **151**, 61–69 (2018)
3. Pudumalar, S., Ramanujam, E., Rajashree, R.H., Kavya, C., Kiruthika, T., Nisha, J.: Crop recommendation system for precision agriculture. In: 2016 Eighth International Conference on Advanced Computing (ICoAC), pp. 32–36. IEEE (2017)
4. Akshatha, K.R., Shreedhara, K.S.: Implementation of machine learning algorithms for crop recommendation using precision agriculture. Int. J. Res. Eng. Sci. Manag. (IJRESM) **1**(6), 58–60 (2018)
5. Gupta, S., Bharti, V., Kumar, A.: A survey on various machine learning algorithms for disease prediction. Int. J. Recent Technol. Eng. **7**(6c), 84–87 (2019)
6. Tarakeswara Rao, B., Patibandla, R.S.M., Ramakrishna Murty, M.: A comparative study on effective approaches for unsupervised statistical machine translation. In: Embedded Systems and Artificial Intelligence, pp. 895–905. Springer, Singapore (2020)

7. Bondre, D.A., Mahagaonkar, S.: Prediction of crop yield and fertilizer recommendation using machine learning algorithms. Int. J. Eng. Appl. Sci. Technol. **4**(5), 371–376 (2019)

8. Savla, A., Israni, N., Dhawan, P., Mandholia, A., Bhadada, H., Bhardwaj, S.: Survey of classification algorithms for formulating yield prediction accuracy in precision agriculture. In: 2015 International Conference on Innovations in Information, Embedded and Communication Systems (ICIIECS), pp. 1–7. IEEE (2015)

9. Rajak, R.K., Pawar, A., Pendke, M., Shinde, P., Rathod, S., Devare, A.: Crop recommendation system to maximize crop yield using machine learning technique. Int. Res. J. Eng. Technol. **4**(12), 950–953 (2017)

10. Data set 1: https://www.kaggle.com/atharvaingle/crop-recommendation-dataset

11. Priyadharshini, A., Chakraborty, S., Kumar, A., Pooniwala, O.R.: Intelligent crop recommendation system using machine learning. In: 2021 5th International Conference on Computing Methodologies and Communication (ICCMC), pp. 843–848. IEEE (2021)

12. Kumar, A., Sarkar, S., Pradhan, C.: Recommendation system for crop identification and pest control technique in agriculture. In: 2019 International Conference on Communication and Signal Processing (ICCSP), pp. 0185–0189. IEEE (2019)

Chapter 49
Technology-Assisted Mental Healthcare: A Novel Approach

Rupsa Rani Sahu, Anjana Raut, and Swati Samantaray

Abstract Mental health issues are rising globally and the present healthcare system has to keep up with the dramatic fluctuations of mental health. Patients' denial to express needs and assort to professional help are a risk factor for declining emotional balance in the society. Technology driven approach can play an integral role in identification, prevention, and treatment of this global health crisis. Access to personalized mental healthcare management is approaching a new frontier with emerging technological tools and applications. Moreover, the healthcare industry is a complex regulatory network and its inability to share and exchange information across patient and providers results in error and disputes. Newer technologies are opening avenues to replace the outdated healthcare payment process aiding in secure transfer of data and cutting down associated risk. This paper focuses on the existing state of the art applications of telemedicine, tricorder, blockchain technology, device tracking, and many more newer ways for management of mental health issues and strongly recommends their increased application to enhance patient care.

49.1 Introduction

Healthcare, with the aid of technology has influenced the world by the transformations that have an effect on the various facets of healthcare such as diagnosis of diseases, clinical appointments, as well as prescription for drugs. Experts in health informatics insinuate that technology has been a major factor for breakthroughs in research, data collection, and treatment. Today, many patients are getting allured by the ease of affordability and accessibility of telehealth services owing to the massive technological expansion in healthcare as discussed in this paper. New tools are being tested by medical providers in order to find innovative ways of practicing medicine

R. R. Sahu · A. Raut
Kalinga Institute of Dental Sciences, Bhubaneswar, India
e-mail: anjana.raut@kids.ac.in

S. Samantaray (✉)
School of Humanities, KIIT Deemed to be University, Bhubaneswar, India
e-mail: ssamantrayfhu@kiit.ac.in

in the future. IT market analysts have reported that technology has led to increased accessibility to treatment. Health Information Technology (or Health IT) unfolds fresh methodologies for research approach and data exploration; hence, it is necessary to empower experts to regulate healthcare in a way that is effective and consumer driven. The convergence of technology, machine learning, and digitalization have a significant impact on the patient experience. So, the healthcare industry is set to deliver breakthrough innovations at unimaginable rates and costs.

49.2   Review of Literature

Gannon et al. executed a research across six therapeutic areas in a major behavioral health system and concluded that direct-to-consumer telemental health was a frequently applied method during the COVID-19 pandemic [1]. Fairchild et al. conducted a study to assess the attributes of telemental health services and concluded that additional resources and support are necessary to access care along with law enforcement measures [2]. Hensel et al. review research described digital healthcare utilization for indigenous mental health, with a focus on existing facts and examples mentioned in the literature. He also explored remote telemental health programme for indigenous kids, as well as presented their specific potential and problems [3]. Armontrout et al. conducted a literature review of judicial guidelines for smartphone health apps claiming mental well-being and associated lawsuit [4]. They did not find any regulatory actions related to adverse outcomes from the use of mental health apps. Yellowlees et al. conducted a study to evaluate the practicality of employing video-based asynchronous mental health examinations through online platform. In this study, psychiatric interviews were performed using the Mini-International Neuropsychiatric Interview by a primary care practitioner and videotaped for subsequent examination by a psychiatrist, who then offered a psychiatric diagnosis. The authors discovered that the procedure worked effectively and allowed psychiatrists to watch and assess individuals who are often unable to visit such doctors due to a shortage of mental health providers in their region [5].

49.3   Advanced Telemedicine

49.3.1   Telemedicine with Artificial Intelligence

A complex procedure like diagnosis necessitates the need for a variety of tests, including clinical, imaging, blood, and gene sequencing. Furthermore, to overcome broad mental health examinations, neurobehavioral and neurocognitive assessments may be necessary [6]. Some of these tests may be inaccessible in a telemedicine setting, while others may be too expensive. These concepts have the capacity to grab straight from information without any prior statistical modeling, resulting in more balanced findings for diagnostic intents over a wide range of different people. In context of the COVID-19 outbreak, multinational efforts to anticipate, manage, and diagnose COVID-19 are ongoing, using data-driven techniques and acquired information. Furthermore, the ML model includes a critical analysis that aids in the quest for low-cost treatments [7].

According to Wahl et al., widespread use of smartphone, together with significant investments in current digital era (like mHealth, EMR, and cloud computing), contributes to a plethora of chances to employ AI applications [8]. AI has demonstrable dominance over humans in scientific thinking and redesigning (particularly when massive dataset is present), and can constructively take care of limitations of human mind. During COVID-19, telemedicine provided a key patient flow conduit during times of disaster in the domain of health service. In addition, it aids in the protection of healthcare institutions and the reduction of dangers to health professionals in the event of pandemics [9]. Telemedicine has the ability to reduce the economic costs of healthcare services. Treatment via telecommunications can be supplied directly to the customer by isolated physicians in medical facilities with remote access, allowing prompt aid to desired patients. Licensed physicians may use teleconsultations to screen patients, identify symptoms of illness, and screen those with greater risk of sickness. Telemedicine solutions save time by filling out insurance paperwork, allowing physicians to dedicate their whole focus to their patients. Health setups are implementing their telemedicine services to make it easier to scrutiny patients for COVID-19, their vigilant identification and subsequent reduction in exposure. Another important factor in aiding holistic care is by combining several specializations, such as teleradiology, teleoncology, and telepathology [10]. The integration of learning models for doctors and paramedical staff, to demonstrate how these innovations resolve affairs and impart imperial education especially for surgical procedures, are important considerations in the telemedical framework discussion [11].

49.3.2 Telemedicine with the Aid of Robotics

Robotics is a potential cutting-edge telemedicine tool with possible applications in changing examination and treatment approach, as well as remote patient monitoring. Both Ebola epidemic and the COVID-19 pandemic have comprehensively explored robotics and telemedicine applications therefore, in these kind of infectious illnesses, robotics can be used for decontamination, distribution of medications and food, assessing vital parameters, improving border control, and automated disinfection [12]. Telepresence robots are a type of software that helps for two-way communication and also lends support to distant places by connecting individuals with their families and physicians [10]. During COVID-19 pandemic in China (Wuhan), a hospital named 'Smart Field' used robots and Internet of Things (IoT) to reduce the exposure to health care workers and patients and also helped in providing medical services. Robots can be used in case of infectious diseases and medical workforce crisis to improve the dimensions of health system and its receptiveness [13]. Lifestyles changes and extensive working hours have brought mental health to the forefront. Software and gadgets related to mental health make use of information technology to keep mental and behavioral diseases under control.

49.3.3 Nanomedicine

Nanomedicine includes advancements in the fields of bioengineering, proteomics, genetics, material science, molecular, and cellular biology. It represents union of medicine with nanotechnology. Cells which are present inside our body also work in nanoscale level as it contains biologically significant molecules like proteins, glucose, enzymes, hemoglobin, receptors, antibodies, and water also have dimensions within the nanoscale limit. As nanoscale technology became popular in medical field, several researches are being carried out in the field of medicine for ease of treatment and also for development of instruments and devices by using nanotechnology to escalate the productiveness, safety, sensitivity, and providing patients with better care. An advantage of nanomedicine is enhanced bioavailability, lessens the toxicity, increases the dose response, and improves solubility in comparison with conventional medicines [14].

49.4 Tricorder

A tricorder is a device that is used to self-diagnose conditions that are related to medical conditions within seconds, and it also helps in taking basic vital measurement. It is a portable screening device. Several researches are going on to create this device and enhance the properties of this device so that it can be used for welfare of the patient. Several studies were done which prove that it is a common device which is similar to Swiss army knife which helps in measuring blood pressure, temperature, and blood flow through a non-invasive way, and it can also be used for diagnosis of a patient's health status by evaluating the information, either autonomously or via linked internet sources. The famous show 'Star trek' had a fictional character called Dr. McCoy who used tricorder for examining the patient's medical condition instantly. The features of the medical tricorder include, easy detachability, high-resolution, portable scanner to send vital data to the tricorder itself. It can be used for various purposes like, for checking the functions of the vital organ, to detect pathogens and helps in evaluation of human physiology [15].

49.5 Blockchain

Blockchain technology is a very secure and tamperproof data storage system that works on Proof of Work (PoW) mechanism using P2P network. It can safeguard health-related records from information loss, cyber threat, and misuse. The use of cryptographic algorithms for data safety and storage allows the patients to have ownership of their health information by virtue of legitimate data access permission. Blockchain has been adopted as a reminder tool for patients on regular medication and health-related products. Also, it allows self-served health status for remote monitoring. Mental disorders are considered social stigma leading to massive under-reporting. Blockchain technology is a boon in maintaining privacy protection and more personalized attention to mentally distressed patients. The technology has been very well adopted by crypto currencies like Bitcoin. Digital ledger technology (DLT) is another name for this technology [16].

In healthcare, deployment of cutting edge technology like 'blockchain' may be used in broad manners to revolutionize the patent care approach and facilitate research by easy retrieval of data. Followings are the important applications of this technology [8]:

- Managing data from electronic medical records (EMR)
- Data security in the healthcare field
- Data management for personal health records
- Management of genomics at the point of care
- Electronics health records data management.

49.6 AI and Smartphone Based Interventions

Healthcare applications of smartphones present enormous technological opportunities and can change the quality of life with positive outcome. Smartphone-based interventions addressing to depression, anxiety, trauma, addiction, and self-inflicted injury need to be explored for the treatment of mental health disorders. These tools can work as an assistant who are available for 24/7 to aid the patient and also help to alert the medical officer when an involvement is needed. AI and smartphone-assisted apps will take over predominantly the next generation of teletherapy. Teletherapy 1.0 was used in 2020 where they provided only live zoom sessions. According to recent studies, teletherapy 2.0 is being used most commonly, where they utilize data from various sources like instantaneous symptom tracking, software usage, and artificial intelligence signals. This form of AI-assisted treatment is especially beneficial when dealing with complicated issues like the recent pandemic-induced mental stress or opiate or benzo weaning. When health problems are both acute and long-term, success demands learning new coping skills, and greater support and assistance [2, 17].

49.7 Prescription Video Games

In the month of June (2017), the first prescription video game was approved by FDA. Several researches have been done for this kind of revolutionary treatment, but finally it has been proved that it can be used in reality as treatment for children of age between 8 and 12 with ADHD. EndeavorRX is gaming app in which the children are emphasized to perform multiple tasks concurrently within the game habitat, and it is also a perfect example of technical tool that supports contemporary kids habit and also enhances the ability of their mental health [17].

49.8 Digital Pills

Digital pills (DP) are one of the most advanced and modern technological innovation in digital medicine, recently discovered by the researchers. Digital pills are drug-device as they can capture and transfer patients individual information that can be used for both clinical and research purposes, and can also be used to track health-related lifestyle behaviors, including medication-taking habits, it consists of a ingestible sensor that gets activated by acid secreted in the stomach after being consumed by patients and produces a signal that can be identified by the wearable patch. All the data received by the wearable patch is immediately redirected to a patient's phone through an application. The program subsequently transfers the information to a web-based interface, making it potentially available to the patient and healthcare professionals. The ingestible sensor can be coated with conventional pharmaceuticals to enable gathering of information on all other health-related lifestyle behaviors [18]. In 2016, the European Medicines Agency (EMA) recognized DP as an eligible scale for assessing compliance in clinical trials, and the Food and Drug Administration (FDA) in the United States allowed the marketing of the first DP blended conventional drug in 2017. The first DP to be authorized as a medication combined the device with aripiprazole, a drug used to treat mental diseases including schizophrenia and bipolar disorder [19].

49.9 Digital Symptom Tracking Device

With the introduction of the smartphone, almost everyone now has a capable note-taking gadget in their pocket. App-based diaries allow for updates to be made from anywhere at any time, and they may be combined with notifications to ensure that data are not neglected. Digital tracking of people with mental health disorder has the ability to metamorphose medical diagnostics and treatment, however this requires careful scrutiny by the caregiver. It enables a new way to detect and monitor health-related disturbances throughout the day.

49.10   Transcranial Magnetic Stimulation

Repetitive Transcranial Magnetic Stimulation (rTMS) is a non-invasive medical procedure for a number of neuropsychiatric illnesses. rTMS is a brain stimulation technology that uses an insulating coil placed over particular locations of the scalp to generate brief magnetic fields. Magnetic fields similar to those employed in magnetic resonance imaging are applied here (MRI). Magnetic pulses provide a small electrical current that stimulates neural circuits in a specific area of the brain. This medication has been proven to be safe and quite well accepted in individuals with depression or any other neuropsychiatric diseases, and it can be beneficial. rTMS can decrease or raise cortex activity in certain parts of the frontal lobe, based on the stimulation frequency. Low-frequency rTMS (LF-rTMS) (less than equal to 1 Hz) has been demonstrated to reduce cortex excitation and be inhibitory, whereas High frequency rTMS (HF-rTMS) (more than equal to 5 Hz) was shown to enhance cortex excitation and be exuberant in most persons [20].

49.11   Remote Patient Monitoring

Remote patient monitoring (RPM) is a platform that allows patients to be observed at places other than clinical settings, such as at residence or at a distant place, hence possibly boosting health insurance and cutting costs. RPM comprises doctors offering patients with continuous remote care, often to monitor medical problems, chronic illnesses, or post-hospitalization recovery [21]. This form of healthcare management is crucial when patients are dealing with tough self-care treatments like emotional disturbances. Remote monitoring and trend analysis of physiological signs, for instance, are crucial components of RPM that enable for early sickness diagnosis, reducing clinical visits, and shorter hospital stays. While new technology is being created to address this sort of health care, clinicians may also use basic communication techniques like Skype, Whatsapp, or even landlines.

49.12   Technological Components

RPM's many applications have resulted in a variety of RPM technology architectures. Most RPM technologies, on the other hand, have a basic structure that comprises of four components: Sensors on a gadget that can assess physiological indicators via wireless connections. Based on the manufacturer, sensors can link to a centralized system through Wi-Fi or cellular connection methods. Local data storage at the patient's location with interfaces to centralized datasets and/or healthcare providers. Sensor information, local data storage, diagnostic apps, and/or healthcare providers is stored in a centralized database. Based on the findings of obtained data, diagnostic application software provides therapy suggestions and intervention warnings. Various sets of sensors, storage, and applications may be used depending on the illness and the parameters being monitored [22].

49.13 Wearables and Digital Biomarker Apps

Biomarker is a trait that is assessed as an indication of regular cellular mechanisms, pathogenic processes, or reactions to a stimulus or intervention, including treatment modalities. This might include genetic, histopathological, radiological, or physiological features. Biomarkers have emerged as a promising new technique for precision medicine and clinical trial assistance. Digital biomarkers are objective, measurable physiological and behavioral data obtained via portable, wearable, implantable, or consumable digital technologies [23]. These tools are widely used to describe, impact, and foretell clinical outcomes. Personalized health baselines may now be created using individualized metrics. The addition of digital biomarkers has had a particularly positive influence in the field of neurophysiology, where objective and non-invasive biomarkers are in high demand. A variety of digital biomarkers are now being evaluated for practicality and accuracy in the diagnosis and treatment of Parkinson's and Alzheimer's disease, as well as clinical outcome evaluations [24]. Wearables and applications connected to the Internet of Things can measure symptoms (such as insomnia and elevated heart rates) that may be shared with trusted specialists. Such real-time data can help doctors make treatment decisions. Professionals can also utilize this information to follow therapy progress, monitor a patient's drug reaction, and choose when to intervene if an appointment is not scheduled in the near future. According to a systematic review published in 2021 the use of sensors for monitoring mental health has increased significantly in past five years and has been well documented in various publications hence validating the use of sensors for assessing emotional state of human mind (as shown in Fig. 49.1) [25].

49.14 Global Market and Regulatory Guidelines

Global market statistics reveal that the smartphone-assisted applications are gaining popularity as healthcare apps are being utilized widely to address mental health well-being (as shown in Fig. 49.2) [26]. They have enormous potential to reach the needy at any corner of the world and initiate timely intervention at a low cost. But, judicial guidelines are necessary to prevent false claims and fake marketing by smartphone apps and devices. A statutory body lays guidelines to protect the public from being misled by inappropriate treatment offered by health apps. Moreover, it also provides insight into the credibility of such software, as well as investigates their efficacy and safety. It is necessary to include mental health guidance in primary health care at

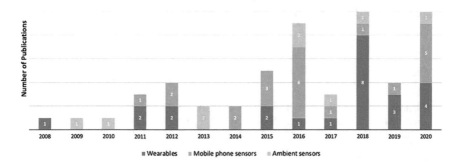

Fig. 49.1 Number of publications investigating monitoring systems in mental health from 2008 to 2020 [25]

Fig. 49.2 Growing global market for health apps [26]

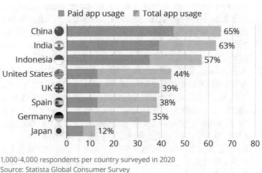

every hospital in blocks and districts and reinforce opportunities for self-care and monitoring.

49.15   Conclusion

Exponential technological growth has been able to offer evidence-based care based on computer software and smart devices. They are user friendly and help distressed patients with adverse mental health events communicate with their healthcare providers to monitor activity levels and social patterns. However, it is still beyond the reach of many needful patients who are not comfortable to use technology-assisted therapy and rely more on conventional manner of care. In such a scenario, technological tools can assist the traditional mode of treatment as well. Long-term

clinical research is necessary to determine the efficacy and outcome of such technological applications for wider application in the future and for framing policies for
technology industry and healthcare providers in the wider interest of the society.

References

1. Gannon, J.M., Schlesinger, A., Glance, J., Sujata, M., Fredrick, N., Wyler, J., Perez, G.: Rapid
   expansion of direct-to-consumer telemental health during the COVID-19 pandemic: a case
   series. Ann. Clin. Psychiatry. **33**(1), 27–34 (2021). https.//doi.org/10.12/88/acp.0020. PMID:
   33529285
2. Fairchild, R., Ferng-Kuo, S.F., Rahmouni, H., Hardesty, D.: An observational study of telemental care delivery and the context for involuntary commitment for mental health patients in
   a group of rural emergency departments. Telemed. Rep. **1**(1), 22–35 (2020). https://doi.org/10.
   1089/tmr.2020.0005
3. Hensel, J.M., Ellard, K., Koltek, M., Wilson, G., Sareen, J.: Digital health solutions for indigenous mental well-being. Curr. Psychiatry Rep. **21**(8), 68 (2019). https://doi.org/10.1007/s11
   920-019-1056-6.PMID:31263971, PMCID:PMC6602981
4. Armontrout, J.A., Torous, J., Cohen, M., McNiel, D.E., Binder, R.: Current regulation of mobile
   mental health applications. J. Am. Acad. Psychiatry Law. **46**(2), 204–211 (2018). https://doi.
   org/10.29158/JAAPL.003748-18. PMID: 30026399
5. Yellowlees, P.M., Odor, A., Parish, M.B., Iosif, A.M., Haught, K., Hilty, D.: A feasibility study
   of the use of asynchronous telepsychiatry for psychiatric consultations. Psychiatr. Serv. **61**(8),
   838–840 (2010). https://doi.org/10.1176/ps.2010.61.8.838. PMID: 20675845
6. Huys, Q.J., Maia, T.V., Frank, M.J.: Computational psychiatry as a bridge from neuroscience
   to clinical applications. Nat Neurosci. **19**(3), 404–413 (2016). https://doi.org/10.1038/nn.
   4238.PMID:26906507, PMCID:PMC5443409
7. Luo, E.M., Newman, S., Amat, M., Charpignon, M.L., Duralde, E.R., Jain, S., Kaufman,
   A.R., Korolev, I., Lai, Y., Lam, B.D., Lipcsey, M., Martinez, A., Mechanic, O.J., Mlabasati,
   J., McCoy, L.G., Nguyen, F.T., Samuel, M., Yang, E., Celi, L.A.: MIT COVID-19 Datathon:
   data without boundaries. BMJ Innov **7**(1), 231–234 (2021). https://doi.org/10.1136/bmjinnov-
   2020-000492. Epub 31 Aug 2020. PMID: 33437494, PMCID: PMC7799368
8. Wahl, B., Cossy-Gantner, A., Germann, S., Schwalbe, N.R.: Artificial intelligence (AI) and
   global health: how can AI contribute to health in resource-poor settings? BMJ Glob Health.
   3(4), e000798 (2018). https://doi.org/10.1136/bmjgh-2018-000798.PMID:30233828, PMCID:
   PMC6135465
9. Bhaskar, S., Bradley, S., Israeli-Korn, S., Menon, B., Chattu, V.K., Thomas, P., Chawla, J.,
   Kumar, R., Prandi, P., Ray, D., Golla, S., Surya, N., Yang, H., Martinez, S., Ozgen, M.H.,
   Codrington, J., González, E.M.J., Toosi, M., Hariya Mohan, N., Menon, K.V., Chahidi, A.,
   MedererHengstl, S.: Chronic neurology in covid-19 era: clinical considerations and recommendations from the REPROGRAM consortium. Front Neurol. **24**(11), 664 (2020). https://
   doi.org/10.3389/fneur.2020.00664.PMID:32695066, PMCID:PMC7339863
10. Bakas, T., Sampsel, D., Israel, J., Chamnikar, A., Ellard, A., Clark, J.G., Ulrich, M.G.,
    Vanderelst, D.: Satisfaction and technology evaluation of a telehealth robotic program to opti-
    mize healthy independent living for older adults. J Nurs Scholarsh. **50**(6), 666–675 (2018).
    https://doi.org/10.1111/jnu.12436. Epub 15 Oct 2018 PMID: 30320967
11. Lucey, C.R., Johnston, S.C.: The transformational effects of COVID-19 on medical education.
    JAMA **324**(11), 1033–1034 (2020). https://doi.org/10.1001/jama.2020.14136
12. Yang, G.Z., Nelson, B.J., Murphy, R.R., Choset, H., Christensen, H., Collins, H.S., Dario,
    P., Goldberg, K., Ikuta, K., Jacobstein, N., Kragic, D., Taylor, R.H., McNutt, M.: Combating
    COVID-19-the role of robotics in managing public health and infectious diseases. Sci. Robot.
    25;5(40), eabb5589 (2020). https://doi.org/10.1126/scirobotics.abb5589. PMID: 33022599

13. Hornyak, T.: CNBC. What America can learn from China's use of robots and telemedicine to combat the coronavirus. (2020). Available online at: https://www.cnbc.com/2020/03/18/how-china-is-using-robots-and-telemedicineto-combat-the-coronavirus.html. Accessed 14 Sep 2020
14. Bawa, R.: Regulating nanomedicine—can the FDA handle it? Curr. Drug Deliv. **8**(3), 227–234 (2011). https://doi.org/10.2174/156720111795256156. PMID: 21291376
15. Frank Simons of Reuters. Scientist beams up a real Star Trek tricorder. Chicago Tribune (April 13, 2012) Retrieved 01 Dec 2012
16. Khezr, S., et al.: Blockchain technology in healthcare: a comprehensive review and directions for future research. Appl. Sci. (2019). https://doi.org/10.3390/app9091736, https://www.mdpi.com/2076-3417/9/9/1736/htm
17. https://www.forbes.com/sites/forbestechcouncil/2020/11/25/five-tech-innovations-that-changed-mental-health-in-2020/?sh=278d4a3b1e9c
18. DiCarlo, L., Moon, G., Intondi, A., Duck, R., Frank, J., Hafazi, H., Behzadi, Y., Robertson, T., Costello, B., Savage, G., Zdeblick, M.: A digital health solution for using and managing medications: wirelessly observed therapy. IEEE Pulse **3**(5), 23–6 (2012). https://doi.org/10.1109/MPUL.2012.2205777. PMID: 23014703
19. FDA approves pill with sensor that digitally tracks if patients haveingested their medication. https://www.fda.gov/newsevents/newsroom/pressannouncements/ucm584933.htm. Accessed 20 Nov 2018
20. Mikellides, G., Michael, P., Tantele, M.: Repetitive transcranial magnetic stimulation: an innovative medical therapy. Psychiatriki (2021). https://doi.org/10.22365/jpsych.2021.012
21. Wicklund, E. (ed.): How COVID-19 affects the telehealth, remote patient monitoring landscape. M Health Intell. (4 May 2021) Retrieved 19 Aug 2021
22. Smith, T., Sweeney, R.: Fusion trends and opportunities medical devices and communications. Analyst Report (Report). Connecticut: NERAC Publication (September 2010)
23. Dorsey, E.R. (ed.): Digital biomarkers, Basel: Karger. Available from: https://www.karger.com/Journal/Home/271954
24. Lipsmeier, F., Taylor, K.I., Kilchenmann, T., Wolf, D., Scotland, A., Schjodt-Eriksen, J., Cheng, W.Y., Fernandez-Garcia, I., Siebourg-Polster, J., Jin, L., Soto, J., Verselis, L., Boess, F., Koller, M., Grundman, M., Monsch, A.U., Postuma, R.B., Ghosh, A., Kremer, T., Czech, C., Gossens, C., Lindemann, M.: Evaluation of smartphone-based testing to generate exploratory outcome measures in a phase 1 Parkinson's disease clinical trial. Mov. Disord **33**(8), 1287–1297 (2018). https://doi.org/10.1002/mds.27376. Epub 2018 Apr 27. PMID: 29701258; PMCID: PMC6175318
25. Sheikh, M., Qassem, M., Kyriacou, P.A.: Wearable, environmental, and smartphone-based passive sensing for mental health monitoring. Front Digit Health. **7**(3), 662811 (2021). https://doi.org/10.3389/fdgth.2021.662811.PMID:34713137;PMCID:PMC8521964
26. Buchholz, K.: Where health app usage is most common. Digital Health (Report) (October 2020). https://www.statista.com/chart/23161/health-app-usage-country-comparison/

Chapter 50
Mizo News Classification Using Machine Learning Techniques

Mercy Lalthangmawii, Ranjita Das, and Robert Lalramhluna

Abstract News is a worldwide resource that is always expanding. Manually categorising texts/data has become tiresome and unproductive as the volume of internet material continues to grow. As a result, there is a compelling need to organise this data systematically and in a personalised manner. This paper presents a case study on identifying MIZO news categories from the news articles collected from the 'Zonet' website. Those articles are divided into three categories, namely Tualchhung (local news), Ramchhung (national news) and Rampawn (international news). In this paper, various machine learning techniques, namely random forest, SVM, KNN, decision tree, Naïve Bayes and neural network, have been explored for the classification of Mizo news, and their accuracy is being measured. The result reflects the superiority of the SVM model over the other models.

50.1 Introduction

The Mizo language or Kuki-Chin-Mizo language belongs to the Tibeto-Burman family of languages, spoken natively by the Mizo people in the Mizoram state of India and Chin State of Myanmar. Although some text classification research has been done on various Indian languages with good accuracy results [1], there is no paper previously published in this field of text classification for the 'Mizo' language. Moreover, since the language is not properly available on a large scale in the internet, a particularly popular Mizo news website https://www.zonet.in/ has been used for the data source.

Information that individuals and businesses exchange online has expanded and continues to grow fast in recent decades, paralleling an exponential increase in the number of people who use the internet. As a result, unstructured and jumbled data

M. Lalthangmawii (✉) · R. Das · R. Lalramhluna
National Institute of Technology Mizoram, Aizawl, Mizoram, India
e-mail: mercy2049@gmail.com

R. Das
e-mail: rdas@nitmz.ac.in

circulate on the internet, posing a difficulty for computers that can only interpret well-defined algorithms. To extract valuable patterns and information from the unstructured text, specialised processing and preparation procedures are necessary.

Text classification is the process of automatically classifying a group of texts into one or more predetermined classes based on their themes or categories [2]. Motivated by linguistic principles and current breakthroughs in deep learning [3], this work studies automated algorithms for news article categorisation for the MIZO language. This entails a review of existing news categorisation standards, textual extraction and classification tools and data.

50.2 News Classification Workflow

Proper classification is the first and most critical stage in this procedure. The operation's following phases are significantly reliant on reliable classification techniques. Instead of manually constructing rules, a model is constructed automatically by learning from examples when we have an example database with pre-classified instances.

There are multiple steps involved in the workflow as shown in Fig. 50.1. Classification is a difficult activity as it requires preprocessing steps to convert the textual data into structured form from the unstructured form. The text classification process involves the following main steps—data collection, preprocessing, feature selection, build classifier and performance evaluation.

50.2.1 Data Collection

The initial stage is where we collect all the necessary and appropriate data that are needed for processing and ultimately training the models.

All the data is collected from the Zonet website (https://www.zonet.in), one of the most famous news providers in Mizoram. The news itself is collected using a Python web scraping tool and is mined by taking the HTML tags. Every news article is then labelled with their appropriate category (Tualchhung/Local news, Ramchhung/National news, Rampawn/International news) into a text file in the order they were mined and are put into their respective folders. The news samples are shown in Fig. 50.2 in which the first column represents the news data collected and the second column represents the category to which the news belong to, respectively.

The total number of news articles mined in each category are as follows:

1. Tualchhung news—2989
2. Ramchhung news—2989
3. Rampawn news—2989.

Phase 1 : Data Collection

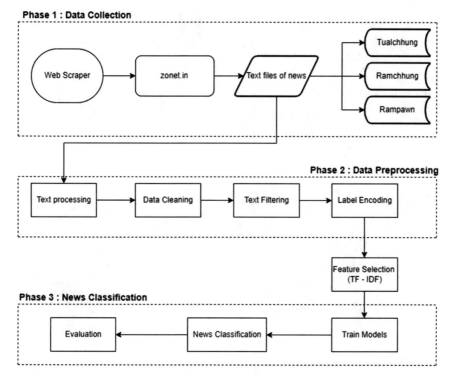

Fig. 50.1 News classification workflow

Fig. 50.2 News sample

	News	Category
0	Education Minister Pu Lalchhandama Ralte in pa...	tualchhung
1	Mizorama hri kai hmuh tawh 284 zinga 189 chu m...	tualchhung
2	Special Court (Prevention of Corruption Act) J...	tualchhung
3	Serchhip khawpui leh a chhehvela mi in rawk ch...	tualchhung
4	Information & Public Relation Minister Pu Lalr...	tualchhung
5	Mizorama bungraw phurh luh phalna, online-a di...	tualchhung
6	Mipui kharkhipna thupek kenkawh mekah Assam la...	tualchhung
7	State Referral Hospital Falkawn, Zoram Medical...	tualchhung
8	Mizoram chhunga chengte atangin Covid 19 a dar...	tualchhung
9	Mizoram sorkarin State Disaster Response Fund,...	tualchhung

50.2.2 News Preprocessing

This phase is the most important and time-consuming step, which results in each document being represented by a small number of index words. It is a technique for cleaning text data and getting it ready to input into a model. It aids in the removal of

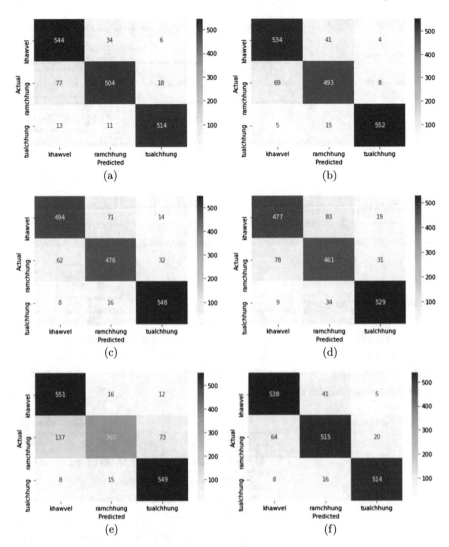

Fig. 50.3 **a** Confusion matrix of random forest, **b** confusion matrix of SVM, **c** confusion matrix of K-neighbor, **d** confusion matrix of decision tree, **e** confusion matrix of Gaussian Naive Bayes, **f** confusion matrix of multilayer perceptron

useless data or noise and the extraction of significant characteristics from texts, as well as the improvement of word-to-document and word-to-category relevance [4].

Text preprocessing is done to the data sets once the relevant data have been obtained. Data cleaning is also essential to verify that there are no corrupt or null records. Unrelated words, such as semicolon, commas, double quotes and brackets, as well as special characters, extraneous texts, quotations, exclamation marks and dates, are all removed from the data.

Table 50.1 Label encoding of news categories

Label	Encoding
Khawvel (International)	0
Ramchhung (National)	1
Tualchhung (Local)	2

50.2.3   Label Encoding

In order to be machine readable, the news categories must also be transformed to a numeric format. In supervised learning, this is a vital step in preparing the structured data set. Table 50.1 shows how the label encoding is done in this experiment.

50.2.4   Feature Selection

The goal of feature selection is to prioritise the significance of the data set's current characteristics and remove the ones that aren't significant. It minimises the number of input variables to our model by only utilising relevant data and removing noise from the data [5].

Term frequency-inverse document frequency (TF-IDF) is a standard approach used to translate text into a meaningful representation of numbers [6], and it is used to identify the features in this study since TF-IDF model contains information on the more important words and the less important ones as well as compared to bag of words that just creates a set of vectors containing the count of word occurrences in the document.

TF-IDF The TF-IDF (term frequency-inverse document frequency) examines the relevance of a word to a document in a collection of documents. This is accomplished by multiplying two metrics: the number of times a word appears in a document and the word's inverse document frequency over a collection of documents. The TF-IDF score of a word in a document is calculated by multiplying these two quantities. The greater the score, the more important the term in that document is

TF-IDF is calculated according to the formula:

$$tfidf_{i,j} = tf_{i,j} \times \log\left(\frac{N}{df_i}\right) \tag{50.1}$$

where

$tf_{i,f}$ total number of occurrences of i in j
df_i total number of documents containing i
N total number of documents.

50.2.5 News Classification

Following feature selection, the news is processed with the goal of assigning them to their appropriate classes. Naïve Bayes, support vector machine, random forest, decision tree, and neural network were utilised to classify the news in this study.

Random Forest Random forest is a supervised machine learning algorithm that is used widely in classification and regression problems [7]. It constructs decision trees from several samples, using the majority vote for classification and the average for regression. While the trees are developing, random forest adds more randomisation to the model. When splitting a node, it looks for the greatest feature from a random group of characteristics instead of the most essential feature. As a result, there is a lot of variety, which leads to a superior model in most cases.

Support Vector Machine A support vector machine (SVM) is a supervised machine learning algorithm that can be employed for both classification and regression purposes [8]. An SVM algorithm's goal is to figure out which category a new data point fits in. As a result, SVM is classified as a non-binary linear classifier. The margins between items on a graph are as wide as feasible in an SVM algorithm, which not only places objects into categories but also places them into categories. They are primarily in charge of determining the decision boundary that will divide the various groups and maximise the margin.

Decision Tree A decision tree is a supervised learning technique that uses a pre-determined goal variable to solve classification problems [9]. We divide the sample into two or more homogeneous groups using this strategy, which is based on the most significant splitter/differentiator in the input variables. A decision tree is a tree-like structure in which each internal node represents an attribute test; each branch indicates a test outcome, and each leaf node (terminal node) stores a class label. By learning decision rules inferred from prior/training data, this is utilised to forecast the classes of target variables.

Naïve Bayes The Bayes' theorem underpins the Naïve Bayes classification approach [10]. It is a collection of probabilistic algorithms that forecast categories using probability theory and Bayes' theorem. Every pair of characteristics being categorised in Naïve Bayes is independent of the other. They are probabilistic, in the sense that they compute the likelihood of each category for a given text and then output the one with the highest probability.

KNN The k-nearest neighbours (KNNs) technique is a straightforward supervised machine learning approach that may be applied to classification and regression applications [11]. The KNN method presupposes that comparable objects exist in close proximity, i.e. that they are nearby to one another. The structure of the data has no bearing on this method. It predicts the values of new data using 'feature similarity', where the new data point is allocated a value based on how closely it resembles the points in the training set. This is accomplished by determining the distance between the test data and all of the training points.

Neural Network The behaviour of the human brain is reflected in neural networks. The basic goal of a neural network is to discover the association between characteristics in a data collection, and it is made up of a series of algorithms that imitate the human brain's functions [12]. In a neural network, a 'neuron' is a mathematical function that gathers and categorises data using a specified design. These layered groupings of neurons make up a neural network. The fundamental benefit of adopting neural networks is that they can adapt to changing input, resulting in the network producing the best possible result without the need to rethink the output criteria. In this study, a multilayer perceptron neural network with 43 hidden layers was used to train and predict the news categories.

50.3 Results and Discussion

In the classification phase, the training size is allotted 80% of the sample size, and the remaining 20% are used as the testing samples.

The classification results and the confusion matrix of the different models are shown in Tables 50.2, 50.3, 50.4, 50.5, 50.6 and 50.7 and in Fig. 50.3a–f, respectively.

Table 50.2 Random forest classification result

Category	Precision	Recall	F-1 score	Support	Accuracy
Khawvel/International	0.88	0.92	0.90	579	0.91400
Ramchhung/National	0.89	0.86	0.88	570	
Tualchhung/Local	0.97	0.97	0.97	572	

Table 50.3 Support vector machine classification result

Category	Precision	Recall	F-1 score	Support	Accuracy
Khawvel/International	0.88	0.92	0.90	579	0.91748
Ramchhung/National	0.90	0.86	0.88	570	
Tualchhung/Local	0.98	0.97	0.97	572	

Table 50.4 K-neighbour classification result

Category	Precision	Recall	F-1 score	Support	Accuracy
Khawvel/International	0.88	0.85	0.86	579	0.88204
Ramchhung/National	0.85	0.84	0.84	570	
Tualchhung/Local	0.92	0.96	0.94	572	

Table 50.5 Decision tree classification result

Category	Precision	Recall	F-1 score	Support	Accuracy
Khawvel/International	0.85	0.82	0.83	579	0.85421
Ramchhung/National	0.80	0.81	0.80	570	
Tualchhung/Local	0.91	0.92	0.92	572	

Table 50.6 Naïve Bayes classification result

Category	Precision	Recall	F-1 score	Support	Accuracy
Khawvel/International	0.79	0.95	0.86	579	0.84834
Ramchhung/National	0.92	0.63	0.75	570	
Tualchhung/Local	0.87	0.96	0.91	572	

Table 50.7 Multilayer perceptron classification result

Category	Precision	Recall	F-1 score	Support	Accuracy
Khawvel/International	0.88	0.92	0.90	584	0.91051
Ramchhung/National	0.90	0.86	0.88	599	
Tualchhung/Local	0.95	0.96	0.95	538	

Table 50.8 Accuracy of models

S. No.	Model name	Accuracy in %
1	Random forest	91.40
2	Support vector machine	91.74
3	K-nearest neighbour	88.20
4	Decision tree	85.42
5	Gaussian Naive Bayes	84.83
6	Multilayer perceptron	91.05

50.3.1 Accuracy Report

From Table 50.8, we can see that support vector machine has the highest accuracy rate of 91.74% for predicting the news articles category. It is followed closely by random forest 91.40% and multilayer perceptron 91.05% while the other models perform below 90%.

50.4 Conclusion

This study explores various machine learning techniques for the classification of Mizo news articles into three categories: Tualchhung/Local news, Ramchhung/National news and Rampawn/International news. The classification results show that the various models, namely random forest, support vector machine, KNN, decision tree, Gaussian Naive Bayes and multilayer perceptron all have good accuracy results well above 80% out of which support vector machine has the highest accuracy.

We can conclude from the findings that text classification using various machine learning algorithms is effective and has promising results.

References

1. Pal, K., Patel, B.V.: A study of current state of work done for classification in Indian languages. Int. J. Sci. Res. Sci. Technol. **3**(7), 403–407 (2017)
2. Kadhim, A.I.: Survey on supervised machine learning techniques for automatic text classification. Artif. Intell. Rev. **52**(1), 273–292 (2019)
3. Hartmann, J., Huppertz, J., Schamp, C., Heitmann, M.: Comparing automated text classification methods. Int. J. Res. Mark. **36**(1), 20–38 (2019)
4. Srividhya, V., Anitha, R.: Evaluating preprocessing techniques in text categorization. Int. J. Comput. Sci. Appl. **47**(11), 49–51 (2010)
5. How, B.C., Narayanan, K.: An empirical study of feature selection for text categorization based on term weightage. In: IEEE/WIC/ACM International Conference on Web Intelligence (WI'04), pp 599–602. IEEE (2004)
6. Liu, C., Sheng, Y., Wei, Z., Yang, Y.-Q.: Research of text classification based on improved tf-idf algorithm. In: 2018 IEEE International Conference of Intelligent Robotic and Control Engineering (IRCE), pp. 218–222. IEEE (2018)
7. Akinyelu, A.A., Adewumi, A.O.: Classification of phishing email using random forest machine learning technique. J. Appl. Math. **2014** (2014)
8. Tong, S., Koller, D.: Support vector machine active learning with applications to text classification. J. Mach. Learn. Res. **2**, 45–66 (2001)
9. Rajaram, R., Balamurugan, A.: Suspicious e-mail detection via decision tree: a data mining approach. J. Comput. Inf. Technol. **15**(2), 161–169 (2007)
10. Dai, W., Xue, G.-R., Yang, Q., Yong, Y.: Transferring Naive Bayes classifiers for text classification. In: AAAI, vol. 7, pp. 540–545 (2007)
11. Khamar, K.: Short text classification using kNN based on distance function. Int. J. Adv. Res. Comput. Commun. Eng. **2**(4), 1916–1919 (2013)
12. Wang, T., Liu, L., Liu, N., Zhang, H., Zhang, L., Feng, S.: A multi-label text classification method via dynamic semantic representation model and deep neural network. Appl. Intell. **50**(8), 2339–2351 (2020)

Chapter 51
Software Defect Prediction Using ROS-KPCA Stacked Generalization Model

Bhaskar Marapelli, Anil Carie, and Sardar M. N. Islam

Abstract Software quality assurance is an area that deals with software defect prediction also. Identifying and eliminating defects is a crucial task that helps organizations deliver quality software products to customers. Machine learning approaches help in identifying software modules that are defective and which are not defective. The existing software defect prediction datasets contain data with features that could classify projects are defective or not. The machine learning model's performance will be degraded with the existence of noisy attributes and class imbalance problems. In this work, we propose a ROS-KPCA-SG model (Random Over Sampling-Kernel Principal Component Analysis-Staked Generalization Model) model to solve the noisy dataset and class imbalance problems and to improve the software defect prediction accuracy. The performance of the ROS-KPCA-SG model is compared with individual models with different combinations of sampling techniques. The results show the proposed ROS-KPCA-SG model solves the problems and gives better performance than other models. The AUC-ROC score is between 0.9 and 1 for the ROS-KPCA-SG model on all the datasets, and the accuracy is near to 90% and above which is a higher value than other models. The proposed model gives accuracy on datasets CM1 is 98%; JM1 is 89%; PC1 is 98%; KC1 is 92% and with KC2 dataset is 89%.

B. Marapelli (✉)
Software Engineering Department, College of Computing and Informatics, Wolkite University, Wolkite, Ethiopia
e-mail: bhaskarmarapelli@wku.edu.et

A. Carie
School of Computer Science and Engineering, VIT-AP University, Amaravati, India

S. M. N. Islam
ISILC and Decision Sciences and Modelling Program, Victoria University, Melbourne, Australia
e-mail: sardar.islam@vu.edu.au

© The Author(s), under exclusive license to Springer Nature Singapore Pte Ltd. 2023   587
V. Bhateja et al. (eds.), *Evolution in Computational Intelligence*, Smart Innovation, Systems and Technologies 326, https://doi.org/10.1007/978-981-19-7513-4_51

51.1   Introduction

In the software development process, software quality plays a crucial role. The software development process should follow the schedule and budget and deliver the product without errors/defects to the customers. Defective software affects the team as well as the organization's reputation, so developers should focus on identifying and eliminating defects in the early stage. Identifying and rectifying defects in the early stage is cost-effective otherwise maintaining defective software is costly according to the U.S National Institute of Technology (NIST) the average cost of such failures and faults of software systems per year is around $59 billion and the cost of software maintenance if 40% more than the cost of software development [1]. So, to reduce quality assurance costs and improve the quality of the software product, prediction of defective modules in the early stage of the software development life cycle is mandatory [2].

Building machine learning models with available datasets to classify software modules as defective or not can help developers in improving software quality by identifying defective modules early and fixing them [2–5]. Based on the machine learning model predictions, project managers can distribute the resources for testing [6]. The machine learning model's performance will be degraded with the existence of noisy attributes and class imbalance problems. For example, the software defect prediction datasets available in the PROMISE repository have problems like there are a number of non-defective modules greater than defective modules; there are attributes that have less relation with defective or non-defective modules, i.e., the class attribute is not balanced, and some attributes in the dataset are with missing or duplicate data [4, 5, 7–9]. With more non-defective modules in datasets, machine learning model results may lead to overfitting, and with noisy attributes, performance of the classification model is reduced [4].

In this work, we propose a ROS-KPCA-SG model to solve the noisy dataset and class imbalance problems and to improve the software defect prediction accuracy. We used the T-link algorithm to remove noise, and then, we combine T-link with other sampling methods like random oversampling (ROS), random under-sampling (RUS), and synthetic minority technique (SMOTE). To reduce the number of features in the dataset, we have used kernel principal component analysis (KPCA), which reduces the dimension of the data. From the suggestions made by Ref. [10], "it is important to balance the classes before reducing the dimensionality of hyperspectral data." We have applied different sampling techniques like ROS, RUS, and SMOTE to the datasets before applying KPCA. The dimensionality reduction method helps in reducing the number of attributes in the dataset while keeping as many of the variations in the original dataset. When there are many attributes in the data, dimensional reduction helps machine learning models from the overfit trap on training data. Kernel-based principal component analysis transforms nonlinear data into a lower-dimensional space of data which can be used with linear classifiers [7, 11, 12].

The class imbalance issue can be handled by ensemble-based machine learning algorithms; there are previous researches for software defect prediction using boosting, bagging, and voting techniques, but less research is done with stacked generalization (SG) models [13–15]. The organization of the remaining part of the paper contains Related Work in Sect. 51.2; Sect. 51.3 contains Proposed Methodology; Sect. 51.4 contains Experimental Setup, Results and Discussion in Sect. 51.5, and finally, Sect. 51.6 contains Conclusion.

51.2   Related Work

Most of the previous researchers employed machine learning models on datasets available for software defect prediction. In the following literature, we briefly discuss related work on software defect prediction. Software defect prediction with machine learning models depends on datasets; the aim will be to reduce prediction error [3]. The problems with software defect prediction datasets that are available in the PROMISE repository are discussed in [4]; there are two problems discussed one because of irrelevant features there are noisy attributes, two there are very few data samples with defective classes compared to non-defective classes. These problems may lead to overfitting and may reduce the performance of machine learning models. The noisy attributes problem can be handled by dimensionality reduction methods, and the overfitting problem can be solved by ensemble machine learning models [4, 7, 11, 12]. Some of the research are briefly described.

In [12], author presented solutions for device defect prediction; he proposed a hybrid feature reduction scheme to handle datasets having enormous dimensions and an artificially dependent neural network strategy for the prediction of software defects. The author used five datasets CM1, KC1, MW1, PC3, and PC4 with a weka simulation tool to handle dimensionality reduction author proposed hybridized approach that incorporates the PCA, random forest, Naïve Bayes, and the SVM software framework. The proposed method is analyzed with parameters like confusion, precision, recall, and recognition accuracy.

In [13], author mentions that finding defects before they occur is a crucial task. The author performed software defect prediction analysis with machine learning and deep learning models. He presented analysis on datasets PC1, CM1, KC1, JM1, and KC2 from the PROMISE repository using 10 machine learning algorithms Naive Bayes, support vector machine, decision tree, K-nearest neighbor, extra trees, gradient boosting, random forest, bagging, multi-layer perceptron, and AdaBoost.

In [11], author presents a comparison of open-source programming frameworks written in java with feature extraction methods utilizing support vector machine (SVM) as the base AI classifier. The feature extraction methods, for example, kernel-based principal component analysis (KPCA), principal component analysis (PCA), linear discriminant analysis (LDA), and autoencoders are assessed utilizing different execution measures like accuracy and ROC-AUC.

In [16], author performed experiments using Naive Bayes, linear regression, and SVM classifiers with the NASA dataset. The author addresses software defect prediction as a division or contrast between two things like defective and non-defective classes. The author used a confusion matrix as a classification metric for determining the performance of the model. In his future work, he suggests performing in-depth research to solve problems of noise and class imbalance.

In [14], author proposed prediction model based on change metric dataset. The author used four open-source software systems datasets and has studied five machine learning algorithms. He used datasets from Eclipse JDT, Eclipse platform, Mozilla, and Bugzilla software systems and studied XGBoost, AdaBoost, random forest, logistic regression, and deep neural network machine learning algorithms.

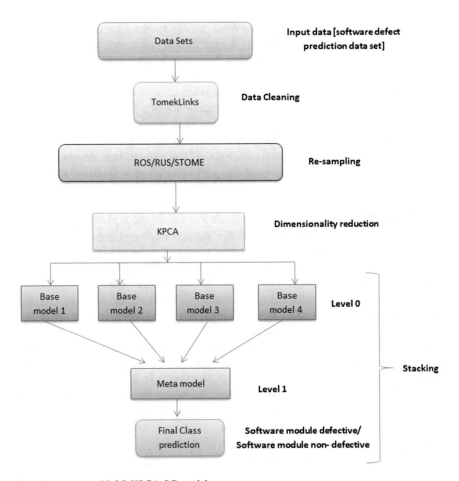

Fig. 51.1 Proposed ROS-KPCA-SG model

51.3   Proposed Methodology

In this study, the first phase deals with data reduction, i.e, removal of noise data; we used the T-link method for noise data removal from the majority class; this helps in improving the classification process. After this to reduce data, we used different re-sampling techniques like RUS, ROS, and SMOTE. For feature selection, we used KPCA a traditional dimension reduction algorithm. KPCA is a type of PCA method that builds a nonlinear mapping between input space and feature space in a natural way, and it is based on kernel functions. In the second phase, different classification algorithms are used to predict defective and non-defective modules. Algorithms like logistic regression (LR), K-nearest neighbors (KNNs), decision tree (DT), Gaussian Naïve Bayes classifier (GB), and multi-layer perceptron classifier (MLP) are used to analyze the prediction. Finally, a stacked generalization (SG) model is used to determine whether ensemble predictors improve the prediction accuracy. In the third phase model, performances were evaluated using the AUC-ROC score and accuracy score. In this stage, by examining distinctive strategies, the improvement of model affectability toward minority class was surveyed. The AUC-ROC curve measures the probability of a data point belonging to an actual class, which is a good measure of model performance when there is a class imbalance in datasets. The proposed ROS-KPCA-SG model framework is shown in Fig. 51.1.

51.3.1   Stacked Generalization Model

The prepared stacking or stacked generalization (SG) model contains logistic regression as metamodel (level-1) and the weak learners or base models (level-0) as decision tree, K-neighbors classifier, and logistic regression, and multi-layer perception. In level-0, we have used 4 classifiers decision tree, K-neighbors classifier and logistic regression, and multi-layer perception; these classifiers are trained using datasets that are divided using k-fold cross-validation. The output of level-0 classifiers is used to train a classifier in level-1; we have used a logistic regression classifier in level-1. The metamodel is trained in this SG using K-fold cross-validation of the base models (Model arrangement shown in Fig. 51.2). The SG model is helpful when the base models or weak learners have different predictions on the dataset [17].

51.4   Implementation

51.4.1   Datasets Used

In this work, we have used five software defect prediction datasets, namely CM1, JM1, KC1, KC2, and PC1 collected from the PROMISE repository. Table 51.1

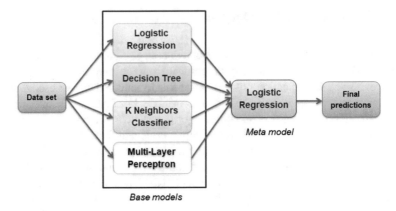

Fig. 51.2 Stacked generalization model

Table 51.1 Software defect prediction dataset description

Dataset	#Attributes	#Rows	Defective (%)	Non-defective (%)
CM1	22	498	9.83	90.16
JM1	22	10,885	19.34	80.65
PC1	22	1109	6.94	93.05
KC1	22	2109	15.45	84.54
KC2	22	522	20.49	79.5

describes the number of attributes, rows, and defective and non-defective data sample percentage. The five datasets were available in .ARFF format before we applied machine learning models we converted them into .CSV format.

51.4.2 Experimental Setup

We have prepared the datasets by cleaning the data using TomkeLinks after that we have applied sampling techniques like RUS, ROS, and SMOTE. Before applying machine learning models, we have used the dimensionality reduction technique KernelPCA for reducing the number of features from the input data. We compared models with different combinations of sampling techniques, with and without using KernelPCA. We have used Python 3.0, and Python machine learning packages like imblearn, sklearn to prepare the models and evaluate them. Since the datasets are imbalanced, we have used sampling techniques to balance the datasets; datasets contain 20 input features to reduce them; we used KernelPCA. As the datasets are imbalanced, the accuracy score of the model is not enough to judge the models, so we have used model evaluation techniques area under the curve (AUC) of receiver characteristic operator (ROC) (AUC-ROC).

51.4.3   Performance Measure

51.4.3.1   AUC-ROC Score

The AUC (Area Under the curve) gives a measure of the classifier's ability to separate classes, and it gives a summary of the ROC (Receiver Operator Characteristic) curve which is a binary classification problem evaluation metric [2]. The probability of a data sample belonging to an actual class is measured when we use AUC as the performance metric measure of the classification model. When AUC = 1, the performance of the model is high since it is able to predict positive and negative classes correctly [2, 8, 11].

AUC value for a machine learning model will be as follows

0.9 < AUC ≤ 1 for excellent model
0.8 < AUC ≤ 0.9 for good model
0.7 < AUC ≤ 0.8 for a fairly performing model
0.6 < AUC ≤ 0.7 for poorly performing model
AUC ≤ 0.6 for failed model.

$$\text{AUC} = \frac{1 + \text{TP}_{\text{rate}} - \text{FP}_{\text{rate}}}{2} \tag{51.1}$$

where TP_{rate} True Positive Rate, FP_{rate} False Positive Rate [4].

51.4.3.2   Accuracy

Accuracy score of machine learning model is a widely used measure to evaluate models; it shows the ratio of all correctly classified samples. Because of the class imbalance in software defect prediction datasets only accuracy is not a proper measure for evaluating the model performance [11, 13].

$$\text{Accuracy} = \frac{n_{\text{TP}} + n_{\text{TN}}}{n} \tag{51.2}$$

where

n_{TP} is the number of samples correctly predicted for defective class, i.e., 'defective' as 'defective'
n_{TN} is number of samples correctly predicted for non-defective class, i.e., 'non-defective' as 'non-defective',
n is the total number of samples in a data set.

Table 51.2 AUC-ROC score on all data sets

Dataset	Classification method	Data preprocessing technique					
		Tlink + ROS	Tlink + RUS	Tlink + SMOTE	Tlink + RUS + KPCA	Tlink + SMOTE + KPCA	Tlink + ROS + KPCA
CM1	Logistic regression	0.860	0.842	0.875	0.741	0.753	0.742
	K nearest neighbors	0.997	0.850	0.994	0.798	0.941	0.998
	Decision tree classifier	1.000	1.000	1.000	1.000	1.000	1.000
	GaussianNB	0.761	0.758	0.749	0.749	0.761	0.752
	MLP classifier	0.953	0.907	0.963	0.744	0.791	0.773
	Stacking classifier	1.000	0.951	1.000	0.973	0.986	1.000
JM1	Logistic regression	0.738	0.732	0.724	0.693	0.678	0.703
	K nearest neighbors	0.960	0.862	0.949	0.843	0.865	0.953
	Decision tree classifier	0.999	0.999	1.000	0.999	1.000	0.999
	GaussianNB	0.710	0.706	0.686	0.697	0.681	0.707
	MLP classifier	0.806	0.772	0.824	0.704	0.697	0.718
	Stacking classifier	0.997	0.927	0.995	0.904	0.896	0.998
KC1	Logistic regression	0.830	0.842	0.841	0.798	0.782	0.797
	K nearest neighbors	0.975	0.892	0.976	0.894	0.925	0.974
	Decision tree classifier	0.998	0.998	0.999	0.998	0.999	0.998
	GaussianNB	0.805	0.815	0.791	0.802	0.782	0.792
	MLP classifier	0.876	0.887	0.898	0.809	0.799	0.808
	Stacking classifier	0.996	0.930	0.994	0.903	0.965	0.997
KC2	Logistic regression	0.882	0.892	0.873	0.879	0.835	0.856
	K nearest neighbors	0.959	0.923	0.960	0.912	0.920	0.961
	Decision tree classifier	0.994	0.995	0.998	0.995	0.998	0.994
	GaussianNB	0.861	0.880	0.843	0.880	0.835	0.855
	MLP classifier	0.900	0.897	0.911	0.882	0.836	0.862
	Stacking classifier	0.984	0.953	0.986	0.903	0.940	0.986
PC1	Logistic regression	0.882	0.884	0.882	0.884	0.727	0.753
	K nearest neighbors	0.997	0.899	0.997	0.899	0.950	0.999
	Decision tree classifier	1.000	1.000	1.000	1.000	1.000	1.000
	GaussianNB	0.728	0.713	0.706	0.713	0.709	0.723
	MLP classifier	0.971	0.940	0.979	0.942	0.767	0.770
	Stacking classifier	1.000	0.987	1.000	0.992	0.977	1.000

Table 51.3 Comparison with baseline models

Model compared	Accuracy scores on datasets				
	CM1 (%)	JM1 (%)	PC1 (%)	KC1 (%)	KC2 (%)
Zhou 2019	88.64	81.11	90.79	84.35	–
Guvenaydin 2020	97	81	94	85	–
Proposed model	**98**	**89**	**98**	**92**	**89**

51.5 Results and Discussion

The experiments were carried out on six machine learning techniques using a 5-fold cross-validation process for assessment. The process we employed splits the dataset into five subgroups of equal size. From these subgroups, one subgroup is used for testing, and the remaining subgroups' are used for training; it is continued until all subgroups have been used for testing. As the datasets are highly imbalanced instead of using only the accuracy score as a measure we have used the AUC-ROC score as the measure. From the AUC-ROC score on all datasets Table 51.2, we observe that the stacked generalization model with ROS-KPCA gives an AUC value high, i.e., using random oversampling as a sampling technique and applying KPCA as feature reduction technique has improved the performance of the classier. The proposed model ROS-KPCA-SG model has an AUC-ROC score of equal to 1 which is the highest of all other individual models for all datasets, the decision tree classifier also shows similar performance.

We have compared our proposed model (ROC-KPCA-SG model) with some baseline models [5, 7], and the results are presented in Table 51.3. In [5], author verified different machine learning algorithms on four datasets, and he concluded that the random forest algorithm got the highest accuracy score; in [7], author used KPCA for dimensionality reduction and applied the SVM classification algorithm for defect prediction he concludes that KPCA-SVM reduces the problem of data redundancy and improves the defect prediction accuracy. Our proposed model (ROC-KPCA-SG model) solves the problem of class imbalance by the use of ROC-KPCA, and overfitting problems by the use of the ensemble machine learning model (SG). The comparison of accuracy scores of baseline models and our proposed model shows that our proposed model outperforms these models and gives high accuracy our model accuracy with dataset CM1 is 98%, JM1 is 89%, PC1 is 98%, KC1 is 92%, and with KC2 dataset it is 89%.

51.6 Conclusion

In this work, we proposed a ROS-KPCA-SG model (Tomek links—Random Over Sampling-Kernel Principal Component Analysis—Staked Generalization Model) to solve the noisy dataset and class imbalance problems and to improve the software defect prediction accuracy. Machine learning algorithms like logistic regression (LR), K-nearest neighbors (KNN), decision tree (DT), Gaussian Naïve Bayes classifier (GB), and multi-layer perceptron classifier (MLP) are used to analyze the prediction. A stacked generalization (SG) model is used to determine whether ensemble predictors improve the prediction accuracy. Two measures are used AUC-ROC score and accuracy score, from both measures it is observed that the proposed model (ROS-KPCA-SG model) outperforms all other individual machine learning models. The AUC is between 0.9 and 1 for the ROS-KPCA-SG model on all the datasets, and accuracy is near to 90% and above which is a higher value than other models. The comparison with other baseline models also shows our proposed model outperforms them with high accuracy.

References

1. Reddivari, S., Raman, J.: Software quality prediction: an investigation based on machine learning. In: 2019 IEEE 20th International Conference on Information Reuse and Integration for Data Science (IRI), pp. 115–122. IEEE (2019)
2. Samir, M., El-Ramly, M., Kamel, A.: Investigating the use of deep neural networks for software defect prediction. In: 2019 IEEE/ACS 16th International Conference on Computer Systems and Applications (AICCSA), pp. 1–6. IEEE (2019)
3. Pradhan, S., Nanniyur, V., Vissapragada, P.K.: On the defect prediction for large scale software systems-from defect density to machine learning. In: 2020 IEEE 20th International Conference on Software Quality, Reliability and Security (QRS), pp. 374–381. IEEE (2020)
4. Bahaweres, R.B., Suroso, A.I., Hutomo, A.W., Solihin, I.P., Hermadi, I., Arkeman, Y.: Tackling feature selection problems with genetic algorithms 18 in software defect prediction for optimization. In: 2020 International Conference on Informatics, Multimedia, Cyber and Information System (ICIMCIS), pp. 64–69. IEEE (2020)
5. Aydin, Z.B.G., Samli, R.: Performance evaluation of some machine learning algorithms in NASA defect prediction data sets. In: 2020 5th International Conference on Computer Science and Engineering (UBMK), pp. 1–3. IEEE (2020)
6. Zhang, T., Du, Q., Xu, J., Li, J., Li, X.: Software defect prediction and localization with attention-based models and ensemble learning. In: 27th Asia-Pacific Software Engineering Conference (APSEC), vol. 2020, pp. 81–90. IEEE (2020)
7. Zhou, Y., Shan, C., Sun, S., Wei, S., Zhang, S.: Software defect prediction model based on KPCA-SVM. In: IEEE SmartWorld, Ubiquitous Intelligence and Computing, Advanced and Trusted Computing, Scalable Computing and Communications, Cloud and Big Data Computing, Internet of People and Smart City Innovation (SmartWorld/SCALCOM/UIC/ATC/CBDCom/IOP/SCI), vol. 2019, pp. 1326–1332. IEEE (2019)
8. Soe, Y.N., Santosa, P.I., Hartanto, R.: Software defect prediction using random forest algorithm. In: 12th South East Asian Technical University Consortium (SEATUC), vol. 1, pp. 1–5. IEEE (2018)
9. Thant, M.W., Aung, N.T.T.: Software defect prediction using hybrid approach. In: 2019 International Conference on Advanced Information Technologies (ICAIT), pp. 262–267. IEEE (2019)
10. García, V., Sánchez, J.S., Mollineda, R.A.: Classification of high dimensional and imbalanced hyperspectral imagery data. In: Iberian Conference on Pattern Recognition and Image Analysis, pp. 644–651. Springer, Berlin (2011)
11. Malhotra, R., Khan, K.: A study on software defect prediction using feature extraction techniques. In: 2020 8th International Conference on 19 Reliability, Infocom Technologies and Optimization (Trends and Future Directions) (ICRITO), pp. 1139–1144. IEEE (2020)
12. Prabha, C.L., Shivakumar, N.: Software defect prediction using machine learning techniques. In: 2020 4th International Conference on Trends in Electronics and Informatics (ICOEI) (48184), pp. 728–733. IEEE (2020)
13. Cetiner, M., Sahingoz, O.K.: A comparative analysis for machine learning based software defect prediction systems. In: 2020 11th International Conference on Computing, Communication and Networking Technologies (ICCCNT), pp. 1–7. IEEE (2020)
14. Tessema, H.D., Abebe, S.L.: Enhancing just-in-time defect prediction using change request-based metrics. In: 2021 IEEE International Conference on Software Analysis, Evolution and Reengineering (SANER), pp. 511–515. IEEE (2021)

15. Raju, K.S., Ramakrishna Murty, M., Varaprasad Rao, M., Satapathy, S.C.: Support vector machine with k-fold cross validation model for software fault prediction. Int. J. Pure Appl. Math. **118**(20), 321–334 (2018)
16. Mi, W., Li, Y., Wang, S.: Empirical evaluation of the active learning strategies on software defects prediction. In: 6th International Symposium on System and Software Reliability (ISSSR), pp. 83–89. IEEE (2020)
17. Mahaweerawa, A., Nilnumpetch, C., Kraipeerapun, P.: Applying stacked generalization with the difference of truth and falsity data to predict student's performance. In: 2020 5th International Conference on Computer and Communication Systems (ICCCS), pp. 148–152. IEEE (2020)

Chapter 52
Combination of Hamming Distance and Entropy Measure of Picture Fuzzy Sets: Case Study of COVID-19 Medicine Selection

Xuan Thao Nguyen, Quoc Hung Nguyen, Duy Dong Le, and Hai Van Pham

Abstract The picture fuzzy set (PFS) is a general form of the fuzzy and intuitionistic fuzzy set for solving real-world problems. Entropy and distance measures play significant measures in solving problems involving fuzzy environments. In multi-criteria decision-making (MCDM) problems, entropy is used to determine the weights of each criterion. Distance measures are used to rank alternatives using fuzzy picture sets. In this paper, we consider relationship between Hamming distance and entropy measures on picture fuzzy sets (PFS) and apply them to solving MCDM problems. Firstly, some distance of picture fuzzy sets are constructed by using the axioms. Secondly, a method to build the entropy measure of the picture fuzzy sets is defined as new Hamming distance measure. Finally, the distance measures apply to solve the MCDM problems. To illustrate the Hamming distance measure, this method is used to construct an entropy based on distance measures of PFS for selection of COVID-19 medicine.

X. T. Nguyen
International School, Vietnam National University, Hanoi and Vietnam National University of Agriculture, Hanoi, Vietnam
e-mail: nxthao@vnua.edu.vn

Q. H. Nguyen · D. D. Le
University of Economics Ho Chi Minh City (UEH), Ho Chi Minh City, Vietnam
e-mail: hungngq@ueh.edu.vn

D. D. Le
e-mail: dongld@ueh.edu.vn

H. Van Pham (✉)
School of Information Technology and Communication, Hanoi University of Science and Technology, Hanoi, Vietnam
e-mail: haipv@soict.hust.edu.vn

52.1 Introduction

Recently, many studies explore using picture fuzzy sets. A picture fuzzy set [1] consists of three components: the positive membership, the negative membership, and the neutral membership function. PFS is defined as a general form of fuzzy sets and intuitionistic fuzzy sets [2]. In 2020, Ganie et al. [3] have investigated in Singh's correlation coefficient which is given in the value [0, 1]. The studies of correlation coefficient in the intuitionistic fuzzy sets of Hung and Yang [4] to build two new correlation coefficients on the picture fuzzy sets with these pattern recognition, medical diagnosis, and clustering analysis. PFS's clustering method is proposed by Son [5]. Furthermore, picture fuzzy sets clustering and reasoning algorithms have been investigated by Hong Lan et al. [6]. In related works [3, 7–14], some PFS aggregation operators are investigated to use in MCDM. Picture linguistic sets were also proposed [15]. Some distance and similarity measures of PFS are initiated to apply in MCDM or pattern recognition problems [16–22]. However, these measures show the restriction analyzed in [14, 17]. In 2019, Thao et al. [23] studied the concept divergence measure of PFSs and applied them in the medical diagnosis.

In real-world problems, entropy measure on fuzzy sets, intuitionistic fuzzy sets can be applied to evaluate uncertain information under dynamic environments [4, 24, 25]. In PFS environment, Thao [17] has indicated the concept of PFS's entropy in sense of probability and proposed some formulas to determine entropies on picture fuzzy sets. Fan and Xi [24], Selvachandran et al. [25] investigated the fuzzy/intuitionistic entropy measures can be induced from the distance measure. Following these ideas of Fan and Xie [24], Selvachandran et al. [25], we built some entropy measures of picture fuzzy sets based on the distance measures. Moreover, the topics related to COVID-19 have also been studied by many people recently [26, 27]. These are the motivations for us to write this paper, the contribution of this paper as follows:

1. Construct some new entropy measures of picture fuzzy sets based on the Hamming distance measures.
2. Give some remark on the entropy measure of picture fuzzy set.
3. Apply our proposed entropy and distance measures to construct a TOPSIS model to select COVID-19 medicine. Here, entropy measure paves to compute the weight of each criterion, distance to rank alternatives.
4. Compare to some other measures on PFSs of the proposed model.

In this paper, entropy measure based on the normal Hamming distance measure has been proposed. The proposed model combined entropy and distance measures to construct the TOPSIS model that solved the MCDM problem. In the medical diagnostic dealing with multi-criteria decision-making problem, the proposed new measures are also proved to be feasible with good results when compared to the other previous measures.

The remainder of this study is described as follows: some concept as picture fuzzy set, distance measure, and entropy of them is presented in Sect. 52.2. Section 52.3 give out the TOPSIS based on the proposed measures and apply to select COVID-19 medicine. Conclusion is presented in Sect. 52.4.

52.2   Methodology

Definition 1 [1]. A PFS A in X is defined by the triple form as $A = \{(x, \mu_A(x), \eta_A(x), \nu_A(x))|x \in X\}$. In which, μ_A is the positive-membership function, η_A is the neutral-membership function, and ν_A is the negative-membership function such that $\mu_A(x)$, $\eta_A(x)$, $\nu_A(x) \in [0, 1]$ and $\mu_A(x) + \eta_A(x) + \nu_A(x) \leq 1$ for all $x \in X$.

Denote PFS(X) is the collection of PFS on X.
For all A, $B \in$ PFS(X), then

- $A^C = \{(x, \nu_A(x), \pi_A(x), \mu_A(x))|x \in X\}$, where $\pi_A(x) = 1 - \mu_A(x) - \eta_A(x) - \nu_A(x)$ for all $x \in X$, is the complement of A.
- $A \subseteq B$ iff $\mu_A(x) \leq \mu_B(x)$, $\eta_A(x) \leq \eta_B(x)$ and $\nu_A(x) \geq \nu_B(x)$ for all $x \in X$.
- $A = B$ iff $A \subseteq B$, $B \subseteq A$.

52.2.1   Terms of Distance Measures of Picture Fuzzy Sets

Definition 2 [18]. A distance measure on PFS(X) is a real function $d :$ PFS$(X) \times$ PFS$(X) \to R$, satisfying all following conditions.

- (M1) $d(A, B) \geq 0$ for all A, $B \in$ PFS(X), $d(A, B) = 0$ if only if $A = B$,
- (M2) $d(A, B) = d(B, A)$ for all $A, B \in$ PFS(X),
- (M3) $d(A, C) \leq d(A, B) + d(B, C)$ for all A, B and $C \in$ PFS(X).

Distance d is called a normal distance measure if only if $0 \leq d(A, B) \leq 1$ for all A, $B \in$ PFS(X).
Now, we give the normal Hamming distance measure:

$$d_H(A, B)$$
$$= \frac{1}{n} \sum_{i=1}^{n} \frac{|\mu_A(x_i) - \mu_B(x_i)| + |\eta_A(x_i) - \eta_B(x_i)| + |\nu_A(x_i) - \nu_B(x_i)| + |\pi_A(x_i) - \pi_B(x_i)|}{3}$$

$$(52.1)$$

Theorem 1 The real functions $d_H :$ PFS$(X) \times$ PFS$(X) \to R$ are normal distance measures on PFS(X).

Proof It is obviously that $0 \leq d_H(A, B) \leq 1$ for all $A, B \in \text{PFS}(X)$. We easy verify that d_1 satisfies conditions (M1), (M2) in definition 3. Applying the inequality $|a - c| \leq |a - b| + |b - c|$ for all real numbers a, b and $c \in R$ to four membership-functions of picture fuzzy sets, we obtain four distances d_1 satisfy condition (M3).

So that, d_1 is normal distance measure on $\text{PFS}(X)$.

Proposition 1 . Put $Q = \{(x_i, 0.25, 0.25, 0.25)|x_i \in X\}$. We have $0 \leq d_H(A, Q) \leq 0.5$ for all $A \in \text{PFS}(X)$.

Proof Indeed, for all $A = \{(\mu_A(x_i), \eta_A(x_i), \nu_A(x_i))|x_i \in X\} \in \text{PFS}(X)$. Only one of the following four situations occurs:

- Case 1. $0 \leq |\mu_A(x_i) - 0.25| \leq |1 - 0.25| = 0.75, |\eta_A(x_i) - 0.25| \leq 0.25,$ $|\nu_A(x_i) - 0.25| \leq 0.25, |\pi_A(x_i) - 0.25| \leq 0.25.$
- Case 2. $|\mu_A(x_i) - 0.25| \leq 0.25, |\eta_A(x_i) - 0.25| \leq 0.75, |\nu_A(x_i) - 0.25| \leq 0.25,$ $|\pi_A(x_i) - 0.25| \leq 0.25.$
- Case 3. $|\mu_A(x_i) - 0.25| \leq 0.25, |\eta_A(x_i) - 0.25| \leq 0.25, |\nu_A(x_i) - 0.25| \leq 0.75,$ $|\pi_A(x_i) - 0.25| \leq 0.25.$
- Case 2. $|\mu_A(x_i) - 0.25| \leq 0.25, |\eta_A(x_i) - 0.25| \leq 0.25, |\nu_A(x_i) - 0.25| \leq 0.25,$ $|\pi_A(x_i) - 0.25| \leq 0.75.$

These imply that $0 \leq d_H(A, Q) \leq 0.5$ ∎

52.2.2 Entropy Induced by Normal Hamming Distance of PFSs

Definition 3 [17]. The function $E : \text{PFS}(X) \rightarrow [0, 1]$ is an entropy on $\text{PFS}(X)$ if it satisfies all following conditions.

- (E1) $E(A) = 0$ if $\mu_A(x_i), \eta_A(x_i), \nu_A(x_i) \in \{0, 1\}$ for all $x_i \in X$.
- (E2) $E(A) = 1$ if only if $A = Q$,
- (E3) $E(A) = E(A^C), \quad \forall A \in \text{PFS}(X)$.
- (E4) $E(A) \leq E(B)$ for all $A, B \in \text{PFS}(X)$ such that either if $\mu_A(x_i) \leq \mu_B(x_i), \eta_A(x_i) \leq \eta_B(x_i)$ and $\nu_A(x_i) \leq \nu_B(x_i)$ if $\max\{\mu_B(x_i), \eta_B(x_i), \nu_B(x_i)\} \leq 0.25$ or $\mu_A(x_i) \geq \mu_B(x_i), \eta_A(x_i) \geq \eta_B(x_i)$ and $\nu_A(x_i) \geq \nu_B(x_i)$ if $\min\{\mu_B(x_i), \eta_B(x_i), \nu_B(x_i)\} \geq 0.25$ for all $x_i \in X$.

Remark 1 Definition entropy of picture fuzzy sets in [17] having properties:

- A crisp set is not fuzzy.
- Q is PFS having highest fuzzy property, it means that $E(Q) = 1$.
- A picture fuzzy set and its complement have the same fuzzy property.

- $E(A) \leq E(B)$ for all A, $B \in$ PFS(X) satisfy the condition $d(A, Q) \geq d(B, Q)$ when either $\max\{\mu_B(x_i), \eta_B(x_i), \nu_B(x_i)\} \leq 0.25$ or $\min\{\mu_B(x_i), \eta_B(x_i), \nu_B(x_i)\} \geq 0.25$.

Theorem 2 We have $E(A) = 1 - 2d_H(A, Q)$, for all $A \in$ PFS(X), is an entropy on PFSs.

Proof First of all, according to proposition 1 we have $0 \leq d_H(A, Q) \leq 0.5$ for all $A \in$ PFS(X). So that $E(A) \in [0, 1]$ for all $A \in$ PFS(X).

- (E1) If $A \in$ PFS(X) has $\mu_A(x_i), \eta_A(x_i), \nu_A(x_i), \in \{0, 1\}$ for all $x_i \in X$, then $d_H(A, Q) = 0.5$. So that $E(A) = 0$ if A has $\mu_A(x_i), \eta_A(x_i), \nu_A(x_i), \in \{0, 1\}$ for all $x_i \in X$.
- (E2) It is easy to check that $E(A) = 1$ if $A = Q$.
- (E3) Because of the symmetry of the positive-membership and negative-membership, neutral and refusal, respectively, in four distance measures, we have $d_H(A, Q) = d_H(A^C, Q)$. So that $E(A) = E(A^C)$, for all $A \in$ PFS(X).
- (E4) For all A, $B \in$ PFS(X), if $d_H(A, Q) \geq d_H(B, Q)$ then $E(A) = 1 - 2d_H(A, Q) \leq E(B) = 1 - 2d_H(B, Q)$. ∎

Apply Theorem 1, with the normal Hamming distance measure, we get the entropy of picture fuzzy sets as follows:

$$E_H(A) = 1 - \frac{2}{n}\sum_{i=1}^{n}\frac{|\mu_A(x_i) - 0.25| + |\eta_A(x_i) - 0.25| + |\nu_A(x_i) - 0.25| + |\pi_A(x_i) - 0.25|}{3}$$

(52.2)

52.3 Application of Normal Hamming Distance and Entropy Measures of Picture Fuzzy Sets to COVID-19 Medicine

Now, we remind some other measures of picture fuzzy sets. They use to the pattern recognition to compare with our proposed distance measures.

Similarity measure of Thao in [17]

$$S_T(A, B)$$
$$= \sum_{i=1}^{n}\omega_i\left\{1 - \frac{|\mu_A(x_i) - \mu_B(x_i)| + |\eta_A(x_i) - \eta_B(x_i)| + |\nu_A(x_i) - \nu_B(x_i)|}{3}\right\}$$

(52.3)

Cosine similarity measure [20].

$$SC_1(A, B) = \sum_{i=1}^{n} \omega_i \frac{\mu_A(x_i)\mu_B(x_i) + \eta_A(x_i)\eta_B(x_i) + v_A(x_i)v_B(x_i)}{\sqrt{\mu_A^2(x_i) + \eta_A^2(x_i) + v_A^2(x_i)}\sqrt{\mu_B^2(x_i) + \eta_B^2(x_i) + v_B^2(x_i)}}$$

(52.4)

Set-theoretic similarity measure [22]

$$SC_3(A, B) = \sum_{i=1}^{n} \omega_i \frac{\mu_A(x_i)\mu_B(x_i) + \eta_A(x_i)\eta_B(x_i) + v_A(x_i)v_B(x_i)}{\max\left\{\begin{array}{l}\mu_A^2(x_i) + \eta_A^2(x_i) + v_A^2(x_i), \\ \mu_B^2(x_i) + \eta_B^2(x_i) + v_B^2(x_i)\end{array}\right\}}$$

(52.5)

Cosine function-based similarity measures [21].

$$SCC_1(A, B) = \sum_{i=1}^{n} \omega_i \cos\left\{\frac{\pi}{2}\max\left\{\begin{array}{l}|\mu_A(x_i) - \mu_B(x_i)|, \\ |\eta_A(x_i) - \eta_B(x_i)|, \\ |v_A(x_i) - v_B(x_i)|\end{array}\right\}\right\}$$

(52.6)

Cotangent function-based similarity measure [21]

$$SCT_1(A, B) = \sum_{i=1}^{n} \omega_i \cot\left\{\frac{\pi}{4} + \frac{\pi}{4}\max\left\{\begin{array}{l}|\mu_A(x_i) - \mu_B(x_i)|, \\ |\eta_A(x_i) - \eta_B(x_i)|, \\ |v_A(x_i) - v_B(x_i)|\end{array}\right\}\right\}$$

(52.7)

Divergence measure of picture fuzzy sets [23]

$$div(A, B) = \frac{1}{n}\sum_{i=1}^{n}[D_\mu^i(A, B) + D_\eta^i(A, B) + D_v^i(A, B)]$$

(52.8)

where

$D_\mu^i(A, B) = \mu_A(x_i)\ln\frac{2\mu_A(x_i)}{\mu_A(x_i)+\mu_B(x_i)} + \mu_B(x_i)\ln\frac{2\mu_B(x_i)}{\mu_A(x_i)+\mu_B(x_i)},$

$D_\eta^i(A, B) = \eta_A(x_i)\ln\frac{2\eta_A(x_i)}{\eta_A(x_i)+\eta_B(x_i)} + \eta_B(x_i)\ln\frac{2\eta_B(x_i)}{\eta_A(x_i)+\eta_B(x_i)},$

and $D_v^i(A, B) = v_A(x_i)\ln\frac{2v_A(x_i)}{v_A(x_i)+v_B(x_i)} + v_B(x_i)\ln\frac{2v_B(x_i)}{v_A(x_i)+v_B(x_i)}.$

Correlation coefficient [28]

$cc(A, B)$

$$= \frac{\sum_{i=1}^{n}\left[\begin{array}{l}(\mu_A(x_i) - \overline{\mu_A})(\mu_B(x_i) - \overline{\mu_B}) + (\eta_A(x_i) - \overline{\eta_A})(\eta_B(x_i) - \overline{\eta_B}) \\ +(v_A(x_i) - \overline{v_A})(v_B(x_i) - \overline{v_B}) + d_i(A)d_i(B)\end{array}\right]}{\sqrt{\sum_{i=1}^{n}\left[\begin{array}{l}(\mu_A(x_i) - \overline{\mu_A})^2 + (\eta_A(x_i) - \overline{\eta_A})^2 \\ +(v_A(x_i) - \overline{v_A})^2 + d_i(A)^2\end{array}\right] \times \sum_{i=1}^{n}\left[\begin{array}{l}(\mu_B(x_i) - \overline{\mu_B})^2 + (\eta_B(x_i) - \overline{\eta_B})^2 \\ +(v_B(x_i) - \overline{v_B})^2 + d_i(B)^2\end{array}\right]}}$$

(52.9)

where

$$\overline{\mu_A} = \frac{1}{n}\sum_{i=1}^{n}\mu_A(x_i),\ \overline{\eta_A} = \frac{1}{n}\sum_{i=1}^{n}\eta_A(x_i),\ \overline{\nu_A} = \frac{1}{n}\sum_{i=1}^{n}\nu_A(x_i)$$

$$d_i(A) = (\mu_A(x_i) - \overline{\mu_A}) - (\eta_A(x_i) - \overline{\eta_A}) - (\nu_A(x_i) - \overline{\nu_A})$$

52.3.1 Application in MCDM to Select COVID-19 Medicine

In this section, we have applied to our proposed entropies and distances to construct TOPSIS model to select COVID-19 medicine. In which, the entropies paved to calculate the weights of criteria in the model, and the distance is used to rank the alternatives.

The MCDM problem: There is a set of alternatives $A = \{A_1, A_2, ..., A_m\}$. In which, each alternative $A_i = \{(C_j, \mu_{ij}, \eta_{ij}, \nu_{ij}) | C_j \in C\}$ $(i = 1, 2, ..., m)$ is a PFS on the universal set of criteria $C = \{C_1, C_2, ..., C_n\}$. Request the ranking of alternatives to select the most optimal option based on the set of criteria $C = \{C_1, C_2, ..., C_n\}$. To solve this MCDM problem, we use the TOPSIS model based on our proposed entropies and distances.

52.3.2 The TOPSIS Model Based on Our Proposed Entropies and Distances

Step 1. Consider $C_j = \{(A_i, \mu_{ij}, \eta_{ij}, \nu_{ij}) | A_i \in A\}$ is a PFS on the universal $A = \{A_1, A_2, ..., A_m\}$, for all $(j = 1, 2, ..., n)$. Using entropy to compute the weights ω_j of criteria C_j, for all $(j = 1, 2, ..., n)$

$$\omega_j = \frac{1 - e_j}{n - \sum_{j=1}^{n} e_j} \tag{52.10}$$

where $e_j = e(C_j)$ is the entropy of criteria C_j, for all $j = 1, 2, ..., n$.

Step 2. Determine the Picture Positive Ideal (PPI): $A^+ = \{(C_j, \mu_{ij}^+, \eta_{ij}^+, \nu_{ij}^+) | C_j \in C\}$.

$$\text{where } \mu_{ij}^+ = \begin{cases} \max\limits_{i=1,2,...,m} \{\mu_{ij}\} & \text{if } C_j \text{ is a benifite criterion} \\ \min\limits_{i=1,2,...,m} \{\mu_{ij}\} & \text{if } C_j \text{ is a non-benifite criterion} \end{cases}, \tag{52.11}$$

$$\eta_{ij}^+ = \begin{cases} \min\left\{ \max_{i=1,2,...,m} \{\eta_{ij}\}, 1 - \mu_{ij}^+ - v_{ij}^+ \right\} & \text{if } C_j \text{ is a benifite criteria} \\ \min_{i=1,2,...,m} \{\eta_{ij}\} & \text{if } C_j \text{ is a non-benifite criteria} \end{cases}$$

$$(52.12)$$

$$v_{ij}^+ = \begin{cases} \min_{i=1,2,...,m} \{v_{ij}\} & \text{if } C_j \text{ is a benifite criterion} \\ \max_{i=1,2,...,m} \{v_{ij}\} & \text{if } C_j \text{ is a non-benifite criterion} \end{cases}$$

$$(52.13)$$

+ Picture Negative Ideal (PNI): $A^- = \left\{ \left(C_j, \mu_{ij}^-, \eta_{ij}^-, v_{ij}^- \right) | C_j \in C \right\}$.

where

$$\mu_{ij}^- = \begin{cases} \min_{i=1,2,...,m} \{\mu_{ij}\} & \text{if } C_j \text{ is a benifite criterion} \\ \max_{i=1,2,...,m} \{\mu_{ij}\} & \text{if } C_j \text{ is a non-benifite criterion} \end{cases}$$

$$(52.14)$$

$$\eta_{ij}^- = \begin{cases} \min_{i=1,2,...,m} \{\eta_{ij}\} & \text{if } C_j \text{ is a benifite criteria} \\ \min\left\{ \max_{i=1,2,...,m} \{\eta_{ij}\}, 1 - \mu_{ij}^- - v_{ij}^- \right\} & \text{if } C_j \text{ is a non-benifite criteria} \end{cases}$$

$$(52.15)$$

$$v_{ij}^- = \begin{cases} \max_{i=1,2,...,m} \{v_{ij}\} & \text{if } C_j \text{ is a benifite criterion} \\ \min_{i=1,2,...,m} \{v_{ij}\} & \text{if } C_j \text{ is a non-benifite criterion} \end{cases}$$

$$(52.16)$$

Step 3. Compute distances $d\left(A_i, A^+\right)$, $d\left(A_i, A^-\right)$ from A_i to A^+, A^-, respectively, for all $j = 1, 2, ..., n$.

Step 4. Determine the closer coefficient of each alternative A_i, for all $i = 1, 2, ..., m$ as

$$CC_i = \frac{d\left(A_i, A^-\right)}{d(A_i, A^+) + d(A_i, A^-)} \tag{52.17}$$

Step 5. Ranking $A_i \succ A_k$ if $CC_i > CC_k$ for all $i, k = 1, 2, ..., m$.

Next, we apply the proposed model with four entropies and distances to select COVID-19 medicine. Data is collected in [26].

Example 1 There are a few drugs such as Remdesivir (A_1), Convalescent plasma (A_2), Tocilizumab (A_3), and Hydroxychloroquine (A_4) being used in a specified subgroup of patients. Doctors in a hospital have selected four major factors as follows: Antiviral activity (C_1), Coolify (C_2), Ease breathing (C_3), and Side effect (C_4) as criteria for evaluation of performance in therapies. Table 52.1 gives us the PFS relation of drugs and four major factors.

Table 52.1 PFS decision matrix

	C_1	C_2	C_3	C_4
A_1	(0.21, 0.48, 0.3)	(0.36, 0.23, 0.35)	(0.33, 0.35, 0.2)	(0.32, 0.34, 0.3)
A_2	(0.25, 0.4, 0.25)	(0.22, 0.35, 0.33)	(0.26, 0.23, 0.45)	(0.39, 0.26, 0.32)
A_3	(0.23, 0.33, 0.31)	(0.61, 0.22, 0.17)	(0.4, 0.1, 0.3)	(0.4, 0.3, 0.3)
A_4	(0.58, 0.13, 0.28)	(0.1, 0.2, 0.6)	(0.1, 0.41, 0.45)	(0.2, 0.3, 0.2)

Step 1. The weight vector of $C = \{C_1, C_2, C_3, C_4\}$, in this case, is $\omega^1 = (0.2428, 0.2872, 0.2663, 0.2037)$ *by using* the Eqs. (52. 2) and (52.10).

Step 2. Using the Eqs. (52.11)–(52.16), we get Picture Positive Ideal (PPI) A^+ and Picture Negative Ideal (PNI) A^- as given in Table 52.2.

Step 3. Using the normal Hamming distances to compute distances $d(A_i, A^+)$, $d(A_i, A^-)$ from A_i to A^+, A^-, respectively, for all $i = 1, 2, \ldots, m$, (see Table 52.3).

Step 4. Using the Eq. (52.17) to determine the closer coefficient of each A_i to A^+, A^-, respectively, for all $i = 1, 2, \ldots, m$ (see Table 52.4).

Step 5. The ranking result is. $A_3 \succ A_1 \succ A_2 \succ A_4$ (see in Table 52.3), i.e., Tocilizumab (A_3) is applicable for a particular patient while processing the treatment.

Table 52.2 PPI A^+ and NPI A^-

	C_1	C_2	C_3	C_4
A^+	(0.58, 0.13, 0.25)	(0.61, 0.2, 0.17)	(0.4, 0.1, 0.3)	(0.4, 0.26, 0.2)
A^-	(0.21, 0.13, 0.31)	(0.1, 0.2, 0.6)	(0.1, 0.1, 0.45)	(0.2, 0.26, 0.32)

Table 52.3 Closer coefficient of A_i and ranking using hamming distance

	A_1	A_2	A_3	A_4
CC_i	0.5483	0.4208	0.6539	0.3679
Ranking	2	3	1	4

Table 52.4 Comparison of the ranking using other measures with the weight vector ω^1

Measures	Ranking	Measures	Ranking
d_1 in the Eq. (52.1)	$A_3 \succ A_1 \succ A_2 \succ A_4$	SCC_1 in the Eq. (52.6)	$A_3 \succ A_1 \succ A_2 \succ A_4$
S_T in the Eq. (52.3)	$A_3 \succ A_2 \succ A_1 \succ A_4$	SCT_1 in the Eq. (52.7)	$A_3 \succ A_1 \succ A_2 \succ A_4$
SC_1 in the Eq. (52.4)	$A_3 \succ A_1 \succ A_2 \succ A_4$	div in the Eq. (52.8)	$A_3 \succ A_2 \succ A_1 \succ A_4$
SC_2 in the Eq. (52.5)	$A_3 \succ A_1 \succ A_2 \succ A_4$	cc in the Eq. (52.9)	$A_3 \succ A_1 \succ A_2 \succ A_4$

52.4   Conclusions

This paper has considered an entropy measure based on the normal Hamming distance measures of picture fuzzy sets. It is indicated that mostly based on four membership-functions of picture fuzzy sets: positive, neutral, negative, and refusal membership functions. The meaningful of these measures are expressed by the TOPSIS model built on these measures to apply to solve the COVID-19 medicine selection problem in fuzzy pairs of knowledge graph [29]. The result of using the new measurement has been as good as the majority of previous measurements, even solving some cases that some previous measures, such as the classification problem as in Example 1. It is then applied to compute the weights of the criteria in the construction of the TOPSIS model to select COVID-19 medicine. In the future, we will investigate this research approach to some types of fuzzy set as complex fuzzy set [6], or Group Decision Support [30, 31], etc.

Acknowledgements This work was supported by the University of Economics Ho Chi Minh City (UEH), Vietnam under project CS-2021-51.

References

1. Cuong, B.C.: Picture fuzzy sets. J. Comput. Sci. Cybern. **30**(4), 409–420 (2014)
2. Atanassov, K.T.: Intuitionistic fuzzy sets. Fuzzy Sets Syst. **20**(1), 87–96 (1986)
3. Ganie, A.H., Singh, S., Bhatia, P.K.: Some new correlation coefficients of picture fuzzy sets with applications. Neural Comput. Appl. 1–17 (2020). https://doi.org/10.1007/s00521-020-04715-y
4. Hung, W.L., Yang, M.S.: Fuzzy entropy on intuitionistic fuzzy sets. Int. J. Intell. Syst. **21**(4), 443–451 (2006)
5. Son, L.H.: Generalized picture distance measure and applications to picture fuzzy clustering. Appl. Soft Comput. **46**(C), 284–295 (2016)
6. Hong Lan, L.T., et al.: A new complex fuzzy inference system with fuzzy knowledge graph and extensions in decision making. IEEE Access **8**, 164899–164921 (2020). https://doi.org/10.1109/ACCESS.2020.3021097
7. Ateş, F., Akay, D.: Some picture fuzzy Bonferroni mean operators with their application to multicriteria decision making. Int. J. Intell. Syst. **35**(4), 625–649 (2020)
8. Garg, H.: Some picture fuzzy aggregation operators and their applications to multicriteria decision-making. Arab. J. Sci. Eng. **42**(12), 5275–5290 (2017)
9. Jana, C., Senapati, T., Pal, M., Yager, R.R.: Picture fuzzy Dombi aggregation operators: Application to MADM process. Appl. Soft Comput. **74**, 99–109 (2019)
10. Le, N.T., Van Nguyen, D., Ngoc, C.M., Nguyen, T.X.: New dissimilarity measures on picture fuzzy sets and applications. J. Comput. Sci. Cybern. **34**(3), 219–231 (2018)
11. Liu, P., Zhang, X.: A novel picture fuzzy linguistic aggregation operator and its application to group decision-making. Cogn. Comput. **10**(2), 242–259 (2018)
12. Wei, G.: Picture fuzzy aggregation operators and their application to multiple attribute decision making. J. Intell. Fuzzy Syst. **33**(2), 713–724 (2017)
13. Wei, G.: Picture fuzzy Hamacher aggregation operators and their application to multiple attribute decision making. Fundamental Informatic **157**(3), 271–320 (2018)

14. Xu, Y., Shang, X., Wang, J., Zhang, R., Li, W., Xing, Y.: A method to multi-attribute decision making with picture fuzzy information based on Muirhead mean. J. Intell. Fuzzy Syst. **36**(4), 3833–3849 (2019)

15. Phong, P.H., Cuong, B.C.: Multi-criteria group decision making with picture linguistic numbers. VNU J. Sci. Comput. Sci. Commun. Eng. **32**(3), 39–53 (2017)

16. Peng, X., Dai, J.: Algorithm for picture fuzzy multiple attribute decision-making based on new distance measure. Int. J. Uncertain. Quantif. **7**(2), 177–187 (2017)

17. Thao, N.X.: Similarity measures of picture fuzzy sets based on entropy and their application in MCDM. Pattern Anal. Appl. 1–11 (2019). https://doi.org/10.1007/s10044-019-00861-9

18. Van Dinh, N., Thao, N.X., Chau, M.N.: Distance and dissimilarity measure of picture fuzzy sets. FAIR'10 104–109 (2017)

19. Van Dinh, N., Thao, N.X.: Some measures of picture fuzzy sets and their application in multi-attribute decision making. Int. J. Math. Sci. Comput. (IJMSC) **4**(3), 23–41 (2018)

20. Wei, G.: Some cosine similarity measures for picture fuzzy sets and their applications to strategic decision making. Informatica **28**(3), 547–564 (2017)

21. Wei, G.: Some similarity measures for picture fuzzy sets and their applications. Iran. J. Fuzzy Syst. **15**(1), 77–89 (2018)

22. Wei, G.W., Gao, H.: The generalized Dice similarity measures for picture fuzzy sets and their applications. Informatica **29**(1), 107–124 (2018)

23. Thao, N.X., Ali, M., Nhung, L.T., Gianey, H.K., Smarandache, F.: A new multi-criteria decision-making algorithm for medical diagnosis and classification problems using divergence measure of picture fuzzy sets. J. Intell. Fuzzy Syst. **37**(6), 7785–7796 (2019)

24. Fan, J., Xie, W.: Distance measure and induced fuzzy entropy. Fuzzy Sets Syst. **104**(2), 305–314 (1999)

25. Selvachandran, G., Maji, P.K., Faisal, R.Q., Salleh, A.R.: Distance and distance induced intuitionistic entropy of generalized intuitionistic fuzzy soft sets. Appl. Intell. **47**(1), 132–147 (2017)

26. Si, A., Das, S., Kar, S.: Picture fuzzy set-based decision-making approach using Dempster–Shafer theory of evidence and grey relation analysis and its application in COVID-19 medicine selection. Soft Comput. 1–15 (2021)

27. Dong, L.D., Nguyen, V.T., Le, D.T., Tiep, M.V., Hien, V.T., Huy, P.P., Hieu, P.T.: Modeling transmission rate of COVID-19 in regional countries to forecast newly infected cases in a nation by the deep learning method. In: International Conference on Future Data and Security Engineering, pp. 411–423. Springer, Singapore (2021, November)

28. Singh, P.: Correlation coefficients for picture fuzzy sets. J. Intell. Fuzzy Syst. **28**(2), 591–604 (2015)

29. Long, C.K., Van Hai, P., Tuan, T.M. et al.: A novel fuzzy knowledge graph pairs approach in decision making. Multimed. Tools Appl. **81**, 26505–26534 (2022). https://doi.org/10.1007/s11 042-022-13067-9

30. Van Pham, H., Moore, P., Tran K.D.: Context matching with reasoning and decision support using hedge algebra with Kansei evaluation. In: Proceedings of the Fifth Symposium on Information and Communication Technology (SoICT '14), pp 202–210. Association for Computing Machinery, New York, NY, USA (2014)

31. Van Pham, H., Khoa, N.D., Bui, T.T.H., Giang, N.T.H., Moore, P.: Applied picture fuzzy sets for group decision-support in the evaluation of pedagogic systems. Int. J. Math. Eng. Manage. Sci. Dehradun **7**(2), 243–257 (2022). https://doi.org/10.33889/IJMEMS.2022.7.2.016

Chapter 53
Analysis of Blood Smear Images Using Dark Contrast Algorithm and Morphological Filters

Sparshi Gupta, Vikrant Bhateja, Siddharth Verma, Sourabh Singh, Zaid Omar, and Chakchai So-In

Abstract In recent years, Biomedical Imaging has emerged as an effective tool in diagnosis of various diseases. In order to perform anatomy or histology of cells, Blood Smear Images are used. To process these images, enhancement plays a major role in order to increase visual quality of the image and for accurate segmentation of Region of Interest (ROI). The motive of this work is to perform enhancement using the Dark Contrast Algorithm (DCA) since it increases the intensity of darker regions, which in case of Blood Smear Images are nucleus. Further, the quality of enhanced image is evaluated using suitable Image Quality Assessment (IQA) metric. This enhanced image is segmented using Morphological Filters with appropriate structuring element to extract ROI which is nucleus and cell periphery. This helps to identify irregularities in cell periphery to detect various blood disorders. The performance of segmentation technique is assessed using Jaccard Coefficient (JC).

53.1 Introduction

Abnormalities in count and shape of leukocytes commonly known as White Blood Cells (WBCs) help to identify various disorders. The conventional methods to detect various blood disorders and inflammation include bone marrow biopsy, manual

S. Gupta · V. Bhateja (✉) · S. Verma · S. Singh
Department of Electronics and Communication Engineering, Shri Ramswaroop Memorial College of Engineering and Management, Faizabad Road, Lucknow, U.P 226028, India
e-mail: bhateja.vikrant@gmail.com

Dr. A.P.J. Abdul, Kalam Technical University, Lucknow, U.P, India

Z. Omar
School of Electrical Engineering, Universiti Teknologi Malaysia, UTM Johor, 81310 Bahru, Malaysia
e-mail: zaidomar@utm.my

C. So-In
Department of Computer Science, College of Computing, Khon Kaen Universty, 123 Midtaparb Rd., Naimuang, Khon Kean 40002, Muang, Thailand
e-mail: chakso@kku.ac.th

© The Author(s), under exclusive license to Springer Nature Singapore Pte Ltd. 2023
V. Bhateja et al. (eds.), *Evolution in Computational Intelligence*, Smart Innovation, Systems and Technologies 326, https://doi.org/10.1007/978-981-19-7513-4_53

counting by hematologists, and through laser machines [1]. These methods have their fair share of limitations like inaccurate cell count, time taking results, and unobtainable machinery. In order to rectify these problems, Computer Aided Diagnosis method can be used which includes analysis of Blood Smear Images. These images are obtained by mounting digital camera on microscope, either optical or electron, used for observing blood slides. These Blood Smear Images are processed to extract cells characteristics to categorize them into various diseases [2]. An approach of nucleus segmentation by Dark Contrast Algorithm (DCA) and Median Filter by Harun et al. [3] enhances the leukocytes in Blood Smear Images by DCA. DCA stretches the contrast of darker regions which in Blood Smear Images is nucleus and improves its visibility. This approach is applied to different malignant cell images. Here, the segmentation is said to be done in the same process using various DCA parameters. Rahman et al. [4] proposed a method to bifurcate between Red Blood Cells (RBCs) and WBCs using Balance Contrast Equalization Technique for enhancement of Blood Smear Images and logical operators for segmentation. The use of logical operator XOR for segmentation segments RBCs in the images to predict various diseases. This approach uses ensemble learning and group of classifiers for classification and gave high accuracy to differentiate the cells. A similar approach to categorize different types of WBCs is discussed in [5] where tissue quant method is used as enhancement technique while segmentation is done using Morphological Operators. For primary steps here, the RGB image is converted to GGB with enhanced green channel. The tissue quant method here converts the enhanced GGB image to grayscale. This enhanced image is further passed through Morphological disk operator which is used to segment the nucleus. This technique of enhancement and segmentation was efficient to categorize the WBCs. An analysis of Blood Smear Images for disease detection by Acharya et al. [6] focuses on segmentation. Although this technique further used Gaussian Filter for noise removal after primary classification. This approach justified preprocessing as an important step in image analysis. Classification and disease detection done by Dasariraju et al. [7] use thresholding for segmentation of cell components without prior enhancement. This technique uses shape features to characterize the cell types and Random Forest Algorithm for classification. The accuracy of this technique ranged from medium to high. These approaches conclude that enhancement preceded by segmentation gives best results in order to identify cell components Experimentally, the accuracy of Morphological Operators for segmentation is quite high. The approach used in this paper uses DCA for contrast enhancement as it works on darker regions of the image and Morphological Filter for segmentation of nucleus and cell periphery. The organization of the paper in the following parts is as follows: Sect. 53.2 contains the proposed approach of enhancement for analysis of Blood Smear Images, Sect. 53.3 contains proposed approach for segmentation, Sect. 53.4 contains Experimental Results, and Sect. 53.5 consists of Conclusion.

53.2 Proposed Enhancement Approach Using Dark Contrast Algorithm (DCA)

In order to perform microscopic anatomy/histology on these Blood Smear Images, the proposed approach focuses on contrast enhancement of these images using DCA. DCA works on contrast stretching which is one of the primary techniques of image enhancement. The enhancement algorithm works in a way that it enhances the area occupied by the leukocytes and their nucleus which are in foreground while compressing the erythrocytes present in background. This enhancement and compression of pixel intensity is dependent on the parameters of DCA, which are Threshold Value (TH) and Dark Stretching Factor (NTH). For DCA, the value of TH is less than that of NTH. The intensity of a pixel varies between the range 0 and 255, where 0 represents the lowest intensity or black and 255 represents brightest intensity or white. The control function on which DCA works is given in Eq. (53.1) [3].

$$O\ (x, y) = \begin{cases} \frac{I(x,y)-\min\_I}{TH-\min\_I} \cdot \text{NTH} & \text{if } I(x, y) < \text{TH} \\ \left[\frac{I(x,y)-\text{TH}}{\max\_I-\text{TH}} \cdot (255 - \text{NTH})\right] + \text{NTH} & \text{if } I(x, y) \geq \text{TH} \end{cases} \tag{53.1}$$

Here, $I(x, y)$ and $O(x, y)$ represent the input and output pixel, respectively, at coordinates x and y. The maximum and minimum intensity of input image are represented by $\max\_I$ and $\min\_I$, respectively. The first case of Eq. (53.1) gives the function for intensity stretching process while the second case gives the function for intensity compressing process. The quality of this enhanced image is assessed on parameters like Measure of Enhancement (EME) [8] and Logarithmic Michelson Contrast Measure (AME). These parameters evaluate image quality by assessing brightness and contrast of the enhanced image and comparing them with that of the input image [9].

53.3 Proposed Segmentation Approach

Morphological Filters use different dimensional shapes such as disk, diamond, line, square, etc., to process the image. The mathematical representation of Morphological Operations is expressed in Eq. (53.2) [10].

$$O\ (x, y) = f\ \{a_1, a_2, a_3, \ldots, a_k\} \tag{53.2}$$

where $O\ (x, y)$ is the output pixel at co-ordinates (x, y), k is the number of pixels in structuring element, and f is any logical operation performed by the structuring element. The matrix which has the pixel to process the image is known as structuring element. In the proposed approach, disk is being taken as the structuring element.

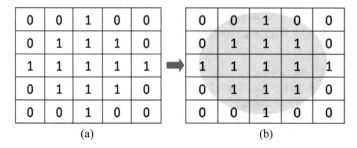

Fig. 53.1 Matrix showing **a** disk structuring element of size 2 and **b** visualization of size 2 disk structure

The structuring element works as the reference pixel to its neighbors in the process of computation of any operation [11]. In the proposed work, disk-shaped structuring element is used to segment both nucleus and cell periphery. Figure 53.1 shows the matrix and visualization of disk-shaped structuring element.

Erosion and Dilation are two methods on which Morphological Operators work. These when performed simultaneously on any image give Opening or Closing mask [12]. Erosion followed by Dilation gives the Morphological Closing mask while Dilation followed by Erosion gives the Morphological Opening mask. The mathematical representation of Morphological Closing Mask is given by Eq. (53.3).

$$O\ (x,\ y) = \text{ERODE}_{SE}\ [\text{DILATE}_{SE}\ \{A\ (x,\ y)\}] \tag{53.3}$$

where $A(x, y)$ is the input pixel and ERODE_{SE} and DILATE_{SE} refer to Erosion and Dilation around the same structuring element. Morphological Closing helps in filling lines and removing unnecessary holes in the binary image [13]. In the proposed approach, Morphological Closing mask is used with disk structuring element as mentioned above with a size of 2 to segment both the nucleus and size 10 to segment cell periphery. The flowchart given as Fig. 53.2 shows the proposed approach of enhancement and segmentation for analysis of Blood Smear Images.

53.4 Experimental Results

53.4.1 Dataset

The dataset of Blood Smear Images used for this experiment is obtained from online database of American Society of Hematology (ASH) [6], for multi-cell images and from Cancer Imaging Archive [7], for single-cell images. These images are acquired by a digital camera mounted on optical microscope. The dimensions for multi-cell images are 720×960 and that of single-cell images are 400×400. Two images were chosen at random, one from each dataset to analyze the results of the

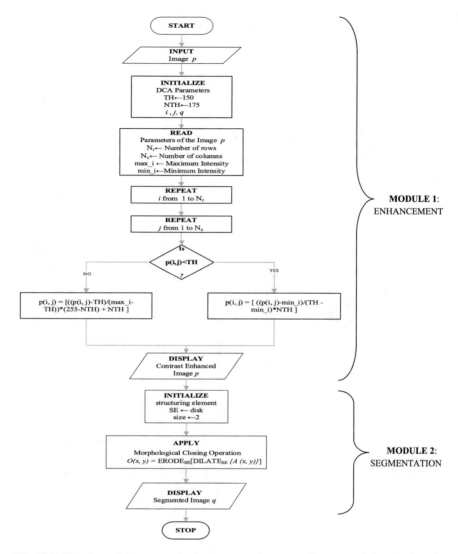

Fig. 53.2 Flowchart of the proposed enhancement and segmentation approach for Blood Smear Images

proposed approach. As a preprocessing step, these images were converted from RGB to grayscale in order to proceed with the proposed approach.

53.4.2   Simulation Results and Analysis of Enhancement Approach

Two random images chosen from the dataset were subjected to the enhancement approach using DCA. The effectiveness of the DCA enhanced image was compared with the original image on the basis of parameters like Measure of Enhancement (EME) and Logarithmic Michelson Contrast Measure (AME) [8, 9]. The results from both the images are shown in Fig. 53.3.

The visual quality of images (c) and (f) is much better than that of (b) and (e) as shown in above Fig. 53.3. We can observe that both the nucleus and cell periphery are more visible in images (c) and (f). The Image Quality Assessment (IQA) metrics of both these images are given in Table 53.1.

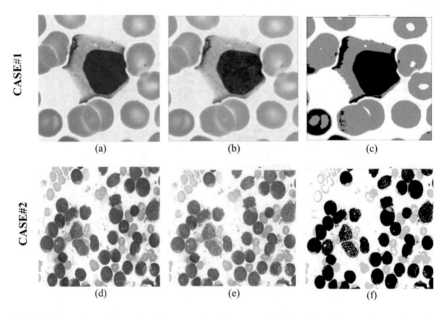

Fig. 53.3 Experimental results of images showing: **a** and **d** input Blood Smear Image, **b** and **e** grayscale converted image and **c** and **f** DCA enhanced image

Table 53.1 IQA metrics for input and enhanced images

Metric	Single-cell image (CASE#1)		Multi-cell image (CASE#2)	
	Input image	Enhanced image	Input image	Enhanced image
EME	5.68	6.074	4.455	5.047
AME	6.087	8.856	7.087	8.009

Observing the table, it is clear that since the value of EME and AME are greater for the enhanced image when compared with the input image, the enhancement technique is effective. The enhancement can be justified both visually and qualitatively.

53.4.3   Simulation Results and Analysis of Segmentation Approach

The enhanced images are further subjected to Morphological Filters [10]. The effectiveness of the segmentation technique is calculated using parameters like Jaccard Coefficient (JC) [14]. The binary masks of nucleus and cell periphery after segmentation are shown in Fig. 53.4.

The segmented binary mask images show that the segmentation is successful for both nucleus and cell periphery. The granules present in images (b) and (e) and the irregularities of cell periphery in images (c) and (f) can be further used to describe blood disorders. To further check the segmentation, these images are assessed qualitatively. The performance of segmentation technique is done using Jaccard coefficient (JC) as given in Table 53.2.

Experimentally observing the value of JC is nearing to 1 in both cases which is a clear indication that segmentation is done successfully. So, visually and qualitatively the performance of segmentation technique is evaluated. The segmentation technique,

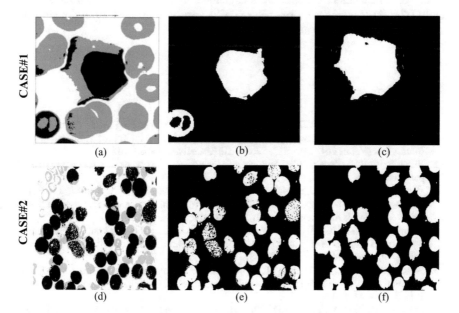

Fig. 53.4 Experimental results of images showing: **a** and **d** enhanced Blood Smear Image, **b** and **e** binary mask of nucleus and **c** and **f** binary mask of cell periphery

Table 53.2 Performance evaluation of segmentation technique

	Single-cell image (CASE#1)		Multi-cell image (CASE#2)	
Metric	Nucleus	Cell periphery	Nucleus	Cell periphery
JC	0.85	0.81	0.92	0.896

therefore, is able to segment the ROI. As observed from both Tables 53.1 and 53.2, the enhancement and segmentation are not only assessed visually but also qualitatively and both gave significant results.

53.5   Conclusion

This paper presents an approach to enhance and segment Blood Smear Images to obtain ROI. The DCA used in enhancement fulfills the motive of enhancing the nucleus in Blood Smear Images. The IQA metrics of the enhanced image also show favorable results. With the nucleus being enhanced which is our ROI, it is easy to segment it from the enhanced image. Morphological Closing with disk as structuring element when varied from a certain radius of 2–10 gives us the binary mask of nucleus and cell periphery. The size for nucleus being 2 and size for cell periphery being 10. This segmented image can be further used to perform microanatomy. This approach of image enhancement using DCA is dependent on selection of its parameters TH and NTH. Further, the performance of DCA can be improved by optimizing the algorithm to get the best values of its parameters.

References

1. Jagadev, P., Virani, H.G.: Detection of leukemia and its types using image processing and machine learning. In: Proceedings of International Conference on Trends in Electronics and Informatics (ICTEI 2017), pp. 522–526. IEEE, Tirunelveli, India (2018)
2. Rezatofighi, S.H., Zadeh, H.S.: Automatic recognition of five types of white blood cells in peripheral blood. Comput. Med. Imaging Graph. 35(4), 333–343 (2011)
3. Harun, N.H., Bakar, J.A., Hambali, H.A., Khair, N.M., Mashor, M.Y., Hassan, R.: Fusion noise—removal technique with modified algorithm for robust segmentation of acute leukemia cell images. Int. J. Adv. Intell. Inf. 4(3), 202–211 (November 2018)
4. Rahman, S., Azam, B., Khan, S.U., Awais, M., Ali, I., Khan, R.J.H.: Automatic identification of abnormal blood Smear Images using color and morphology variation of RBCS and central pallor. Comput. Med. Imaging Graph. 87, 101813 (2021)
5. Hegde, R.B., Prasad, K., Hebbar, H., Singh, B.M.K.: Development of a robust algorithm for detection of nuclei and classification of white blood cells in peripheral blood Smear Images. J. Med. Syst. 42, 110 (2018)
6. Acharya, V., Ravi, V., Pham, T.D., Chakraborty, C.: Peripheral blood smear analysis using automated computer-aided diagnosis system to identify Acute myeloid leukemia. IEEE Trans. Eng. Manage. 1–14 (2021)

7. Dasariraju, S., Huo, M., McCalla, S.: Detection and classification of immature leukocytes for diagnosis of acute myeloid leukemia using random forest algorithm. Bioengineering **7**(4), 120–131 (2020)
8. Trivedi, M., Jaiswal, A., Bhateja, V.: A No-reference image quality index for contrast and sharpness measurement. In: 3rd IEEE International Advance Computing Conference (IACC), pp. 1234–1239. IEEE, India (2013)
9. Prajapati, P., Narmawala, Z., Darji, N.P., Moorthi, S.M., Ramakrishnan, R.: Evaluation of perceptual contrast and sharpness measures for meteorological satellite images. In: Soni, A.K., Lobiyal, D.K. (eds) 3rd International Conference on Recent Trends in Computing (ICRTC), Procedia Computer Science, vol. 57, pp. 17–24. Springer, India (2015)
10. Batchelor, B.G., Waltz, F.M.: Computer vision handbook, 1st edn. Springer, USA (2011)
11. Bhateja, V., Nigam, M., Bhadauria, A.S., Arya, A., Zhang, E.D.: Human visual system based optimized mathematical morphology approach for enhancement of brain MR images. J. Ambient Intell. Human Comput. (2019)
12. Srivastava, A., Raj, A., Bhateja, V.: Combination of wavelet transform and morphological filtering for enhancement of magnetic resonance images. In: Snasel, V., Platos, J., El-Qawasmeh, E. (eds.) Digital Information Processing and Communications. ICDIPC 2011, Part I Communications in Computer and Information Science, vol. 460–474, pp. 460–474. Springer, Heidleberg (2011)
13. Bhateja, V., Urooj, S., Mehrotra, R., Verma, R., Lay-Ekuakilli, A., Verma, V.D.: A composite wavelets and morphology approach for ECG noise filtering. In: Maji, P., Ghosh, A., Murty, M.N., Ghosh, K., Pal, S.K. (eds.) Pattern Recognition and Machine Intelligence. International Conference on Pattern Recognition and Machine Intelligence, vol. 8251, pp. 361–366. Springer, Heidelberg (2013)
14. Kumar, S.N., Lenin Fred, A., Ajay Kumar, H., Sebastin Varghese, P.: Performance metric evaluation of segmentation algorithms for gold standard medical images. In: Sa, P., Bakshi, S., Hatzilygeroudis, I., Sahoo, M. (eds.) Recent Findings in Intelligent Computing Techniques. Advances in Intelligent Systems and Computing, vol. 709. Springer, Singapore (2018)

Chapter 54
Automatic Detection of Diabetic Eye Disease Using Convolutional Neural Network

M. Shanmuga Sundari⊙, Ch Deekshitha, V. Esthar Rani, and D. SriChandana

Abstract Diabetes is a chronic health disease where the pancreas fails to produce insulin, or when the body cannot make good use of the insulin it produces. It affects the eyes that induce other problems like diabetic retinopathy, diabetic macular edema, cataracts, and glaucoma. We use Convolutional Neural Network (CNN), a branch of deep learning in the AI/ML domain to help us detect and predict vision-loss problems with the most effective machine learning methods. We deal this with the screening of Diabetic Retinopathy (DR) and its stages of severity by using the color fundus retinal photography as input. The deep CNN model has been proposed to analyze the fundus images and automatically distinguish between controls no DR, mild DR, moderate DR, severe DR, and proliferative DR. This detection results in early treatment and helps the patient from suffering from blindness.

54.1 Introduction

Diabetic Retinopathy (DR) is a disease developed by the injury of blood vessels in the tissue at the back of the eye, i.e., retina, also called Diabetic Eye Disease (DED). Diabetic Retinopathy is a severe and most common disease all over the world. DED is a disease in which a human body fails to produce insulin or does not produce sufficient insulin. The World Health Organization (WHO) has declared that 347 million people are affected by diabetes and this count might increase to 552 million over the year 2030.

M. Shanmuga Sundari (✉) · C. Deekshitha · V. E. Rani · D. SriChandana
BVRIT Hyderabad College of Engineering for Women, Bachupally, Hyderabad, India
e-mail: sundari.m@bvrithyderabad.edu.in

C. Deekshitha
e-mail: 18wh1a05b8@bvrithyderabad.edu.in

V. E. Rani
e-mail: 18wh1a0571@bvrithyderabad.edu.in

D. SriChandana
e-mail: 18wh1a05c0@bvrithyderabad.edu.in

Diabetic Retinopathy also can show no symptoms or sometimes only mild vision (sight) problems which may also lead to blindness. The compilation of eye that affects about 75% of diabetic patients which leads to vision loss within the age group of 20–64. Using automated methods such as support vector machines and K-NN classifiers, significant effort has been done to recognize the features of DR. The majority of these classification techniques are on two-class classification for DR and no DR. This condition can develop in anyone who has type 1 or type 2 diabetes and a patient is more likely to have eye complications if he affects by diabetes for a long time and has less control on blood sugar level.

The symptoms of Diabetic Retinopathy may include the following:

Spots or dark strings floating in the vision (floaters)
Blur vision
Fluctuating vision
Dark of empty areas in vision
Vision loss.

Convolutional Neural Network (CNN) approach is used for automating Diabetic Retinopathy using the color fundus images (retinal) in earlier stage of detection. CNNs will apply individually to solve certain problems. CNN is the method of deep learning which has a magnificent record of advantages for image analysis and interpretation. CNN is considered one of the important methods to fix an automatic analysis of retinal images. It is used for the classification of retinal injury to an appropriate degree and also to extract the features of retinal fundus images. A human eye retina is the main cause to analyze the light and give the signals to the brain. According to the International Clinical Diabetic Retinopathy disease severity scale, the existence of Diabetic Retinopathy is categorized into five divisions which consist of (0) None, (1) Mild, (2) Moderate, (3) Severe (Extreme), and (4) Proliferative. There are mainly two stages of Diabetic Retinopathy: (i) Early Diabetic Retinopathy (Early DR) and (ii) Advanced Diabetic Retinopathy (Advanced DR). In Early Diabetic Retinopathy, new blood vessels will not develop the Proliferating and it is also known as Non-Proliferative Diabetic Retinopathy (NPDR). NPDR is mainly divided into three types: (1) Mild, (2) Moderate, and (3) Severe. Advanced Diabetic Retinopathy is termed as Proliferative Diabetic Retinopathy in which damaged blood vessels will leak the transparent jelly-like fluid, and it fills the center of the eye.

54.2   Literature Survey

Diabetic Retinopathy [1] is a leading cause of blindness among working-age adults. In this study, it is shown how to recognize the stages of Diabetic Retinopathy using a Convolutional Neural Network using color fundus images. Diabetic Eye Disease (DED) [2] is a collection of eye problems that affect diabetic people. The retinal fundus image plays a vital role in the early DED classification and identification. The quality and amount of the fundus picture are the most important factors in this classification and identification model. The proposed DED automatic categorization framework was developed in multiple steps, including (i) Enhancement of image quality (ii) Image Augmentation (iii) Image Classification, and (iv) Image Segmentation, designed with CNN produced the best results.

Diabetic Eye Disease is a set of ocular issues [3] that affect diabetic people. Screening on time increases the odds of prompt treatment and prevents permanent vision impairment [4]. This study focused on patient situations to develop an automated classification system based on two parameters: (1) Mild multi-class Diabetic Eye Disease (DED) and (2) Multi-class DED. Diabetic Retinopathy [5] is caused due to damaged blood vessels of the eye. This disease in turn causes blurred vision [6], sometimes also leading to blindness. DR mainly consists of five stages.

Diabetic Retinopathy is a rapidly developing condition caused by diabetes that affects people all over the world. This DR [7] offered a technique for automatically identifying and grading DR. This work provides a thorough examination of the identification of DR in three areas: retinal datasets, DR detection algorithms, and performance evaluation metrics. DR is an incurable eye disease that develops as a result of long-term diabetes mellitus [8]. It causes blindness and impairs the veins in the retina. Using the data gathered during the eye exam, the idea of ensemble learning is applied to forecast the blindness in retinal problems in the eyes. The process [9] has to mine properly to get the activity flow. The classification method [10] in CNN can also predict brain tumor problems also.

54.3   Proposed system

54.3.1   Dataset

There are various publicly available retina datasets for DR detection and vasculature [11], commonly used for training and testing the model and comparing performance metrics. Fundus color pictures and Optical Coherence Tomography are two methods of retinal imaging (OCT) [12]. The OCT images are two- and three-dimensional retina datasets obtained with low cohesiveness light and provide a lot of information on the thickness and structure of the retina; however, the fundus images are two-dimensional retain datasets collected with reflected light (Fig. 54.1).

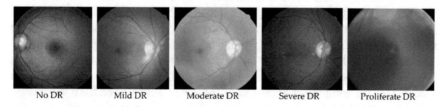

No DR   Mild DR   Moderate DR   Severe DR   Proliferate DR

Fig. 54.1 Dataset images

This work uses the Kaggle dataset [13] that contains 88,702 high-resolution images with various resolutions, ranging from 433 × 298 pixels to 5184 × 3456 pixels taken under a variety of imaging conditions. It is collected from different cameras, and all the images are classified into five DR stages. A clinician has rated the presence of diabetic retinopathy in each image on a scale of 0–4, according to the scale 0—No DR, 1—Mild, 2—Moderate, 3—Severe, and 4—Proliferative DR. A left and right filed is provided for every subject. Images labeled with a subject id with left or right (e.g., 3_right.jpg is the right eye of patient id 3).

54.3.2 Data Preprocessing

Data preprocessing [14] refers to the manipulation or dropping of attributes. Images show the various photo preprocessing steps for visualization Enhancement. If the images are brighter and cleaner, the community can extract more salient and particular functions.

In a maximum of the image preprocessing strategies, experienced channel extraction is used. The green channel images (photograph) will produce the extra records than blue and crimson channels. The other famous image preprocessing technique is "CONTRAST ENHANCEMENT." The application of assessment enhancement further improves the comparison on a green channel image. To improve the contrast of the image, assessment Enhancement is hired to the inexperienced channel of the image. Many DED datasets contain images with a black border, with researchers choosing to segment the useless black border to concentrate on the region of interest (ROI) [15].

Image augmentation is done if there is an imbalance in the images. In comparison to DED retina images [16], photos are mirrored, rotated, scaled, and cropped to make cases of the selected images for a class where the number of images is lower than the proportion of healthy retina images. Augmentation [17] is a frequent method for improving results and avoiding overfitting.

54.4 Architecture

A CNN model was developed in order to classify the five stages of the Drin deep learning model, CNN model is applied to the Kaggle dataset for DR classification. Firstly, we feed the model with an input color fundus image and then preprocess the input image. The preprocessing on the image starts under two phases, wherein the first stage, the ROI Detection (region of interest) is done. ROI Detection detects the affected part of the eye. In phase two, Image Enhancement is done where the affected area is highlighted and the rest of the image is eliminated from the picture. Now this preprocessed image goes through DR Classification where it involves the classification stages data and the CNN model to detect and predict the severity and

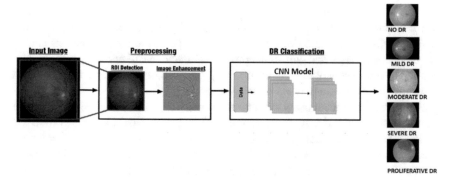

Fig. 54.2 Architecture of proposed system

stage of the Diabetic Retinopathy (DR). Here, the image received is examined and classified accordingly and gives us the output analyzing the severity of the disease. It involves five stages, namely No DR, Mild DR, Moderate DR, Severe DR, and Proliferative DR. So after examination gone through under DR classification stage, the disease detected will fall under any one of these categories. No DR depicts as the name itself suggests that there is no diabetic retinopathy detected in the image. Mild DR depicts that the patient might suffer from little or mild stages of DR which can be treated easily. Moderate DR depicts that the patient is suffering from an intermediate stage of the DR, and a cure can be done. Severe DR is a stage where the disease has reached a level more than the intermediate stage and has to be considered seriously. Proliferative DR is a stage where the disease has reached a higher level and immediate action has to be taken before the patient might go blind. So, these are the five stages that it classifies the image into and gives an accurate result (Fig. 54.2).

54.5   Implementation

In this project, a CNN architecture was proposed for the automatic detection of Diabetic Eye Disease. This model is used to predict the severity of the eye damage of a patient who is suffering from diabetes. The entire dataset is divided into train and test folders with 500 and 100 images, respectively, where each classification consists of 100 images. As the images are of high resolution, data augmentation is performed every image is resized to 224 × 224 pixels. These resized images are given as input to the model.

Every image is thoroughly observed for features that are used to classify the stages of Diabetic Retinopathy. Each image was randomly augmented with random rotation 0°–90° at each epoch. There are random yes or no horizontal and vertical flips, as well as random horizontal and vertical shifts.

The visibility of using fundus images to diagnose and categorize DR with a deep Convolutional Neural Network was proven in the given article. Compared to the existing CNN-based and SVM-based strategies reported in the literature as above, the suggested strategy had greater diagnostic accuracy, sensitivity, and specificity. In addition to this excellent diagnostic accuracy, the proposed technique can also have a great degree of accuracy when it comes to grading DR.

Even though the accuracy of validation is regarded as an appropriate performance measure because the dataset with mismatched category sizes was used to train the model. It may not accurately describe the performance of the suggested architecture. Another method for evaluating classifier performance and measuring agreement between two raters: The quadratic weighted kappa score is the difference between anticipated and ground truth labels. The score will range from -1 to $+1$, -1 score indicates the complete disagreement between forecasts and ground truth and a $+1$ score indicates the complete agreement between the two labels. The score can be zero if the agreement between labels occurred by chance. The quadratic weighted kappa score is then calculated using the five steps as listed below:

- Calculate the confusion matrix and then normalize it (C)
- Create the weights W matrix, in which predictions with bigger deviations with the actual label gives more weight. The following below formula (54.1) is used to calculate weights:

$$w(i, j) = \frac{(i - J)^2}{L - 1)^2} \tag{54.1}$$

- Create a histogram for both the actual and predicted labels vectors and then normalize it.
- Calculate and normalize the outer product (P) of the two histograms.
- The quadratic weighted kappa (K) is calculated as follows (54.2):

$$K = 1 - \frac{\sum_{i=0}^{L} \sum_{j=0}^{L} w(i, j) c(i, j)}{\sum_{i=0}^{L} \sum_{j=0}^{L} w(i, j) p(i, j)} \tag{54.2}$$

54.6 Result

The suggested model was evaluated on the Kaggle dataset with 80% of the retina dataset being used for training and the remaining 20% for testing. The studies acquired 90% of accuracy when it is tempted to believe that more data might acquire a good result. The signals used for classification lies in the part of the image which is visible to the person according to the visualization of the characteristics learned by CNN. Overall, we want to improve the eye illness diagnosis and migrate to more complex and advantageous multi-grade disease detection in the future (Fig. 54.3).

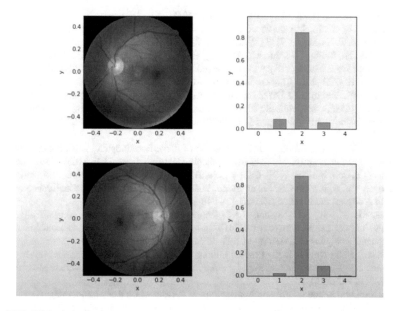

Fig. 54.3 Diabetic analysis

54.7   Conclusion

Our research has shown that a CNN method can use to solve the five-class challenge for countrywide DR screening. Our network has shown hints of learning the features needed to categorize fundus images, correctly classifying the majority of proliferative and non-DR cases. As the dataset research, great specificity has occurred at the expense of sensitivity. Without any feature-specific detection and utilizing a considerably larger dataset, our method provides accurate results.

Our trained CNN has the potential benefit of categorizing hundreds of photos for about a minute, which allows it to use in real time whenever a new image is collected. All the images are provided to physicians for categorizing in practice, but they are not appropriately rated when the patient is bought for screening. The trained CNN makes it possible to diagnose and respond quickly to a patient. The network was able to achieve these results with only one image per eye.

References

1. Zago, G.T., Andreão, R.V., Dorizzi, B., Salles, E.O.T.: Diabetic retinopathy detection using red lesion localization and convolutional neural networks. Comput. Biol. Med. **116**, 103537 (2020)
2. Kwasigroch, A., Jarzembinski, B., Grochowski, M.: Deep CNN based decision support system for detection and assessing the stage of diabetic retinopathy In. International Interdisciplinary

Ph.D. Workshop (IIPhDW) **2018**, 111–116 (2018)
3. Saman, G., et al.: Automatic detection and severity classification of diabetic retinopathy. Multimed. Tools Appl. **79**(43), 31803–31817 (2020)
4. Kaushik, H., Singh, D., Kaur, M., Alshazly, H., Zaguia, A., Hamam, H.: Diabetic retinopathy diagnosis from fundus images using stacked generalization of deep models. IEEE Access **9**, 108276–108292 (2021). https://doi.org/10.1109/ACCESS.2021.3101142
5. Patil, U.A., Wagh, S.J.: Detection of diabetic retinopathy (DR) using convolutional neural network (CNN) and multiple classifier techniques in machine learning. In: Handbook of Research on Applied Intelligence for Health and Clinical Informatics, pp. 201–209. IGI Global (2022)
6. Liu, Z., Wang, C., Wang, J.: Discrimination of diabetic retinopathy from optical coherence tomography angiography images using machine learning methods. **9**, 51689–51694 (2021)
7. Chen, Q., Sun, X., Zhang, N., Cao, Y., Liu, B.: Mini lesions detection on diabetic retinopathy images via large scale CNN features. In: 2019 IEEE 31st International Conference on Tools with Artificial Intelligence (ICTAI), pp. 348–352 (2019)
8. Jena, M., Mishra, S.P., Mishra, D.: Detection of diabetic retinopathy images using a fully convolutional neural network. In: 2018 2nd International Conference on Data Science and Business Analytics (ICDSBA), pp. 523–527 (2018)
9. Sundari, M.S., Nayak, R.K.: Process mining in healthcare systems: A critical review and its future. Int. J. Emerg. Trends Eng. Res. **8**(9), 5197–5208 (2020). https://doi.org/10.30534/ijeter/2020/50892020
10. Khatoon Mohammed, T., Shanmuga Sundari, M., Sivani, U.L.: Brain tumor image classification with CNN perception model. In: Soft Computing and Signal Processing, pp. 351–361. Springer (2022)
11. Nunes, F., Moutinho, R., Oliveira, T., Soares, F.: A mobile tele-ophthalmology system for planned and opportunistic screening of diabetic retinopathy in primary care. **9** (2021). https://doi.org/10.1109/ACCESS.2021.3085404
12. Khan, S.H., Abbas, Z., Rizvi, S.M.D.: Classification of diabetic retinopathy images based on customised CNN architecture. In: Amity International Conference on Artificial Intelligence (AICAI) **2019**, 244–248 (2019)
13. Dataset. https://www.kaggle.com/c/diabetic-retinopathy-detection/data
14. Tripathy, R., et al.: Spectral clustering based fuzzy C-means algorithm for prediction of membrane cholesterol from ATP-binding cassette transporters. In: Intelligent and Cloud Computing, pp. 439–448. Springer (2021)
15. Lakshmi, L., Purushotham Reddy, M., Praveen, A., Suniha, K.V.N.: Identification of diabetes with recursive partitioning algorithm using machine learning. Int. J. Emerg. Technol. **11**(3) (2020)
16. Padmaja, B., Prasad, V.V.R., Sunitha, K.V.N., Reddy, N.C.S., Anil, C.H.: Detectstress: a novel stress detection system based on smartphone and wireless physical activity tracker. In: Advances in Intelligent Systems and Computing, vol. 815 (2019). https://doi.org/10.1007/978-981-13-1580-0_7
17. Sundari, M.S., Nayak, R.K.: Efficient tracing and detection of activity deviation in event log using ProM in health care industry. In: 2021 Fifth International Conference on I-SMAC (IoT in Social, Mobile, Analytics and Cloud)(I-SMAC), pp. 1238–1245 (2021)

Chapter 55
Fusion Approach for Enhancement of Microscopy Images Using Morphological Filter and Wavelets

Vikrant Bhateja, Disha Singh, Ankit Yadav, Peter Peer, and Milan Tuba

Abstract The microscopic view of blood sample examined under a microscope is termed as microscopy images. The visual quality of such images is not so reassuring due to its acquisition via lens of the microscope. Contrast Limited Adaptive Histogram Equalization (CLAHE) has been useful for improving the visibility of foggy images; the contrast enhancement and edge sharpening can be done using Morphological Filter (MF). In this paper, an approach of fusing the responses of the aforementioned filters is proposed using Discrete Wavelet Transform (DWT). The Image Quality Assessment (IQA) of the image fused is estimated to adjudge the quality characteristics of different responses provided by the filter. Increased values of the IQA parameters depict that the output images are noises free, have better sharpness, and contrast so as to differentiate the bacterial clusters properly from the background.

V. Bhateja (✉)
Department of Electronics Engineering, Faculty of Engineering and Technology, Veer Bahadur Singh Purvanchal University, Jaunpur, Uttar Pradesh 222003, India
e-mail: bhateja.vikrant@gmail.com

D. Singh · A. Yadav
Department of Electronics and Communication Engineering, Shri Ramswaroop Memorial College of Engineering and Management , Lucknow, Uttar Pradesh 226028, India

Dr. A. P. J. Abdul, Kalam Technical University (AKTU), Lucknow, Uttar Pradesh 226031, India

P. Peer
Computer Vision Laboratory, Faculty of Computer & Information Science, University of Ljubljana, Ljubljana, Slovenia
e-mail: peter.peer@fri.uni-lj.si

M. Tuba
Singidunum University, Belgrade, Serbia
e-mail: tuba@np.ac.rs

© The Author(s), under exclusive license to Springer Nature Singapore Pte Ltd. 2023
V. Bhateja et al. (eds.), *Evolution in Computational Intelligence*, Smart Innovation, Systems and Technologies 326, https://doi.org/10.1007/978-981-19-7513-4_55

55.1 Introduction

The bacterial species and their genera should be identified properly to gain biological information about the microorganisms. This information plays a very important role in food industry, medicine, veterinary science, and farming. Most of the microorganisms are very useful in the day-to-day life of the human beings such as for curd formation. These microorganisms can be in the same way very harmful to human life as they are the cause of many diseases which can be life taking. Therefore, these microorganisms should be properly identified for the proper diagnosis of the disease [1]. The organisms which are too small to be seen through the naked eye are known as microorganisms. Microscopes are used for seeing the microorganisms. The view through the microscope is also captured which is known as microscopy images. For capturing such images, the sample (which can be a blood sample, urine sample or any kind of sample obtained from the body) is placed on a slide and put under a microscope, and the image is captured. This image is known as the microscopy images. The pathologist who examines the blood samples use these images for a clearer examination [2]. But these images have certain shortcomings such as poor visibility, contrast, and noise is also acquired while they are captured. To remove these shortcomings, the image enhancement is very necessary which not only increases the accuracy of the pathologist but also makes the process less time consuming. To achieve the required image enhancement, image preprocessing techniques are used for improving the quality of the images. Multi-scale Retinex [3–5] has been used in these works for improving the quality of the images. Multi-scale Retinex produces color constancy as well as compression of the dynamic range in the images. These works have mainly focused on modifying the MSR techniques by using color correction and changing the local and the global contrast in the images. CLAHE [6] is used for the visibility improvement in the images. This method produces bright images but has a shortcoming as it produces gradient reversal artifact effect in the images. Contrast enhancement has been carried out by using MF [7] but it blurs the images. A pre-defined illumination factor is used for the correction of illumination in [8]. UM [9] which is highly noise sensitive filter has been used by the authors for sharpening of the images, but this filter produces noisy images. For producing a brighter and edge enhanced image, the responses of CLAHE and MF can be fused together. Such fused image response would have better visibility, increased sharpness, reduces noise, and enhanced contrast which can be considerable result. The next sections of the paper are organized in the following way: Sect. 55.2 discusses the general overview of CLAHE and MF along with proposed fusion-based enhancement approach; In Sect. 55.3, discussion is on the IQA, summary of the results, performance metrics, and discussion of the outcomes of this work. In Sect. 55.4, the conclusion of the work is discussed.

55.2 Proposed Image Fusion-Based Enhancement Approach

55.2.1 CLAHE

CLAHE is a methodology that is used for the improvement of the low contrast problems in the microscopy images as well as for the medical images [10]. CLAHE is superior than Adaptive Histogram Equalization (AHE) as well as Histogram Equalization (HE) because of its outperforming results with the microscopy images. CLAHE is used for the limiting of the level of contrast enhancement in an image which is not limited in the HE method which finally leads to enhancement of the noise as well. Thus, by limiting the level of contrast enhancement, CLAHE avoids noise enhancement in the image, and the contrast is enhanced successfully [11]. Therefore, CLAHE produces desirable results for contrast improvement as well as the visibility of the image is improved. But, the edges of the bacterial cells in the microscopy images are not improved. Contrast enhancement is limited by CLAHE by limiting the slope of the function which relates to input image intensity and the desired result image intensity. Contrast improvement can also be limited by clipping the value of the histogram obtained at these intensity values.

55.2.2 Morphological Filter (MF)

A group of nonlinear filtering techniques that rely upon the structure and shape of the objects are termed as MF [12]. Morphological Filtering technique is carried out by using a combination of morphological operators which are dependent on the shape of the objects in the image. These operators are used separately on structuring element and the input image. A layout named structuring element determines the neighborhood of the pixel and selects its shape and size which is greatly dependent on the information obtained from the image [13]. The most rapidly used morphological operators are Erosion, Dilation, Closing, and Opening [14]. Erosion is used to reduce the area of the image, whereas Dilation is used to expand the image area [15]. For a given input image, a disk-shaped structuring element of size 7 (Fig. 55.1) is selected in this work. Following this, Top-hat and Bottom-hat filters are applied on this image. Finally for achieving the contrast enhanced response of this filter, the Bottom-hat response is subtracted from the Top-hat response, and the original image is added to it.

Fig. 55.1 Disc structuring
element of size 7

55.2.3 *Fusion-Based Enhancement Filter Using DWT*

The obtained results from CLAHE and MF are fused together to get the desired output. For achieving this, firstly, CLAHE is applied on the input microscopy image followed by MF. These results are decomposed using DWT [16] to achieve the Approximate Coefficient (AC) and Detailed Coefficient (DC). AC consists of the low frequency information, and DC consists of high frequency components of the image. Further, AC of the CLAHE and MF responses are fused using Particle Component Analysis (PCA) [16]. Then, DC of the filter responses are fused using PCA. Inverse Discrete Wavelet Transform (IDWT) [17] is applied on the fused results of AC and DC, respectively. The obtained result is again fused using PCA for achieving the desired results as shown in Fig. 55.2.

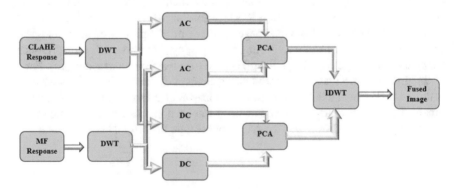

Fig. 55.2 Proposed design methodology for fusion-based enhancement of microscopy images

DWT is used for decomposing a signal into a large number of sets which represents the coefficient into time series which describes the timely evolution of the signal in the frequency domain. In DWT, the image is decomposed into various components in a discrete manner. DWT is very advantageous as it captures the information in both frequency as well as time domain. The weighted sum of the source image is used for fusion by PCA. The weighted sum of the images if found by normalizing the Eigen vector of the covariance matrix of the image. This weighted sum is used for fusion of the images.

55.3 Results and Discussions

55.3.1 Experimental Results

DiBAS dataset [18] of bacterial microscopy images is used for evaluating the performance of the above discussed simulation. The IQA is carried out using EME, PSNR, and SD [19, 20]. Initially, some test images are selected from the database and converted to grayscale. After this, CLAHE is applied on these grayscale images. Then, MF is applied on the grayscale images. Further, these results are fused as discussed in Sect. 2.3. Response of the proposed work is depicted in Figs. 55.3 and 55.4. In Fig. 55.3, Test Image#1 is shown along with the results at all the stages of preprocessing while Fig. 55.4 shows rest of the test images along with the final output after the fusion process.

EME, PSNR, and SD are calculated for IQA of the images as discussed in the Sect. 3.1. The respective values of the parameters are tabulated in Table 55.1:

55.3.2 Discussions

Candida albicans bacteria species is shown in Fig. 55.3 along with its response images at all the stages of the preprocessing, whereas in Fig. 55.4, enterococcus faecalis and micrococcus spp species of the bacterial images are depicted with the final fusion response of the images. In the Test Image#1 which shows the candida albicans bacteria cells, it can be observed that the original image is not so clear, and the appearance of the cells is hazy so it is difficult to predict the species. When MF is applied to this image, it can be observed that the image becomes noise free, and the image is also sharpened. MF also enhances the edges of the images as shown in Fig. 55.3b. The original Test Image#3 is also passed through CLAHE. The response of the CLAHE filter is brightened, and the visibility of the images is also improved as shown in Fig. 55.3c. To get the properties of both these filters in a single response,

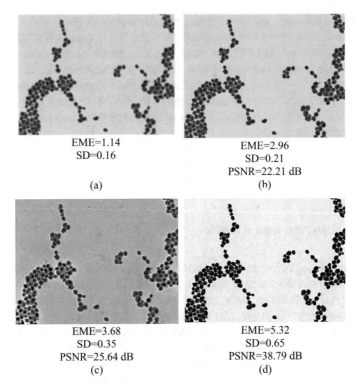

Fig. 55.3 **a** Original test image#1, **b** response of MF, **c** response of CLAHE, and **d** response of fusion-based enhancement approach test image#1

these responses are fused using DWT. The resultant response is depicted in Fig. 55.3d. It can be clearly observed that the fused image is clearer and the bacterial cells can be properly differentiated from the background. In the Test Image#2, enterococcus faecalis species of bacterial cell is shown and compared with the final fusion response of the proposed filter. The final response has better visual characteristics as compared to the original image. The Test Image#3 shows the micrococcus spp and the response of the fusion of the MF and CLAHE filters. The final response has better sharpness, less noise, and more brightness than the original image as shown in Fig. 55.4c and d. These observations are supported by the IQA parameters calculated and given in Table 55.1.

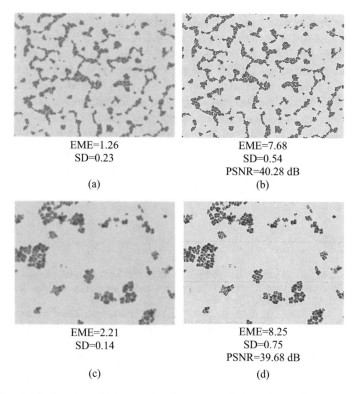

EME=1.26
SD=0.23

(a)

EME=7.68
SD=0.54
PSNR=40.28 dB

(b)

EME=2.21
SD=0.14

(c)

EME=8.25
SD=0.75
PSNR=39.68 dB

(d)

Fig. 55.4 **a** Original test image#2, **b** response of fusion-based enhancement approach test image#2, **c** original test image#3, and **d** response of fusion-based enhancement approach test image#3

Table 55.1 Values of PSNR, SD, and EME for the proposed method

Images	PSNR (in dB) of result	SD		EME	
		Original	Fused	Original	Fused
Test image#1	38.79	0.16	0.65	1.14	5.32
Test image#2	40.28	0.23	0.54	1.26	7.68
Test image#3	39.68	0.14	0.75	2.21	8.25

55.4 Conclusion

This paper presents an improved approach of image enhancement by the DWT-based fusion of the responses of contrast enhancement and edge sharpening filters. The contrast enhancement is carried out by CLAHE. MF is used for the noise reduction and the edge sharpening of the image. The image obtained after fusion shows very improved visual characteristics, and the evidence of the same is provided by the IQA parameters calculated and used for comparison which can be seen in Table 55.1. IQA parameters have shown an increased value at each stage of the preprocessing which concludes the improvement of the test images at each stage of the preprocessing. The value of PSNR is also comparable to the results of various other techniques. Therefore, it can be concluded that in this DWT-based fusion approach for the enhancement of the bacterial images, the useful information in the images (in this case the bacterial cells) are conserved, and their visual quality is improved. They can be clearly distinguished from the background which can be used for the disease diagnosis. Hence, this work provides with a significant and dependable result. The technique used in MF can be further improved by the optimization of the filter for the selection of an appropriate structuring element.

References

1. Hiremath, M.: Segmentation and recognition of E. coli bacteria cell in digital microscopic images based on enhanced particle filtering framework. Emerg. Res. Comput. Inf. Commun. Appl. **3**(1), 503–512 (2019)
2. Panicker, R.O., Soman, B., Saini, G., Rajan, J.: A review of automatic methods based on image processing techniques for tuberculosis detection from microscopic sputum smear images. J. Med. Syst. **40**(1), 1–13 (2015)
3. Jobson, D.J., Rahman, Z., Woodell, G.A.: Spatial aspect of color and scientific implications of retinex image processing. Vis. Inf. Process. Int. Soc. Opt. Photonics **4388**, 117–128 (2001)
4. Biswas, B., Roy, P., Choudhuri, R., Sen, B.K.: Microscopic image contrast and brightness enhancement using multi-scale retinex and cuckoo search algorithm. Proc. Comput. Sci. **70**, 348–354 (2015)
5. Barnard, K., Funt, B.: Analysis and improvement of multi-scale retinex in color and imaging. Soc. Imaging Sci. Technol. **1997**(1), 221–226 (1997)
6. Pandey, A., Yadav, A., Bhateja, V.: Design of new volterra filter for mammogram enhancement. In: Proceedings of the International Conference on Frontiers of Intelligent Computing: Theory and Applications (FICTA), pp. 143–151. Berlin, Heidelberg (June 2013)
7. Bhateja, V., Nigam, M., Bhadauria, A.S., Arya, A., Zhang, E.Y.: Human visual system based optimized mathematical morphology approach for enhancement of brain MR images. J. Ambient Intell. Humanized Comput. 1–9 (July 2019)
8. Tek, F.B., Dempster, A.G., Kale, I.: Parasite detection and identification for automated thin blood film malaria diagnosis. Comput. Vis. Image Underst. **114**(1), 21–32 (2010)
9. Bhateja, V., Misra, M., Urooj, S.: Human visual system based unsharp masking for enhancement of mammographic images. J. Comput. Sci. **21**, 387–393 (2017)
10. Reza, A.M.: Realization of the contrast limited adaptive histogram equalization (CLAHE) for real-time image enhancement. J. VLSI Signal Process. Syst. Signal Image Video Technol. **38**(1), 35–44 (2004)
11. Sahu, S., Singh, A.K., Ghrera, S.P., Elhoseny, M.: An approach for de-noising and contrast enhancement of retinal fundus image using CLAHE. Opt. Laser Technol. **110**, 87–98 (2019)
12. Alankrita, A.R., Shrivastava, A., Bhateja, V.: Contrast improvement of cerebral MRI features using combination of non-linear enhancement operator and morphological filter. In: Proceedings of IEEE international conference on network and computational intelligence (ICNCI), pp. 182–187. Zhengzhou, China (May, 2011)
13. Somasundaram, K., Kalaiselvi, T.: Automatic brain extraction methods for T1 magnetic resonance images using region labeling and morphological operations. Comput. Biol. Med. **41**(8), 716–725 (2011)

14. Benson, C.C., Lajish, V.L.: Morphology based enhancement and skull stripping of MRI brain images. In: Proceedings of IEEE International Conference on Intelligent Computing Applications (ICICA), pp. 254–257. Coimbatore, India (March 2014)
15. Tiwari, D.K., Bhateja, V., Anand, D., Srivastava, A., Omar, Z.: Combination of EEMD and morphological filtering for baseline wander correction in EMG signals. In: Proceedings of 2nd International Conference on Micro-Electronics, Electromagnetics and Telecommunications, pp. 365–373. Singapore (September 2018)
16. A. Krishn, V. Bhateja and A. Sahu, "Medical image fusion using combination of PCA and wavelet analysis. In: Proceedings of International Conference on Advances in Computing, Communications and Informatics (ICACCI), pp. 986–991. Delhi, India (September 2014)
17. Srivastava, A., Bhateja, V., Shankar, A., Taquee, A.: On analysis of suitable wavelet family for processing of cough signals. In: Proceedings of Frontiers in Intelligent Computing: Theory and Applications (FICTA), pp. 194–200. Singapore (October 2020)
18. The bacterial image dataset (DIBaS) is available online at: http://misztal.edu.pl/software/databases/dibas/. Last visited on 10 Dec 2020
19. Singh, D., Bhateja, V., Yadav, A.: Stationary wavelet based fusion approach for enhancement of microscopy images. In: Proceedings of 9th International Conference on Frontiers of Intelligent Computing: Theory and Applications (FICTA 2021), pp. 1–10. Mizoram, India (June 2021)
20. Yadav, A., Bhateja, V., Singh, D., Chauhan, B.K.: Segmentation of microscopy images using guided filter and otsu thresholding. In: Proceedings of 6th International Conference on Microelectronics, Electromagnetics and Telecommunications (ICMEET 2021), pp. 1–7. Bhubaneshwar, India (August 2021)

Author Index

Printed in the United States
by Baker & Taylor Publisher Services